History of Printing Techniques in China

Pan Jixing

History of Printing Techniques in China

中国科学技术大学出版社
UNIVERSITY OF SCIENCE AND TECHNOLOGY OF CHINA PRESS

Springer

Pan Jixing
The institute for history of natural sciences
Chinese Academy of Sciences
Beijing, China

Translated by
Chen Jiliang
Interdisciplinary Center for Language Science
University of Science and Technology of China
Hefei, China

Mu Yuanyuan
School of Foreign Languages
Chaohu University
Hefei, China

Chen Fumei
Foreign Language Learning Center
University of Science and Technology of China
Hefei, China

ISBN 978-981-96-6118-3 ISBN 978-981-96-6119-0 (eBook)
https://doi.org/10.1007/978-981-96-6119-0

Jointly published with University of Science and Technology of China Press
The printed edition is not for sale in China Mainland. Customers from China Mainland please order the print book from University of Science and Technology of China Press.
ISBN of the Co-Publisher's edition: 978-7-312-05291-0

Supported by the Chinese Fund for the Humanities and Social Sciences

This Springer imprint is published by the registered company Springer Nature Singapore Pte Ltd.
The registered company address is: 152 Beach Road, #21-01/04 Gateway East, Singapore 189721, Singapore

Preface

Driven by the thriving of Buddhism in the Sui and Tang dynasties and prosperity of imperial examinations in the Song dynasty, printing, one of the four great inventions of ancient China, had undergone rapid development in ancient times ever since its emergence in China. It was this invention that promoted popularization of books or written records in Chinese society at a greatly higher rate, and thus made an important contribution to cultural transmission in ancient China and even in the whole world. If papermaking is deemed as an economical and widespread carrier for cultural transmission, then, printing is supposed to make this easier and faster. It is also the invention of printing to which China owes its title as one of the greatest ancient civilizations and a magnificent country with glamorous culture. For this reason, after doing research on papermaking technology, I have always wanted to systematically summarize my research results on the printing technology and compile them in a monograph like this.

Since the 20th century, studies of ancient Chinese printing have been conducted in China, Japan, and Korea, and outstanding results have been achieved with regard to research on the origin of printing, application and process of wooden, copper and other movable-type printing techniques. During this period, there was, of course, a big debate among people that led to further concerns from academic experts on printing, as well as prevailing research on ancient Chinese printing. Moreover, as new archaeological discoveries have constantly been made, new ideas or thoughts have emerged, enriching people's understanding of the ancient Chinese printing and promoting increasing research activities in this field.

I have been focusing on studying the printing techniques for decades, and have discussed my findings in *History of Chinese Science and Technology: Papermaking and Printing* (1998). Later, I expounded and analyzed several aspects of ancient Chinese printing in *History of Metal Movable Type Printing Technology in China* (2001) and *On the History of Scientific and Technological Exchange between China and Foreign Countries* (2012). The problems that I found in the course of my study, accompanied by new archaeological discoveries, aroused my keen interest in the history of Chinese printing as well as my growing thirst for overall and systematic study. However, most current researches on printing are subject to the existing paradigms of studies in the field of papermaking. The history of printing is thus simply classified as the history of printing books, or the two mingle with each other, which leads to people's blurred understanding of ancient Chinese printing. Therefore, it has been my greatest desire in recent years to have a monograph on ancient Chinese printing published, and fortunately, I have finally completed this monograph based on years of experience.

The book is divided into three main parts:

Part I (Chaps. 1–3) introduces the definition of printing technology and the research objects, significance, and methodology of the history of printing. It summarizes the research history of printing, followed by a description and analysis of the origins of printing and papermaking respectively. In this part, the origin of ancient Chinese papermaking that has caused a lot of controversies over many years has been defined, supported by a conclusion that paper had appeared in China before Cai Lun's inventive attempts.

Based on archaeological discoveries and literature reviews, the second part (Chaps. 4–7) discusses the findings on several research fields of ancient Chinese printing, including the origin and development of woodblock printing, the origin and development of copperplate printing, mono-printing, and color printing, the origin and development of non-metal moveable-type printing such as wooden-type and clay-type printing, and the origin and development of moveable metal-type printing, such as copper-type printing. Among the above-mentioned research topics, movable metal-type printing, in my opinion, dated back to the Northern Song dynasty and further developed in the Yuan, Ming, and Qing dynasties.

The third part, including Chaps. 8 and 9, focuses on the spread of ancient Chinese printing and its significance for the development of world civilization. This part begins with the spread of Chinese printing in the Korean Peninsula and Japan, followed by a discussion on the spread of this technique in other parts of Asia, Africa, Europe and America. It ends with a description of the important contribution of Chinese printing to the world civilization and its important effects on politics and economy.

I feel very gratified that my dream for many years has finally been realized, and sincerely hope to get your support and criticism. I hope that this book can serve as a link between the past endeavors and the future efforts, which is my greatest desire!

Beijing, China

Pan Jixing

Brief Introduction

This book presents a systematic discussion of the history of Chinese printing technology, encompassing the origins and evolution of writing, the invention and refinement of papermaking techniques, and the emergence and progression of various printing techniques. It meticulously examines related processes, including papermaking, woodblock printing, movable type printing, and mono-printing, alongside their respective impacts on printing itself. Furthermore, the book systematically explores the profound influence of Chinese printing technology on neighboring countries and its contribution to global culture since its inception. It serves not only as a history of printing but also as a concise narrative of the history of civilization. By delving into the history of printing, readers are afforded insights into the intricate interplay between printing technology and various societal aspects, including economy, legal systems, art, culture and education, as well as its essential role in facilitating social advancement.

Contents

Chapter 1
Introduction

1.1 Definition of Printing Technology and Research Objects, Significance, and Methodology of Printing History

1.1.1 Definition of Printing Technology

It's essential to clearly define printing technology before studying the history of printing. As the connotation of printing varies at different periods, printing technology is supposed to be defined in diversified ways. Printing emerged after the advent of paper and presupposes the existence of paper, which must be taken into account when it comes to the definition of printing technology. In fact, the history of printing cannot be equated to the history of printing books, as printed materials, mostly in the form of books, also include tickets, banknotes or paper money, and census registers as research objects. Thus, it is improper to simply classify the history of printing as the history of printing books. Printing is related to use of seals, which, in fact, does not entail printing activities per se since it had existed before the invention of paper. Modern printing evolves from ancient printing, but there are differences between the two. Modern printing is also different from both ancient and early modern printing.

Since we are investigating ancient or traditional printing first, it is necessary to provide a definition accordingly. As different definitions of printing have been given in various Chinese and foreign works, with significant discrepancies in its denotation and connotation, it is useful to make a review before defining the term. *Cihai* ("Sea of Words", 1979), a well-known unabridged comprehensive Chinese dictionary, classifies printing as a technique of making printed materials according to original texts or images, and adds that in early China, images or texts were carved on the wooden boards and printed with ink. This definition is not inclusive enough in terms of its denotation, and therefore it does not reflect the characteristics of printing.

P. Jixing, *History of Printing Techniques in China*,
https://doi.org/10.1007/978-981-96-6119-0_1

The New Encyclopaedia Britannica (1980) states that "Printing has traditionally been defined as a technique for applying under pressure a certain quantity of colouring agent onto a specified surface to form a body of text or an illustration. "However, there is no discussion about what special surface it refers to. In *The Encyclopaedia Americana* (1980), "printing refers to a process of reproducing images or texts on paper, cloth, or other surfaces. Although there are considerable variants of the printing technique, printing typically refers to a process that transfers characters carved in reverse form from a plate or a similar surface containing the reversing characters to the material to be printed." This definition is comparatively more specific, in terms of technological features.

The Encyclopedia of Japan (1980) defines printing as a process of producing printed materials by virtue of reproducing methods, in which ink was applied to a plate containing text and images and transferred to paper and other materials. However, the said "paper and other materials" include paper, polymer composites, wood, glass, ceramics, and other materials that are applicable to modern times rather than ancient times. Qian Cunxun defined printing as a process of reproducing a reverse form of the object on paper and other surfaces with ink, and pointed out that this process was at least composed of three basic steps: (1) carving a flat plate in advance with the convex and reverse form of the object to be printed; (2) applying ink to the reverse form of the object; (3) transferring the reverse form to the surface of the material to be printed in a standardized form.[1] This definition is given from the technical point of view. The above-mentioned definitions can generally reflect some ideas in Chinese and foreign works, some of which are based on modern printing. However, as modern printing is supposed to be distinguished from ancient printing due to the difference in printing process, the term "traditionally" must be emphasized when ancient printing is defined, just as the case in *The New Encyclopaedia Britannica.* Definition of modern printing, which can not be fully applicable to traditional printing, should take into consideration ancient printing practices to accurately determine connotation and denotation of printing.

When defining printing, it is important to specify the printing materials, printing process and purpose so as to avoid confusion of concepts. If cloth and clay are regarded as carriers, calico and lute (or sealing clay) used in ancient times can be classified as printed materials, which will make the origin of printing more complicated. If printing process and purpose are not specified, affixing a seal can be regarded as a printing activity. Thus, we can see that some of the definitions above are ambiguous while others show enlarged denotation by including the non-printing activities. For that reason, we believe that traditional printing is composed of mono-block printing and movable-type printing. Mono-block printing is a technique in which original texts or images are carved in a convex and reverse form on a monolithic block where a coloring agent is applied, and then reproduced in a standardized form on paper to form a reading material. Movable-type printing means that after original characters are separately carved in a convex and reverse form onto hard materials, the

---

[1] Needham J. *Science and Civilisation in China: Vol. 5.* Cambridge: Cambridge University Press, 1985: 132–133.

separate characters are integrated into a whole according to the original text, and the following process is similar to that of mono-block printing. The difference between the two is that the plates are made in a different way, and that the characters can be recycled and re-integrated after movable-type printing while the contents to be printed should be the same in mono-block printing. Mono-block printing includes woodblock printing and copperplate printing, while movable-type printing includes non-metal and movable metal-type printing.

In the definition we proposed, there are three basic elements: (1) printing materials. The main carrier of a printed material is paper. The monolithic block is made of wood or metal alloys, while movable types are made of clay, wood, and metal alloys; the ink is mainly used as the coloring agent, and dyes and pigments are used for color printing. (2) Process and method. Reproduce original texts or images in a convex and reverse form on a monolithic block or metal plate, or create individual blocks in a convex and reverse form on a hard material before integrating them into a whole. Then, apply a coloring agent onto the plate, cover it with printing paper and brush the paper. (3) Purpose. The finished product is used primarily as a reading material, and secondarily as a decorative material, banknote, instrument or document, note, etc. The definition of printing, which is composed of the above three elements, can basically reflect the essence of traditional printing or legacies at home and abroad. As we defined, the carrier is limited to paper, because it is the primary material used by traditional printing. For any doubt about this definition in a narrow sense, we are open to these questions since printing in the broad sense is not within the scope of this book.

As we have stated above in the definition of printing technology, printing is different from papermaking, because the former mainly involves mechanical processes on the raw materials through more physical force and less chemical force. Therefore, only morphological changes (no essential changes) happen in the course of material processing. The same is true for use of the metal alloy to produce the movable types, except for clay casting, in which chemical change will occur. Therefore, differing from the papermaking industry, the printing industry creates no pollution to the environment, but causes a large consumption of hard wood. The printing work has the characteristics of individual operation, as operations including cutting and brushing are often completed by individual workers. Printing also sets high requirements for labor skills. Cutting of movable types must be completed indoors by an educated worker, while production of plates should be completed indoors or in the courtyard by a skilled worker. Printing plants are often located in cities or towns owing to little demand for water. Instead, paper mills must be situated near the river and raw materials, and the papermaking process is often completed outdoors. The papermaking industry requires more workers, while fewer workers are required but well-paid by the printing industry. The difference between the two lies in the facts above.

1.1.2 Research Objects and Significance

Both monolithic blocks and movable types are ancestors of the modern large printers, and movable-type printing is the origin of modern printing. Both of them originated from China and spread to other countries. Papermaking, printing, gunpowder and compass, known as the Four Great Technological Inventions of ancient China, have played a major role in advancing the development of human civilization. It is important to understand the history of printing culture, especially for those who are or will be engaged in printing, because development of the printing technique relies on historical inheritance to some extent, and today's printing technology is developed from the printing techniques of the past. Thus, studying and understanding its past help us get to know its present, which is conducive to prediction of its future. We can learn from the history of printing and know well the development patterns and rules of printing techniques over time to broaden our horizons and improve our historical wisdom.

Studying the history of printing also provides us with an interesting insight into the interaction between printing techniques and social economy, regulations, religion, academy, art, culture and education, as well as the role of printing in promoting the development of social civilization. This is not accessible through professional printing tasks and is useful for enriching the humanistic knowledge of practitioners. Hard work and creative spirits of ancient printing craftsmen and technicians, as well as their successes and failures, will enlighten our minds, just as the old Chinese saying goes, learning from ancient people, we can weigh the successes and failures. Chinese printing culture, with a history of more than 2000 years, is a rich knowledge treasure that constitutes a significant part of Chinese culture. Studying the history of Chinese printing culture is critical to understanding the Chinese cultural heritage and carrying forward its essence. Knowing the long and outstanding printing culture created by our ancestors is beneficial to enhancing practitioners' patriotic and professional spirit. The history of Chinese printing is a part of the history of the world printing, and people in every country have made their own contributions to the development of printing. As the Chinese people learn from foreign printing techniques and spread their own technical fruits to other countries, the history of printing, from a professional perspective, represents scientific and cultural exchanges between China and foreign countries.

In this book, there are a total of nine chapters. Chapters 3 through 7 discuss the origins of paper, woodblock printing, copperplate printing, mono-printing, color printing, and movable-type printing and their formation and development in China until the end of the Qing dynasty in the nineteenth century. Thus, history of ancient printing is mainly involved in these chapters. Chapter 8 summarizes the spread of Chinese printing in Asia, Africa, Europe, and the United States, as well as the early printing situation in these countries. In Chap. 9, the influence of paper and printing on the development of world civilization is discussed. The above constitute the research objects mentioned in this book. They have a span of 1400 years chronologically from the Sui and Tang dynasties in the sixth century to the twentieth century, and stretch from China to the five continents; They reach across many fields of printing, ancient or modern. These objects are distributed so widely that we can only make an overview in this book.

1.1.3 Methodology

When studying the history of printing, we attempt to employ a new approach that integrates literature review, archaeological excavation, study of ancient physical specimens, simulation experiments, investigation of traditional techniques, comparison of Chinese and foreign techniques, and exploration of technical principles while taking into account relevant socio-historical background. This approach has proven to be effective and will be described below.

To study the history of printing, it's necessary to acquire a large number of Chinese and foreign literature, especially primary literature. However, these materials are rather scattered. Thus, it is important to connect these scattered historical data, carry out textual proofreading, technical judgement and date identification, discern the false from the genuine, and tease out the meaning of the technical terms contained. Historical data can be adopted and trusted, but not necessarily followed blindly. We must review one by one the historical data listed by the predecessors before citing them, and if not, we may make mistakes. For example, after referring to *Dhāraṇī of the Great Protectress* translated by Amoghavajra (AD 705–774), a monk in the Tang dynasty in AD 758, some stated that the practice of carrying along with mantra started from the middle and late Tang dynasty, and that matrix-shaped character layout that was modeled on the shape of maṇḍala appeared not earlier than that time. However, after studying historical data, we found that this practice came into fashion in the early Tang dynasty, and that *Dhāraṇī of the Great Protectress* was translated as early as 693. Thus, the statement that the practice of carrying along with mantra started from the middle and late Tang dynasty is not factual. Others claimed that the Tibetan version of *The Tripitaka* was published by the Tibetan Jiamuxiang in the reign of Emperor Renzong (1311–1320) of Yuan.[2] However, according to historical data, the term Jiamuxiang was not the name of a person, but meant *Huofo* (a Buddha) or khatuktu in Mongolian. It appeared since the Qing dynasty and did not exist in the Yuan dynasty. It is found from further investigation that the Tibetan version of *Tripitaka* was first published in the ninth year of Yongle era (1411) of Ming Emperor Chengzu, as evidenced by the physical possession of the Potala Palace in Lhasa.

While documentary records are important, research on the history of printing requires on-site collection and scientific analysis of handed-down documents and unearthed material data, instead of simply staying in the study to look for documentary records. As these handed-down documents and unearthed material data are more authoritative, it's necessary to pay close attention to the information about archaeological excavation and relic collection. For example, according to *Hou Han Shu* •

[2] In this book, an emperor's calendar is based on his tenure.

Cai Lun Zhuan ("Biography of Cai Lun in Book of Later Han"), the papermaking technique was invented by Cai Lun, a eunuch of the Eastern Han dynasty in 105 AD, which was in contradiction to physical evidence of Western Han paper from the second century B.C. to the first century B.C. Such paper has been unearthed since the twentieth century and later proven as hemp paper, some of which still have the handwriting. Thus, the statement that the papermaking technique was invented by Cai Lun should be corrected. For another example, whether there were wooden movable-type components in the Song dynasty had once sparked a hot debate among scholars, some of whom claimed they could be dated back to the Yuan dynasty by referring to *Nong Shu* ("Book on Agriculture") written by Wang Zhen in the Yuan dynasty. But in recent decades, the discovery of the wooden movable-type-based Buddhist texts in Tangut script has resolved this debate, and the existing findings have proved the existence of wooden movable-type-based books since the Northern Song dynasty in the twelfth century. The question about when metal movable-type first appeared in China remained unresolved for a long time, but findings of study on Song and Jin coin plates and other unearthed cultural relics showed that metal movable-type components appeared at the same time as wooden movable-type components.

There are still workshops in Chinese mainland and ethnic minority areas, which produce books using traditional techniques and production tools to varying degrees. Investigation of their production techniques will provide clues to the actual state of ancient technology. For example, both the Tianjin Yangliuqing Woodblock Printing Factory and the Sichuan Derg Tibetan Printing Institute have still utilized the traditional methods to produce books and cultivated technology successors, in addition to continuing to use printing plates and tools of the past. Investigation of their tools and techniques can provide a variety of valuable information that are certainly inaccessible from ancient books and records. Just as investigation of modern plant and animal fossils provides clues about ancient plant and animal morphology, the study of technical "fossils" is also an effective method.

In order to make accurate and comprehensive judgement for any inaccuracy or omission in ancient records about technical process, it is necessary to conduct simulated experiments according to the ancient records and through hands-on practice, and decide the essential steps that must be taken to produce the required products. Then, compare the experimental results with ancient objects, and finally explore and recover the details of the lost ancient technologies. Our study on the hemp papermaking technique of Han is a good example of such methods. Another good example is the discovery of the Dou printing, a kind of woodblock watermarking technique in the Ming dynasty, by Beijing Studio of Glorious Treasures, which lays the foundation for reproduction of ancient art masterpieces. When studying the evolution of the binding forms of ancient printed books, we used waste paper to simulate the books bound in different forms, and then made comparison. Our findings showed that books bound in a cylindrical form would certainly be replaced by those bound in a flat or square form, because the latter met characteristics of the printing technique and was easy to read and keep. This is the case with all countries.

When studying the history of Chinese printing, it is advisable to obtain a global view, understand the history of printing in other countries and make comparison rather than just focus on Chinese printing. Only in this way, can the printing technique at home and abroad be well understood, which provides us with an insight into Sino-foreign technical exchange. Take the early European woodblock printings in the fourteenth and fifteenth centuries as an example, which were similar to Chinese printings in format and binding form, but significantly deviated from the traditional European practices. The engraver cut the plate at an oblique angle in the direction of the grain of the plate itself, just as the Chinese engraver did. Such convergence may result from the westward spread of Chinese technology in the era when east–west traffic routes were opened up. Likewise, the copper movable-type printings that appeared in the Goryeo dynasty in the fourteenth century were the same as those used by the Song dynasty in terms of format, binding form, typecasting, typesetting and even fonts. This suggests the spread of Chinese printing technology, because the Korean Peninsula was, at that time, under the rule of the Yuan dynasty, and it is highly unlikely that the Goryeo invented such technology independently. Based on numerous historical data around the world, the recognized theory of technology communication was proposed: during the period from the first millennium after A.D. to the rise of modern science (first to fifteenth centuries), the more complex the technology, the less likely it is to be repeatedly invented in several different regions within a certain period of time. Emergence of a technology invented by a country in another country or region can only be explained by technology transmission.[3]

Finally, study on the history of printing should focus on the printing technology rather than book printing or publishing. Therefore, it is necessary to evaluate the technical activities of the ancients according to theories of modern sciences, and to interpret the words and sentences in ancient books technically not just literally. That's exactly what traditional historiography is missing. For example, some people misinterpreted the term "cast copper-based tin movable- type printing" (implies the "cast copper plate and tin types"), invented by Hua Sui (1439–1513) in the Ming dynasty, as a technique of carving hundreds of thousands of tin characters one by one and then affixing them to copper plates. Such interpretation that is literally compliant violates the printing principle. In fact, the metal movable-type components created by Hua Sui were made of copper-tin alloy, and the so-called "movable-type copper plate should refer to "copper movable-type printing". Thus, we should make corresponding correction for improper usage of words of literati in the Ming dynasty. Another example is the term "Taishan porcelain clay movable-type printing"created by Xu Zhidin of Qing in 1719. Some interpreted it as a process in which the glazed porcelain clay types were assembled to form a printing plate, or fired to form the whole porcelain clay plate. Such interpretation was not conducted from a modern scientific point of view. In fact, glazed porcelain clay types are very expensive and easy to deform and difficult to repair after being fired. In addition, it is technically unfeasible to produce the whole porcelain clay plate using fire. The only possibility is to produce movable types using porcelain clay or kaolinite at the temperature of 900 degrees centigrade. As these types are white, they are also called "porcelain

---

[3] Needham J. *Science and Civilisation in China*: *Vol. 1* [M]. Cambridge: Cambridge University Press, 1954: 229.

clay types". Research methods mentioned above should complement each other to produce a synthesis effect.

1.2 Past and Present of Research on History of Printing

1.2.1 Research Review

In China, the printing technology came out approximately 800 years later than the paper-making technology, but this is not the case for our study of the printing history. In ancient times, the origin of the printing technology was occasionally mentioned when it came to book editions. For example, Song dynasty scholars occasionally mentioned the origin of printing technology, but there were many misunderstandings due to limited historical materials. Yuan dynasty scholars who followed the views of Song dynasty offered limited in-depth analysis. Hu Yinglin (1551–1602), a Ming dynasty bibliographer, concluded in his *Shaoshishan Fang Bicong* ("Notes from Shaoshishan Studio", ca.1589) after investigating relevant historical records that "the woodblock printing originated from the Sui dynasty, flourished in the Tang dynasty, spread during the Five Dynasties, and perfected in the Song dynasty." He thus became one of the most insightful early scholars studying the Chinese printing history owing to his broader mind. Even more than 400 years later, his penetrating insight remains profound and reasonable, and can be used as an outline for compiling the history of woodblock printing. However, Qing dynasty scholars fell behind Ming scholars when it came to study of the printing origins.

The study of Chinese printing history using modern historical methods began in the nineteenth century. In 1847, the French sinologist Stanislas Julien (1799–1873) published a paper titled *Materials on Woodblock, Lithography, and Typography Printing*,[4] where he cited some of Chinese historical materials and made a relatively comprehensive introduction to Chinese printing history. In this paper, Julien traced woodblock printing back to the Sui dynasty, and movable-type printing to the Northern Song dynasty. This paper produced a significant influence in the West. In the early twentieth century, both *Shulin Qinghua* ("A Plain Talk on Books", 1911) by Ye Dehui (1864–1927) and *Zhongguo Diaoban Yuanliu Kao* ("Research on the Origin of Chinese Block Printing", 1918) by Sun Yuxiu (1864–1930) systematically discussed Chinese printing history, but followed traditional practices in research methods and were seldom related to modern methods. However, both books are commendable, as they have accumulated a wealth of historical materials. Laterin the 1920s, Wang Guowei (1877–1927) published specialized essays on editions from the Five dynasties, Song, and Western Xia, which were then incorporated into *Haining Wang Jingan Xiansheng Yishu* ("Posthumous Papers of Wang Guowei", 1930) owing to cautious research approaches. The classic work *The Invention of Printing in China and Its*

[4] Julien S. Documents sur l'artd'imprime, al'aide des planches au bois, des planches au Pierre et des types mobiles [J]. *Journal Asiatique*, 1847, 4(9): 508.

Spread Westward by American sinologist Thomas Francis Carter (1882–1925), first published in 1925 and subsequently reprinted multiple times, marked the authoritative beginning of systematic and scientific research on Chinese printing history. Carter became a leading figure in this field, from whom scholars around the world benefited.

Shortly after the appearance of his masterpiece, Carter passed away, and new materials continued to emerge. French sinologist Paul Pelliot (1878–1945) attempted to revise Carter's masterpiece but failed. Another American sinologist, Luther Carrington Goodrich (1894–1986) filled Pelliot's shoes and had the second revised edition of Carter's masterpiece published in New York in 1955. This second revised edition was the most widely recognized and has been translated into Chinese and Japanese. In addition to revising Carter's masterpiece, both Pelliot and Goodrich also wrote some papers on printing history. In the United States, scholars such as Wu Guangqing and Richard Rudolph (1909–2003) also conducted similar researches. Chinese scholars in the 1950s to 1980s, including Liu Guojun, Zhao Wanli, Qu Wanli, Chang Peter, Li Shuhua, Shi Meicen, and Zhang Xiumin, also published works related to printing history, especially *The Invention of Printing in China and Its Influence* (1958) by Zhang Xiumin and *History of Printing in China* (1989) and other relevant papers that represent the most recent systematic research on printing history following Carter's book. In 1985, Professor Qian Cunxun of the University of Chicago compiled the "Volume of Papermaking and Printing" in English for Dr. Joseph Needham's *Science and Civilisation in China*, which set a precedent for combining the history of papermaking and printing in a single work. This volume offered a more extensive range and richer historical sources than Carter's masterpiece, and made many innovative contributions by drawing upon all sorts of Chinese and overseas materials. The works of Lu Qian, Jiang Yuanqing, Qian Cunxun, Zhang Binglun, and others dealt with specific printing issues, marking the beginning of research in the history of technology.

1.2.2 Current Studies on Printing History

Japanese scholars have cited abundant materials about Chinese printing history in their works, especially in Nagasawa's *History of Chinese Printing*, which offers valuable insights by integrating the printing histories of China and Japan. China is also mentioned in Korean works on printing history, of which *A Comparative Study of Ancient Movable Type Printing Techniques in China and Korea* (1986) by Choeng Kyong-jin is worth reading. These works can serve as references for comparative studies between China and Korea. Since the 1990s, China entered its glory days for studies of printing history, when Niu Dasheng, Shi Jinbo, and Wushou Er conducted more in-depth studies on the movable-type printing technologies in the Tangut script and the Uighur, and provided many new insights on printing technologies, identification of printed text, movable-type components, the exchange of printing technologies among various ethnic groups, and the transmission of Chinese printing technology to the West. Concurrently, a more in-depth academic discussion on the origin of printing took place. Most scholars considered it must be too late to trace back the woodblock printing to the middle of the Tang dynasty. Studies on the history of metal movable-type printing in China were inadequate till 1998 when PanJixing, based on the study of excavated artifacts, asserted in his book *On the Origins of Metal Movable Type Technology* (written in both Chinese and English), that movable-type copper plates originated from the Northern Song dynasty in the early twelfth century. Pan made a significant breakthrough in this field. He also investigated the influence of Chinese metal movable-type printing on Korea and Europe, and published both Chinese and English versions of *A Comparative Study of Early Printing in China, Korea, and Europe.* Based on the above, Pan had another publication titled *A History of Chinese Movable Metal-Type Printing Technology* in 2001, which became the first academic monograph in this field. In 1998, Pan added a volume of "Papermaking and Printing" to Dr. Joseph Needham's *Science and Civilisation in China*, which combined the histories of papermaking and printing and offered another work following Qian Cunxun, and plenty of new information and ideas are added to the book.

Since the 1990s, scholars from the mainland of China and Taiwan region have cooperated extensively in exploring Chinese printing history. This unprecedented collaboration has been highly effective and productive, bringing out many classic works including *Essays on the History of Chinese Printing* (1997), edited by Xu Yingjian. The book extensively revised the data provided by Zhang Xiumin, adding new materials, exquisite illustrations, and well-designed layouts. It stands as a successful example of academic collaboration between the mainland of China and Taiwan region. In another representative work, *A General History of Chinese Printing* (1998) revised by Li Xingcai and authored by Zhang Shudong and others, modern Chinese printing history and technology were also explored in addition to ancient ones. Following *Essays on the History of Chinese Printing,* it is another successful example of academic collaboration between the mainland of China and Taiwan region. Woodblock printing and polychrome printing are products that combine printing and art. Works compiled by Zheng Zhenduo (1898–1958), Wang Bomin, and others collected a wealth of related materials. In 1999, Feng Pengsheng from Beijing Studio of Glorious Treasures provided an overview of woodblock printing and polychrome printing and summarized the latest research in this area from both historical and technical perspectives in his book *An Overview of Woodblock Watermarking Technology in China.* For domestic and foreign publications on the printing history from 1900 to 1992, please refer to Qian Cunxun's *A Concise Bibliography of Chinese Printing History.*[5] Due to limited space, these publications can not be listed one after another but will be mentioned and cited appropriately in the following sections.

Over the past century, the study of Chinese printing history has become well-established under the joint efforts of Chinese and foreign scholars. Nearly 400 papers

[5] Qian Cunxun. *A Concise Bibliography of Chinese Printing History* [M] // Qian Cunxun. *Collection of Papers on the History of Books, Paper, Ink and Printing in China.* Hong Kong: The Chinese University of Hong Kong Press, 1992: 205–326.

and more than 60 monographs in Chinese, as well as nearly a hundred papers and over 20 monographs in English, have been published mostly after the 1950s. Of these publications, most obviously investigate woodblock printing, particularly from the perspectives of editions, catalogs, and publishing, while the development of printing technology is relatively scarcely analyzed. The study of the printing origin and early printed materials remains underexplored, and no consensus has been reached among scholars. Investigation of early historical materials is still ongoing, and there is much more to explore. As archaeological work progresses, more early printing materials are expected to be unearthed, necessitating scientific research and thus diminishing some disagreements. In the past, there was little research on the history of movable-type printing. However, with the discovery of new movable-type books in the 1980s and 1990s, research on wooden movable-type printing, clay movable-type printing, and ceramic wooden movable-type printing has taken on a new look. The study of metal movable-type printing should continue to expand. Previously it was generally believed that the movable-type components appear in the order of clay movable-type, wooden movable-type and metal movable-type. However, recent evidence shows they are not, and metal movable-type components did not emerge as late as previously thought, which urges us to change our ideas. It is essential to establish a new theory in which the relationship and evolutionary mechanism between woodblock printing and movable-type printing, non-metal movable-type and metal movable-type components should be comprehensively analyzed from both technical and historical perspectives. During the study, we should consider the characteristics of the printing industry and its development patterns, and printing history cannot be equated with the history of printing books. Although it's essential to study the ancient book publishing history from the bibliographic point of view, after all, it is not the history of printing technology, and as the history of the printing technology has been less studied, more efforts should be made in the future.

It is essential to analyze the impact of social norms, economic factors, Buddhism, academia, art, cultural education and the papermaking industry on the development course of printing technology because of their close relation with printing technology. However, the existing publications reveal that there are relatively few studies addressing these issues. Most studies limit themselves to describing the publication of books during different historical periods and in various regions, without conducting a humanistic analysis of the printing history. This makes it difficult to get rid of the limitations caused by traditional historiographical methods and to fully understand the development characteristics of printing technology in different eras and regions. Over the past century, scholars have accumulated a vast amount of historical materials. The challenge now lies in how to synthesize these materials to provide theoretical explanations. Research on the history of Chinese-foreign exchange in printing technology and comparisons between Chinese and foreign printing technologies have begun since the Carter era, but often focused more on the Western transmission of woodblock printing and less on Eastern transmission, not to mention the outward transmission of the movable type printing technology. Therefore, research gaps still exist in this area. In recent years, along with the opposition to confusing the printing history and the book printing history, another ideological tendency has emerged: expanding the extension of the concept of printing activities. As a result, some non-printing activities have been considered as printing ones, leading to academic disagreement in the printing archology and the diversification of the birthplaces. Therefore, it is still necessary to clearly define the printing technology before the study. While considerable progress has been made in Chinese printing history over the past century, there are still many unresolved issues, and much more work remains to be done for researchers in the twenty-first century. Early researches were primarily conducted by foreign sinologists, and it was not until after the 1950s that Chinese scholars and overseas Chinese scholars took the lead in this field. However, as many of them are now elderly or deceased, there is an urgent need for young scholars to fill in the research vacancy.

1.3 Source of Primary Materials for History of Printing

1.3.1 Physical and Documentary Materials in China

The materials related to the history of printing in China are abundant but rather scattered. Over the past century, through the efforts of Chinese and overseas scholars, a considerable amount of research materials has been collected, which can generally be divided into the following categories:

The first comes to printed materials of various dynasties, which serve as tangible evidence reflecting the state of printing technology during different historical periods. Museums, libraries, and private collections house these materials, with the earliest originating from the Song and Yuan dynasties, followed by those from the Ming and Qing dynasties. The numbers of these materials, mostly dated back to the Ming and Qing dynasties, range from hundreds of thousands to millions. Since the early twentieth century, tens of thousands of hand-copied materials from the Six Dynasties to the Tang dynasty and a small number of block-printed copies from the Tang and Five Dynasties have been discovered in the Dunhuang Library Cave in Gansu. Among these, the most famous is *Diamond Sutra* printed in the ninth year of Emperor Xiantong (868) of the late Tang dynasty, making it possible to study printed materials from the Tang and Five Dynasties period. Subsequently, a batch of block-printed Buddhist scriptures, mantras, and Buddhism collections from the early Tang dynasty to the middle and late Tang dynasty have been unearthed in various places in China, of which some are kept in China while others have been taken overseas. Wood engraved editions and clay movable-type editions from Northern Song, wood engraved editions, clay movable-type editions and wooden movable-type editions in Tangut script have occasionally been excavated. There are also some movable type works from Ming and Qing dynasties. Wood engraved editions, woodcuts and color overprints from the Liao, Jin, and Yuan periods, as well as printed Yuan dynasty paper money, have also been extensively unearthed, with some in remarkably preserved condition. These tangible materials constitute the most authoritative primary sources for the study of printing history.

The second category consists of printing tools and equipment excavated or handed down from one generation to another, including woodblock plates, cutting tools, copper plates with copper movable type, wooden movable type, clay movable type, etc. These items can date back to the Northern Song, Western Xia, and Jin dynasties. In some museums, there are engraving tools and coir scrub brushes from the Qing dynasty. There are still a considerable number of woodblock plates from the Ming and Qing dynasties, which are engraved in either Chinese or Tibetan.

The third category includes original textual records related to printing, such as books, rubbings from stone tablets or printing plate rubbings, etc. Important documents studying the history of printing movable-type components include Volume 18 of *Mengxi Bitan* ("Dream Pool Essays", 1088) by Shen Kuo, a scientist in the Song dynasty, *Zao Huozi Yinshu Fa* ("The Technology of Printing Movable Type", 1298) by Wang Zhen in the Yuan dynasty and *Wuyingdian Juzhenban Chengshi* ("Regulations on Movable Type Printing Adopted at the Hall of Martial Valor", 1776) by Jing Jian in the Qing dynasty. In addition, early records of woodblock printing activities can also be found scattered in the works of the Sui and Tang dynasties, and inscriptions or vows attached to various printed materials also hold historical value by providing information about the publication time, location, and the names of publishers or bookshops. Rubbings from various kinds of plates can offer textual information for the study of copperplate printing and copper movable type printing. Stone tablets can provide clues about publishers, engravers, and the printing status in the area at that time. Records about printing in history books of various dynasties can also be considered original. Descriptions of printing activities by printers themselves should be regarded as the most reliable first-hand information. In addition to the works from Wang Zhen, Jin Jian, and the others mentioned above, this category also contains memories from Zhou Bida, a Song dynasty scholar, Hua Sui, a Ming dynasty scholar and Zhai Jinsheng, a Qing dynasty scholar. The so-called original records refer to events that occurred during the author's lifetime or writings about their activities and are deemed authoritative as the authors are witnesses or participants. However, among the above-mentioned records, those recording events of the preceding dynasties are called non-original, and a clear distinction should be made. For example, records from Shen Kuo about his contemporary Bi Sheng who invented the porcelain clay movable-type printing technology were deemed as original, but his saying that "engraved copies that did not prevail in the Tang dynasty began with copies of Five Classics engraved by Feng Dao in the Five dynasties" is not considered original, and must be validated by physical evidence and original records from the Tang dynasty. Some people have misunderstood Shen Kuo's statement about the absence of printing in the Tang dynasty as original and wrongly concluded that the printing began in the Five Dynasties. Knowledgeable as he is, Shen Kuo surprisingly did not investigate original records related to woodblock printing in the Tang dynasty, which can serve as a lesson for us.

Sometimes, original records related to printing exist in the hand-copied form, making them only available to a few scholars and less accessible to the wider society.

Or they were published in a period rather than during the author's lifetime and scattered in various books. Or the printing activities might be recorded not by printers themselves, but by their friends or relatives who then passed them down to future generations. At this point, information related to these original records can only be found in subsequent literature. These records belong to the fourth category including local chronicles, genealogies, notes from literati reading, novels, anthologies, unofficial histories, memoirs, sketchbooks, and travelogues. Although the range is quite extensive, valuable findings can be made only when long-term meticulous search is conducted. For example, *DaTang Xinyu* ("New Anecdotes of Tang Dynasty") by Liu Su of Tang dynasty contains records about the printing of gate pass under an order from Empress Wu Zetian in 693. These records may be sourced from the archival materials about the court at that time. Similarly, Volume 608 of *Cefu Yuangui* ("Prime Tortoise of the Record Bureau") by Wang Qinruo in the Song dynasty cited a memorial made by the Chancellor of Later Tang, Feng Dao, in 932, that "We observed that the engraved Confucian classics with abundant styles mostly came from Wu and Shu (two kingdoms in the Three Kingdoms period)." This quote is deemed reliable as it comes from a chancellor in the former dynasty. Volume 15 of *Mu'an Ji* ("Collections of Yao Sui", Yao Sui also styled himself as Mu'an) by Yao Sui in the Yuan dynasty contains a record that Yao Shu used clay movable types to print the works of Song dynasty scholars including Zhu Xi during the early Yuan period. Since Yao Sui was Yao Shu's nephew, and he either witnessed the printing activity or heard it from his uncle, such records included in his uncle's tombstone can be considered reliable. From *Fenghua Xianzhi* ("Fenghua County Chronicles") of Zhejiang, we also learned that Ma Chengde, the then head of the county, printed books using 100,000 pieces of wooden movable-type components in 1322. This printing activity was not recorded in other books, and was only known to the locals as cited from a tablet inscription authored by Deng Wenyuan, a writer from the Yuan dynasty. Research on the history of printing requires a laborious collection of basic historical materials. The lifetime efforts of one person can only contribute to the discovery of a few new pieces of information on the basis of the foundation laid by previous researchers. More information awaits exploration by many.

1.3.2 Physical and Documentary Materials from Foreign Countries

When we conduct research on the history of exchanges of printing technologies between China and foreign countries and also research on comparisons of printing techonlogies between China and foreign countries, significant efforts are expected to collect original documents written in different languages abroad and recently published primary materials. Books from ancient Japan, Korea and the Democratic People's Republic of Korea are easy to understand, because the Chinese characters were also widely adopted in writing in those countries at that time. However, where some ancient books are still difficult to access, materials published in Japanese, Korean, and other languages must be used as references. Latin books from the Western world are hard to comprehend or obtain. We should refer to works written in English, French, German, and other languages in addition to frequent visits to museums and libraries overseas. Some ancient Arabic books originally existed only in hand-copied form and have had many language versions over time, some of which have been translated into Chinese.The most important is *Jāmi' al-Tawārikh* or *Collected Histories* (1311) from Rashid al-Din (1247–1318), a Persian author. Some records about Chinese printing that are not contained in Chinese historical materials were collected and written down by foreign visitors and thus fill the gaps in Chinese records. Studing Chinese history without understanding the histories of other Eastern and Western countries seems to be impossible, and meanwhile it is not enough to use records in the Chinese language as the single source. Simultaneously, attention must be paid to the latest information on Chinese prints collected and discovered overseas. Therefore, scholars who study printing history need to be proficient in not only ancient Chinese but also Japanese and several Western languages. Without direct access to original materials through foreign-language works, researching Chinese printing history would be quite challenging. Today, different from the time when Ye Dehui and Sun Yuxiu lived, the information age in which we live sets higher requirements on research.

Chapter 2
Historical Origin of Printing Techniques

2.1 Invention and Development of Chinese Characters

2.1.1 From Oracle Bone Script of Shang and Yin Dynasties to Bronze Script of Zhou Dynasty

The printing technology is the result of social and cultural development and also the driving force for the progress of human civilization. The carriers of culture refer to classics with developed characters, and it's self-evident that any country or nation can only develop printing after it has a fully functional writing system. China, as one of the oldest civilizations in the world, has a written history of over 5000 years. Before the invention of characters, people exchanged ideas through spoken language and gestures, passed stories from mouth to mouth, and relied on their memories day in and day out. Later, they began to make knotted cords to keep track of events. A set of various colored strings with knots that came in different sizes, shapes, and numbers were used for recording and "passing on" the events, serving the purpose of language.

According to *Zhouyi Xici* ("Attached Verbalizations of the Book of Changes"), "in the remotest antiquity, the government was carried on successfully by the use of knotted cords (to preserve the memory of things). In subsequent ages, the sages substituted written characters and bonds for them." *Zhuangzi Quqie* ("Prying Open Valises") recorded, in the times of Fuxi and Shennong, that is, the Early Neolithic (4800–4400 BC), "people made knots on cords in carrying on their affairs". Knotting, however, had to be supplemented by memory, because it could hardly record complex events, and it was also uncertain whether others could accurately understand the original intention of the knotter. Therefore, human beings created characters as symbols to express thoughts and languages. The creation of characters is a milestone in the history of human civilization.

P. Jixing, *History of Printing Techniques in China*,
https://doi.org/10.1007/978-981-96-6119-0_2

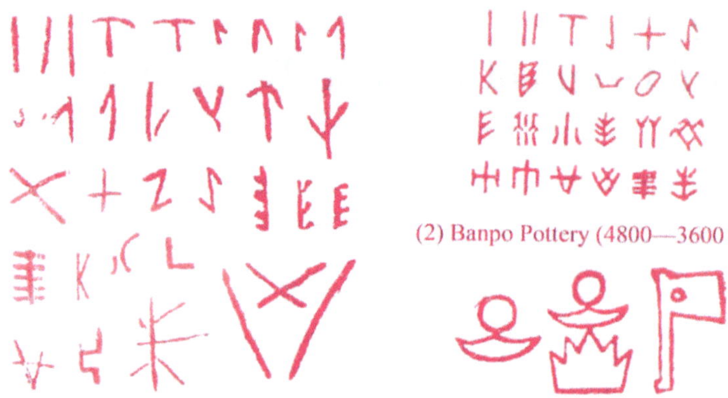

(2) Banpo Pottery (4800—3600 BC)

(1) Yangshao Pottery (5000—4000 BC) (3) Longshan Pottery (3000—2300 BC)

Fig. 2.1 Graphs and symbols on Chinese ancient pottery

The original characters were abstract and simplified graphs carved on wood, stone, or pottery. For example, the graphs found on the painted pottery unearthed in Banpo, Xi'an were used to express thoughts or record certain events (Fig. 2.1). These graphs could date back to around 4000 BC. Much more advanced than knotting, they are deemed as the origins of Chinese writing. The graphs on the unearthed objects are relatively simple, and there should be more complex ones to be discovered.

Later, the graphs evolved into pictographs that developed into today's Chinese characters. According to historical records, Chinese writing was invented by Cangjie. Xun Kuang (313–238 BC) described in his *Jiebi* ("Free from Obsessions"), "there were many who liked to write, but there was only one whose writing was passed down through the ages, Cangjie, because of the soleness of his creation." Cangjie is a legendary ancient Chinese figure said to have been an official historian for the Yellow Emperor. He might have collected and organized but never created Chinese characters, because characters evolved over a long period of time and cannot be created by any individual. The pictographs used in the Xia dynasty (2070–1600 BC)[1] were more advanced than those in the era of Cangjie. We can recognize and decode the pictographs used from Xia to Shang (1600–1300 BC) and Yin (1300–1046 BC). From the end of the nineteenth century to the twentieth century, archaeologists found at the ancient capital site of the Yin dynasty in Anyang, Henan, many inscriptions carved on tortoise shells and ox scapulae, and thus called them *jiagu wen*, or oracle bone script (Fig. 2.2). Since then, more oracle bone inscriptions were unearthed successively in other places. Around 160,000 pieces of oracle bone have been found so far. Among the 4600 characters inscribed on them, only 2000 have been decoded. Oracle bone script can be roughly divided into early, middle, and late periods. The

[1] The chronology of ancient Chinese history in this book is based on *Chinese History Chronology* written by Fang Shiming, and the chronology before 841 BC is based on *Xia, Shang, and Zhou Chronology*.

Fig. 2.2 Oracle bone inscriptions of the Shang King Wu Ding period in the Yin dynasty[2] (1250–1192 BC, 21 cm × 19 cm). *Note* The divination inscriptions were carved on an animal bone, containing 129 characters in three parts

early period is the King Wu Ding period in the Yin dynasty (thirteenth–twelfth century BC, or 1250–1192 BC). The late period is the King Di Xin period at the end of the Yin dynasty (1075–1046 BC). In 1950, a small number of oracle bones were found in the Erligang early Shang site (1620 BC), Zhengzhou, providing more information on the oracle bone script in early Shang dynasty, and giving clues for the pictographs of the Xia dynasty. We can tell from the unearthed inscriptions that the oracle bone appeared and were used no later than 4000 years ago. In terms of character structure, the oracle bone script had already become mature. In addition to pictographs, it used more advanced methods such as "signific-phonetic", "combined meaning", and "wrongly borrowing" to form characters.

People of the Shang dynasty believed in ghosts and spirits. Before traveling, hunting, and fighting, they performed divination for good and bad luck. A *wushi*, meaning shaman-historian or diviner, drilled or chiseled pits on the bone or shell that had been prepared in advance, and put fire in the pit until it cracked in various directions and depths. The diviner would read the cracks to learn the answer to the divination, and then use a sharp knife to engrave the divination results near the cracks. This is *buci*, meaning divination inscription. Most of the oracle bone inscriptions are *buci*, so they are also called "*jiagu buci*", or oracle bone divination inscription. The oracle bones with inscriptions were tied together for preservation, which were called *ce* (archive). *Zhoushu Duoshi* ("Many Scholars of Book of Zhou") recorded that "ancestors of Yin had their archive and books." The oracle bone script of *dian* (book) looks like two hands holding a book, indicating that the Shang dynasty witnessed the earliest form of books. Each oracle bone generally contained over 50

[2] Dong Zuobin, *Yin Li Pu* ("Calendar of the Yin Dynasty") II, *Volume 9*, Lizhuang, Institute of History and Philology, Academia Sinica, 1945: 43.

characters, and sometimes more than 100 characters.[3] These inscriptions provided rich information about social economy, science, and technology, etc., and have been sorted out and studied by experts. They also represent the technical skills and early origin of engraving. The columns of characters are straight, the strokes are clear and delicate, arranged straight up and down, and the thickness of each stroke is basically even. The rounded forms are often converted to rectilinear ones, which is convenient for cutting, laying the foundation of the square-shaped Chinese characters. Each line is engraved word by word from top to bottom, which also forms the basis for straight Chinese writing. The engraving tool is a sharp knife with a thin blade. After one line is engraved, another line is engraved. It is worth noting that sometimes characters are engraved in reverse form on oracle bone tablets, which is the origin of reversing characters for later generations. Inscriptions can also be found on bronzes cast in the Shang dynasty. There are more than 1000 Shang bronzes with inscriptions, but the number of inscriptions is relatively small.

Shang (seventeenth–thirteenth century BC) and Yin (thirteenth–eleventh century BC) lasted for about 600 years. In 1046 BC, King Wu of Zhou established the Western Zhou dynasty (1046–771 BC) with the capital located in Haojing (Chang'an District in Xi'an, Shaanxi). In 770 BC, King Ping of Zhou (770–720 BC) moved the capital to Luoyang, which brought about the beginning of the Eastern Zhou dynasty. Eastern Zhou dynasty is divided into two periods, the Spring and Autumn period (770–476 BC) and the Warring States period (475–221 BC). After the relocation of the capital, Zhou emperors lost their ability to control the feudal lords, and China entered a period of political division and hegemony. The characters used in the Western Zhou is basically the same as those in the Yin dynasty. The Zhouyuan oracle bone script of the eleventh century BC is very similar to the *buci* of the Shang Yi and Di Xin (King Zhou) periods in the late Yin dynasty. The Western Zhou bronzes and their inscriptions are also the same as those in the late Yin dynasty. Inscriptions on the bronze wares are also known as *jin wen* or "bronze script". Inscriptions were found on most of the 4000 existing Western Zhou bronze objects. On each object, the number of characters varies from a few to over 100. The characters of the Shang, Yin, and early Western Zhou dynasties, sometimes complex, sometimes simple, have no unified way of writing, as the radicals are sometimes on the left and sometimes on the right, and most characters are pictographs. However, since the middle and late Western Zhou (ninth–eighth century BC), the characters have evolved significantly. Among the bronze artifacts of the mid-Western Zhou that were unearthed in Zhuangbai Village, Fufeng County, Shanxi Province in 1976, the Shi Qiang Pan (plate for Qiang the official historian) has an inscription of 284 characters (Fig. 2.3), which contains the largest number of inscriptions found in the past 50 years. This plate was made during the reign of King Gong of Zhou (922–900 BC).[4] Comparing the mid-Western Zhou

[3] To find more about the oracle bone script, please read: Wu Jiekun and Pan You, *History of Study on Oracle Bone Inscription in China*, Shanghai People's Press, Shanghai, 1985; and Xia Nai, *Archaeological Discoveries and Research in China*, Cultural Relics Press, Beijing, 1984.

[4] Archaeological Team of Zhouyuan, Shaanxi. Brief Report on the Excavation of Western Zhou Bronze Vessel Cache at Zhuangbai No. 1, Fufeng, Shaanxi [J]. *Cultural Relics*, 1978 (3): 1–16.

Fig. 2.3 Bronze script on the Shi Qiang Pan of the Western Zhou

bronze script with the Shang and Yin oracle bone script, we can find that the strokes are better-proportioned and have more rounded turns, and the pictograph is gradually decreasing.

2.1.2 From Large Seal Script of Zhou Dynasty to Small Seal Script of Qin Dynasty

If we say that Chinese characters represented by the oracle bone script during the Shang dynasty are already mature, they have become even more succinct and enriched after hundreds of years of utilization and consummation. Later characters, represented by the Western Zhou bronze script, were capable of recording everything people think and say, and thus became advanced ideographs which have one syllable for each character. Written materials expanded the space and time range of information transmission, recorded human activities into faithful historical accounts, thus cultivating the spiritual civilizations. Historiographers in Western Zhou systematically collected and standardized the existing characters, and compiled and popularized standard handbooks of characters, which is significant to the development of

Chinese characters. *Yiwen Zhi* ("Literature Records"), the 30th Scroll of *Qian Han Shu* ("Book of Former Han"), completed in 83 AD, recorded that the *Shizhoupian Textbook* has 15 chapters and "it is the book that historiographers used to teach schoolchildren to read and write." Yan Shigu, a philologist from the Tang dynasty, annotated that "in the times of King Xuan of Zhou, the *taishi* (Grand Historian) wrote 15 chapters in large seal style." That is to say, during the late period of King Xuan of Western Zhou (827BC–782BC), there were historiographers who compiled *Shizhou*, a book of characters, and used it as the textbook for schoolchildren to learn characters. The Chinese character *Zhou* originally means "to read aloud, to recite", and its meaning can be transferred to "textbook". *Shizhou* refers to the textbook compiled by the historiographer for learning characters, and these standardized characters given by *shizhou* were called *zhou wen* by later generations. According to paleographer Tang Lan, the word *shiliu* found on the Yueding vessel of the 19th year of King Li of Western Zhou (860 BC), which is now collected in Museum of Chinese History, is actually the *shizhou* mentioned in *Qian Han Shu*.[5] It can be estimated that the writers of this earliest literature on grammatology are the historiographers in the times of King Li and King Xuan of the Western Zhou (ninth century BC). The *zhou wen* was the emerging character prevalent during the Western Zhou, which is known as *da zhuan* or large seal script, the intermediate between Shang dynasty oracle bone script and Qin dynasty seal script. Bronze inscriptions of the Zhou dynasty were mostly characters of large seal script.

During the Spring and Autumn Period and the Warring States Period, with the decline of the Eastern Zhou royal family, although the large seal script was still used by the states, different varieties had come into being in different regions. Large seal of different varieties could be found on bamboo slips or bronze wares collected from different places in this period. The development of Chinese characters necessitated the conclusion of this situation where numerous varieties of large seal style coexisted. It required the unification and standardization of the characters, just as how Chinese people at that time required the ending of an age of division and the return to a unified power. When Qin Shi Huang, the first emperor of China, conquered all the other six states and unified the country in 221 BC, he unified the characters, which is another milestone in the development of Chinese characters. According to *Annals of Qin Shi Huang* in *Shiji* ("Records of the Grand Historian", 93 BC), in the 26th year of Qin Shi Huang (221 BC), after the unification of the country, the prime minister Li Si (284 BC–208 BC) reported to the throne that "we'd better have a unified measurement for *dan*, *zhang* and *chi*, unified roads for carriages, and unified characters for books". *Shuowen Jiezi Xu* ("Preface to Analytical Dictionary of Characters") recorded that during the Spring and Autumn Period and the Warring States Period, the country was not unified by the Zhou emperors, but divided into seven countries, and "they all used their own characters with different shapes. When Qin Shi Huang had just unified the country, prime minister Li Si suggested their unification, and abolished the characters that were different from Qin's. Li Si wrote a book, *Cangjie Pian*

[5] Liu Qiyi. Inscription on the Bokuifu's Vessel and the Reign of King Li [J]. *Cultural Relics*, 1979 (11): 16–20.

("Compilation by Cangjie")... in which characters were all selected from *shizhou*, and the large seal underwent major change or modification to form what we call *xiao zhuan*, or small seal." Li Si standardized and simplified the large seal style that had existed in Qin, reduced the pictograph, and added linear symbols. He compiled *Cangjie Pian*, popularized it by law, and abolished different characters that had existed in the other six states, making Qin seal script, or small seal script, the official characters. We can see the appearance of small seal on the Qin decrees (Fig. 2.4), measuring instruments, stones carved for extolling, etc. handed down or excavated. The strokes of small seal are concise, and the radicals are in the fixed position, and there is only one way to write one character. The writing or carving of small seal style is easier than that of large seal. After this great revolution, Chinese characters were basically finalized. Later developments were just minor changes in styles or further simplification of structure. Despite the loss of *Cangjie Pian*, *Shuowen Jiezi* of the Han dynasty includes 9300 characters of small seal characters which should have no great difference from those included in *Cangjie Pian*.

Fig. 2.4 Qin decree in small seal script (from the Qin decree of the 26th year of Qin Shi Huang)

2.1.3 From Clerical Script of Han Dynasty, Clerical-Regular Script of Six Dynasties, to Regular Script of Sui and Tang Dynasties

The development of Chinese characters is from complex to simple, making them easy to write and learn. Although small seal script was simplified, its shapes and strokes were still very similar to those of the large seal, which made it hard to write. As a result, people modified and simplified them again, developing the *li shu*, or clerical script. Clerical script first appeared in the Qin dynasty, and was also called *zuo shu*. Wei Heng (245–291) of the Western Jin recorded in his *Siti Shushi* ("Calligraphy Postures of Four Styles") that "people in the Qin dynasty use the small seal script. Since there were many things to deal with and seal script was difficult to write, petty officials were then commanded to write in *zuo shu*, which was called clerical." *Li* refers to inferior clerical officers who were engaged in paperwork in feudal authorities. Since they needed to deal with piles of files, they used a convenient way to enhance efficiency, and thus developed a new style of writing. According to the legend, clerical script was created by Cheng Miao, a prison officer of a county government during the times of Qin Shi Huang. However, the formation of a character style is a gradual process and is impossible to be created by an individual in a short period of time. We could only say that someone or some people made a summary of the already existing character styles. The clerical script of the Qin dynasty still shared the style of seal script, but in the early Western Han dynasty (206 BC–25), the strokes of small seal were gradually diminishing, and the clerical script soon became the official script. During the Eastern Han dynasty (25–220), clerical script entered the mature stage (Fig. 2.5). Clerical script inherited the style of small seal, but changed the curve stroke into angular stroke. The most apparent feature of the clerical script is that the beginning of horizontal lines is akin to the head of a silkworm while the finish is akin to the tail of a swallow and that its form appears flat and square. The characters were no longer with a trace of pictographs, and the six traditional methods of constructing the form of characters were not evident. Clerical script was followed by regular script, which was identical to the characters commonly used nowadays, and that was another revolution in the history of Chinese characters. When Chinese people were writing clerical script, they formed a convenient and cursive way, i.e., the clerical-cursive script or the cursive form of clerical script.

Clerical-cursive script further simplified the clerical script, and developed into the regular script (*kai shu*) or real script (*zhen shu*). Regular script could be traced back to the beginning of the third century, from the late Eastern Han dynasty to the Three Kingdoms Period. Early regular script was found on the rubbings of stone inscriptions of the Wei calligrapher Zhong Yao (151–230) and the Eastern Jin calligrapher Wang Xizhi (321–379), as well as the Buddhist scriptures excavated in Dunhuang stone rooms. It replaced clerical script and became commonly used during the Wei, Jin, and Northern and Southern dynasties (third–sixth century). Since the style of clerical script was partially preserved in the new script, scholars named it *li kai*, meaning clerical-regular script (Fig. 2.6), which is a transition from clerical script to

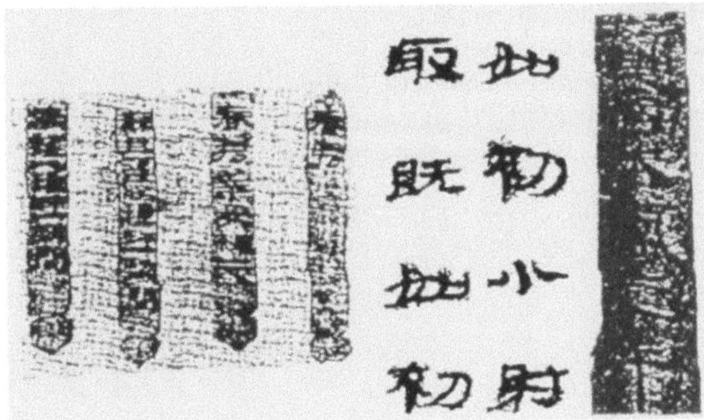

Fig. 2.5 Clerical script of the Han dynasty (from the unearthed bamboo slips)

regular script. After the unification of China during the Sui and Tang dynasties, as the economy and culture turned out to be more advanced, the characters also became more unified and standardized. Clerical-regular script was completely replaced by the regular script that was used across the whole country (Fig. 2.7). The regular script written by famous calligraphers, such as Zhi Yong in the Sui dynasty, and Yu Shinan, Ouyang Xun, Chu Suiliang, Yan Zhenqing, Liu Zongyuan in the Tang dynasty, was imitated by schoolchildren when they were learning to write. The regular script used in the Sui and Tang dynasties has reached a mature stage. It has become the commonly used formal style for about 1400 years and has remained in use until now. The end of horizontal strokes is not raised but rather retracted. The left-falling stroke slopes downwards and has a sharp ending. The hook stroke does not form a turning but a solid hook. The word turns from being flat to square and regular. Compared with oracle bone script, large seal, small seal, and clerical script, regular script is the easiest to write, carve, recognize, and read (Table 2.1), which makes it suitable for woodblock printing. Thanks to regular script, printing techniques began to bud in the Sui dynasty, gaining momentum in the Tang and the Five dynasties, and reaching their prime in the Song dynasty. As calligraphers wielded the writing brush in different manners, the style of regular script could vary creatively and calligraphy thus became an art. Characters carved on the woodblocks usually imitated the handwriting of famous calligraphers, so printing techniques were combined with the art of calligraphy, rendering printed books artworks with aesthetic value. Regular script could also be written fast, thus semi-cursive script and cursive script derived from it, and each person could form his or her own writing style. Chinese characters sufficiently demonstrate the extensive, profound, long-standing and colorful characteristics of Chinese culture.

Fig. 2.6 Clerical-regular script (from the manuscript of *Mahāparinirvāṇa Sūtra* in the Northern Wei)

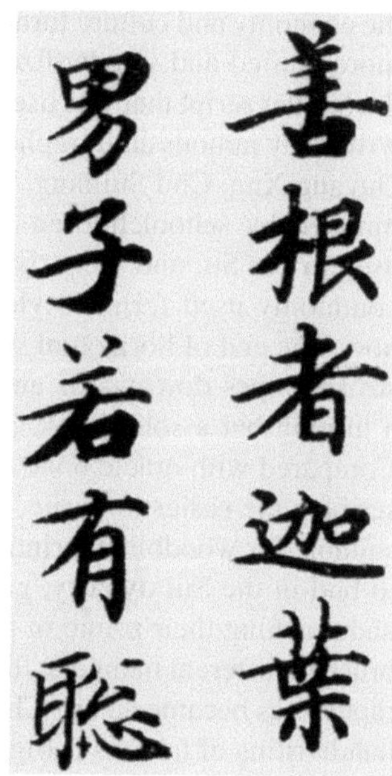

Fig. 2.7 Regular script (from the manuscript of *Mahāparinirvāṇa Sūtra* in the Tang dynasty)

Table 2.1 Comparison table of various fonts

Font	Character					
Oracle bone script	典	册	千	年	印	史
Large seal	典	册	千	年	印	史
Small seal	典	册	千	年	印	史
Clerical script	典	册	千	年	印	史
Regular script	典	册	千	年	印	史

2.2 Development of Brush, Ink, and Ancient Books

2.2.1 *History of Writing Brush and Pen*

Pen is a tool for writing. People use pens to write classics so that civilization can be passed on. Pen is also an essential material prerequisite for developing ancient printing techniques. All the characters were written on the block using pens before they got carved and printed. China has a history of over 5000 years of developing pens in tandem with characters. Pens in ancient times can be divided into hard-pens and ink-brushes. According to *Dongtian Qinglu Ji* ("Records of the Pure Registers of the Cavern Heaven", ca. 1240) written by Zhao Xigu, "in ancient times people dipped bamboo stalks into the paint to write." On the pottery of Yangshao culture which is about 5000 years before present, excavated in Banpo and Jiangzhai Town, Xi'an City, Shaanxi, there were both primitive characters and colorful pictures. The characters were written with hard-pens that were made of bamboo stalks or wooden sticks with a sharpened end, while the pictures were painted by ink-brushes. Ancient hard-pens produced the same effect as the fountain pens now, and they were durable and easy to make. In 1900–1901, Aurel Stein (1862–1943) discovered wooden pens from the third century in Niya ruins near Hotan Prefecture, Xinjiang. The pens are 15–23.5 cm long.[6] They are sharpened at one end, and some of them split at point. Bamboo pens and wooden pens were not only used by the Han Chinese but also by Tibetan and Uyghur people. For example, the eighth century transcript of the *Infinite Life Sutra* in Tibetan script excavated in Dunhuang, Gansu, was written with hard-pens.

People in the Shang and Yin dynasties continued to use hard-pens and brushes as writing tools. The original form of the Chinese character *Bi* (筆, "pen", simplified Chinese: 笔) is *Yu* (聿). Its form in oracle bone script is , which assembles a hand holding an ink brush to write. Its form in small seal is . Oracle bone script written with ink-brush dipped with cinnabar or graphite was found on animal bones, pottery, and jade pieces excavated in Yinxu, Anyang. In most cases, people firstly wrote on the oracle bones with ink brushes or hard-pens, and then carved words with sharp knives. Written but not carved characters could sometimes be found on oracle bones that contained carved characters. After the Spring and Autumn period and the Warring States period, ink brushes became the major writing tools. They were not only recorded by literatures but also testified by physical evidence unearthed (Fig. 2.8). In 1954, ink brushes were excavated in the No. 15 Chu tomb of the Warring States period in Zuojiagongshan, Changsha, Hunan. They are 21 cm long, and their sheaths are 23.5 cm long. They are all made of bamboo. The hair of the brushes is twined by thread at one end of the pen (No. 1, Fig. 2.8).[7] In 1958, bamboo slips and ink brushes were unearthed from two Chu tombs of the late Spring and

[6] Stein A. *Ancient Khotan*: *Vol. 1* [M]. Oxford: Clarendon Press, 1907: 398, 403.

[7] Hunan Provincial Cultural Relics Management Committee. Wooden Coffin Tomb of the Warring States in Zuojiagongshan, Changsha [J]. *Cultural Relics Reference Materials*, 1954 (12): 3–19.

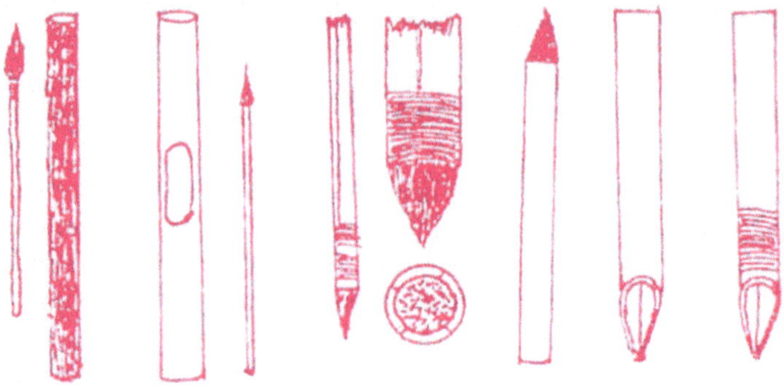

Fig. 2.8 Brushes and pens in ancient China

Autumn period (6 BC) in Changtai, Xinyang, Henan. In 1975, 1150 bamboo slips and 3 ink brushes were unearthed from a Qin tomb (217 BC) in Shuihudi, Yunmeng, Hubei. The ink brushes are with the length of 21.5 cm and the diameter of 0.5 cm. The stalks are made of bamboo and are thin at one end and thick at the other. The middle of the handle is hollowed out at two sides so that the brush could be easily taken. The end of the brush is a tube securing the bundle of hair to the stalk, which looks the same as ink brushes nowadays (No. 2, Fig. 2.8).[8] The producing techniques of ink brushes unearthed in Zuojiagongshan are not so advanced as those in Shuihudi. Legend wrongly credits the invention of the ink brush to the Qin general Meng Tian (ca. 265–210 BC).[9] We cannot say that Meng Tian invented brushes, but he might ameliorate the techniques of producing brushes. After the unification of characters in the Qin dynasty, the techniques of producing brushes have been improved, which could be testified by the unearthed Qin brushes. In 1975, a 24.9-cm long bamboo ink brush was excavated in the tomb of Sui Shaoyan of the early Western Han dynasty (167 BC) in Fenghuangshan, Jianglin, Hubei. Its tip is made of a tube filled with a bundle of animal hair, which is the same as the ink brush from the Qin dynasty. Ink blocks, ink stones, and wooden slips were also found in the tomb.[10]

In northern regions, due to the lack of bamboo, brush was made of wood. In 1931, a wooden ink brush of the Han dynasty was excavated in Juyan, Banner of Ejina in Inner Mongolia. The total length of the brush is 23.2 cm and the tip, broken, is 1.4 cm long. The wooden stalk was chopped into four parts that were then joined together with a bundle of hair put in between and then bound tight with twine. The other

---

[8] Hubei Museum. *Tomb of the Qin Dynasty in Shuihudi, Yunmeng* [M]. Beijing: Cultural Relics Press, 1981: 20.

[9] If the year of birth or death of the people mentioned in this book is unknown, it will be represented by "?"; if the year of birth or death is uncertain, it will follow the word "ca."; if both the year of birth and death are uncertain but an event is recorded, it will be noted as "alive during" the start and end time of the event that is recorded.

[10] Fenghuangshan No. 167 Han Tomb Excavation Team. Report on the Excavation of the No. 167 Han Tomb at Fenghuangshan, *Jiangling, Hubei* [J]. *Cultural Relics*, 1976 (10): 31–35.

end of the brush was also bound tight (No. 3, Fig. 2.8).[11] However, the bamboo ink brushes excavated in the tomb of Eastern Han in Mozuizi, Wuwei, Gansu, are similar to those excavated in Changsha and Jiangling, with a diameter of 0.6 cm, a stalk of 21.9 cm, and a tip of 1.6 cm. The stalk is a painted tube with weasel's hair inserted into one end and bound with silk thread. The other end of the stalk is sharpened and the intaglio *Baimazuo* (Made by Baima) is carved in the middle of the stalk. On another brush is the intaglio *Shihuzuo* (Made by Shihu). These should be the names of the makers of the brushes.[12] The excavation of the above-mentioned ink brushes from the Warring States, Qin, and Han dynasties has provided physical materials for us to know the shape, structure, size, ingredient and producing method of ancient brushes, based on which the ink brushes in the Northern and Southern, Sui and Tang dynasties evolved. In 1973, three wooden ink brushes from the Tang dynasty were excavated in Astana, Turpan, Xinjiang. They were placed on a wooden brush-holder that is 7.8 cm high, composed of five columns and two bars, with the same style as the Central Plain. The tip of the brush is short and blunt. It is notable that the stalk, no longer bound with thread and painted, is more like the ink brushes used nowadays, which reflects the progress of techniques of producing brushes in the Tang dynasty. In 1967, a reed pen from the Tang dynasty was excavated in Astana, with the length of 10.6 cm.[13] In 1972, people discovered a set of transcripts in Chinese script, Tangut script, and Tibetan script from the Western Xia regime in the twelfth century in a cave in Zhangyi, Wuwei, Gansu. There are also two wooden hard-pens, with the length of 13.6 cm and 9.5 cm. These wooden hard-pens are cut to a slope at the thinner end and then cleaved to form the tip of the pen.[14] So far, there is physical evidence unearthed for both hard-pens and ink brushes in ancient times. Hard-pens were not only used by ethnic minorities but also by the Han people. The main material for producing brushes in ancient times was the hair of rabbit, deer, wolf, and sheep. Hard hair and soft hair were often mixed to make brushes.

1. Bamboo brush from the Warring States period excavated in Changsha, 1954. The brush is 21 cm long with a bundle of hair fixed at one end of the bamboo tube. The outer tube is 23.5 cm long, made of bamboo and painted in black.
2. Bamboo brush from the Qin dynasty excavated in Yunmeng, Hubei, 1954. The brush has a length of 21.5 cm and a diameter of 0.5 cm. The hole in the middle of the outer tube is 8 cm long.
3. Wooden brush from the Han dynasty excavated in Juyan, 1931. The brush is 23.2 cm long. The wooden stalk is hollow and split into four pieces. The tip is 1.4 cm long and bound with twine.

---

[11] Ma Heng. Report on Juyan Brushes of the Han Dynasty [J]. *Chinese Studies Quarterly*, 1932 (1): 67–73.

[12] Gansu Museum, A Brief Report on the Excavation of Three Han Tombs in Mozuiz, Wuwei. [J]. *Cultural Relics*, 1972 (12): 9–19.

[13] Xinjiang Uygur Autonomous Region Museum, *Artifacts Unearthed in Xinjiang* [M]. Beijing: Cultural Relics Press, 1975:129.

[14] Gansu Museum, Western Xia Relics Discovered in Wuwei, Gansu [J]. *Archaeology*, 1974 (3): 200–204.

4. (null)
5. Bamboo hard-pen from the Western Xia excavated in Wuwei, Gansu, 1972. It is 13.6 cm long and cut to a slope at the thinner end as the tip of the pen.
6. Hard-pen from the Tang dynasty excavated in Turpan, 1967. It is 10.6 cm long. The tip is bound with twine.

2.2.2 History and Production of Chinese Ink

Ink is a colorant used in the printing process and also an indispensable writing material in ancient times. Chinese ink has a history of about 3000 years. It shows a color of pure black and maintains an artisan sheen that never fades. This feature makes it superior to the ink produced in other countries, and thus it is spoken highly of in the world and modelled on by the Western and Eastern countries. The main ingredient of Chinese ink is carbon black, which is the light and loose black powder produced by carbonaceous matter after incomplete combustion when the oxygen supply is insufficient. Carbon black is amorphous carbon, composed of many tiny graphite crystals, with complicated microstructures. Chinese people originally used carbon black and gum to make liquid ink, and later, on the basis of that, they made solid ink blocks. When using the ink blocks, people dipped them into water and ground them on inkstones whose development was closely connected with the utilization of solid ink. Other countries also used ink to write in ancient times, but the ingredients and techniques of production were different, and we will not go into details here. Speaking of the history of Chinese ink, we have already mentioned that the oracle bone script from the Yin dynasty in the fourteenth century BC was written in black or red. The microchemical analysis of the black pigment shows that its ingredient is carbon black.[15] It is evident that people had already entered the prehistory of using ink at that time. In the Western Zhou and the Spring and Autumn period, with the development of technology, people made a breakthrough in ink manufacturing techniques. In 1964, seven wares with characters written in black ink were excavated in the tomb of a Western Zhou aristocrat in Beiyao, Luoyang, Henan. The characters were written respectively at the bottom of a copper *gui* (簋, a square container) and at the base of copper spears and lead spears. All characters are perfectly legible.[16] Copper *gui* is from the times of King Kang of the Western Zhou (1020 BC–996 BC). Spears and halberds with characters written in ink are also from the early Western Zhou. The characters on metal artefacts written in ink are still intact after about 3000 years (Fig. 2.9).

Tian Zi-Fang, Chapter 21 of the Outer Chapters of Zhuangzi, compiled in the Warring States (BC 290), recorded that "the ruler Yuan of Kingdom Song wishing to

[15] Benedetti-Pickler A. Microchemical Analysis of Pigments Used in the Fossae of the Incisions of Chinese Oracle Bones [J]. *Industrial & Engineering Chemistry: Analytical Edition*, 1937 (9): 149–152.

[16] Cai Yunzhang. A Brief Discussion on the Ink Writing in the Western Zhou Tombs at Beiyao, Luoyang [J]. *Cultural Relics*, 1994 (7): 64–69.

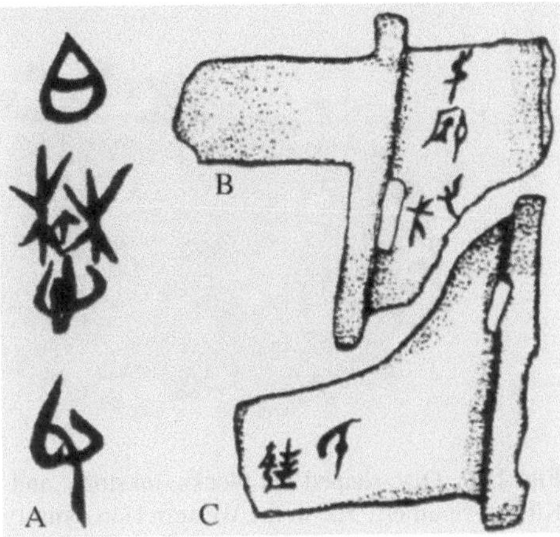

Fig. 2.9 Characters written in black ink[17] (on bronze wares in the tomb of an aristocrat from the times of King Kang of the Western Zhou)

have a map drawn, the masters of the history all came (to undertake the task). Having received his instructions and made their bows, they stood, licking their brushes to prepare their ink…" The ruler Yuan of Song refers to Duke Yuan of the State of Song (530 BC–516 BC). The event that his liegemen licked the brush took place in the sixth century BC. Literature from the Spring and Autumn and the Warring States was usually written in ink on silk wadding and bamboo slips, which were frequently unearthed. In 1975, a disc-shaped ink block, an inkstone, and two wooden slips were excavated in the No. 4 Qin tomb (223 BC) at Shuihudi, Yunmeng, Hubei.[18] They were produced in the late Warring States period before the unification of China by Qin Shi Huang, so the ink should be the same as that in the times of Duke Yuan. Speaking of the manufacturing techniques of ink blocks, the even and small particles of carbon black were firstly acquired, then blended with glue, and finally made in the shape of discs or balls. Ink block is a great technological breakthrough in ink production. Ink block is more storable and portable than liquid ink. People only need to put some water on the inkstone and grind the ink block using a grindstone, so that they can get the ink and write. This kind of ink should be the early soot ink that could be seen now. Since it entails the usage of grindstones, it seems that it had not been processed by shaping moulds.

Ink produced after the Western Han dynasty has been excavated in various regions. For example, in 1975, a set of stationery was excavated in the No. 168 Han Tomb at Fenghuangshan, Jiangling, Hubei, which includes bamboo brushes, ink, inkstones, blank wooden slips and writing knives. The tomb owner is Sui Shaoyan, who was buried in the thirteenth year of King Wen (167 BC). There are five black ink blocks in the tomb. The biggest one is 1.5 cm long, 1.1 cm wide and 0.4 cm thick, in the shape

17 Cai Yunzhang. A Brief Discussion on the Ink Writing in the Western Zhou Tombs at Beiyao, Luoyang [J]. *Cultural Relics*, 1994 (7): 64–69.

18 Hubei Xiaogan Region Phase II Archaeological Training Class for Both Industrial and Agricultural Artifacts. Brief Report on the Excavation of Eleven Qin Tombs at Suihudi, Yunmeng, Hubei [J]. *Cultural Relics*, 1976 (9): 53.

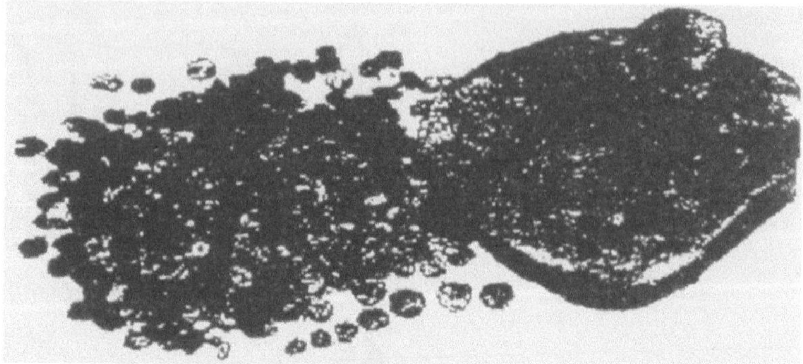

Fig. 2.10 Disc-shaped ink blocks, inkstone, and grindstone excavated in the tomb of Zhao Mei, King of Southern Yue in the Western Han dynasty, in Guangzhou[21]

of melon seeds. The inkstone is round, with a diameter of 9.8 cm at the bottom and a thickness of 1.8 cm. There is a 3.5-cm-high grindstone on the inkstone, and ink mark was found on the inkstone and under the grindstone.[19] In 1983, 4386 disc-shaped ink blocks were excavated in the tomb of Western Han in Xianggang, Guangzhou (Fig. 2.10). They are not equal in size. Large ones are 1.31 cm in diameter and 0.42 cm in thickness; medium ones 1.16 cm in diameter and 0.33 cm in thickness; and small ones 0.81 cm and 0.23 cm. The ink blocks are black and slightly reddish, having a smooth and fine texture. They are flat at the bottom and rise around the perimeters, with a shape like a water drop. It can be estimated that people mixed particles of carbon black and glue to form a paste-like substance and dropped it onto the board, so they are of different sizes. Excavated along with them in the same tomb were inkstones and grindstones. The inkstones are in the shape of squares with rounded corners, without edges. Their sides are 12.5–13.2 cm long and 2.8 cm thick. Grindstones are cylindrical, concave in the middle, 3.2–3.3 cm in diameter and 2.2 cm in height.[20] The tomb owner is Zhao Mei (137 BC–122 BC), the second King of Southern Yue, buried in the first year of Yuanshou (122 BC).

In the Eastern Han dynasty (25–220), new development has been made in the techniques of ink production. According to *Baiguan Zhi* ("Treatise on the Officials") in *Hou Han Shu* ("Book of Later Han"), the official *Shougongling* "manages the paper, brush, and ink used by the emperor, properties of *Shangshu* and the sealing clay", and the *Shangshu You Cheng* (right aide of *Shangshu*) "manages the silk

[19] Jinan Cheng Fenghuangshang No. 168 Han Tomb Excavation Group. Brief Report on the Excavation of Western Han Tomb No. 168 at Fenghuangshan, Jiangling, Hubei [J]. *Cultural Relics*, 1975 (9): 1–8.

[20] Mai Yinghao, Huang Zhanyue. *Tomb of the Nanyue King of the Western Han Dynasty: Vol. 1* [M]. Beijing: Cultural Relics Press, 1991: 128, 142. Mai Yinghao, Huang Zhanyue. *Tomb of the Nanyue King of the Western Han Dynasty: Vol. 2* [M]. Beijing: Cultural Relics Press, 1991: 76.

[21] Mai Yinghao, Huang Zhanyue. *Tomb of the Nanyue King of the Western Han Dynasty: Vol. 1* [M]. Beijing: Cultural Relics Press, 1991: 128, 142. Mai Yinghao, Huang Zhanyue. *Tomb of the Nanyue King of the Western Han Dynasty: Vol. 2* [M]. Beijing: Cultural Relics Press, 1991: 76.

ribbon and other properties in storage such as paper, brush, and ink". Officials that specially managed brush, paper, and ink were set up in the Office of the Emperor, which was rarely seen in the times before. In the Eastern Han dynasty, Ying Shao (140–206) recorded in his *Han Guan Yi* ("Etiquettes of the Officialdom of the Han Dynasty", 197) that "each of officials, *Shangshuling* (Director of *Shangshu*), *Pu* (Vice Director), *Cheng* (Aides), and *Lang* (Attendant Gentlemen), was offered a big block and a small block of ink from Yumi every day." In the Jin dynasty, Zhang Chang described the etiquette of the Han and Wei dynasties in his *Dong Gong Jiu Shi* ("History of Eastern Palace") and recorded that "when a crown prince salutes for the first time, he would receive four balls of fragrant ink." There are two notable things in these records. Firstly, it has been mentioned that the high-quality ink used by officials was produced in Yumi, i.e., Qian Yang, Shaanxi nowadays. Secondly, ink was no longer amorphous tiny block at that time, but rather a large block with fixed shape, and thus people used "block" and "ball" as units of measurement. The ink from Yumi was made from pine soot which was burned from pines in Zhongnan mountain. This ink was famous around the whole country, and later generations used "Yumi" as a substitute for ink. Cao Zhi (192–232) from the Wei dynasty wrote in his Yuefu poem, *Changge Xing* ("Song of Changge"), that "ink comes from the soot of green pines, and brush comes from the feathers of crafty rabbits." Chao Guanzhi from the Song dynasty mentioned in his *Mo Jing* ("Book of Ink", ca. 1110) that "people in the Han dynasty treasured pines on Zhongnan mountain in Yumi, Fufeng," since they are the first-class materials for making ink. Ink in the Eastern Han was made with spice and thus had a fragrance. Since ink was made into large block, people could put some water onto the inkstone and grind the ink block directly on it, so that the grindstone could be omitted. The large ink block could also be used for a longer time, which is very convenient. Ink was moulded into the shape of cylinder during the Six Dynasties period and cuboid in the Tang dynasty, and there are characters or patterns on the ink blocks.

In 1970, a gilt bronze inkstone case was unearthed from an Eastern Han tomb in Xuzhou, Jiangsu. The inkstone case is a cuboid with lid. Decorated with red coral and turquoise, it is beautifully crafted in the shape of a crouching toad.[22] The tomb owner, dressed in a silver thread jade burial suit, is Liu Gong, Duke of Pengcheng, the fifth son of the Emperor Ming of Han, who was buried in the sixth year of Yongchu (112). The unearthed inkstones after the Eastern Han dynasty are without grindstones and put into cases, which suggests that ink could be ground directly on the inkstones at that time. In 1965, three ink blocks were excavated from a tomb of the Eastern Han dynasty in Liujiaqu, Shan county, Henan. They are in the shape of cylinder, which was designed to be directly ground on the ink stones.[23] In the mural in No. 1 Eastern Han tomb at Wangdu, Hebei, discovered in 1952, an official in charge of recording history is sitting by a three-leg inkstone writing with an ink brush, and there

---

22 Wu Xuewen. Silver Threaded Jade Garment, Bronze Ink Box, Carved Stone [N]. *Guangming Daily*, April 7, 1973.

23 Yellow River Reservoir Archaeological Team. Han Tomb of Liu Jiaqu, Shaan County, Henan [J]. *Archaeology*, 1965 (1): 107–168.

is a cylindrical black ink on the inkstone and a water pot put beside it.[24] In 1958, a three-leg celadon inkstone was excavated in a tomb of the Western Jin dynasty in Laohushan, Nanjing. On the inkstone, there is a block of ink with a length of 6 cm and a width of 2.5 cm. The assay shows that its ingredient is similar to the ink used nowadays.[25] The usage of porcelain inkstones suggests that ink was directly ground on inkstones, while the shape of the inkstone is the same as that from the Eastern Han dynasty in Wangdu. The ink excavated in Dantu, Zhenjiang in 1961, collected by Zhenjiang Museum in Jiangsu, is in the shape of a long elliptical cube, 3 cm in length, 4.2 cm in width and 1.9 cm in thickness, made during the Six Dynasties.[26] The ink produced by the great master ink-maker, Wei Dan (179–253) during the Han and Wei dynasties (second–third century) was the most famous. He has summarized the techniques of producing ink since the Han dynasty, and recorded them in his *Bi Mo Fang* ("Prescription of Pen and Ink").

Wei Dan's ink manufacturing method was quoted by Jia Sixie (473–545), an agronomist in the Northern Wei dynasty, in the 91st chapter, *Brush and Ink*, of his *Qimin Yaoshu* ("Important Arts for the People's Welfare", ca. 538), and became the model followed by ink technicians during the Six Dynasties. According to Wei Dan' method, 1 *jin*[27] of carbon black from pine soot and 5 *liang* of animal glue are combined, meaning the ratio of their weight is 100:31, i.e., 100 *jin* of ink is composed of carbon black of 67%–77% and animal glue of 23%–33%. This ratio has been employed for a long time in history. Later, the amount of glue fluctuated from time to time, but the ratio remains 100:30–100:50. This ink manufacturing method requires six materials, carbon black from pine soot, animal glue, juice of ash bark, albumen, cinnabar, and musk. Animal glue is the binder for carbon black and ink, while the other materials are additives. Musk is the dry secretion in the musk pod at the abdomen of *Moschus moschiferus*. It is a first-class perfume, which makes the ink more fragrant and has antibacterial and antiseptic effects. Ash bark is the bark of *Fraxinus bungeana* (also called "small-leaved ash tree"). The bark is greyish-brown or greyish-black, while it would turn yellowish-green when immersed in water. It has an antibacterial effect and can affect the color of the ink. Cinnabar (HgS) is bright red and gives the black ink a little red glimmer. Albumen contains protein which can improve the wettability of particles of carbon black and make them more dispersed in the glue solution. During the Sui, Tang, Song, Yuan, Ming and Qing dynasties, additives in ink became increasingly diverse and ingredients became increasingly complicated, which has become another characteristic of Chinese ink.

How to acquire carbon black from pine soot? The technique is rarely mentioned in the documents in the Han and Jin dynasties. In the Northern Song dynasty, Chao

[24] Hebei Provincial Museum, Beijing History Museum. *Wall Paintings of the Wangdu Han Tomb* [M]. Beijing: China Classical Art Publishing House, 1955: 13–14.

[25] Nanjing Cultural Relics Management Committee. Jin Tomb of Laohushan Nanjing [J]. *Archaeology*, 1959 (6): 295.

[26] Guo Ruoyu, Zhu Shuyi, et al. *Illustrated Catalogue of Writing Brushes, Ink, Paper, and Inkstones* [M]. Shanghai: Shanghai Education Publishing House, 1979: 3.

[27] In ancient China, one *jin* was 16 *liang*, which is about 224 grammes.

Guanzhi (1050–1120) introduced in his *Mo Jing* that there are two ways of firing pine wood in the kiln, that is, fired in the vertical kiln and the horizontal kiln.[28] Before the Song dynasty, the vertical method was commonly used. The vertical kiln is about 1 *zhang* high.[29] It has a broad interior but a narrow opening. There is no chimney and the kiln is covered by a big urn, on top of which five urns of different sizes are piled up. Hence there are altogether six urns, the upper, the smaller, put onto another one. The upper urn has a hole at the bottom that connects the lower one, and the joint is sealed up with mud. People put pine wood into the kiln and ignited it. Airflow and pine smoke would move upward to each urn, going through the holes so that the rate of flow could be controlled appropriately and cooled particles of pine soot would stay in the urns. The whole kiln forms an oxygen-poor environment that will make combustion incomplete. The airflow would be impeded and cooled after going through six urns. After a thick layer of pine soot particles was formed in the urns, people would stop the fire, and after the urns cooled down, they would get the carbon black by sweeping the urns with a chicken feather. Carbon black in the uppermost urn is the finest and of the best quality, while the particles are relatively coarser in the lower urns. The particles in the lowest urn are the biggest, which can be used to make second-class ink or black paints, or be used in printing. This method only requires simple equipment, but the productivity is limited since the flue is relatively short and particles of carbon black are easy to disperse.

After the Song dynasty, people generally used horizontal kilns to make pine soot. The aforementioned *Mo Jing* recorded that people built up a sloping horizontal kiln on a hill according to the topography. The kiln has a total length of 100 *chi*, a ridge height of 3 *chi*, and a width of 5 *chi*. It is composed of several soot chambers with baffle boards put in between. The interior place for burning is at the bottom of the kiln, while the opening is a square with the length of each side of 1 *chi*, through which pine wood is put into the kiln. There are holes connecting the burning place and the soot chambers. The holes are squares with the length of each side of 2 *chi*. The soot rises gradually from the lower flues and reaches the end through each chamber. People added 3–5 sections of pine wood from the opening at the bottom and ignited them, and then continued to add pine wood according to the condition of combustion. The whole process would last 7 days, which is called *yihui* (a phase). After that, people would enter the kiln and sweep out the pine soot. The particles of carbon black from the chambers nearest to the fire are large and inferior, while those from the farther end are finer. Li Xiaomei (1055–1115) from the Northern Song dynasty also described the horizontal kiln and the techniques of making ink in his *Mo Pu* ("Ink Manual", 1095) with illustrations (Fig. 2.11).[30] Compared with vertical kilns, horizontal kilns have greater output and higher productivity. Apart from pine soot ink, people in the Song dynasty also used lampblack to produce ink. However, due to the high cost of

---

28 Chao Guanzhi. *Book of Ink*. Changshu: Ji Gu Ge, 1628.

29 In ancient China, *cun* (寸), *chi* (尺), *zhang* (丈) were slightly different from that nowadays. Nowadays 1 *cun* ≈ 0.033 m, 1 *chi* ≈ 0.33 m, 1 zhang ≈ 3.33 m.

30 Li Xiaomei. *Ink Manual* [M]. Beijing: Palace Museum Facsimile Editions of Ming Dynasty Publications, 1930.

Fig. 2.11 Ancient techniques for producing pine soot and ink making[31]

lampblack ink, pine soot ink was still mainly used for printing. Lampblack ink will not be mentioned in details here.

2.2.3 Development of Ancient Books

Now let's turn to the evolution of the material carriers of texts or images. Based on archaeological findings, in primitive societies, people used pens to write or draw symbols on bark, stones, or inscribed them on pottery, such as the colored pottery unearthed in Banpo, Xi'an, more than 6000 years ago. Although this was an advancement from knot tying for record-keeping, it still could not fully express ideas. When drawings, scripts, or characters evolved into pictographic writing, China entered the era of written history. Pictographic writing began in the Xia dynasty, flourished in the Shang dynasty, and because such characters were found extensively on oracle bones and animal bones in the ruins of Shang dynasty capitals, it was named "oracle bone script". Oracle bone script was initially written on bones with a pen and then carved into them with a sharp tool, with each bone typically containing over 50 characters, sometimes more than 100. Bones with characters were also perforated and strung together with a rope, forming the earliest form of books. This type of text carrier was also called a *shuqi*, with *qi* meaning carving. Therefore, the earliest Chinese documents were carved on oracle bones. In this sense, *shuqi* in the Shang dynasty were

[31] Li Xiaomei. *Ink Manual* [M]. Beijing: Palace Museum Facsimile Editions of Ming Dynasty Publications, 1930.

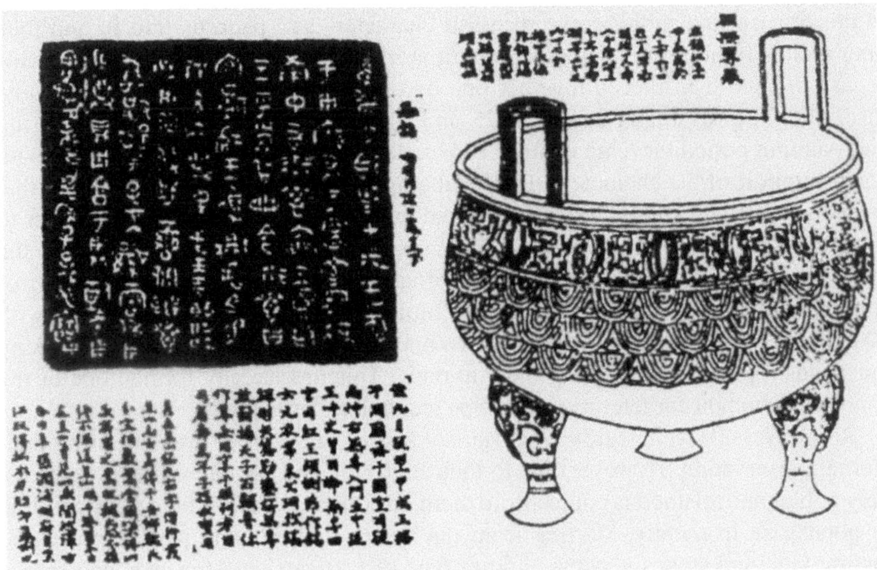

Fig. 2.12 *Zhou Wu Zhuan Ding* and rubbings of the inscription[32] (cast in the 16th year of King Xuan of Zhou)

books carved with characters. However, such characters were also cast on bronze vessels, making metal another important text carrier. The Western Zhou dynasty inherited the Shang tradition, still using oracle bones and metals as text carriers, but with more bronze artifacts unearthed than in the Shang dynasty, many inscribed with texts documenting historical events. This made Western Zhou bronze inscriptions significant written records for understanding that period of history (Fig. 2.12).

During the Spring and Autumn and Warring States periods, due to the development of economy, culture, and scientific technology, there were new changes in the carriers of written texts. The trend of recording historical events on bronze vessels with inscriptions, which began in the Western Zhou dynasty, further expanded. Some inscriptions on bronze vessels contained as many as three to four hundred characters. Besides being cast on bronze vessels, inscriptions were also cast on iron objects, reflecting advancements in casting technology. For instance, the state of Jin once cast legal statutes on an iron tripod, known as the "punishment tripod". *Zuozhuan Zhaogong Ershijiu Nian* ("The 29th year of Duke Zhao from Zuo's Commentary") records in 513 BC, "Zhao Yang of Jin then gave an iron drum to cast a punishment tripod, and cast the law made by Fan Xuanzi on it". Large inscribed tripods even became treasures passed down through generations and symbols of political power, hence the term *wending* (seeking the tripod) to refer to attempts to seize political power. In ancient times, when casting characters on bronze vessels, they first had

32 Feng Yunpeng, Feng Yunyuan. *Jinshi Suo* [M]. Woodblock Edition. Ziyang: Ziyang County Office, 1821 (Qing Daoguang First Year).

to create a casting mold, carve recessed characters and patterns into it, and then pour molten bronze into the mold. Sometimes molds with characters had to be made separately, and then pieced together one by one to form a complete mold with an inscription. For example, around 1925, in Gansu, a bronze vessel from the Spring and Autumn period (seventh century BC) called Qin Gong Gui was unearthed with an inscription of 50 characters. From rubbings of this vessel, it can be seen that each character was carved individually, and multiple molds were pieced together to form the entire inscription, with clear boundaries between each character.[33] By the late nineteenth century, inscriptions were found on a decree from the 26th year of Emperor Qin Shi Huang (221 BC) on clay molds unearthed in Linzi, Shandong, with 40 characters, two characters per line, two lines (four characters) a mold, combining ten molds together and pouring them at once. This has already formed one of the sources of thought for later movable-type technology principles.

Bronze vessels were sturdy as carriers of historical events in texts, ensuring their eternal preservation. However, due to their high cost and complex crafting process, only nobles and aristocrats could afford them, making these heavy objects impossible to popularize in society. Starting from the Spring and Autumn period, processed natural jade and stones became another type of text carrier. After jade and stones were polished into flat plates, characters could be directly inscribed on them using cinnabar, creating red characters known as "cinnabar books". *Zuozhuan Xianggong Ershisan Nian* ("The 23th year of Duke Xiang from Zuo's Commentary") records in 550 BC, "Fei Bao is a slave whose sin was written with cinnabar." From 1965 to 1966, thousands of documents written with ink and cinnabar on jade stones were unearthed in Houma, Shanxi. These documents dated back to 497 BC, documenting agreements and oaths made by Zhao Yang, the state minister of Jin, after defeating the Zhao clan of Handan. There are over 600 complete pieces, some with up to 220 characters on thin jade tablets, with legible inscriptions to this day.[34] Stone materials were cheaper and easier to obtain and process than jade, so characters were first written in cinnabar on stone and then carved, becoming another way to record historical events since the Spring and Autumn period. The earliest existing stone inscriptions are from the early Tang dynasty, unearthed in Fengxiang, Shaanxi, in the shape of cylinders, varying in height from 45 to 90 cm, with an average circumference of 210 cm, totaling 10 pieces engraved in large seal script. Due to their drum-like appearance, they are called *shiguwen* or "stone drum inscriptions" and are now housed in the Palace Museum in Beijing. These inscriptions contain about 700 characters and were made in the 8th year of Duke Xiang of Qin (770 BC), recording Duke Xiang's military achievements in assisting King Ping of Zhou to combat the Western Rong tribes, with four characters per rhyme.[35] Afterward, the

[33] Luo Zhenyu. *Songxi Jingao: Vol. 1* [M]. Stone Lithograph Edition. Shangyu: Luo Family, 1925: 32–33. Rong Geng. *Comprehensive Study of Shang Ritual Bronzes: Vol. 1* [M]. Beijing: Harvard-Yenching Institute, Yenching University, 1941: 88, 158.

[34] Tao Zhenggang, Wang Kelin. The Eastern Zhou Alliance Oath Site at Houma [J]. *Cultural Relics*, 1972 (4): 27.

[35] Guo Moruo. *Studies on Stone Drum Inscriptions* [M]. Beijing: People's Publishing House, 1955: 9–10.

tradition of carving stone continued to flourish through successive dynasties. Early stone carvings were cylindrical, but they changed to rectangular shapes after the Han dynasty. Such artifacts and their rubbings are countless, with Xi'an Beilin Museum alone housing over 1000 stone tablets from the Han to Qing dynasties. The tradition of recording events by stone carving continues to this day.

After the Spring and Autumn period, oracle bone script evolved into large seal script. Oracle bones were no longer used as text carriers, replaced by metal and stone. However, metal and stone were hard and heavy materials, suitable only for storage at fixed locations and impractical for portable use due to their large size and the limited amount of information they could store. To address this situation, ancient people turned to bamboo and wood to make tablets for writing and recording events. Rectangular bamboo strips smoothed and prepared for writing were called *zhujian* (bamboo slips), or simply *jian* (slips). Wooden pieces were called *du* (tablets), and together they were referred to as *jiandu* (slips and tablets). People could write and record on these slips and tablets with a brush. Each slip had specified dimensions,[36] typically 24–50 cm in length and capable of holding 22–25 characters. After writing on one slip, another could be added, and once a certain length was reached, they were bound together with strips of leather or hemp thread, finally rolled up into bundles. When reading, they were unfurled, and when not in use, rolled up and stored horizontally on shelves. The practice of writing on slips and tablets dates back at least to the Zhou dynasty. *Shiji* ("Records of the Grand Historian"), in the *Biography of Confucius*, mentions: "Confucius (551–479 BC) was fond of *Yi* ("Book of Changes") in his later years… he read *Yi* written on bamboo slips, with the strips of leather binding the slips having broken three times." Slips and tablets were true books, and it was thanks to them that Chinese cultural classics were passed down to later generations. Although no slips from the Spring and Autumn period have survived, a large number of slips from the Warring States and Qin and Han dynasties have been unearthed in recent decades. In 1953, 43 bamboo slips were unearthed from a Warring States Chu tomb at Yangtian Lake, Changsha, Hunan. Dating back to the fourth century BC, each slip measures 22 cm in length, 1.2 cm in width, and 0.1 cm in thickness, with each slip containing between 2 to 21 characters.[37] In 1975, 1150 Qin bamboo slips were excavated from Tomb No. 11 in Shuihudi, Yunmeng, Hubei (217 BC), containing books such as *Biannian Ji* ("Chronicles"), *Yushu* ("Language Studies"), *Qinlv Shiba Zhong* ("Eighteen Types of Qin Laws"), *Falv Dawen* ("Legal Questions and Answers"), and *Ri Shu* ("Daily Records").[38] In 1972, 4942 bamboo slips were unearthed from the Western Han Tomb No. 1 in Yinqueshan, Linyi, Shandong, each approximately 27.6 cm long and 0.5–0.9 cm wide, containing ancient texts like *Sun Zi Bingfa* ("The Art of War by Sun Zi"), *Sun*

---

36 Wang Guowei. *Studies on Bamboo Slips and Tablets* [M] // *The Collected Works of Mr. Wang Jing'an of Haining: Vol. 26*. Shanghai: Zhonghua Book Company, 1936.

37 Shi Shuqing. *Research on Chu Bamboo Slips Excavated at Yangtianhu, Changsha* [M]. Shanghai: Qunlian Publishing House, 1955: 2, 6–18.

38 Hubei Xiaogan Region Phase II Archaeological Training Class for Both Industrial and Agricultural Artifacts. Brief Report on the Excavation of Eleven Qin Tombs at Suihudi, Yunmeng, Hubei [J]. *Cultural Relics*, 1976 (6): 1–10, 95–99.

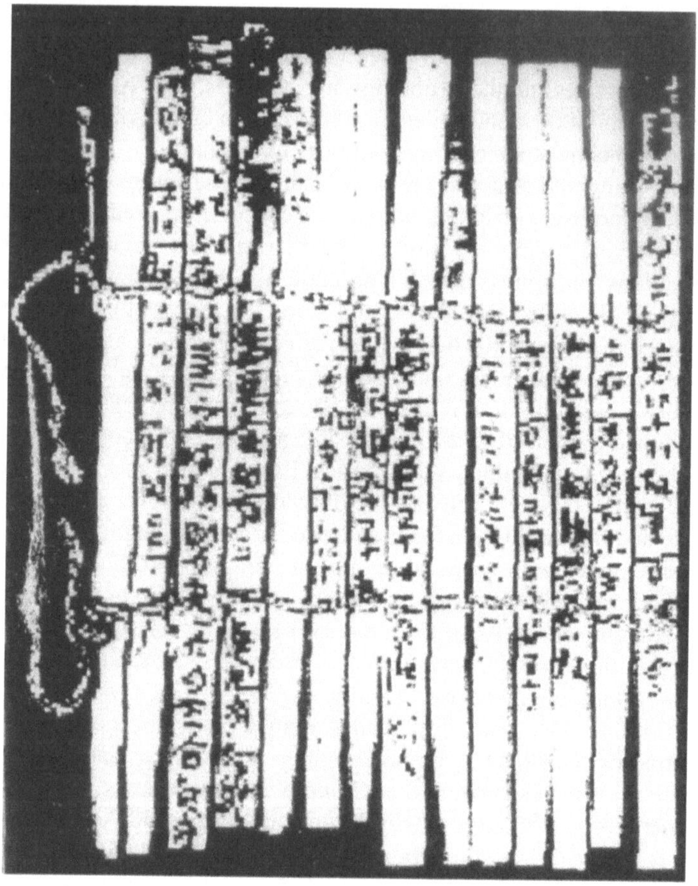

Fig. 2.13 The wooden slips from the Han Dynasty, dating from 93–95, unearthed at the Juyan site[40]

Bin Bingfa ("The Art of War by Sun Bin"), Liu Tao ("The Six Secret Teachings"), *Weiliaozi*, *Guanzi*, *Yanzi*, *Mozi*, as well as pre-Qin classics such as *Xianggou Jing* ("The Book of Dogs") and *Yinyang Shu* ("The Book of Yin and Yang").[39] In recent decades, tens of thousands of slips and tablets have been unearthed in provinces such as Gansu, Shandong, Hunan, and Hubei (Fig. 2.13), holding significant academic research value and allowing us to witness the original appearance of ancient books and documents.

[39] Wu Jiulong, Bi Baoqi. Report on the Discovery of Bamboo Slips of The Art of War by Sun Zi and The Art of War by Sun Bin in a Western Han Tomb in Linyi, Shandong [J]. *Cultural Relics*, 1974 (2): 15–26, 71–78.

[40] Lao Han. *Study and Explanation of the Han Bamboo Slips from Juyan* [M]. Shanghai: Commercial Press, 1949.

China is the birthplace of sericulture and silk spinning techniques. By the latest in the Spring and Autumn period, silk fabric was already used as a material for writing and painting. The fabric was cut into appropriate lengths, marked with red grid lines, and then written on. These segments could be joined together to form a long scroll, with a roller at the end for rolling up the entire scroll. This is the scroll form of books, hence ancient books were measured in "scrolls". In 1942, silk books were unearthed from a Chu tomb of the Warring States period in Changsha, containing both text and illustrations, now housed abroad. In 1973, the same site yielded the *Renwu Yulong* ("Human Riding A Dragon") silk painting, also a Chu artifact. From 1972 to 1973, a large quantity of items were unearthed from Western Han Tomb No. 3 in Changsha, dating to the 12th year of Emperor Wen of Han (168 BC), including silk paintings, silk books such as *Yijing*, *Laozi*, and *Zhanguo Ce* ("Strategies of the Warring States"), as well as 20 other fragmentary scrolls of astronomy, calendars, Five Elements, miscellaneous divinations, and maps, totaling 100,000 characters, with an additional 18 bamboo slips of medical books written in early Han clerical script.[41] This illustrates that from the Spring and Autumn period to the Qin and Han dynasties, slips and silk were used side by side. Thus, *Mozi*'s chapter *Minggui* ("On Ghost", fourth century BC) stated: "In ancient times, the sage kings… feared that future generations would not understand them, so they wrote on bamboo and silk, passing down to future generations. They feared that these would decay and be destroyed, and future generations would not know, so they engraved them on trays and bowls, carved them in metal and stone to preserve them." The chapter *Jian'ai* ("Universal Love") also recorded: "Knowing the wisdom of the six sage kings… Master Mozi said, 'I was not born at the same time as they were, nor did I hear their voices or see their appearance. I have written what they wrote on bamboo and silk, carved it in metal and stone, and engraved it on trays and bowls, so that future grandchildren can know it.'" Here, Mozi (468–376 BC) listed several types of text carriers used in the pre-Qin period, apart from oracle bones, among which only slips and silk (Fig. 2.14) are materials that can be used to write books. Their emergence had remarkable historical significance and greatly contributed to the development of science and culture.

However, after five or six hundred years of use, the limitations of bamboo slips and wooden tablets became apparent by the time of the Qin and Han dynasties. With the development of science and culture, longer literary works continued to emerge, and the variety and quantity of classics expanded significantly. For instance, *Hanshu Yiwen Zhi* ("Literature Records in Book of Han", 100 AD) recorded 596 works by 13,269 scrolls. Each bamboo slip could hold only a small number of characters, requiring about 400 slips to transcribe a ten thousand-word book. Compiling them into multiple volumes increased their volume and weight significantly. In the *Tian Xia* chapter of *Zhuangzi*, it is mentioned that the Warring States scholar Huishi was "multi-talented", with his collection of bamboo and wooden slips filling five carts,

[41] Hunan Provincial Museum and Institute of Archaeology, Chinese Academy of Sciences. Excavation Report of the Han Tombs No. 2 and No. 3 at Mawangdui, Changsha [J]. *Cultural Relics*, 1974 (7): 39–48, 95–111.

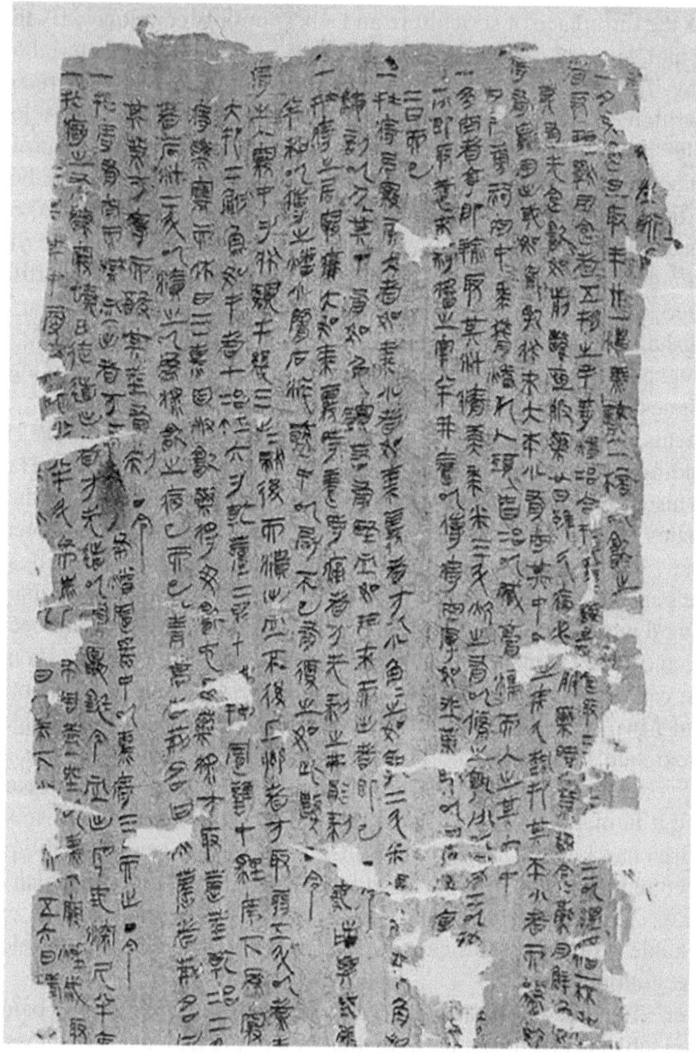

Fig. 2.14 Partial content from the Western Han dynasty silk manuscript *Wushier Bingfang* ("Fifty-Two Diseases Formula")[42]

all of which he thoroughly read. This led later generations to use the phrase "five cart loads of knowledge" to describe profound knowledge. According to *Annals of Qin Shi Huang* in *Shiji*, when Emperor Qin Shi Huang handled affairs, the bamboo

[42] Research Office of the History of Medicine and Literature, China Academy of Traditional Chinese Medicine. Introduction to Four Ancient Medical Texts from the Mawangdui Silk Manuscripts [J]. *Cultural Relics*, 1975 (6): 16–19.

slips submitted for review often amounted to one *dan* (about 60 kg), "with submissions day and night, leaving no room for rest until matters were resolved". In the *Biographies of Humorists* in *Shiji*, during the reign of Emperor Wu of the Western Han dynasty, the Qi scholar Dongfang Shuo (154–93 BC) "arrived in Chang'an and immediately submitted memorials, totalling three thousand slips. The carriage commander instructed two people to carry them for presentation… and it took two months to read them all." These examples illustrate that under new historical conditions, bamboo slips and wooden tablets had become insufficient in terms of character capacity and information volume, causing inconvenience in their use. Silk fabrics were light, smooth, and could accommodate more characters written in ink, but they were too expensive. During the Han dynasty, one piece of silk was worth six *dan* of rice, which ordinary people could not afford, hence the saying "the poor could not afford plain silk". Since the Zhou and Qin periods, bamboo slips and silk replaced oracle bones and metal stones as the main media for writing. However, they themselves faced the fate of being replaced by a completely new writing medium during the Qin and Han periods with the invention and application of plant fiber paper. This invention fundamentally changed the medium of writing, marking the transition of books into manuscript form, and providing the material foundation for the subsequent invention of printing. This will be discussed in detail in Chap. 3 of this book.

Although oracle bone inscriptions from the Shang and Zhou dynasties were replaced by silk and bamboo after the Zhou dynasty, they marked the beginning of the Chinese people's engraving on hard materials to record events, a practice that has lasted for more than 3000 years. During the Shang and Zhou periods, divination was carried out using turtle shells and animal bones, with the majority of turtle shells coming from the Yangtze River Basin in the south, offered as tribute by local lords. The plastrons were generally 28 cm long, 20 cm wide, and 0.6–0.7 cm thick. Due to the unevenness of the turtle's carapace, it had to be split in half longitudinally, with each half measuring 27–35 cm long and 11–15 cm wide. Animal bones were mostly cow scapulae, known for their large and smooth surfaces, occasionally using other animal bones. Before divination, other adherents on the carapace and bone were removed, scraped and polished, and then elliptical and circular grooves with a diameter of about 1 cm were chiseled on them, intersecting each other. The number of grooves on each piece of carapace varies. Before divination, the carapace was scorched with fire on the ridge surface with grooves, resulting in longitudinal and transverse cracks around the grooves. Wizards discerned good or bad fortune based on different shapes and historians recorded the results on the carapace, which became divination inscriptions. This is the main content of oracle bone inscriptions. After divination, the oracle bones containing the divination inscriptions were strung together through a hole to form *ce* for archival preservation. Due to the long passage of time, complete archives of oracle bone divination inscriptions have long since scattered into individual pieces, but traces of their stringed connections can still be seen, verifying the reliability of historical records stating that "the ancestors of the Yin dynasty had books and archives".

So, how were the divination inscriptions recorded on the oracle bone pieces? In his compilation *Yinxu Wenzi Yibian* ("Oracle Bone Script Compilation 2"), Dong

Zuobin (1895–1963), included 9105 pieces of oracle bone inscriptions in volume 1 (1948), volume 2 (1949), and volume 3 (1953). Previously, he had compiled *Yinxu Wenzi Jiabian* ("Oracle Bone Script Compilation 1", 1948) containing 3942 oracle bone inscriptions. Most of the divination inscriptions on the oracle bone pieces were engraved, but a very small number of them were written with a brush dipped in cinnabar or ink. In particular, some of the turtle plastrons had inscriptions engraved on their front and inscribed with a sharp and vigorous pen, while the reverse side of the inscription was written without carving, with thick and soft strokes. Some characters (for example *zhen hu*, 贞乎) have strokes that are very thick when handwritten, but are finely engraved, with some characters still unfinished. This explains that the divination inscription on the oracle bone usually first writes the character style on the bone and then carves it with a knife.[43] From the handwriting samples that were not carved in time, we can see that the writing sequence was from top to bottom, left to right, exactly the same as the order of writing today. Since handwritten characters could not be traced, they had to be photographed instead. Sometimes, skilled craftsmen could also carve the characters directly, but this was rare. There were also instances where several lines of stems and branches were repeatedly engraved on bone pieces, appearing very elegant, while the same characters below were crooked and uneven. This was an example of masters teaching their apprentices how to carve characters.[44]

In his work *Jianshoutang Suocang Yinxu Wenzi Kaoshi* ("Explanations of Oracle Bone Inscriptions in the Jianshou Hall Collection of Yin Ruins", 1917), Mr. Wang Guowei also provided examples to illustrate the carving techniques used during the Shang and Zhou dynasties. One piece of oracle bone inscription should read "苦方出", but it was only partially carved as , clearly indicating it was incomplete. The correct and complete carving should have been 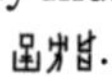. Another inscription with the characters "亞其" was similarly not fully carved out as . The complete carving should have been . This demonstrates that the craftsmen first carved each character with straight lines using a knife, making it easier to chisel. Then, they rotated the oracle bone piece 90 degrees to carve the horizontal lines of the characters vertically.[45] For smaller characters and finer strokes, each stroke was made with a single cut of the knife, while larger characters and thicker strokes required two cuts, removing material from both sides of the stroke. The carving knives were made of hard bronze with 20%-30% tin content, featuring various blade shapes. Characters in oracle bone inscriptions sometimes exhibited mirrored forms, with components on the left or right, resulting in what today appear to be different characters but are actually the same. For example, the character "明" might be carved as or 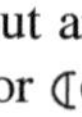, while "如" could appear as or 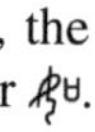. Such examples are numerous. Therefore, it can be said that

[43] Dong Zuobin. Preface to Oracle Bone Script Compilation 2 [J]. *Journal of Chinese Archaeology*, 1949 (4): 255–289.

[44] Guo Moruo. *Dialectical Development of Ancient Script* [M] // *The Slave Society Era*. Beijing: People's Publishing House, 1977: 251–252.

[45] Wang Guowei. *Explanations of Oracle Bone Inscriptions in the Jianshou Hall Collection of Yin Ruins* [M]. Shanghai: Cang Sheng Ming Zhi University, 1917: 46.

the carving techniques of oracle bone script during the Shang and Zhou dynasties laid the foundation for engraving techniques in later woodblock printing. Despite the transition from oracle bones to materials like stone, the technique of carving characters continued to evolve and thrive as an art form. While turtle shells and animal bones could be used for writing, their inscriptions were not durable over time, necessitating the additional step of carving them into more permanent materials. Due to their limited size and capacity for characters, and the impracticality of storing or reading them when linked together, oracle bone pieces were eventually replaced first by bamboo and silk, and later by paper.

In the evolutionary history of writing media from oracle bones, metal and stone, bamboo and silk, to paper, we can observe a common cultural trend: as time progresses, the content and quantity of text on these media become increasingly rich and extensive, while the time required to complete writing tasks decreases. In other words, the pace of work accelerates with the passage of time. To adapt to this trend, Chinese characters underwent several large-scale standardization and simplification processes from oracle bone script, large seal script, small seal script, clerical script, clerical-regular script, to regular script during the periods of Shang and Zhou dynasties, Spring and Autumn period, Warring States period, Qin and Han dynasties, Six Dynasties, and Sui and Tang dynasties. Each reform made Chinese writing easier and more efficient. By the Sui dynasty, regular script had become a stable script suitable for both writing and printing. Alongside these reforms in writing styles, there were corresponding changes in writing materials—from oracle bones, metal and stone, bamboo slips and silk manuscripts, to paper. There existed a mutually reinforcing interaction between the evolution of writing styles and writing materials. The formats of books also underwent numerous changes, making reading and usage increasingly convenient. With the invention of paper in the Western Han dynasty, China became the first to possess writing materials suitable for both writing and printing. Considering the use of brushes and ink, by the later period of the Northern and Southern dynasties, China had already possessed all the necessary materials and technological conditions for the development of woodblock printing. As long as the idea of replicating books was put into practice and there was an appropriate social environment, the emergence of printing technology became inevitable.

Chapter 3
Invention and Development of Papermaking Techniques

3.1 Definition of Paper and Scientific Principles of Papermaking

3.1.1 Definition of Paper

Among the Four Great Inventions of ancient China, papermaking and printing are the most closely related. For over a thousand years, paper has been the primary material for printing, and it has been essential for the advancement of printing technology. Without paper, no country or region could have developed printing. Therefore, a comprehensive study of printing technology necessarily includes an examination of the history of papermaking. Dr. Thomas Francis Carter, a renowned American scholar, emphasized that the invention and development of printing were fundamentally rooted in the development of papermaking. In his world-famous book *The Invention of Printing in China and Its Spread Westward* (1925), the first chapter is dedicated to the invention of papermaking. His approach should serve as a standard model for writing such works.

In Chap. 1, it was noted that before the invention of paper, ancient China utilized a variety of writing materials, including oracle bones, gold, stone, bamboo slips, and silk. Other civilizations employed materials such as clay tablets, palm leaves, bark, papyrus, and sheepskin to record their history and culture. These materials can be categorized into three types based on their properties. The first category consists of heavy and rigid materials like oracle bones, gold, stones, bamboo slips, and clay tablets. Although durable, these materials are bulky, inflexible, and inconvenient to carry and store. The second category includes lightweight and fragile materials such as leaves, bark, and papyrus. Though small, portable, and inexpensive, they are prone to damage from folding and wear. The third category features lightweight and flexible materials like silk and sheepskin. These materials have a smooth surface, absorb ink well, and are easy to roll and bind into books, but they are costly. Oracle bones, gold,

P. Jixing, *History of Printing Techniques in China*,
https://doi.org/10.1007/978-981-96-6119-0_3

stones, and clay tablets were phased out first due to their impracticality. Bamboo slips, silk, sheepskin, papyrus, and palm leaves also had their limitations and were ultimately replaced by paper.

Compared to other ancient writing and note-taking materials, paper offers numerous advantages. First, its smooth, white surface makes it ideal for writing with both soft brushes and hard pens. Second, paper is available in large formats, providing ample writing space, and is lightweight, foldable, and easy to carry. Third, it boasts a long lifespan and is highly amenable to coloring, making it suitable for various advanced processing techniques. Fourth, paper is incredibly versatile, with applications ranging from writing and printing to packaging. It can also be transformed into a wide array of products used in industries, agriculture, and daily life. Finally, the greatest advantage of paper is its combination of high quality and low cost. With abundant raw materials, it can be produced anywhere in the world.

For over 2000 years, paper has been a remarkably versatile material, enjoying widespread and enduring use across the globe. Its introduction revolutionized the history of writing and note-taking, marking a significant milestone. The advent of paper notebooks paved the way for the invention of printing, with paper serving as the essential medium for printed materials. The use of paper in writing and printing has been instrumental in advancing human civilization and societal development.

The word *zhi* (paper) is not found in pre-Qin documents (before the third century BC). It appeared later and was written in small seal script as 紙. It is reported that the character *zhi* was first seen in the wooden slips of the middle and late Western Han dynasty (first century BC) unearthed in ancient Juyan, northwest China. There is a phrase of "official written paper" in it.[1] Juyan was an important garrison site for the Western Han dynasty. It was located in today's Ejina Banner of Inner Mongolia. At that time, paper was already used for writing at the border, so the use of paper in the interior should have been earlier than that. The word "paper" appears to express a new type of writing material that is different from bamboo slips and silk. Studying the Western Han dynasty paper has deepened our understanding of this material. Generally, "paper" is defined as a substance made from dispersed fibers that result from the mechanical and chemical processing of plant materials. These fibers are mixed with wood and turned into a slurry, which is then spread onto a porous mold curtain. The wet layer is filtered and dried to create a strong, durable sheet of interwoven fibers. This material is widely used for writing, printing, and packaging. The definition of paper encompasses five key elements: (1) A smooth surface that accepts ink, suitable for writing, packaging, etc.; (2) The primary component is pure, dispersed plant fiber; (3) The fibers are intricately intertwined in various directions; (4) The fibers are cut into short lengths, and a phenomenon called "fibrillation" occurs when they break; (5) The fibers undergo processes like pulping, papermaking, and drying. Anything lacking these characteristics does not qualify as paper.

[1] Chen Zhi. *New Evidence on the Book of Han* [M]. Tianjin: Tianjin People's Publishing House, 1979: 467–468.

The definition and criteria for paper provided above apply to all hand-made paper, both ancient and modern, across cultures and regions. Historically, misunderstandings about what constitutes paper have led to confusion and misconceptions about the origins of papermaking. For instance, some have mistakenly classified the bark cloth, or tapa, used by Pacific islanders as paper, and consequently, believed that the Maya civilization in ancient Mexico was the birthplace of papermaking.[2] However, tapa is made from beaten tree bark and does not undergo pulping or sheet formation processes; its basic component is fiber bundles, making it fundamentally different from paper. Some have mistakenly considered the papyrus sheets produced by ancient Egyptians as paper, leading to the belief that paper originated in Egypt.[3] In reality, the Egyptians made papyrus by cutting the stalks of the Cyperus papyrus plant into short strips, splitting them, and pressing them into layers, which were then flattened with a hammer after being treated with vinegar. This material, characterized by its horizontal and vertical fibers, differs significantly from paper in terms of composition, physical structure, appearance, and properties. Similarly, palm leaves used in ancient India were not paper. When paper emerged, these materials fell out of use, underscoring their fundamental differences from true paper.

In *Shuowen Jiezi* ("Analytical Dictionary of Characters", 121) written by Xu Shen (circa 58–147) during the Eastern Han dynasty, the character for paper is explained as follows: "*zhi* (paper, 纸) refers to *xu* (fiber, 絮) and *shan* (curtain, 箈). It is composed of the radical *mi* (fine threads, 糸), and phonetic component *zhi* (氏)."[4] Here, *xu* denotes fibers, which typically refer to lower-grade animal fibers like cotton or plant fibers resembling cotton. *Shan* is synonymous with *ze* (簀), meaning bamboo mat or bamboo curtain. Xu Shen thus understood that paper was made by filtering a fiber suspension through a bamboo curtain to form thin sheets. By Xu Shen's time, hemp paper was already in use, so he would have been familiar with the raw materials and methods of papermaking. Therefore, his use of *xu* or fibers likely refers to dispersed hemp fibers. His definition encompasses both the raw material (hemp fiber) and the papermaking tool (bamboo curtain). This aligns well with the 1963 edition of the Encyclopaedia Americana, which describes paper as "a thin sheet formed from plant fibers suspended in water and collected on a screen".[5] Understanding Xu Shen's explanation in this way is consistent with both its original meaning and historical facts.

The Eastern Han scholar Liu Xi, in his work *Shiming* ("Explanations of Appellations", ca. 100), states: "*zhi* (纸) refers to *zhi* (砥), meaning smooth like a grinding stone." Therefore, the character "纸" for paper evolved from "砥" by replacing the "stone" radical with "fiber" (糸), indicating a smooth material made from fibers used as a writing surface, which was the intention behind its creation during the Western Han period. This also explains Xu Shen's interpretation of the character as

[2] Hans L. *Coras del Papel en Mesoamérica* [M]. Mexico: 1984, 74.

[3] Jaroslav C. *Paper and Books in Ancient Egypt* [M]. London: H. K. Lewis, 1952, 31.

[4] Xu Shen. *Shuowen Jiezi Annotated: Vol. 13, Silk Section* [M]. Annotated by Duan Yucai. Shanghai: Wensheng Bookstore, 1908, 9.

[5] *The Encyclopaedia Americana*: *Vol. 21* [M]. New York: Grolier, 1963, 258–259.

"a compound of 糸 (meaning 'fiber') with the phonetic element 氏 pronounced as *zhi*". Additionally, the Eastern Han scholar Fu Qian, in his work *Tongsu Wen* ("Plain Language Text", ca. 180), referred to paper as *fang xu* (方絮), a term that praised its whiteness and fibrous texture resembling cotton. During the Han dynasty, there was no distinction between animal and plant fibers; both were collectively called *xu*, which has led to some confusion. For example, in 232, the scholar Zhang Yi, in his *Gujin Zigu* ("Historical and Modern Etymology of Chinese Characters"), suggested that there were two types of paper: "ancient paper", made from silk and called *fan zhi* (幡纸), and "modern paper", made from used hemp cloth and referred to as *zhi* (帋).[6] This notion, though new, lacks historical evidence. Prior to the Han dynasty, silk fabrics were used for writing but were not called paper. The term *fan zhi* first appeared in *Han Ji* ("Records of Han Dynasty") written by Xun Yue (148–209), a contemporary of Zhang Yi. The concept of "ancient paper" and "modern paper" is therefore erroneous and unnecessary. It is time to correct this misunderstanding.

We can distinguish paper from other ancient materials by examining its raw material processing, manufacturing techniques, and underlying principles, which highlight its superiority. Leaves, tree bark, papyrus, animal skins, oracle bones, stone, and bamboo slips involve only basic processing, with the physical properties, chemical composition, and appearance of the raw materials remaining largely unchanged. These are considered primary processed products. Bronze and silk represent more advanced processing, but metal cannot be written on, and although silk is a suitable writing material, it is prohibitively expensive and does not alter the form of the raw material. In contrast, paper is the result of a sophisticated process that combines both chemical and mechanical treatments. The production of paper involves significant physical and chemical changes in the raw materials. Papermakers convert old, soiled rags into clean, smooth paper, embodying the traditional Chinese philosophy expressed by Zhuangzi (369–283 BC) that "decay and corruption transform into something miraculous". Paper emerged from this traditional perspective. The science behind papermaking involves complex principles that have only been fully understood in the past century, especially in recent decades. Remarkably, the techniques and equipment developed by our ancestors 2000 years ago inadvertently applied these principles.

3.1.2 *Scientific Principles of Papermaking*

Papermaking raw materials consist of dispersed plant fibers, which are not found in nature but are artificially extracted from plants and then subjected to chemical purification and mechanical processing. Unlike other ancient writing materials that were readily available and required minimal processing, papermaking raw materials are diverse and not naturally occurring. Estimates suggest there are between 500 and

[6] Zhang Yi. *Gujin Zigu: Jin Section* [M]. // Li Fang, *Taiping Yulan* ("Taiping Imperial Encyclopedia"): *Vol. 605, Paper Section.* Beijing: Zhonghua Book Company, 1960, 2724.

1000 varieties of papermaking raw materials, including grasses, woods, barks, hemp, and waste materials. Ancient papermakers primarily used tough plant fibers and stem fibers. Tough plant fibers are sourced from the bast or phloem of herbaceous and woody plants. Herbaceous plants like hemp are mostly annual, while woody plants are perennial, such as paper mulberry (Broussonetia papyrifera), mulberry, rattan, Tetrapanax papyrifer, and blue sandalwood (Pteroceltis tatarinowii Maxim.), among others. Stem fibers, found in the vascular bundles of monocotyledons, are difficult to separate from basic tissues. This category includes both annual plants (e.g., rice, wheat, and grasses) and perennial plants (e.g., bamboo). Therefore, papermaking raw materials represent a deeper exploration and understanding of natural resources. Their widespread availability around the world has contributed to paper becoming a universally used material. Details on the microscopic morphology and fiber length-to-width ratios of various papermaking plant fibers are provided in Table 3.1.

Different raw materials have varying fiber lengths and widths, leading to differences in the quality of produced paper. Paper made from long fibers is generally superior to that made from short fibers, with a higher length-to-width ratio resulting in denser tissue and greater tensile strength. China has historically chosen high-quality hemp fibers for papermaking, with average length-to-width ratios of 1000 for hemp and 3000 for ramie (Fig. 3.1). Following hemp, bark materials such as paper mulberry bark fibers have length-to-width ratios of 290, while bark from the mulberry plant has a ratio of 463, and blue sandalwood has a ratio of 276. Bamboo fibers, being shorter, have average length-to-width ratios ranging from 123 to 133. Grass fibers, with the lowest ratios, have lengths ranging from 102 for wheat straw to 114 for rice straw. The length-to-width ratio of ramie fibers is 29 times that of wheat straw. Therefore, in ancient China, hemp paper and bark paper were used for cultural purposes, grass paper for packaging and hygiene, and "fire paper" used in

Table 3.1 Fiber length, width, and length-to-width ratios of common papermaking plants in ancient China

Fiber		Length (mm)			Width (mm)			Length-to-width ratios
		Max	Min	Range	Max	Min	Range	
1	Hemp	29.0	12.4	15.0–25.5	0.032	0.007	0.015–0.025	1000
2	Ramie	231.0	36.5	120.0–180.3	0.076	0.009	0.024–0.047	3000
3	Paper mulberry	14.0	0.57	6.0–9.0	0.032	0.018	0.024–0.028	290
4	Mulberry plant	45.2	6.5	14.0–20.0	0.038	0.005	0.019–0.025	463
5	Giraldi Daphne	5.8	0.95	3.1–4.5	0.030	0.004	0.015–0.019	222
6	Blue sandalwood	18.0	0.72	9.0–14.0	0.034	0.007	0.019–0.023	276
7	Moso bamboo	3.20	0.34	1.52–2.09	0.030	0.006	0.012–0.019	123
8	Omei mountain bamboo	2.85	0.34	1.33–1.90	0.028	0.003	0.009–0.019	133
9	Rice straw	2.66	0.28	1.14–1.52	0.028	0.003	0.006–0.009	114
10	Wheat straw	3.27	0.47	1.30–1.71	0.044	0.004	0.017–0.019	102

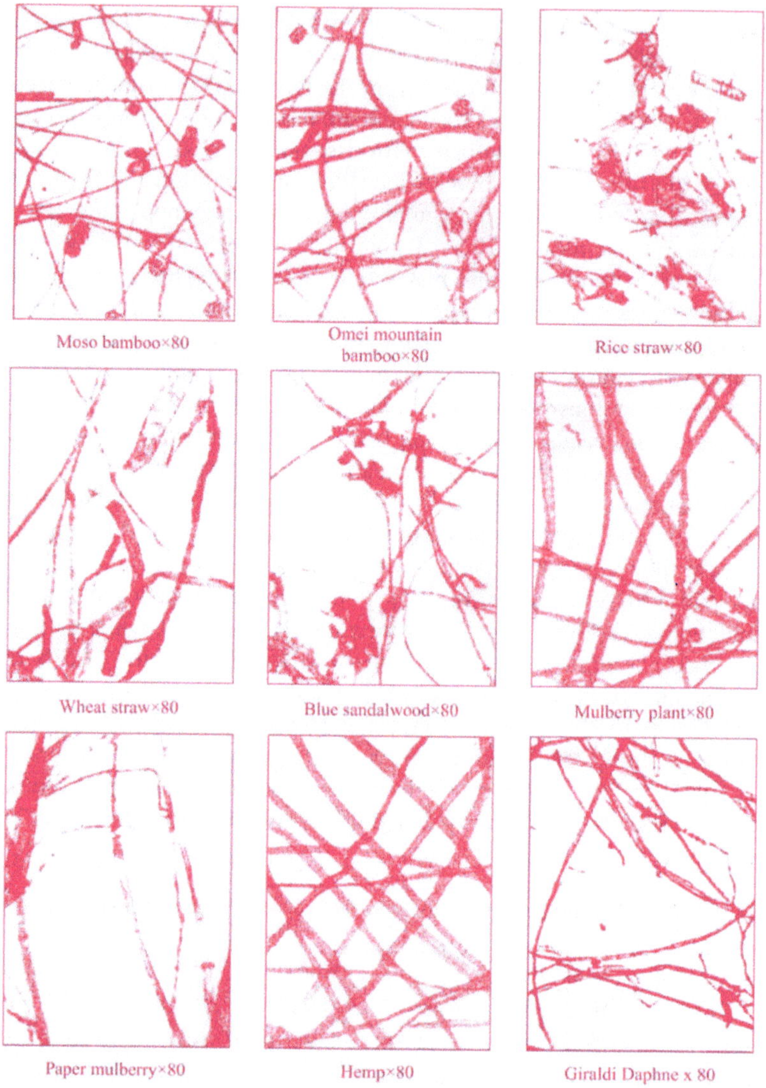

Fig. 3.1 Fiber atlas of common papermaking plants in ancient China[7]

funeral rituals. Bamboo paper, being short-fibered and cost-effective, was widely used. The diversification of papermaking materials is a significant characteristic of China. Another feature is the proportionate blending of long and short fibers to ensure quality while lowering costs.

[7] Institute of Paper Industry Science Research, Ministry of Light Industry. *Fiber Atlas of Chinese Papermaking Materials* [M]. Beijing: Light Industry Publishing House, 1965.

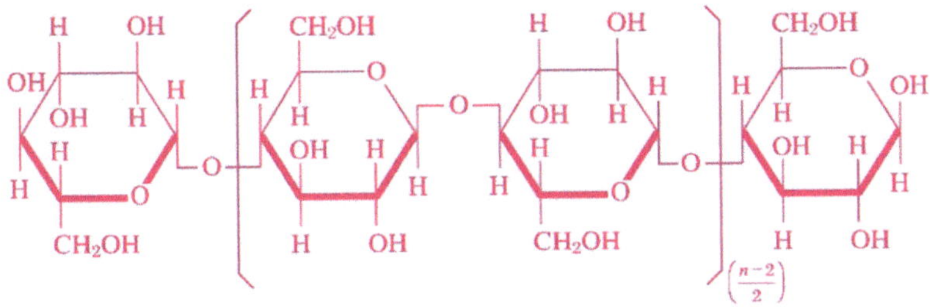

Fig. 3.2 Molecular formula of cellulose

Upon closer microscopic observation, it is evident that plant fibers are primarily composed of cellulose, a fundamental component of cell walls. At the molecular level, cellulose is a polymer polysaccharide formed by the linkage of many α-glucose units (α-glucoside, $C_6H_{10}O_5$) through 1-4-β-glycosidic bonds. Cellulose can be viewed as a long-chain polymer of glucose units, represented by the formula $(C_6H_{10}O_5)_n$. The value of *n*, representing the degree of polymerization, indicates the length of the molecular chain; for example, the degree of polymerization for ramie cellulose is 8580, though it varies among different raw materials. A larger *n* value corresponds to longer fibers. The degree of polymerization is also directly proportional to the molecular weight of cellulose, such as an average molecular weight of 335,000 for flax. The chemical structure of the cellulose molecule can be expressed using various formulas, and the molecular formula provided by British chemist Walter Norman Haworth (1883–1950) is shown in Fig. 3.2.

From the molecular formula of fibers in Fig. 3.2, it can be seen that each glucose unit in the cellulose chain polymer has three hydroxyl groups (OH). Therefore, each cellulose molecule contains three times the degree of polymerization (3n) of hydroxyl groups. For example, the cellulose molecule of hemp fiber contains $3 \times 8580 = 25{,}740$ hydroxyl groups, close to 26,000. It exhibits significant hydrophilicity, and when plant fibers are purified and dispersed in water, the numerous hydroxyl groups in cellulose molecules attract water molecules, causing the fibers to swell. When cellulose molecules come close to each other, the oxygen atoms (O) of the adjacent hydroxyl groups in two molecules pull water molecules (H–O–H) together, acting like a bridge connecting two cellulose molecules. This is referred to as "water bridge" (Fig. 3.3). This is the state when the paper pulp is scooped out with a mold, drained, and excess water is pressed out, forming a wet paper layer on the mold. At this stage, the water bridges between fibers are not very strong, so the wet paper layer has relatively low strength.

However, when the wet paper is dried, and water is evaporated, the fibers are subjected to strong surface tension. The distance between cellulose molecules is greatly reduced to below 2.75 angstroms (Å, 10^{-10} m). At this point, the fibers are no longer connected by water bridges but are bound together by hydrogen bonds formed between the countless hydroxyl groups of adjacent molecules. Hydrogen bonds are chemical bonds formed by the attraction of the oxygen atom in one polar

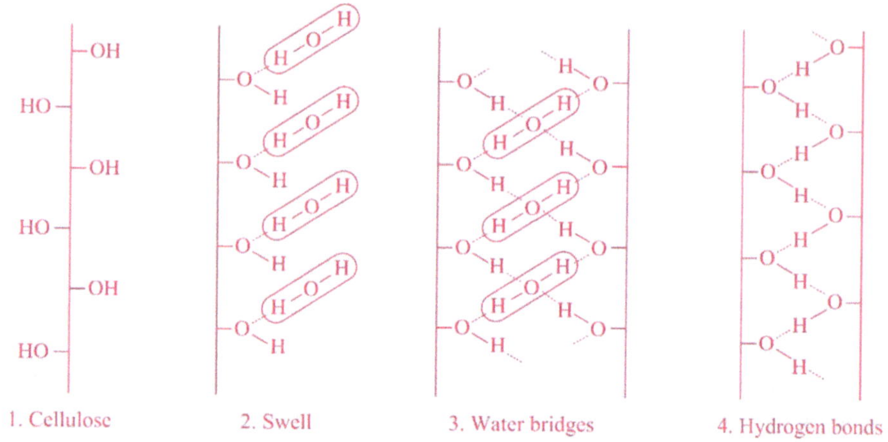

Fig. 3.3 The mechanism of fiber formation into paper[8]

hydroxyl group to the hydrogen atom in another hydroxyl group in a compound. These hydrogen bonds have an energy of 5–8 kcal/mol, which is 2–3 times greater than the energy of typical intermolecular forces, such as Van der Waals forces. Hydrogen bonding is the primary form of interaction between cellulose molecules. All hydroxyl groups in cellulose can potentially form hydrogen bonds. Hydrogen bonding leads to the tight interconnection of cellulose molecules, forming thin sheets of paper with specific strength. Since the dehydration of wet paper involves the formation of hydrogen bonds, the papermaking process is essentially a chemical reaction.

The inclusion of chemical processes in papermaking is also evident in the chemical purification of fiber raw materials. To ensure that the process of bonding cellulose molecules through hydrogen bonds to form paper is not disturbed, harmful impurities must be removed. Impurities such as ash, pectin, lignin, protein, hemicellulose, or poly-pentose, and pigments in the raw material need to be eliminated, as they can have adverse effects on the quality of the finished paper. The common chemical components of papermaking raw materials are shown in Table 3.2.

The higher the cellulose content in the raw material, the better, and the lower the amount of non-cellulosic substances, the better. The main impurities in the raw material are pectin and lignin, with lignin being the most harmful and difficult to remove. According to Table 3.2, hemp has the most ideal composition, with cellulose content ranging from 70% to 83% and lignin content from 2% to 4%. Next is bark, with cellulose content of 40% to 55% and lignin content of 9% to 14%. Bamboo and grass have cellulose content of 36% to 45% and lignin content of 14% to 31%. The chemical composition ranking of various raw materials aligns with physical indicators such as length, width, and length-to-width ratio. Regardless of the criteria used,

[8] *A History of Papermaking Technology in China* [M]. Beijing: Cultural Relics Press, 1979.

Table 3.2 Chemical components of common papermaking materials in ancient China

No	Material	Moisture	Ash content	Extractives				Pentosan	Protein	Pectin	Lignin	Cellulose
				Cold water	Hot water	Ether	1%NaOH					
1	Hemp	9.25	2.85	6.45	10.50		30.76			2.06	4.03	69.51
2	Ramie	6.60	2.93	4.08	6.29		16.81			3.46	1.81	82.81
3	Paper mulberry	11.25	2.70	5.85	18.92	2.31	44.61	9.46	6.04	9.46	14.32	39.08
4	Mulberry plant		4.40		2.39	3.37	35.47	10.42	6.13	8.84	8.74	54.81
5	Blue sandalwood	11.86	4.79	6.45	20.18	4.75	32.45	8.14	4.23	5.60	10.31	40.02
6	Moso bamboo	12.14	1.10	2.38	5.96	0.66	30.98	21.12		0.72	30.67	45.50
7	Omei mountain bamboo	12.56	1.20	2.42	6.78	0.71	31.24	25.41		0.87	31.28	44.35
8	Rice straw	9.87	15.50	6.85	28.50	0.65	47.70	18.06	6.04	0.21	14.05	36.20
9	Wheat straw	10.65	6.04	5.36	23.15	0.51	44.56	25.56	2.30	0.30	22.34	40.40

the ranking remains consistent: hemp is the best, followed by bark, then bamboo, and lastly grass.

Impurities in fibers, such as hemicellulose, consist of various monosaccharides, primarily polymers of xylose in the β-pyranose form. An increase in hemicellulose in the pulp can reduce the strength of the paper. Pectin, which is partially or completely methoxylated galacturonic acid or pectic acid, makes fibers rigid and difficult to break, and consumes alkali solution unnecessarily during the raw material cooking process. Pectin can be decomposed by alkaline solutions or biologically degraded by filamentous fungi through fermentation. Therefore, it is crucial to remove pectin from the raw material before papermaking. In ancient times, raw materials were typically soaked in water tanks and depectinized through fermentation. The process included:

1. Preparation Stage: The raw material absorbs water in the tank, and some organic and inorganic substances dissolve in the water, giving it a light yellow color. At the same time, spore fungi brought in with the raw material begin to ferment, forming bubbles in the water, and the water color gradually deepens while the temperature rises.
2. Pectin Fermentation Stage: Spore fungi reproduce, releasing pectinase to break down pectin, and the temperature continues to rise.
3. Termination Stage: At this point, the raw material becomes soft, and fibers separate, causing the water in the tank to turn brown. Ideally, the pH of the water should be neutral (pH 6–7), and the temperature should be 37–42 °C. The raw material is pressed underwater with stones and should not be exposed. The weight of water is 10 times the weight of the material. If the raw material is pre-cooked with clear water, the effect is better. The fermentation liquid can be recycled or partially drained, and the required time depends on the season, generally taking 7–10 days.

Lignin in the raw material can reduce the strength and lifespan of paper, and it easily oxidizes into pigments, making it challenging to produce white paper. Lignin is a large molecule containing some aromatic benzene rings with a molecular weight of 840 and a complex structure. After delignification of the raw material, it needs to be cooked in an alkaline solution to break down and degrade lignin into soluble substances. This process can also dissolve fats and oils in the raw material, destroy natural pigments, and dissolve substances like tannins, proteins, and starch. Washing can be employed to remove these impurities. Since the Han dynasty, China has used solutions of grass and wood ash water and lime water to cook raw materials, which increases the total alkali content of the solution. After cooking, the lignin content in the raw material decreases from 25%–28% to 1.5%–2%, the hemicellulose content decreases from 25%–28% to 11%–12%, and the cellulose content increases from 55%–58% to 88%–89%. In the production of high-quality paper in ancient times, there were instances of additional cooking and sunlight bleaching. With these methods, lignin could be almost completely removed, resulting in raw

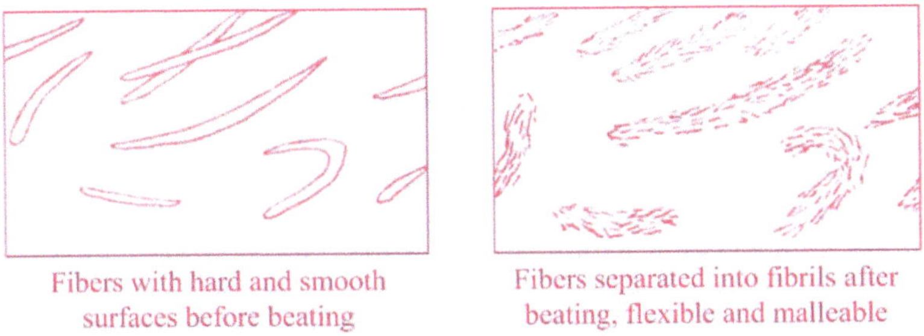

Fig. 3.4 Comparison of fibers before and after beating (Illustrated by Pan Jixing, 1979)

material with cellulose content reaching 99%–100%. The cooking time was typically around 7 days, with continuous firing. The cooking liquid turned black, and after washing, the raw material became white and soft.

Although the purified fiber raw material no longer contains impurities such as pectin, lignin, and pigments, there are still many tangled fiber bundles and a hard outer shell. The hydroxyl groups in cellulose are bound within, unable to be fully exposed and utilized. To produce dense paper, it is necessary to break and brush the excessively long fibers. For this purpose, a beating process is employed, where mechanical force is used to crush the fiber cell walls and fiber bundles, breaking the long fibers. Experimental evidence shows that the bonding force between fibers differs by a factor of 10 before and after beating. In ancient China, various beating tools were used, including pole mortars, treadle mortars, stone mortars, water mortars, etc., powered by human labor, animal power, and hydraulic power. Among them, the equipment with several water-powered stone mortars for pounding material was considered the most advanced. After beating, the fiber surface becomes fuzzy, undergoing fibrillation (Fig. 3.4), sometimes requiring repeated pounding and brushing.

The process of papermaking involves dispersing the crushed fibers in water to create pulp, which can then be used to make paper. Since fibers do not dissolve in water, they can only be formed into a suspension, requiring continuous stirring. However, the inevitable phenomenon of fiber flocculation occurs, resulting in variations in concentration within the pulp, making it challenging to produce uniformly thick paper. To address this issue, ancient papermakers initially added starch to the pulp to improve the suspension properties of the fibers. Later, they introduced plant mucilage as a suspending agent. Commonly used plant mucilage was extracted from the stems of plants such as Chinese gooseberry vine, tamarillo vine, and Abelmoschus manihot L. In water, this mucilage exhibited the characteristics of a filamentous high-molecular-weight electrolyte, preventing fiber settling and avoiding the sticking together of wet paper when one sheet was lifted at a time. Chinese papermakers referred to plant mucilage as *zhi yao* (paper medicine) or *hua shui* (slippery water). After preparing the pulp, a papermaker would insert a paper-making screen, woven from fine bamboo strips, into the pulp. The screen is then shaken to allow the water to drain through the gaps, forming a wet sheet of paper. The remaining water is removed, and the paper is then lifted off the screen. The resulting paper is typically ready for immediate use in writing, packaging, and printing.

However, when using paper for painting or writing small characters, there may be issues like ink spreading or color bleeding. Microscopic studies of paper structure indicate the presence of gaps and capillaries between fibers. To prevent ink spreading, these gaps and capillaries need to be blocked or compressed. The simplest method for this is to rub a smooth stone on the paper, a practice known as "polishing". Another method involves wetting the paper with starch paste and repeatedly beating it with a wooden mallet, a process known as "sizing with starch paste". A more effective method is sizing, which increases the paper's resistance to liquid penetration. Sizing can be applied to the surface of the paper or within the paper itself. The initial sizing agent was starch paste, later replaced by animal glue with alum as a precipitant. The particles of the sizing agent deposit in the gaps between paper fibers (see Fig. 3.5). Additionally, plant dyes could be used to dye paper in various colors, enhancing its aesthetics and improving its performance. To improve the whiteness, smoothness, opacity, and moisture resistance of paper, ancient people used to apply a mixture of white mineral powder and glue on the paper surface, forming a protective layer (see Fig. 3.6). This was known as "powdered paper" or, in modern terms, coated paper. Coating paper with wax enhances its water resistance. Coating colored paper with wax was referred to as colored wax paper.

Applying white powder to paper, followed by coloring and waxing, creates colored powder and wax paper. In ancient times, gold powder or gold foil was fixed onto colored paper using glue, resulting in what was known as "sprinkled gold paper". If patterns were drawn with gold paint on colored paper, it was referred to as "gold flower paper". Another technique involved carving reversed characters or patterns into a wooden board, pressing it onto paper to produce the correct characters or patterns, known as "embossing paper". Creating raised patterns on paper curtains with threads and incorporating them with the curtain patterns was termed "watermarks paper", now known as "watermarked paper". Paper processing in China began

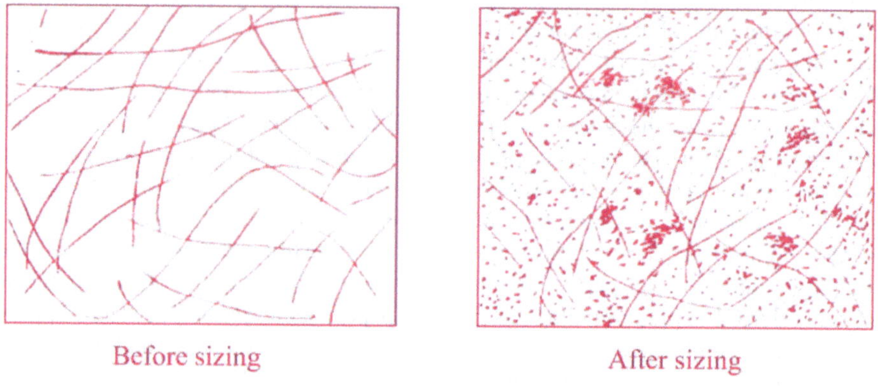

Fig. 3.5 Comparison of paper before and after sizing (Illustrated by Pan Jixing, 1979)

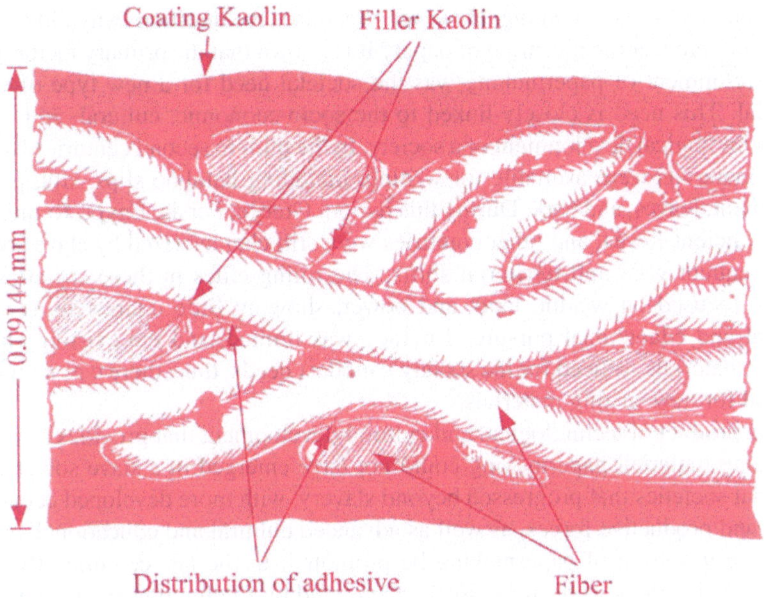

Fig. 3.6 Microscopic image of the cross-section of coated paper[9]

during the Han dynasty with techniques such as polishing and coloring. After the Tang dynasty, various types of processed paper emerged with innovative patterns, contributing to a vibrant paper culture. The purposes of paper processing were to: (1) enhance the visual and intrinsic beauty of paper, moving towards artistic development; (2) improve paper performance and expand its range of applications; and (3) implement protective measures to extend its lifespan.[10]

3.2 Origin of Papermaking Techniques

3.2.1 Why the Origin of Papermaking is in China

The earliest paper was made from hemp rope and discarded cloth, resulting in hemp paper. In China, hemp cloth was woven from hemp or ramie, while in other countries or regions, hemp cloth was woven from flax or jute. Paper could also be made from discarded cloth. From this perspective, both ancient China and other countries had

[9] Pan Jixing. *A History of Papermaking Technology in China* [M]. Beijing: Cultural Relics Press, 1979.

[10] Pan Jixing. Ten Kinds of Modified Paper in Ancient China [J]. *Cultural Relics*, 1979 (2): 38–48. Pan Jixing. Ten Kinds of Modified Paper in Ancient China [J]. *Bulletin of the International Association of Paper Historians*, 1983 (4): 151–155.

the resources for papermaking. However, this raises the question: why did China, in particular, pioneer the invention of paper? It is known that the primary factor driving the development of papermaking was the societal need for a new type of writing material. This need is closely linked to the socio-economic, cultural, educational, and technological development of a society. In the third to second centuries BC, only a few materials were available for writing, including bamboo slips, silk, papyrus, parchment, leaves, and bark. During this period, Greece, Persia, Egypt, Rome, India, Japan, ancient Korea, and other countries were still characterized by slave societies with frequent wars and territorial shifts. The ruling elites in these societies were mainly focused on wealth, land, and power, showing little interest in advancing cultural and educational pursuits. Literacy was limited, and the existing classical writing materials sufficed for the society's modest needs; thus, there was no pressing demand for new writing materials.

As a product of technological and cultural development that provides a new type of writing material, papermaking could not have emerged in a slave society. Only in feudal societies that progressed beyond slavery, with more developed economies, social and productive forces, as well as advanced cultural and educational systems, could the invention of papermaking be promoted. In the last centuries BC, some ancient civilizations were successively conquered by foreign powers, leading to the destruction of their civilizations. These civilizations were still under the rule of slave systems. For example, in 525 BC, Egypt was conquered by Persia, and in 330 BC, Persia was conquered by the Macedonian Kingdom of Greece. In 146 BC, Greece succumbed to the Roman Empire. India, in the East, was ravaged by Greece in 323 BC, and only under the rule of Emperor Ashoka (273–232 BC) did it establish a slave state, which later fell into division after Ashoka's death. In 146 BC, Greece fell to the Roman Empire, marking the decline of Greek civilization. However, the Romans valued military prowess and indulgence, while they disregarded civilization. There was little need for reading materials, and sheepskin books and papyrus scrolls were damaged even before people had the chance to read them, sometimes being used for burning to provide hot water for public baths. Even until the beginning of the Common Era, there was no emergence of new feudal societies in these regions.

On the contrary, China may have entered the slave society later than some countries, but it entered the feudal society earlier, possibly around the fifth century BC. In 221 BC, Qin Shi Huang established a unified feudal empire, standardizing the writing system. This was followed by the Western Han dynasty, which maintained the momentum during the reigns of Emperors Wen and Jing (179–141 BC), creating an era of relative peace and prosperity. Emperor Wu of Han (140–87 BC) demonstrated great talent and strategic thinking, continuing this momentum during his reign. The unified state, along with the economic, cultural, educational, and academic development, required a large quantity of writing materials. Historical records mention that during the reign of Qin Shi Huang, memorials were submitted on stone tablets, with reports presented day and night, and he would not rest until all of them had been reviewed. During the reign of Emperor Wu of Han, Dongfang Shuo submitted a memorial written on 3000 bamboo slips, requiring two people to carry it in, and it took two months to read it completely. Although silk was lightweight, in the Han dynasty, a bolt of silk cost the equivalent of 360 kg of rice, making it unaffordable for common people to use for writing. Hence, there was a saying that "the poor are unable to afford silk". People keenly felt the inconvenience of expensive silk and heavy bamboo slips, leading to the exploration of new materials that were inexpensive, of high quality, and could replace silk and bamboo slips. This paved the way for the invention of paper. China possessed the social background necessary for the development of papermaking.

Among all classical writing materials in the world, only Chinese silk and paper are directly related. The character for paper (纸) itself provides clear evidence, as it incorporates the radical for silk (糸). Paper and silk share several characteristics: both are smooth for writing, lightweight and soft, foldable and easy to roll, and made from pure fibers. The technological concept of papermaking was also inspired by the process of floating and beating silk during the production of silk fabric. This process involved placing silk fibers on a bamboo mat, beating them in water, and then drying and removing the crushed silk, resulting in thin sheets similar to paper. This procedure implied actions similar to pulp beating and scooping. The depiction of the "floating and beating" process (Fig. 3.7) in the illustrated book *Cansang Luosi Zhichou Tushuo* ("Illustrations and Explanations of Silk Culture and Sericulture", 1891) by the Qing dynasty painter Wu Jiayou (styled himself as Youru, 1818–1893) can help us understand this operation.

In ancient times, before weaving silk, cocoons were sorted; the finest cocoons were used to produce high-quality silk fabric, while those of lower quality were repurposed

Fig. 3.7 Illustration of floating and beating process in ancient China (by Wu Jiayou, 1891)

for insulation materials. To prepare these lower-quality cocoons for use, they were first boiled in water with plant ash to remove sericin, then floated, washed, beaten, and dried. The floating and beating process was typically performed by women, as recorded in Zhuangzi's *Xiaoyao You* ("Free and Easy Wanderings").

The floating and beating process demonstrates how dispersed fibers can be retained on a bamboo mat to form thin sheets resembling silk, without requiring a spinning process. By using plant fibers such as hemp instead of silk fibers, paper could be produced. Craftsmen in the Qin and Han dynasties aimed to achieve this, replicating the floating and beating technique with hemp fibers and using bamboo screens for sheet formation. The key to this process was substituting hemp fibers for silk to reduce costs. China, with a rich history of sericulture and silk weaving spanning over 5000 years, was the birthplace of these practices. The tradition of floating and beating is among the oldest, and hemp and ramie have long been used in fabric weaving. Hemp and ramie were native to China, and historical records such as *Shijing* ("Book of Songs") from the fifth century BC document their processing. The retting process, which includes pond retting and steaming with lime water (or plant ash water), is described in Wang Zhen's *Nongshu* ("Book on Agriculture", 1313) and Xu Guangqi's *Nongzheng Quanshu* ("Complete Treatise on Agriculture", 1628). The purpose of retting is to remove impurities chemically, yield pure hemp fibers, and make them soft and white like silk, which are then spun into yarn for weaving. These combined factors ultimately contributed to the invention of papermaking.

3.2.2 On the Origin of Papermaking in Western Han Dynasty

Although China is the birthplace of papermaking, there are differing opinions regarding its origin. The first perspective, represented by Zhang Yi and Fan Ye (397–445), suggests that before the Han dynasty, writing materials like silk were referred to as paper. Due to its high cost, Cai Lun (ca. 61–121) invented paper using plant materials in the first year of Yuanxing (105).[11] However, silk-based materials in ancient times were not paper and cannot be considered as such. Since paper is made from plant materials and records of using paper existed before Cai Lun in the Han dynasty, a second perspective emerged, represented by Zhang Huaiguan (686–758) and Shi Shengzu (1204–1278). They argue that paper had already replaced bamboo slips before Cai Lun, making him an improver rather than the inventor of paper.[12] This viewpoint aligns with a historical development perspective, supported by archaeological discoveries. Papermaking, like metallurgy, is a collective effort that cannot be attributed to a single person or a sudden invention on a particular day. Paper likely emerged during the technological exploration for a new writing

[11] Fan Ye. *Book of Later Han: Vol. 108, Biography of Cai Lun* [M]. // *Twenty-Five Histories: Vol. 2*. Compact Edition. Shanghai: Shanghai Classics Publishing House, 1986: 262.

[12] Zhang Huaizhuan. *Book Criticism* [M]. // Tao Zongyi. *Saying of the Book Collection*. Shanghai: Commercial Press, 1927.

material to replace silk scrolls around the Qin and Han periods. In feudal societies, craftsmen had low social status and were rarely recorded in history. Their technological innovations were often attributed to certain nobles or officials. The Song dynasty scientist Shen Kuo criticized this practice in his work *Changxing Ji* ("Collection of Long-lasting Prosperity"), noting that laborers from various industries contributed to the development of new technologies.

Archaeological excavations in China since the twentieth century have provided abundant physical evidence to address the millennium-long debate on the origin of papermaking. In 1933, Dr. Huang Wenbi (1893–1966) discovered ancient Han dynasty paper in the Han dynasty beacon tower site in Lop Nor, Xinjiang. Describing the finding as hemp paper, he stated that:

Hemp Paper: This paper is made from hemp, white in color, and is produced in square, thin sheets. The edges are irregular, measuring approximately 4.0 cm in length and 10.0 cm in width. It has a very coarse and uneven texture, with visible hemp fibers on the surface. This indicates that it was produced during the early stages of papermaking, thus lacking refinement. This paper was found at the site of the ancient beacon tower of Lop Nor, along with wooden slips dated to the first year of the Huanglong era (49 BC), during the reign of Emperor Xuan of the Han dynasty (73–49 BC). Therefore, this paper is also from the Western Han period. Based on this evidence, it is clear that paper was already in use during the Western Han dynasty. However, the paper from this period was rougher compared to the finer paper made by Cai Lun later on.[13]

Dr. Huang was the first to identify, based on archaeological evidence, that papermaking began during the Western Han dynasty. In the autumn of 1942, Dr. Lao Gan, an archaeologist, discovered a piece of writing paper beneath a Han dynasty beacon tower in what is now Ejina Banner, Inner Mongolia. At this site, 78 Han bamboo slips were found, most of which recorded weapons from the fifth to seventh years of Yongyuan (93–95 AD) and postal records from the tenth year of Yongyuan (98 AD). The paper was located underneath wooden slips and "was buried slightly earlier than the Yongyuan tenth-year slips." Botanist Wu Yinchan from Tongji University identified it as plant-fiber paper.[14] The paper contains 50 surviving characters spread across 8 lines (Fig. 3.8). This text appears to be an official document discussing the transfer of weapons. Given that the Han dynasty was engaged in battles with the Xiongnu around 200 BC, this location served as a garrison point for Han soldiers on the front lines. In 1975, Dr. Lao added the following notes:

The dating of the Juyan paper can be as recent as the Yongyuan era (93–98 AD) and can extend back to the Zhao and Xuan periods (89–49 BC). This is because the wooden tablets discovered in the Juyan region, particularly the Yongyuan Weapons Register, represent the latest compilation from that era. The majority of the other

[13] Huang Wenbi. *Archaeological Study of Lop Nor* [M]. Beiping: National Peking University Press, 1948: 168.

[14] Lao Gan. On the Origins of Chinese Papermaking Technology [J]. *Collected Papers of the Institute of Historical Linguistics*, 1948: 489–498.

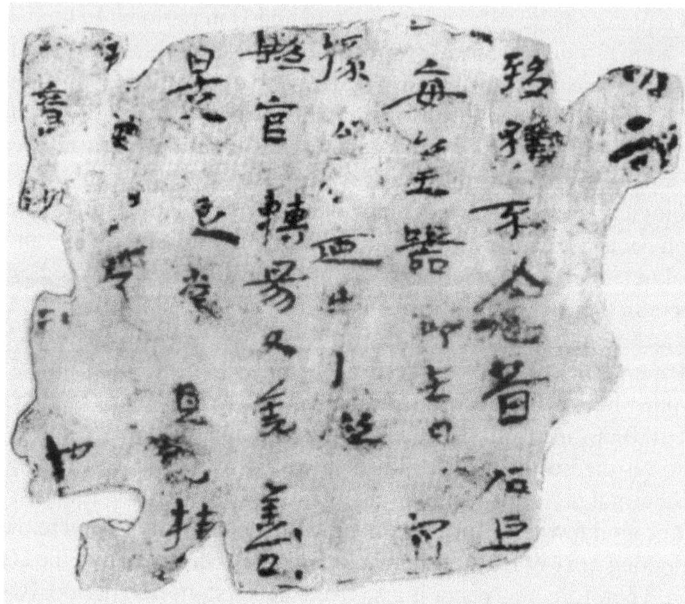

Fig. 3.8 Western Han dynasty paper with Chinese characters unearthed from Chake'ertie in Inner Mongolia in 1942 (Collection of the Institute of History and Philology, Taipei)

tablets were mostly found in the Western Han period, especially during the reigns of Emperor Zhao and Emperor Xuan.[15]

Therefore, both Huang Wenbi and Lao Gan, two esteemed predecessors in the Chinese archaeological community, assert that the origin of papermaking can be traced back to the Western Han period. Dr. Lao even excavated paper from the Western Han era. On May 8, 1957, during an excavation at a brick and tile factory site in Baqiao District, Xi'an, Shaanxi, artifacts such as bronze mirrors and bronze swords were discovered in the soil. The following day, personnel from the provincial museum were dispatched to investigate and found nearly a hundred cultural relics in a north–south oriented tomb. Underneath a trichord knobbed mirror, hemp cloth was found, and beneath it were several layers of paper, which had fragmented, with the largest piece measuring 8 cm × 12 cm. Both the cloth and the paper showed green rust stains. Archaeologists compared these artifacts with those from other tombs of known ages and confirmed that they dated no later than the reign of Emperor Wu of the Western Han dynasty (140–87 BC).[16] Apart from this tomb, no other burial sites or architectural remains were found in the vicinity. Initially, no chemical analysis of

[15] Lao Gan. *Afterword to the History of Ancient Chinese Books* [M]. // Qian Cunxun. *History of Ancient Chinese Books*. Hong Kong: Chinese University Press, 1975: 183–184.

[16] Tian Ye. Western Han Paper Discovered in Baqiao, Shaanxi [J]. *Cultural Relics Reference Materials*, 1957 (7): 78–81. Cheng Xuehua. Dating and Related Information of the Baqiao Paper in Xi'an [J]. *Collected Works on the History of Science and Technology*, 1989 (15): 17–22.

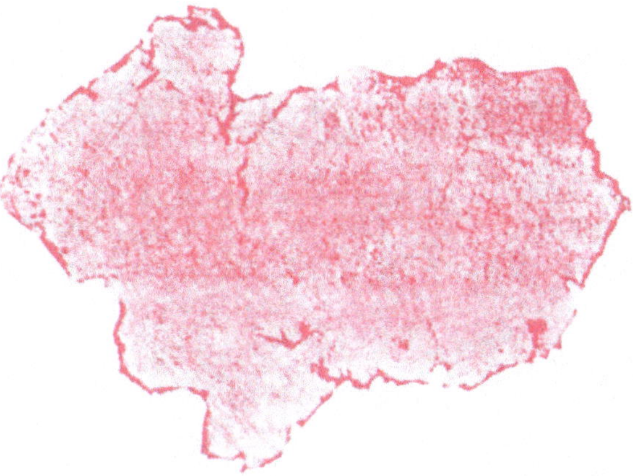

Original state (Image provided by Cheng Xuehua)

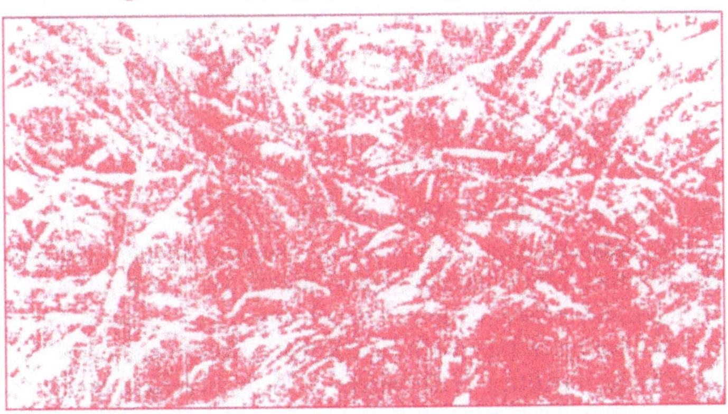

Magnified 10x (Image provided by Pan Jixing)

Fig. 3.9 Western Han hemp paper (140–87 BC) unearthed in Baqiao, Xi'an in 1957

the paper was conducted, and it was believed to be "paper made from fibers similar to silk". However, in 1964, laboratory analysis confirmed that the paper was actually made from hemp fibers (Fig. 3.9).[17]

In 1973, the Gansu Provincial Great Wall Archaeological Team conducted systematic scientific excavations at the Han dynasty military outpost site of Jianshuijinguan on the east bank of the Ejina River in Gansu Province. They uncovered dated wooden slips, silk pieces, hemp cloth, brushes, inkstones, and paper. Two pieces of paper were excavated: the first paper (coded EJF1) was white, measuring 21 cm × 19 cm, and was found alongside wooden slips dating mostly to the periods of Emperor Zhao and Emperor Xuan, with the latest being in the second year of Emperor Xuan's Ganlu era

[17] Pan Jixing. The Earliest Plant Fiber Paper in the World [J]. Cultural Relics, 1964 (11): 48–49. Pan Jixing. Discussion on the Earliest Plant Fiber Paper in the World [J]. *Huaxue Tongbao*, 1974 (5): 45–47.

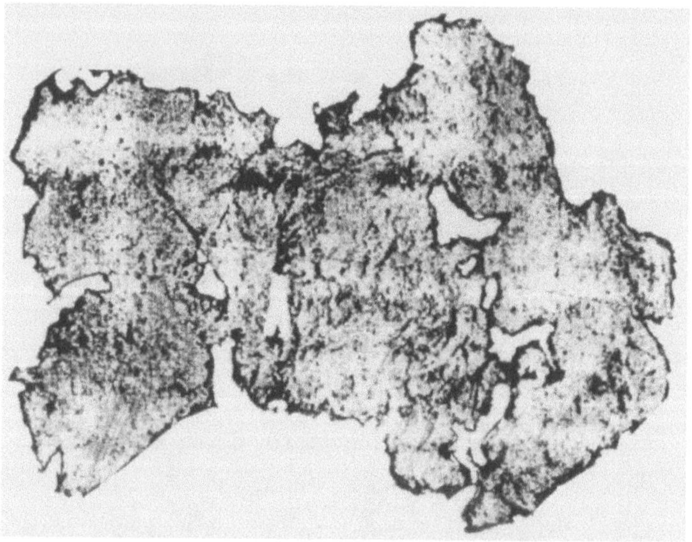

Fig. 3.10 Western Han hemp paper (52 BC) unearthed from the Juyan Jinguan, Gansu in 1973[20]

(52 BC). This paper was thin and abundant (Fig. 3.10). The second paper (EJT30:3) was dark yellow, measuring 11.5 cm × 9 cm, coarser in texture, and dated back to the first year of Emperor Ai's Jianping era (6 BC).[18] The site's various components were well-defined, and the stratigraphic relationships of the cultural deposits and the chronological attribution of the paper were reliable. In December 1978, during the construction of water facilities in Zhongyan Village, Fufeng County, Shaanxi, the remains of a Han dynasty structure were discovered. In a pit beneath the accumulation of roof tiles, a buried pottery jar was found, containing over 90 artifacts, including copper objects and coins. Inside the copper ornaments (round-headed copper nails) of lacquerware decorations, paper was stuffed. The largest piece measured 6.8 cm × 7.2 cm, was white and flexible, exhibiting copper rust stains. Upon examination, the large pottery jar dated to the period around Emperor Xuan, the coins ranged from Emperor Wen to Emperor Ping, and the upper limit for the age of the paper was within the period from Emperor Xuan to Emperor Ping, no later than Emperor Ping (1–5 AD).[19] This pit had never been disturbed and remained in its original state when excavated.

In October 1979, a large-scale excavation was carried out by the Gansu Provincial Great Wall Joint Survey Team at the site of the Western Han military garrison in

[18] Chu Shibin, Ren Buyun. Han Dynasty Relics and Newly Unearthed Wooden Slips from the Juyan Site [J]. *Cultural Relics*, 1978 (1): 6.

[19] Luo Xizhang. Western Han Bronze Ware and Ancient Paper Discovered in Zhongyan Village, Fufeng County, Shaanxi [J]. *Cultural Relics*, 1978 (9): 17–20.

[20] Chu Shibin, Ren Buyun. Han Dynasty Relics and Newly Unearthed Wooden Slips from the Juyan Site [J]. *Cultural Relics*, 1978 (1): 6.

Majuanwan, northwest of Dunhuang. The garrison, established during the reign of Emperor Wu, reached its peak during the reign of Emperor Xuan and was abandoned during the Xin Mang period (8–23 CE). A total of 337 artifacts were unearthed, including silk textiles, coins, copper and iron objects, seals, writing implements, hemp paper, and 1217 wooden slips. There was a total of eight pieces of hemp paper (Fig. 3.11). Paper- I (T12:47), a complete sheet of paper, was yellow and coarse, measuring 32 cm × 20 cm. It dated to around 65–50 BC, based on the wooden slips from the same stratigraphic layer. Paper- II (T10:06) and Paper- III (T9:26), totaling four pieces, had a finer texture, and their chronological context was determined by wooden slips to be from around 32 BC to 5 AD. Paper- IV (T9:25) was white and finely textured, while Paper- V (T12:18), totaling two pieces, dated to the Xin Mang period.[21] In June to September 1986, a paper with a map was unearthed from Tomb No. 5 at Fangmatan in Tianshui, Gansu (Fig. 3.12). The surviving dimensions were 5.6 cm × 2.8 cm, and it was placed on the chest of the deceased. The paper was yellow and stained, featuring fine black lines depicting mountains, rivers, and roads, similar to the silk maps excavated from Tomb No. 3 in Mawangdui, Changsha, in 1973 (168 BC). The burial structure at Fangmatan was similar to Qin tombs, and the burial artifacts were consistent with early Western Han tombs in Shaanxi and Hubei Yunmeng. The tomb's dating corresponds to the Wen and Jing periods (179–141 BC), making it the earliest known paper map to date.[22]

From October 1990 to December 1992, the Gansu Provincial Institute of Cultural Relics and Archaeology conducted extensive excavations at the Han dynasty *Xuanquanzhi* site in the northeast of Dunhuang. *Xuanquanzhi* was an institution for transmitting information and receiving guests, covering an area of 22,500 square meters. A total of 70,000 artifacts were excavated, including 35,000 wooden slips, 470 pieces of ancient paper, as well as leather goods, silk fabrics, writing implements, inkstones, grains, and various daily items. It was recognized as one of China's top ten archaeological discoveries in 1991. The accumulated deposits in the area were divided into five layers with clear stratification, and the dating could be determined by the stratigraphy of the wooden slips found within each layer. Notably, nine pieces of Western Han paper were discovered: T212④:1, which is white, measuring 18 cm × 12 cm, and bears the clerical script "付子". It was found in the fourth layer, dating to the periods of Emperor Wu and Emperor Zhao (140–74 BC); T114③:609, which is yellow and finely textured, with characters such as "持书来……啬" in cursive clerical script, measuring 7 cm × 3.5 cm. It was found in the third layer and belongs to the periods of Emperor Xuan and Emperor Cheng (73–7 BC); F2④:1, a complete sheet of writing paper, light yellow, measuring 34 cm × 25 cm. This excavation yielded the largest quantity of underground ancient paper to date.[23] The summarized information on the excavation of Western Han paper is presented in Table 3.3.

---

21 Yue Banghu, Wu Rengxiang. Brief Report on the Excavation of Han Dynasty Beacon Tower Site at Majuanwan, Dunhuang [J]. *Cultural Relics*, 1981 (10): 1–8.

22 He Shuangquan. Excavation of the Qin and Han Tombs at Fangmatan, Tianshui, Gansu [J]. *Cultural Relics*, 1989 (2): 1–11, 31.

23 He Shuangquan. Brief Report on the Excavation of the Han Dynasty Xuanquanzhi Site in Dunhuang, Gansu [J]. *Cultural Relics*, 2000 (5): 4–20.

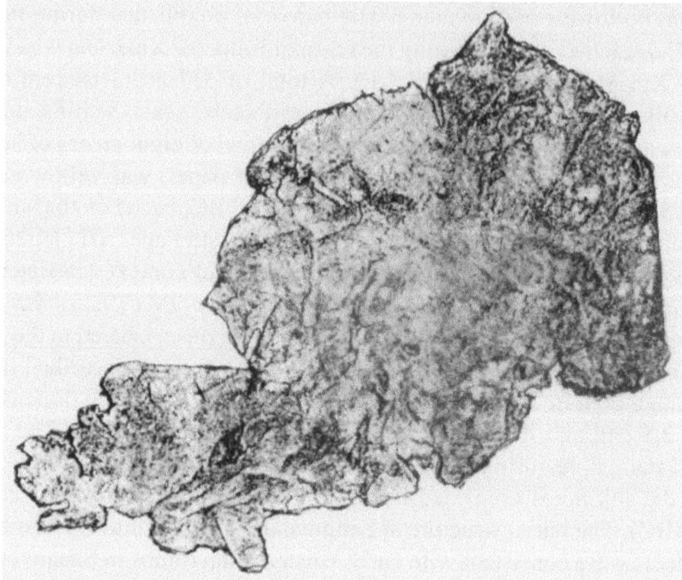

Fig. 3.11 Western Han hemp paper unearthed from Majuanwan, Dunhuang in 1979[24]

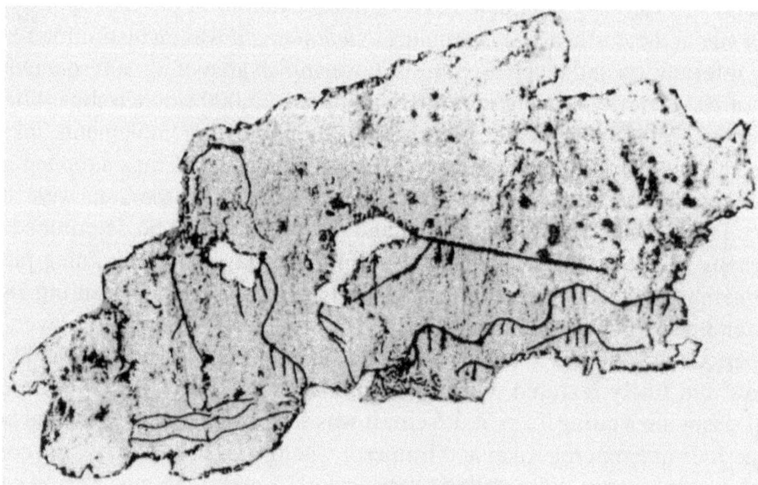

Fig. 3.12 Western Han paper map unearthed from Fangmatan, Tianshui in 1986[25]

[24] Yue Banghu, Wu Rengxiang. Brief Report on the Excavation of Han Dynasty Beacon Tower Site at Majuanwan, Dunhuang [J]. *Cultural Relics*, 1981 (10): 1–8.

[25] He Shuangquan. Excavation of the Qin and Han Tombs at Fangmatan, Tianshui, Gansu [J]. *Cultural Relics*, 1989 (2): 1–11, 31.

Table 3.3 Discovery of Western Han ancient paper (partial)

No	Name	Era (AD)	Discovery year	Discovery location	Dimensions (cm)	Appearance description
1	Lop Nor Paper	73–49 BC	1933	Han dynasty beacon tower site at Lop Nor, Xinjiang	4 × 10	White, thin paper with a rough texture, featuring numerous bundles of fibers and undispersed hemp fibers on the surface
2	Chake'ertie Paper	89–77 BC	1942	Han dynasty beacon tower site at Chake'ertie, East Bank of the Ejina River, Gansu	10 × 11.3	The paper has 8 lines of text, totaling about 50 characters. Over 10 characters are legible. The paper is yellowish-gray in color
3	Baqiao Paper	140–87 BC	1957	Baqiao, Xi'an, Shaanxi	8 × 12	Light yellow, thin paper, layered and compressed under a bronze mirror, torn into 88 pieces. It has many fiber bundles and uneven interweaving. The paper shows green copper rust stains
4	Jinguan Paper-I	52 BC	1973	Jinguan, East Bank of the Ejina River, Gansu	21 × 19	White, fine texture, strong, with fewer fiber bundles
5	Jinguan Paper-II	6 BC	1973	Jinguan, East Bank of the Ejina River, Gansu	11.5 × 9	Dark yellow, with a coarser texture

(continued)

Table 3.3 (continued)

No	Name	Era (AD)	Discovery year	Discovery location	Dimensions (cm)	Appearance description
6	Zhongyan Paper	1–5	1978	Zhongyan, Fufeng, Shaanxi	6.8 × 7.2	White and flexible, of good quality. The paper shows mold marks. This paper, along with other artifacts, was found as part of a cache
7	Majuanwan Paper-I	65–50 BC	1979	Majuanwan, Dunhuang, Gansu	32 × 20	Yellow, relatively coarse, with natural edges around it. This is the most complete piece of paper, measuring to its original size
8	Majuanwan Paper-III	32–1 BC	1979	Majuanwan, Dunhuang, Gansu	9.5 × 16	A total of 2 pieces, originally white but now stained a soil-like color. Some areas remain white. The paper is finely made
9	Majuanwan Paper-IV	1–5	1979	Majuanwan, Dunhuang, Gansu	9 × 15.5	White, fine texture, with few fiber bundles. The paper shows clear mold marks
10	Majuanwan Paper-V	8–23	1979	Majuanwan, Dunhuang, Gansu	17.5 × 18.5	White, fine texture, few fiber bundles, with good strength

(continued)

Table 3.3 (continued)

No	Name	Era (AD)	Discovery year	Discovery location	Dimensions (cm)	Appearance description
11	Fangmatan Paper	179–141 BC	1986	Fangmatan, Tianshui, Gansu	5.6 × 2.8	Originally yellow when unearthed, now faded to a yellowish-gray. The paper is thin and soft, with a map drawn on it, and the surface has stains
12	Xuanquanzhi Paper-I	140–74 BC	1990–1992	Xuanzhiquan Site in Tianshuijing, Dunhuang, Gansu	18 × 12	White, good texture, with the inscription *fu zi*
13	Xuanquanzhi Paper-II	73–7 BC	1990–1992	Xuanzhiquan Site in Tianshuijing, Dunhuang, Gansu	3 × 4	Light yellow, with fine fibers and good texture. The paper bears the inscription *xi xin*. It has mold marks, with mold strips approximately 0.3 mm thick, and the paper is thin, with a thickness of 0.286 mm
14	Xuanquanzhi Paper-III	73–7 BC	1990–1992	Xuanzhiquan Site in Tianshuijing, Dunhuang, Gansu	13.5 × 7	Light yellow, slightly thicker, with the inscription *xun li*. The paper is of good quality

Instances of paper use before Cai Lun are also found in historical records. The *Stories of San Fu*, compiled in the fourth century, mentions an incident during the reign of Emperor Wu. Crown Prince Liu Ju (128–91 BC) has a large nose. When Emperor Wu fell ill, the Crown Prince visited him, and the court official Jiang Chong (141–91 BC) said, "The Emperor dislikes large noses. You should cover your nose with paper before entering." However, Emperor Wu did not become angry because of his nose, and the prince subsequently had Jiang Chong killed.[26] This event occurred in the second year of Emperor Wu's Zhenghe era (91 BC) during the emperor's illness in Ganquan. This is the earliest recorded use of paper in ancient texts. The *Han Shu* ("Book of Han") in volume 57, records that after Emperor Wu read Sima Xiangru's *Zixu Fu*, he summoned Sima Xiangru to the palace in the third year of Jianyuan (138 BC). Xiangru made a *Youlie Fu* and the emperor ordered the Ministry of Personnel to provide writing materials. Yan Shigu (581–645), in his commentary, explained that at that time, paper was not widely used, so they were provided with bamboo slips for writing.[27] This indicates that paper was not universally used at the beginning of the Han dynasty, and the archaeological discoveries mentioned earlier suggest that both bamboo slips and paper were used simultaneously. Records found on wooden slips dating to the mid-late Western Han period (first century BC) from the Juyan excavations mention "official written paper", as mentioned earlier.

In the *History of the Former Han, Volume on the Imperial Relatives*, it is recorded that in the third year of Hongjia (18 BC), Emperor Cheng of the Western Han dynasty appointed Zhao Feiyan (approximately 42–1 BC) as Empress, but she had no children. In the first year of Yuan Yan (12 BC), a concubine named Cao Weineng gave birth to a son. The younger sister of the Empress instructed someone to send Cao Weineng to a cold palace and then had the prison minister Ji Wu use poison to kill her. Afterward, the child was taken away. The poison was wrapped in a material called *he ti*, and Eastern Han scholar Ying Shao (140–206 AD) explained in his *Hanshu Jijie Yinyi* ("Explanations, Pronunciations, and Meanings of Bookof Han") that *he ti* refers to thin small paper.[28] Liu Xiu, the founding emperor of the Eastern Han dynasty (5 BC–57 AD), ascended the throne on August 5, 25 AD in Hao (present-day Gaoyi, Henan). In the winter of the same year, the imperial carriage entered Luoyang, establishing it as the capital. According to Ying Shao's *Fengsu Tongyi* ("Annotation of Customs", 175), Emperor Guangwu transported books to Luoyang using 2000 carriages, carrying books made of silk, bamboo slips, and paper.[29] These 2000 carriages constituted the first batch of imperial collections in the Eastern Han internal library, undoubtedly dating back to the Western Han before the year 25 AD. Emperor Guangwu had a preference for paper. In his accession year, he ordered the establishment of the *Shangshutai* (Ministry of Personnel) as the central administrative institution of the state. The *Shangshu You Cheng* (right aide of Shangshu) was appointed to "manage the silk ribbon and other properties in storage such as paper, brush, and ink". The emperor also directed the official *Shougongling* to "manage the paper, brush, and ink used by the emperor, properties of *Shangshu* and the sealing

---

[26] Zhang Shu. *The Stories of Sanfu* [M]. Eryou Tang Series Edition, 1820: 9.

[27] Ban Gu. *Book of Han: Vol. 57, Biography of Sima Xiangru* [M] // *Twenty-Five Histories: Vol. 1.* Compact Edition. Shanghai: Shanghai Classics Publishing House, 1986: 601.

[28] Ban Gu. *Book of Han: Vol. 97, Biography of Empress Zhao of the Han Dynasty* [M] // *Twenty-Five Histories: Vol. 1.* Compact Edition. Shanghai: Shanghai Classics Publishing House, 1986: 370.

[29] Ying Shao. *Comprehensive Meaning of Customs* [M]. // Qian Yong, Huang Han, Yin Yuanwei, et al. *Collection of Notes and Novels*. Yangzhou: Jiangsu Guangling Ancient Books Printing House, 1983.

clay".[30] This occurred over 30 years before Cai Lun's birth (in 25 AD). In the first year of Jianchu (76 AD) during the reign of Emperor Zhang of Han, scholar Jia Kui (29–101 AD) was ordered to deliver lectures at the Northern Palace White Tiger Hall. The emperor instructed Jia Kui to select twenty students with high abilities to teach them *Zuozhuan* ("Zuo's Commentary") and gave each of them a set of bamboo slips and paper-written classics.[31]

In summary, from the twentieth century onwards, ancient Han paper dating from the second century BC to the first century BC has been unearthed in Xinjiang, Gansu, and Shaanxi in eight separate archaeological excavations in 1933, 1942, 1957, 1973, 1978, 1979, 1986, and between 1990 and 1992. Some of these papers even feature maps or inscriptions. These archaeological discoveries, along with a series of historical records, confirm that writing paper was already in use during the early Western Han period, 200 years prior to Cai Lun's time in 105 AD. It is clear that paper was not invented by Cai Lun. Zhang Yi, Fan Ye, and their followers have recognized that paper existed during the early Han period but have maintained that this paper could only have been made from silk rather than plant fibers. The validity of this claim can be established by analyzing and testing the unearthed ancient paper. Some individual samples were tested as early as the 1940s, and from the 1960s onwards, comprehensive systematic tests were conducted on Western Han papers unearthed between 1957 and 1990. The process involved observing the external morphological characteristics of the paper under high magnification, extracting fibers from the samples, dyeing them, creating slides, and examining the fiber disintegration patterns using high-magnification optical or scanning electron microscopes. Relevant data were measured to determine fiber types, and each test was repeated at least twice. Simultaneously, parallel comparative tests were conducted on known samples of various plant fibers and post-Han papers of known ages. The results of each test are presented in Tables 3.4 and 3.5.

From the microscopic images of the longitudinal and cross-sections, as well as the overall appearance of various fiber samples (Fig. 3.13, 3.14, 3.15, 3.16), it is evident that the features align with those of hemp and ramie fibers. The average width of the fiber samples is 18–26 μm (1 μm = 10^{-3} mm), consistent with the variation range in hemp width (7–32 μm). Experts from the China Textile Academy reexamined the paper samples and found only hemp fibers, with no evidence of silk fibers. We also noted that all samples had undispersed small hemp thread ends, which were consistent with the fiber composition of the paper, indicating that the paper raw material was broken hemp fabric. Under the microscope, it was observed that the purity of hemp fibers was relatively high, with fewer impurities and foreign cells. In contrast to raw hemp and hemp fabric fibers, some paper samples exhibited a whiteness of 25%, while the majority ranged between 40 and 45%. This consistency in purity and whiteness with Han dynasty paper suggests that the paper raw material underwent chemical

[30] Fan Ye. *Book of Later Han: Vol. 36, Records of the Officials* [M]. // *Twenty-Five Histories: Vol. 2*. Compact Edition. Shanghai: Shanghai Classics Publishing House, 1986: 80–81.

[31] Fan Ye. *Book of Later Han: Vol. 66, Biography of Jia Kui* [M]. // *Twenty-Five Histories: Vol. 2*. Compact Edition. Shanghai: Shanghai Classics Publishing House, 1986: 152–153.

Table 3.4 Analysis and testing results of ancient paper found before Cai Lun

No	Name	Era	Raw material	Width (mm)	Basis weight (g/m^2)	Density (g/m^2)	Whiteness (%)	Average fiber length (mm)	Average fiber width (μm)
1	Baqiao Paper	140–87 BC	Hemp	0.10	29.2	0.29	25	0.88	25.55
2	Jinguan Paper-I	52 BC	Hemp	0.22	61.7	0.28	40	2.10	18.73
3	Zhongyan Paper	1–5	Hemp	0.22	61.9	0.28	43	2.12	20.26
4	Majuanwan Paper-V	8–23	Hemp	0.29	95.1	0.33	42	1.93	18.18
5	Simulating Han Dynasty Paper	1965	Hemp	0.14	38.9	0.28	42	2.85	22.10
6	Fengxiang Hemp Paper	1980	Hemp	0.10	38.5	0.45	45	1.56	20.89

Table 3.5 Analysis and testing results of three types of unearthed Western Han dynasty paper

No	Name	Era	Raw material	Width (mm)	Basis weight (g/m^2)	Density (g/m^2)	Whiteness (%)	Average fiber length (mm)	Average fiber width (μm)
1	Baqiao Paper	140–87 BC	Hemp	0.085	21.0	0.25	25	1.05	18
2	Jinguan Paper	52 BC	Hemp	0.25	63.8	0.26	40	1.03	17
3	Zhongyan Paper	1–5	Hemp	0.23	58.4	0.262	40	1.29	19

purification through steaming with wood ash water. Simulation experiments indicate that without such treatment, it is challenging to achieve higher whiteness in hemp materials, and low whiteness is due to inadequate steaming effects.[32]

The average length of fibers in Western Han dynasty paper ranges from 0.9 to 2.2 mm, with the longest fibers (1%) measuring up to 10 mm. Fibers with lengths between 0.5 and 1.9 mm constitute more than 50% of the total. In comparison, untreated raw hemp fibers have an average length of 15–25 mm, and untreated raw ramie fibers have an average length of 120–180 mm. The difference is several dozen

[32] Pan Jixing, Miao Junying, Zhang Jinying, et al. *Comprehensive Analysis and Testing of Four Types of Unearthed Han Dynasty Paper* [M] // Pan Jixing. *History of Chinese Science and Technology: Papermaking and Printing Volume.* Beijing: Science Press, 1998: 64.

Fig. 3.13 Photograph 1 of Western Han Baqiao paper fibers under scanning electron microscope (×100) (Provided by Pan Jixing, 1988)

Fig. 3.14 Photograph 2 of Western Han Baqiao paper fibers under scanning electron microscope (×100) (Provided by Pan Jixing, 1988)

to several hundred times, indicating that the paper fibers underwent cutting and shortening processing. It is unlikely that the fibers naturally shortened due to decay, and this phenomenon was not observed. Although there are undispersed hemp rope ends on the paper, they have been cut short, and the cuts can be seen under the microscope. All samples lacked fiber bundles or rope ends longer than 15 mm.

Fig. 3.15 Photograph of Western Han Jinguan paper fibers under scanning electron microscope (×300) (Provided by Pan Jixing, 1988)

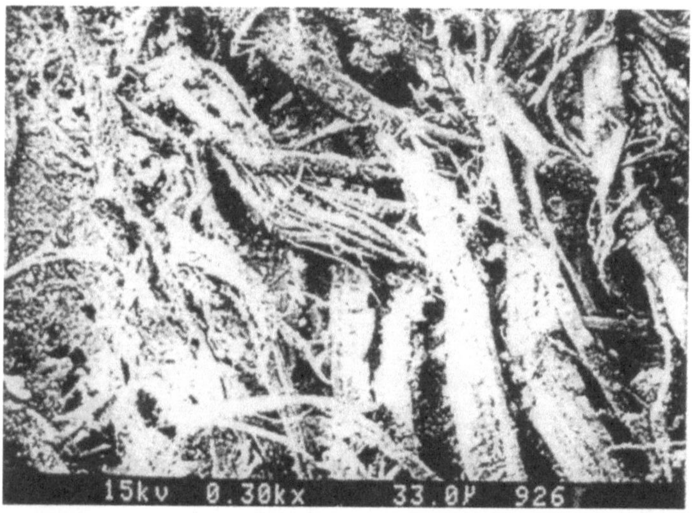

Fig. 3.16 Photograph of Western Han Majuanwan paper fibers under scanning electron microscope (×100) (Provided by Pan Jixing, 1988)

Under the microscope, the basic composition of Western Han dynasty paper samples consists of dispersed individual fibers arranged in a non-directional interweaving pattern, with some fibers aligned in the same direction in certain areas. The average thickness of the papers ranges from 0.1 to 2.9 mm, typically 0.2–0.25 mm, and they all exhibit a certain level of strength and tightness ranging from 0.28 to 0.33 g/

cm3. This aligns with the requirements for handmade paper and is not indicative of naturally accumulated fibers.

All observed Western Han dynasty paper samples exhibit signs of compression and fibrillation under a microscope, with the degree of fibrillation being 40% for Jinguan paper and Zhongyan paper, and the degree of beating reaching approximately 50°SR. The presence of brushing and beating indicates intentional processing of the fibers. If the raw materials were not cut short or beaten, such phenomena would not occur. Only the fibers in the Baqiao paper show a relatively low degree of fibrillation; however, areas that were beaten still retain the structure of paper. The Zhongyan paper, Majuanwan paper, Xuanquanzhi paper, and Baqiao paper all exhibit watermark patterns, each with a width of 2 mm, indicating that they were made using watermark screens. Some paper watermark patterns are less distinct or vary in thickness due to the use of woven moulds. Based on the test results mentioned, it can be concluded that all the unearthed Western Han paper was made from hemp fibers. The raw materials, including hemp stems and old cloth, underwent processes such as shredding, washing, steaming in plant ash water, mechanical beating, and papermaking. The paper surfaces are smooth, accept ink well, and some even retain writing, demonstrating a certain degree of mechanical strength. Microscopic observation reveals that the primary component is purified dispersed fibers, forming thin sheets 0.2–0.25 mm thick through tight interlocking. Most fibers exhibit a fibrillation phenomenon. Therefore, from the perspectives of appearance, physical structure, technical indicators, and performance, the unearthed Western Han paper meets the definition of handmade paper. It is genuine plant fiber paper that replaced silk and bamboo slips for writing. However, there are variations in processing quality, with differences in texture and use. The origin of papermaking technology in the early second century BC Western Han is now confirmed beyond doubt.

3.2.3 Papermaking Techniques in Han Dynasty

Since the beginning of the second century BC, during the early Western Han dynasty, hemp paper was the predominant type of paper in China for nearly a thousand years. Understanding how this paper was manufactured requires detailed investigation. There are scarce records in ancient texts about the technology of producing hemp paper, as writings mostly focus on the production of bark paper and bamboo paper since hemp paper gradually gave way to these alternatives after the Song dynasty. Therefore, researching hemp paper manufacturing involves reconstructing a lost technology. To unravel the mystery, on-site surveys of the few remaining handcraft workshops producing hemp paper are essential. Analyzing Western Han paper samples can reveal insights into raw materials and processing techniques. To uncover these details, we need to conduct simulation experiments in handcraft workshops, compare the experimental products with unearthed artifacts, and explore various experimental methods. Due to space constraints, some details will not be covered here. Instead, we will briefly summarize our research findings. The technological process for making hemp paper should include at least the following steps: (1) Soak and breaking hemp cloth → (2) Shredding → (3) Water washing → (4) Steaming hemp materials with plant ash water → (5) Water washing → (6) Pounding → (7) Water washing → (8) Preparing paper pulp → (9) Papermaking → (10) Sun drying → (11) Peeling paper → (12) Sorting and packaging.

In 1965, we conducted a simulation experiment at the Fengxiang Hemp Paper Factory in Shaanxi following the outlined process, working alongside experienced craftsmen, led by the late papermaker Huang Yansheng (1912–1968). To mimic actual production conditions, we used 20–25 kg of broken hemp products as raw materials, manually operated with real production tools and equipment. However, the papermaking mold was temporarily designed and manufactured to the dimensions of Han dynasty paper. The paper we produced closely resembled the unearthed Western Han paper, with similar technical indicators. The papermaking process we employed likely replicated the process used during the Western Han period. Papermaking procedures could be simpler or more complex than the outlined steps, resulting in coarser or finer paper. Some products might not fall into the category of paper, while others closely resemble later paper. Therefore, the 12 steps mentioned above should reflect the early process of paper manufacturing.[33] Here is a breakdown of the process:

1. Mechanical pretreatment of raw materials (soaking, shredding, water washing): Old hemp cloth, ropes, etc., were weighed and placed in a basket, soaked in water to moisten and remove dust and sediment. The damp hemp material was cut into small pieces using a tool, removing metal, wood chips, feathers, leather, and debris during the process. The shredded hemp material was then washed in river water.
2. Chemical treatment of raw materials (soaking in plant ash water, steaming, water washing): Plants were burned to ash, soaked in hot water, filtered, and plant ash water was obtained (Fig. 3.17), which had weak alkalinity. Taking rice straw ash as an example, its total alkali content (as potassium oxide K_2O) was 57.7 g/L, hydroxide (as potassium hydroxide KOH) was 36.6 g/L, and carbonate (as potassium carbonate K_2CO_3) was 42.5 g/L.

   The shredded hemp material was soaked in grass-wood ash water, then placed in a steaming pot and steamed with grass-wood ash water or lime water poured from top to bottom. The steaming pot, made of iron, contained water, with a rack and two open-ended barrels placed on top. The hemp material was placed on the rack, filling the barrels, and sealed with hemp bag pieces at the top. The pot was heated with firewood underneath (Fig. 3.18). After steaming, the hemp material was taken out using an iron fork, placed in a basket, and washed in river water to discard the dark liquid. Steaming took about 7 days, and the washed hemp fibers were purified, becoming white and soft.
3. Crushing hemp material: The washed hemp material was pounded and flipped in batches in a stone mortar until fine (Fig. 3.19). Using a hand-held pestle to pound is quite laborious, while using a treadle mill is more efficient. After pounding,

[33] Pan Jixing. The Technology of Han Dynasty Hemp Paper Production: Insights from Experimental Simulations. [J] *Cultural Relics*, 1977 (1): 51–58.

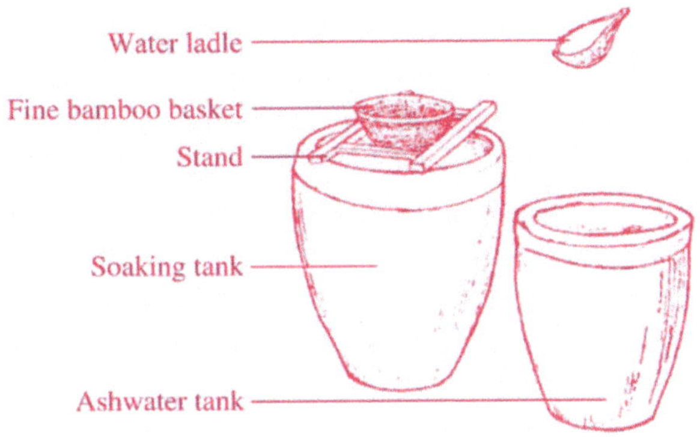

Fig. 3.17 Soaking device for plant ash water (Illustrated by Pan Jixing, 1979)

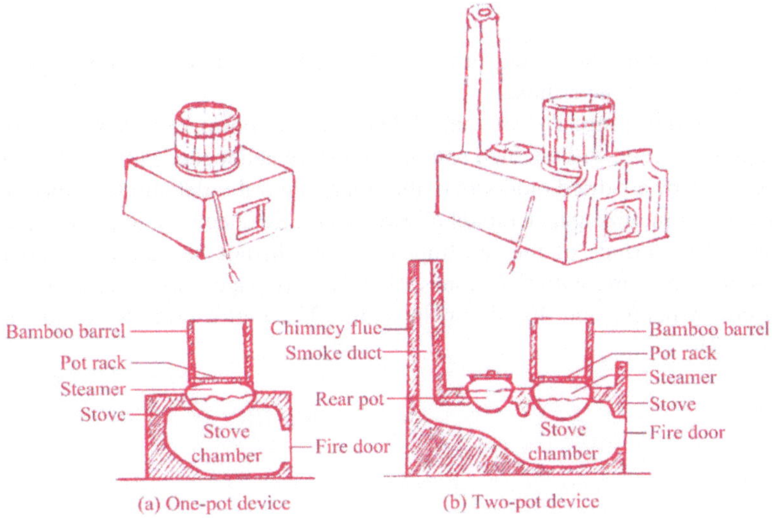

Fig. 3.18 Steaming pot used in Han dynasty papermaking (Illustrated by Pan Jixing, 1979)

the hemp material turns into dispersed fine fibers. Pounding was labor-intensive and took the most time, typically done by strong laborers. After crushing, the material was placed in a fine bamboo basket or cloth bag, then washed in river water to remove ash particles, mud, and other impurities.

4. Pulping and paper scooping: Crushed and washed hemp fibers, resembling white cotton fluff, were placed in a rectangular wooden trough. Clean well water or mountain spring water was added to make a pulp of appropriate thickness. Stirring with a wooden stick was done to ensure the fibers floated in the water. The paper mold was a rectangular (34 cm × 24 cm) wooden frame, with horsehair woven into a sieve fixed on top or a bamboo curtain fixed on top (Fig. 3.20). The paper mold was slanted into the pulp with both hands, shaking it back and forth, and then

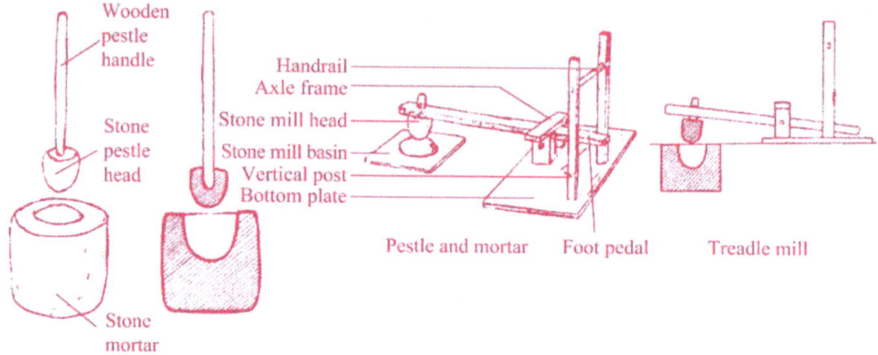

Fig. 3.19 Equipment for pounding hemp material (Illustrated by Pan Jixing, 1979)

lifting it to drain the water, forming a wet paper sheet. This step was performed by experienced papermakers.

5. Drying and peeling: The wet paper sheet, after draining excess water during the shaping process, still retained residual moisture and lacked sufficient strength. It needed to be air-dried outdoors in the sun to naturally dehydrate before peeling. Making numerous paper molds allowed multiple sheets to be made simultaneously. The surface of the resulting paper might not be perfectly smooth and required polishing with fine stones. Han dynasty paper molds were directly 24–25 cm in height and 35–55 cm in width. The entire process is illustrated in Fig. 3.21.

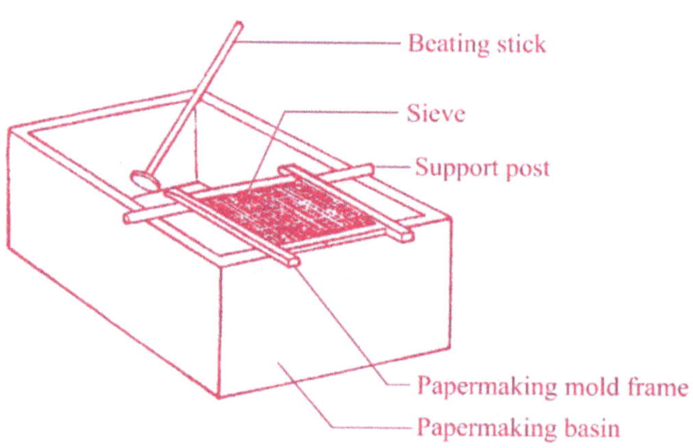

Fig. 3.20 Wooden trough for paper scooping (Illustrated by Pan Jixing, 1979)

Fig. 3.21 Flow chart for Han dynasty hemp paper production (Designed by Pan Jixing and illustrated by Zhang Xiaoyou, 1979)

3.2.4 Historical Evaluation of Cai Lun

Through archaeological discoveries and scientific research on the unearthed ancient paper from the Western Han dynasty, the claim that Cai Lun invented paper cannot be substantiated. Instead, the origin of papermaking can be traced back to the second century BC, indicating the long and rich history of this Chinese invention and giving due recognition to the creative labor of papermaking predecessors in the Western

Han dynasty. Therefore, the academic debates on the origin of paper since the Tang and Song dynasties should come to a conclusion. The remaining question is how to evaluate Cai Lun's historical role. Cai Lun, courtesy name Jingzhong, was born in Guiyang (modern-day Leiyang, Hunan) around the sixth year of the Yongping era of Emperor Ming of the Eastern Han dynasty (63). He entered the palace as a eunuch in the late Yongping era (75) and served as a *Huangmen Shilang* (Yellow Gate Attendant) in the early years of Emperor Zhang of the Han (76–86), handling official matters inside and outside the palace and accompanying various princes in their audiences. When Emperor He ascended the throne in the first year of the Yongyuan era (89), Cai Lun was promoted to *Zhongchangshi* (Central Palace Attendant), attending the emperor closely and participating in military and state affairs, marking the beginning of eunuch involvement in politics in history. In the third year of Yongyuan (91), Cai Lun also took on the role of *Shangfangling* (Superintendent of the Imperial Household), overseeing the manufacturing of knives, swords, and various objects used in the palace. The weapons he supervised were well-crafted and some were unearthed. In the fourteenth year of Yongyuan (102), Empress Deng Sui (81–121) was enthroned, and Cai Lun presented her with fine paper. Deng Sui, not interested in playing with precious gems and jewels, only required the annual supply of paper and ink from various households.[34] Seeing the empress's fondness for paper, in 105, Cai Lun suggested promoting the use of paper, which was accepted. As Emperor He passed away in the same year, and his young son succeeded him, Empress Dowager Deng came into power, so the paper produced under Cai Lun's supervision was, in fact, dedicated to Empress Dowager Deng.

Empress Dowager Deng held power for the rest of her life (105–121). In the first year of Yuan Chu (114), she bestowed the title of *Longting Hou* (Marquis of Longting) upon Cai Lun, granting him land in present-day Yang County, Shaanxi, along with a fief of three hundred households. Later, Cai Lun was further honored as *Changle Taipu* (Grand Master of Changle), equivalent to *Da Changqiu* (Grand Chancellor), reaching the pinnacle of his power. In the first year of Jianguang (121), Empress Dowager Deng passed away, and Emperor An took over the throne. In the third year of the Jianchu era (78 AD), Empress Dou ordered Cai Lun to falsely accuse Empress Dowager Song, the birth mother of Emperor An, which led to her death. Emperor An issued an edict for the Chief Judge to arrest Cai Lun. Knowing that death was inevitable, Cai Lun committed suicide by poisoning himself. The court stripped Cai Lun of his noble title and confiscated his estates. Due to his involvement in the palace power struggles, he met a tragic end. Although Cai Lun, as a eunuch, did not achieve much politically, he made significant contributions to technological advancement during his time as *Shangshuling*, particularly in the development of paper. At that time, silk fabrics were expensive and cumbersome to use, and Cai Lun recognized the advantages of paper. To cater to Empress Deng's preference for paper, he organized skilled craftsmen in Luoyang to produce high-quality paper. Besides using hemp as a raw material, he successfully developed paper made from paper

[34] Yuan Hongjin. *Annals of Later Han: Vol. 14, Reign of Emperor He* [M]. Shanghai: Commercial Press, 1926: 12.

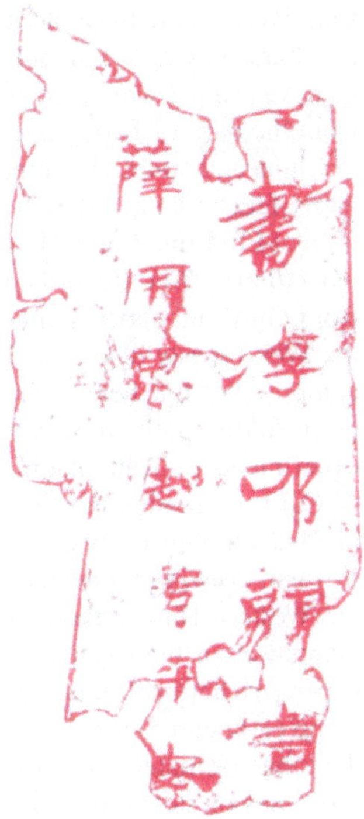

Fig. 3.22 Han dynasty letter unearthed in Xinjiang[35]

mulberry. In 105, after presenting this paper to the court, he received praise. In 114, after Cai Lun was granted the title of *Longting Hou* (Marquis of Longting), the paper produced for use in the palace was known as *Cai Hou Zhi* (Marquis Cai's Paper). A Han dynasty letter unearthed in Xinjiang is shown in Fig. 3.22.

The earliest biography of Cai Lun appears in the Eastern Han official history, *Dongguan Hanji* ("History of Eastern Han"). This book was written during the reign of Emperor Huan (151–166) by Yan Du (ca. 97–167) and Cao Shou. However, the book gradually became lost after the Song dynasty. Tang dynasty references to this biography mention, "Cai Lun, Yellow Gate Attendant, is in charge of papermaking in the Imperial Household, known as Marquis Cai's Paper."[36] This suggests that Cai Lun supervised papermaking in the Imperial Household, and after being titled a marquis, the paper produced was referred to as Marquis Cai's Paper. When his marquis title was later stripped, the name of this paper fell out of use during the Eastern Han period. Dong Ba (200–275), a scholar from the Wei dynasty, mentioned in *Dahan Yufu Zhi* ("Records of Carriages and Clothing in Han Dynasty") that, "in Dongjing (Luoyang), there was Marquis Cai's Paper, which referred to the paper made by Cai

35 Luo Zhenyu, Wang Guowei. *Falling Sands and Lost Bamboo Slips: Vol. 1* [M]. Beijing: Zhonghua Book Company, 1914.

36 Xu Jian. *Beginning of Learning: Vol. 26, Paper* [M]. Beijing: Zhonghua Book Company, 1962: 516.

Lun. Paper made from old hemp was called *ma zhi* (hemp paper), paper made from tree bark was called *gu zhi* (paper hemp paper), and paper made from old fishing nets was called *wang zhi* (net paper)."[37] The term *gu zhi* here denotes paper made from the bark of Broussonetia papyrifera (also known as *gu* or *chu*), a plant native to China and an excellent material for papermaking. Paper made from Broussonetia papyrifera bark did not exist in the Western Han period and likely emerged during the time of Cai Lun. Zhang Hua (232–300), a Jin dynasty scholar, also mentioned in his work *Bowu Zhi* ("Records of the Investigations of Things", ca. 200) that "Cai Lun from Guiyang started pounding old fishing nets to make paper."[38] Dong Ba from the Wei dynasty also talked about paper made from fishing nets. Fishing nets are composed of fine hemp ropes with many knots, making them difficult to pound into pulp. Additionally, they were treated with tung oil and pig's blood. Our simulated experiments indicate that making paper from old fishing nets is challenging.

Cai Lun, though not the original inventor of papermaking, played a crucial role in advancing the early development of the craft.

First, drawing on nearly 200 years of Chinese experience with hemp paper from the Western Han to the early Eastern Han period, he leveraged the resources and manpower of the royal workshops to produce high-quality hemp paper. This new paper set a national standard for paper production. Prior to Cai Lun's innovations, old hemp cloth and rope ends were commonly used as raw materials. Cai Lun expanded the range of materials by experimenting with old fishing nets. He likely refined the processes of steaming and beating, repeating these treatments as needed, and may have improved the papermaking equipment. These changes represented a significant technological advancement in hemp paper production.

Second, Cai Lun promoted the use of paper mulberry bark for papermaking, achieving a breakthrough by utilizing wood-derived fibers. Mulberry bark fiber, being raw, contains more impurities such as pectin and lignin compared to hemp and old fabric. This required an extra boiling process, which was not necessary for hemp paper. The production of mulberry bark paper also demanded a more rigorous steaming process, and careful removal of the outer layer of the bark during boiling was essential to avoid brownish-black spots in the finished paper. The development of mulberry bark paper introduced a more complex set of technological processes for using plant-based fibers. This innovation led to the creation of new varieties of bark paper, including those made from mulberry bark, fragrant sumac, and rattan, marking a significant advancement in the field of bark paper.

Third, as *Zhongchangshi* and *Longting Hou*, Cai Lun, a high-ranking eunuch trusted by the ruling class, proposed the widespread use of paper to the court and gained support from Empress Dowager Deng, who was in charge during the emperor's minority. Paper mills were subsequently established in various regions, facilitating the rapid development of the papermaking industry and the widespread

[37] Dong Ba. *Records of Carriages and Clothing in Han Dynasty* [M] // Li Fang, Li Mu, Xu Xuan. *Taiping Yulan: Vol. 605, Paper*. Beijing: Zhonghua Book Company, 1960: 2724.

[38] Zhang Hua. *Records of the Investigations of Things* [M]. Beijing: Zhonghua Book Company, 1980: 125.

use of paper in society. Therefore, Cai Lun played a pivotal role as an innovator in the evolution of papermaking technology.

We emphasize that papermaking originated in the Western Han dynasty, and acknowledging Cai Lun's contributions during the Eastern Han dynasty is grounded in extensive historical evidence. This perspective is not intended to diminish Cai Lun's achievements. It not only upholds the significance of this Chinese invention but also garners recognition and acclaim from scholars both domestically and internationally. However, this viewpoint has faced criticism from some individuals within the papermaking community in China who habitually consider Cai Lun as the patriarch of the industry. They criticize archaeologists for what they believe is an incorrect dating of the unearthed Western Han papers, demanding that the ages of these papers be adjusted to postdate Cai Lun. Such demands have naturally been rejected by archaeologists. Additionally, there are claims that papers like Baqiao paper and Zhongyan paper "bear characters written in the Eastern Jin style",[39] which, in reality, is purely fictional. These individuals treat the account of Cai Lun inventing paper in *Book of Later Han* as the ultimate truth and reject the archaeological community's dating of Western Han papers, taking a confrontational stance against archaeological findings. This not only fails to uphold the narrative of Cai Lun inventing paper but also generates negative consequences.

Some individuals go so far as to assert, based on analytical examinations, that the unearthed Western Han papers are not plant fiber papers; at best, they claim these are merely "naturally accumulated fibers" or "proto-paper".[40] They argue that only papers made after Cai Lun should be considered true paper. However, such assertions are not tenable. Any objective researcher conducting a microscopic comparison of papers from before and after Cai Lun would find that the physical structure, fiber types, and conditions of both kinds of paper are indistinguishable—they are both papers, with differences only in age and texture. When people observe Western Han papers with writing on them and touch them even briefly, they can easily recognize them as paper without the need for chemical analysis. Therefore, foreign scholars believe that those who still insist on Cai Lun inventing paper are "devotees with a religious sentiment towards Cai Lun". They argue that the criticisms against the existence of paper in the Western Han period lack a basis and are driven by emotional attachment.[41] However, emotions should not influence research into ancient papers, as the microscope is impartial. The resurgence of the claim that Cai Lun invented paper after the repeated discoveries of Western Han papers is merely a fading echo before this claim withdrew from academic discourse.

---

39 Rong Yuankai. Doubts on Western Han Hemp Paper [J]. *Journal of Jiangxi University* (Social Sciences Edition), 1980 (2): 56–60.

40 Wang Juhua, Li Yuhua. A Preliminary Discussion on the Invention of Papermaking in China Based on the Analysis and Identification of Several Han Papers [J]. *Cultural Relics*, 1980 (1): 80–84.

41 Shigeru Nakayama. *Science for the Citizens* [M]. Tokyo: Social Review Press, 1984: 44–47.

3.3 Development of Papermaking Techniques

3.3.1 Papermaking in Wei, Jin, and Northern and Southern Dynasties

The Western Han dynasty laid the groundwork for the papermaking industry, primarily producing hemp paper while still using both paper and bamboo slips. The Wei, Jin, and Northern and Southern dynasties (third–sixth centuries) saw significant advancements in papermaking technology. Systematic analysis of unearthed papers from this period shows improvements in paper quality, including increased whiteness, a smoother surface, more tightly interwoven fibers, fewer fiber bundles, and noticeable horizontal stripes. The beating degree improved, with some Jin dynasty paper reaching a high level of 70°SR. During the Southern and Northern dynasties (420–589), paper thickness decreased to 0.1–0.15 mm, compared to the thicker Han dynasty paper that measured 0.2–0.3 mm. Technological advancements included the use of movable sieve-bed papermaking machines, which were more advanced and efficient than earlier fixed machines. These new machines enabled the production of dense, thin, and uniform paper, and allowed for the continuous production of millions of sheets without changing the paper mold—a significant innovation. The use of movable sieve-bed machines also required a higher beating degree and enhanced steaming processes. Additionally, this period saw the introduction of a wet paper pressing step and changes in the paper drying method.

During the Han dynasty, papermaking involved using a fixed machine that required drying the mold and the wet paper together in sunlight, which meant multiple paper molds were needed. However, the introduction of the movable sieve-bed papermaking machine improved efficiency. With this new method, the bamboo sieve with wet paper was removed and the paper was transferred to a wooden board, where it was stacked layer by layer. The sieve was then put back on the frame for the next batch. This innovation allowed a single worker to operate just one papermaking machine. After pressing the excess water from the stacked wet paper, the papermaker peeled off the partially dried paper and placed it on a wooden board or a wall coated with lime. It quickly dried into single-sided smooth paper. The side facing the board was smooth and referred to as the front side, while the other side was the back side. Studies of the paper from the Wei, Jin, and Northern and Southern dynasties reveal evidence of both front and back sides. The papermaking machine of this period comprised three main components: a wooden bed, a sieve, and side pillars (Fig. 3.23). The paper sieve was made from thin bamboo strips (Fig. 3.24). A sieve with more than 9 bamboo strips per centimeter (ranging from 9 to 15 strips) was considered fine, while a sieve with 5–7 strips (mostly 5) was deemed coarse. In regions without bamboo, such as northern areas, sieves were woven from grass such as Achnatherum splendens or Hemerocallis fulva stems. Extensive testing of various types of paper showed differences in length and width, as detailed in Table 3.6.

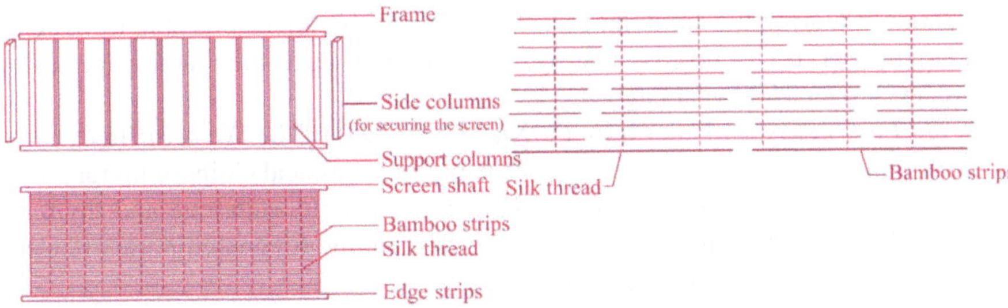

Fig. 3.23 Movable sieve-bed papermaking machine and principle of sieve weaving (Illustrated by Pan Jixing, 1979)

Fig. 3.24 Illustration of sieve weaving operation (Illustrated by Pan Jixing, 1998)

Table 3.6 Dimensions of paper and papermaking tools from the Wei, Jin, and Northern and Southern dynasties

Era	Wei and Jin		Southern and Northern dynasties	
Type	Type A (Small Paper)	Type B (Large Paper)	Type A (Small Paper)	Type B (Large Paper)
Height (cm)	23.5–24.0	26–27	24.0–24.5	25.5–26.5
Width (cm)	40.7–44.5	42–52	36.3–55.0	54.7–55.0

Based on the dimensions of the paper, it can be seen that the size of the paper sieve was somewhat larger than in the Han dynasty. Paper came in different sizes depending on the need. In February 1972, documents written in the 20th year of the Jianyuan era (384) were unearthed from tombs dating back to the Former Qin period (351–394) in Turpan, Xinjiang. The size of the paper used for these documents was 23.4 cm × 35.5 cm, which is roughly the size of a page in the *Beijing Evening News*. Larger paper would have dimensions similar to a page in today's *The New*

York Times.[42] During the Jin dynasty, calligrapher Wang Xizhi was able to write on such large paper. It was no easy feat for a person to produce such large paper using a sieve, even over 1500 years ago.

In *Zhi Fu* ("Essay on Paper") by Fu Xian (239–294) of the Western Jin dynasty, a description of paper is given.[43] It praises paper in a classical Chinese literary style called *fu* that blends poetry and prose, characterized by its elaborate and rhythmic language. However, the language can be quite challenging for contemporary readers. We have endeavored to translate it as follows:

The quality of writing impacts governance, for it influences both advantages and disadvantages. Thus, institutional practices evolve with the times, and writing tools change accordingly. Just as we have replaced cords with oracle bones and written contracts, we have also developed paper to replace bamboo slips. This transition mirrors the shift from simplicity to suitability, adapting to new needs. Paper, as a material, is both beautiful and valuable, embodying simplicity and integrity. It accommodates refined writing and literary works, cherished by scholars and poets alike. Its smooth and versatile nature allows for easy handling and storage. Even from afar, one can express feelings and maintain connections through letters sent over great distances.[44]

Since the Jin dynasty, people began producing clean and smooth paper, eliminating the need for expensive silk and cumbersome bamboo slips. Paper became the dominant writing material, leading to a significant increase in the number of books. During the Eastern Jin dynasty (317–420), the practice of copying classics, historical records, literary works, official and private documents, contracts, and religious scriptures using paper became widespread. Handwritten copies were in scroll format, adhering to a fixed writing style, with each sheet containing 400–500 characters. These sheets were connected into long scrolls, which could be rolled up or spread out. In the eighth year of the Yuanjia era (431) of the Liu Song dynasty, the imperial library contained 64,000 scrolls, and during the Liang dynasty, Emperor Yuan had 70,000 scrolls in Jiangling. By the time of Emperor Wu of Liang (502–549), households across the country possessed literary and historical works. This proliferation of written material significantly promoted the development of Chinese society, culture, education, science, and technology, and also contributed to the flourishing of religion (Fig. 3.25). The extant early paper manuscripts found in Dunhuang, such as

---

[42] Pan Jixing. *History of Chinese Science and Technology: Papermaking and Printing Volume* [M]. Beijing: Science Press, 1998: 1217.

[43] Fu Xian. *Essay on Paper* [M] // Yan Kejun. *Complete Texts from the Ancient Period through the Three dynasties, Qin, Han, Three Kingdoms, and Six dynasties: Vol. 51, Complete Works of the Jin dynasty*. Beijing: Zhonghua Book Company, 1958: 5.

[44] Pan Jixing. *History of Chinese Science and Technology: Papermaking and Printing Volume* [M]. Beijing: Science Press, 1998: 104.

Fig. 3.25 Copying Buddhist scriptures during the Northern and Southern dynasties (Illustrated by Zhang Xiaoyou, 1979)

Avādana-sūtra from the first year of the Ganlu era (359) of the Former Qin dynasty, provide valuable physical evidence. The surviving portion is the thirtieth chapter, *Chu Diyu Pin* ("Departing from Hell"), with a total length of 166 cm, composed of seven connected sheets measuring 23.6 cm × 30.3 cm each. The hemp paper is white with a slight yellowish tint, and the surface is relatively smooth with coarse horizontal stripes (0.2 mm). This manuscript was produced in the northwest and is now housed in the Tokyo Calligraphy Museum (Fig. 3.26).

During the Han dynasty, the dominant script styles were small seal script and clerical script. Small seal script featured uniformly thick strokes, while clerical script introduced variations in stroke thickness, providing more room for artistic expression. In the Wei, Jin, and Northern and Southern dynasties, regular script and semi-cursive script became prevalent. These scripts demonstrated more fluid and expressive strokes, reflecting the transition from bamboo slips to paper. The manuscripts such as *Avādana-sūtra* and the Western Jin dynasty manuscript of *Zhufo Yaoji Jing* ("Essentials of the Buddha") from the sixth year of the Yuan Kang era (296), as well

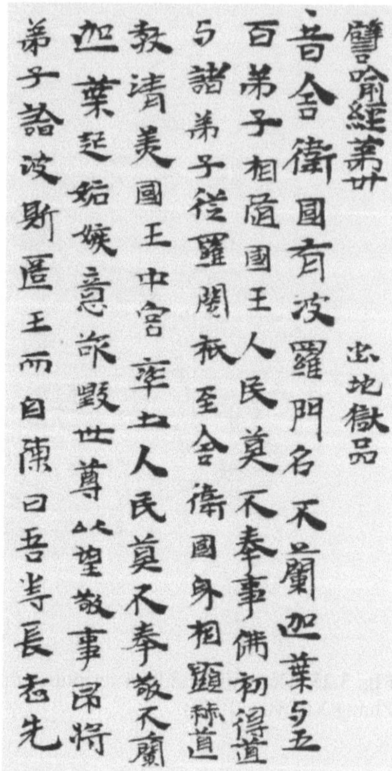

Fig. 3.26 *Avādana-sūtra* written on hemp paper from the first year of the Ganlu era (359) of the Former Qin dynasty (Provided by Nakamura Fusetsu, 1934)

as the Eastern Jin dynasty manuscript of *Sanguo Zhi* ("History of the Three Kingdoms", Fig. 3.27), were all written in clerical and regular scripts, showcasing the popular script styles of the time. The emergence of calligraphic masters like Wang Xizhi and Wang Xianzhi during the Jin dynasty can be attributed to the widespread use of paper. The scripts of the Two Wangs became models for future generations (Fig. 3.28).

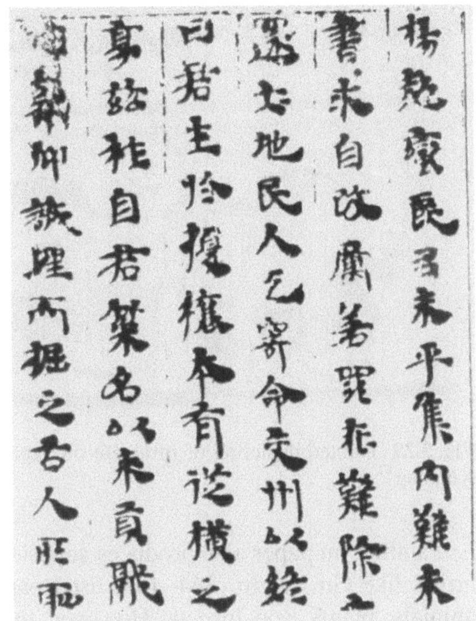

Fig. 3.27 Eastern Jin dynasty manuscript of *Records of the Three Kingdoms* unearthed in Turpan, Xinjiang[45]

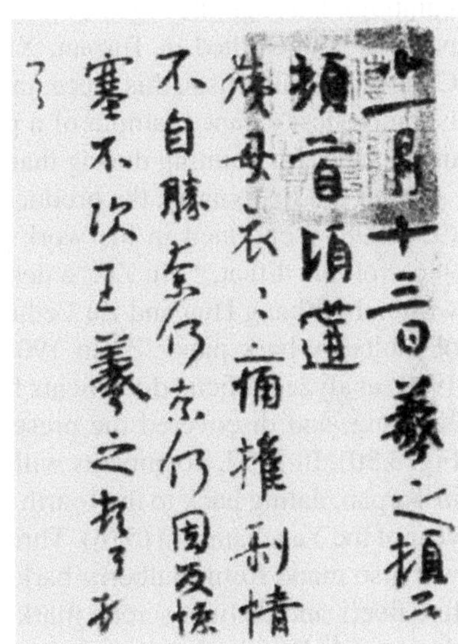

Fig. 3.28 Calligraphy of Wang Xizhi[46]

[45] Guo Moruo. Fragments of the Eastern Jin dynasty Manuscript of 'Records of the Three Kingdoms' Newly Unearthed in Xinjiang [J]. *Cultural Relics*, 1972 (8): 2–6.

[46] Xu Senyu. My View on the Authenticity of 'Preface to the Orchid Pavilion Collection' [J]. *Cultural Relics*, 1965 (11): 1–8.

Fig. 3.29 Painted depiction of rural life on paper in Eastern Jin dynasty, unearthed in Turpan, Xinjiang[47]

Painting on paper also produces superior artistic effects. During the Jin dynasty, artists like Gu Kaizhi (344–405) used paper to create paintings of figures, birds, animals, plants, and insects. However, these works have not survived, and their existence is only recorded in Zhang Yanyuan's *Lidai Minghua Ji* ("Record of Famous Paintings Through the Ages", ca. 870). In 1964, a painted depiction of rural life on paper was unearthed in Turpan, Xinjiang, dating back to the Eastern Jin dynasty. Created by folk artists, this piece, measuring 106.5 cm in length and 47 cm in height, is the earliest extant example of a painted artwork on paper, reflecting the trend of using paper for painting during that time (Fig. 3.29).

After the Jin dynasty, the production of bark paper gradually increased. Su Yijian (958–998) mentioned in his work *Wenfang Sipu* ("Four Treasures of the Study", 986) volume 4 that, "Mu Zhi, a descendant of Lei Kongzhang, still had documents written by Zhang Hua and Zu Leihuan (230–290), and these documents were made of mulberry bark paper."[48] In 1901, Austrian scholar Julius von Wiesner (1853–1913) analyzed official documents from the third to fifth centuries found in Lop Nor, Xinjiang, and discovered the presence of mulberry bark paper (mulberry tree see Fig. 3.30). In 1972, documents with writing were unearthed from Astana Cemetery in Turpan, dating back to the fourth year of the Jianchang era (558) and the sixteenth year of the Yanchang era (576). Through our analysis, it was confirmed that the paper was also made from mulberry bark. The paper is thin and white, with uniform and fine fibers and distinct screen marks. It measures 42.6 cm in length and is partially damaged.[49] We also examined paper made from a mix of hemp and tree bark fibers.

[47] Xinjiang Uygur Autonomous Region Museum. *Artifacts Unearthed in Xinjiang* [M]. Beijing: Cultural Relics Press, 1975.

[48] Su Yijian. *Four Treasures of the Study: Vol. 4, Paper* [M]. Shanghai: Commercial Press, 1960: 51.

[49] Pan Jixing. Research on Ancient Paper Unearthed in Xinjiang [J]. *Cultural Relics*, 1973 (10): 50–60.

Fig. 3.30 Mulberry tree[50]

By the Three Kingdoms period, paper made from mulberry bark had spread from the Yellow River region to the Yangtze River basin and northern Vietnam. Wu scholar Lu Ji (courtesy name Yuanke, 210–279) wrote:

In Youzhou (present-day Beijing), it is called *gu sang* or *chu sang*, while in Jing (present-day Hubei), Yang (present-day Yangzhou, Jiangsu), Jiao (northern Vietnam), and Guang (present-day Guangzhou), it is also referred to as *gu*. In Zhongzhou (present-day Henan), it is called *chu sang*… People in the lower reaches of the Yangtze River now make the bark into cloth and beat it into paper, calling it *gu pi zhi* (mulberry bark paper)[51] (Fig. 3.31).

The regions producing mulberry bark paper, such as Jingzhou, Yangzhou, Jiaozhou, and Guangzhou, were all under the control of Wu (222–280). Mulberry bark paper was also produced within the Wei (220–265) territory. Jia Sixie, an agronomist from the Later Wei dynasty, dedicated a chapter to the cultivation of paper mulberry in his work *Qimin Yaoshu* ("Important Arts for the People's Welfare", ca. 538). This indicates that by the sixth century, there were specialists in cultivating paper mulberry trees and intermediaries known as "mulberry merchants" who dealt in mulberry bark. Jia Sixie noted that those who grew paper mulberry trees and made their own paper could achieve greater profits.[52] In 1973, paper used for *Dabei Rulai Gaoshu* ("Great Compassion Sutra"), unearthed from the Land Temple of the Thousand Buddha Caves in Dunhuang, dating to the third year of the Xing'an era (454) of the Northern Wei, was found to be made from paper mulberry bark. In 1972, paper with inscriptions from the fourth year of the Jianchang era (558) and the sixteenth year of the Yanchang era (576) discovered in the Astana tombs was identified as paper made

50 Chinese Academy of Sciences, Institute of Botany. *Illustrated Flora of China* [M]. Beijing: Science Press, 1987.

51 Lu Ji. *Commentary on Plants, Birds, Beasts, Insects, and Fish Mentioned in Mao's Poems* [M]. Shanghai: Commercial Press, 1935: 29–30.

52 Jia Sixie. *Selected Readings from Qimin Yaoshu* [M]. Annotated by Shi Shenghan. Beijing: Agricultural Publishing House, 1961: 280.

Fig. 3.31 Paper mulberry[53]

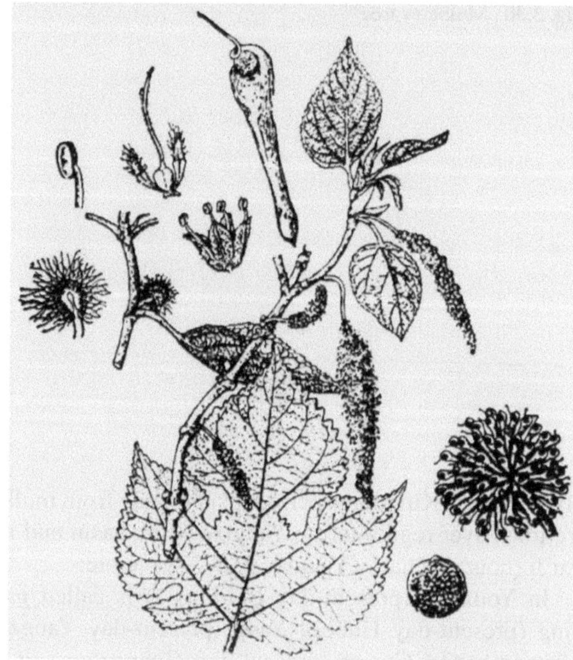

from paper mulberry and mulberry, respectively. To further expand the sources of paper materials and develop new types of paper, by the Jin dynasty, people near the upper reaches of the Cao'e River in Shengzhou, Zhejiang, began making paper from wild vine bark fibers. This practice later spread to other vine-producing regions. Yu Shinan (558–638), in his *Beitang Shuchao* ("Literary and Historical Texts of the Northern Hall", 630), volume 104, quoted Fan Ning (339–401) from the Eastern Jin dynasty, who, when serving as a local official in Zhejiang, instructed his subordinates: "Local paper is unsuitable for writing; all must use vine bark paper." "Local paper" referred to the lower-quality hemp paper produced locally, while "vine bark paper" referred to paper made from vine bark. Surveys indicate that Shengzhou produced various types of vines such as Illigera, Wisteria, and Pueraria, all of which could be used to make paper.

With the increase in paper production, the use of paper expanded beyond writing and began to replace silk in various applications. Paper umbrellas, paper kites, paper flowers, paper-cutting art, paper-folding art, and toilet paper emerged. The surface of the traditional umbrella was made of silk, which was not effective against rain. The use of paper for the umbrella surface, coated with tung oil, became a cost-effective and rain-resistant alternative. The use of paper for umbrellas can be traced back to the Northern Wei dynasty (386–534). *Wei Shu* ("History of Wei") records that in the early years of the reign of Emperor Xuanwu (500), a group of mountain tribes,

[53] Chinese Academy of Sciences, Institute of Botany. *Illustrated Flora of China* [M]. Beijing: Science Press, 1987.

"wearing plain clothes and holding white umbrellas and banners, resisted against the emperor."[54] The *Tongdian Zhiguandian* ("Imperial Encyclopedia", 801) notes that umbrellas were not present during the Jin dynasty, and the practice of making paper umbrellas started with the Xianbei Tuoba tribe ruling the Northern Wei. The paper umbrellas were initially designed for cavalry during rainy military marches and later spread from minority regions to the central plains, becoming widespread throughout China after the Tang dynasty.

In ancient times, kites were made using bamboo sticks and then coated with silk fabric. Paper kites emerged during the Southern and Northern dynasties. In the year 550, after the Northern Qi ruler Gao Yang defeated Eastern Wei, he persecuted the son of the Eastern Wei emperor, Yuan Huangtou. Gao Yang ordered Yuan Huangtou and some prisoners to climb onto the 26-*zhang* (about 86.67 m) tall platform and jump off using paper kites. Surprisingly, Yuan Huangtou survived the fall.[55] Later on, kites were also used for military purposes. Papercutting, as a folk art, originated in the Wei and Jin periods, matured during the Southern and Northern dynasties, and there are archaeological findings associated with it.[56] Sun Fang (330–370), a person from the Jin dynasty, mentioned in the *Xisi Ming* ("Ode to the West Temple") that when he served as an official in Changsha (present-day capital of Hunan Province), he observed children inserting paper flowers into the ground near the West Temple, indicating the paper flower art had a long history. Yan Zhitui (531–591), a person from the Southern Qi dynasty, mentioned in the *Yanshi Jiaxun* ("Family Instructions for the Yan Clan") that there were papers with the Five Classics and the names of virtuous individuals, emphasizing their avoidance for inappropriate uses. The term "inappropriate uses" refers to using paper for bodily waste, such as toilet paper and menstrual hygiene products, which played a crucial role in public health, hygiene, and preventing infections among the general population.

The processing techniques of paper also underwent development, and the application of sizing technology began in the latter half of the third century during the Wei and Jin periods. The earliest sizing agent used was starch paste. In 1973, while studying paper from a tomb in Xinjiang dating back to the first year of the Baique era (384) of Later Qin, we observed that the surface of hemp paper underwent sizing and polishing treatment. Although this sizing improved the writing properties of the paper, it had poor resistance to pests and was brittle. To overcome these drawbacks, surface coating technology emerged. In the early twentieth century, Wiesner discovered that some paper from the Jin and Southern and Northern dynasties were coated with a layer of gypsum. In 1974, a letter written by Wang Zong from the 36th year of the Jianxing era (348) was unearthed from the Hala Hezhuo Tomb in Turpan. Our analysis confirmed that the paper used for this letter is coated hemp paper. In 1965, a

[54] Wei Shou. *History of Wei: Vol. 69, Biography of Pei Yan* [M]. Shanghai: Shanghai Classics Publishing House, 1986: 2346.

[55] Li Yanshou. *History of Northern Dynasties: Vol. 19, Biography of Wang Xie of Pengcheng* [M]. Shanghai: Shanghai Classics Publishing House, 1986: 2967.

[56] Chen Jing. *Research on Chinese Folk Paper-cutting Art* [M]. Beijing: Beijing Arts and Crafts Publishing House, 1972: 146–147.

manuscript of *Sanguo Zhi* from the Eastern Jin period was unearthed in Turpan, and its surface was coated with a layer of white mineral powder and had been polished.

During the Wei, Jin, and Southern and Northern dynasties (third–sixth centuries), yellow paper was the most popular dyed paper, as evidenced in the Dunhuang caves. This type of paper was resistant to pests, and in case of mistakes, orpiment (As_2S_3) could be applied for corrections. The yellow color also conveyed a sense of solemnity. In the Jin dynasty, there were two methods of dyeing yellow paper: either writing first and then dyeing, or dyeing first and then writing. Based on archaeological findings, the latter method was more common, especially in official documents, Confucian classics, and Buddhist scriptures. The dye for yellow paper was derived from the bark of the Phellodendron chinense Schneid tree. It contained effective components such as berberine ($C_{20}H_{19}O_5N$), imparting a yellow color and a bitter taste, soluble in water, and exhibiting insect-repelling properties. The *Qimin Yaoshu*, in Volume 3, provides information on cultivating the Phellodendron chinense Schneid tree and the process of dyeing with yellow. The *Taiping Yulan* ("Taipingg Imperial Encyclopedia"), in Volume 605, quotes Ying Dezhan's *Huan Xuan Weishi* ("False Accounts of Huan Xuan"), stating that after declaring himself emperor, Huan Xuan (369–404) "ordered officials to produce blue, red, light blue, green, and pink paper, emphasizing precision and speed." In 333 AD, the ruler of the Later Zhao, Shi Hu (295–349), also ordered the production of five colored papers in the capital. The dyes used were mostly plant-based, such as red flowers and Chinese redwood for red, indigo for blue, and purple grass for purple, resulting in various intermediate colors formed by combining red, yellow, and blue.

3.3.2 Papermaking in Sui, Tang, and Five Dynasties

During the Sui, Tang, and Five dynasties period (sixth–tenth centuries), China saw significant advancements in papermaking techniques. The range of raw materials expanded to include not only hemp, mulberry bark, and vein bark, but also the bark of fragrant sumac, cotton rose hibiscus, and various types of bamboo. This era also marked the use of wild hemp in papermaking and the initiation of recycling waste paper to produce reborn paper. The traditional practice of blending various raw materials for pulping continued. The production of bark paper significantly increased, compared to previous periods. Due to the high consumption of vine paper, the 250-km-long vine forests along the Shanxi River in Shengzhou, Zhejiang, were completely depleted, leading to ecological deterioration. In response, Shu Yuanyu (ca. 760–835) wrote *Bei Shanxi Guteng Wen* ("Ode to the Ancient Vines of Shanxi River"), urging people to conserve paper, protect plant resources, and practice sustainable harvesting and replanting to prevent such environmental damage from recurring. Mulberry bark paper gained widespread admiration and was celebrated as the "national paper", with the term *chu* (paper mulberry) becoming synonymous with paper, and *chu mo* (mulberry ink) referring to ink used on this paper. Early Tang dynasty calligrapher Xue Ji (649–713) honored mulberry paper as *chu guogong* (Duke Mulberry Paper), ranking top among all papers. The literary giant Han Yu (768–824) referred to it as *chu xiansheng* (Master Mulberry Paper). Mulberry paper became the most esteemed type of paper. According to *Huayan Jingzhuanji* ("Records of the *Avatamsaka Sutra*") by the monk Fa Zang (703), many Buddhist monks cultivated paper mulberry trees and irrigated them with fragrant water. After making white paper, they used it to respectfully write *Avatamsaka Sutra*, signifying its importance and solemnity.

While traditional hemp paper reached its peak during the Sui and Tang periods, being the main type of paper with increased production and quality surpassing that of the Southern and Northern dynasties, bark paper, especially represented by paper made from paper mulberry, demonstrated a rising trend due to its thin, flexible, and durable characteristics. Over 30% of the Dunhuang and Xinjiang Buddhist scriptures we have analyzed from this period were written on bark paper, reflecting the situation in the northwest region. For the inland areas, the percentage of bark paper was likely higher. Examples include *Huguo Bore Boluomi Jing* ("Perfection of Wisdom Sutra for Humane Kings Protecting Their Countries") written in 600 during the Sui dynasty, using yellow mulberry bark paper, and the Daoist scripture *Wushang Miyao* ("Esoteric Essentials of the Most High") from 718 during the Tang dynasty, which was written on yellow waxed mulberry bark paper. A *Miaofa Lianhua Jing* ("Lotus Sutra") from the Sui-Tang period (early seventh century), discovered in Dunhuang, was written on yellow-dyed mulberry bark paper. The Palace Museum in Beijing holds a Tang dynasty painting by Han Huang (723–787) titled *Wu Niu Tu* ("Five Oxen"), which was done on mulberry bark paper. Additionally, a document from the second year of the Linde era (665) found in Xinjiang, known as *Bu Laoshi Jieqian Qi* (Contract for Borrowing Money by Teacher Bu), was written on paper made from a mix of hemp and mulberry bark fibers. Other Tang dynasty manuscripts from Dunhuang were written on paper made from fragrant sumac bark. Based on our research findings, the process for producing bark paper during the Tang dynasty is as follows:

1. Cutting trees
2. Stripping bark
3. Soaking
4. Removing outer green bark
5. Washing
6. Soaking in plant ash and lime water
7. Steaming
8. Rinsing
9. Removing residual green bark
10. Chopping and crushing bark
11. Pounding and beating
12. Washing with water
13. Pulp-making
14. Papermaking
15. Pressing to remove water

16. Drying
17. Lifting the paper
18. Sorting and packaging (Fig. 3.32).

Fig. 3.32 Flow chart for Tang dynasty bark paper production (Designed by Pan Jixing and illustrated by Zhang Xiaoyou, 1979)

At the end of the Tang dynasty, a new type of paper made from bamboo emerged as a significant achievement in the history of papermaking. Li Zhao (791–830) recorded in *Guoshi Bu* ("Supplement to the History of the Nation", ca. 829), that there were various types of paper, including "Shan vine and moss paper from Yue (present-day Zhejiang), hemp paper from Shu (present-day Sichuan), Liuhe paper from Yang (present-day Yangzhou in Jiangsu), and bamboo paper from Shao (present-day Shaoguan in Guangdong)." Duan Gonglu (840–905) mentioned bamboo paper in *Beihu Lu* ("Historical Records of Products in South China", 875) when discussing sandalwood bark paper in Luozhou, Guangdong, noting that it was "not as good as mulberry paper and bamboo paper". In the late Tang dynasty, Cui Guitu, when annotating *Beihu Lu,* pointed out that bamboo paper was produced in Muzhou, which is present-day Chun'an in Zhejiang. This suggests that by the ninth century, bamboo paper had already become prominent in Guangzhou and Zhejiang. Bamboo paper was made from wild bamboo, making raw materials relatively inexpensive. China, being a major bamboo-producing country, had a constant supply of materials. The development of bamboo paper had significant implications in the history of papermaking, representing the pioneering use of stem fibers for paper production. This principle and approach later influenced the successful development of wood pulp paper in Europe.

During the Sui and Tang dynasties, paper production was widespread across regions including present-day Shaanxi, Henan, Beijing, Shanxi, Gansu, Shandong, Xinjiang, Jiangsu, Zhejiang, Sichuan, Hunan, Hubei, Guangdong, Anhui, Jiangxi, Fujian, and Tibet. Provinces such as Jiangsu, Zhejiang, Jiangxi, Anhui, and Sichuan were major production areas, indicating a shift in the economic center from the Yellow River basin to the Yangtze River basin. In addition to large-scale imperial paper mills, there were numerous private workshops, and the total production remained the highest in the world for an extended period. Besides meeting domestic demand, paper was also exported. The cultural paper produced was sufficient to meet technical needs of calligraphy, painting, and printing. Renowned calligraphers like Yan Zhenqing and Ouyang Xun, as well as painters like Wu Daozi and Han Huang, used paper extensively, and scholars and literati were no exception. The consumption of paper for Buddhist scriptures was considerable, with Dunhuang Mogao Caves alone containing tens of thousands of scrolls. In addition to handwritten manuscripts, there were also printed materials. Due to the relatively low cost of paper, the variety of daily paper products gradually increased. New products such as paper fans, paper hats, paper clothing, paper armor (for protection), paper tents, playing cards, business cards, paper screens, paper lanterns, and window papers emerged. Paper was extensively used, even during funeral ceremonies where large quantities of paper money were burned. Paper was also used for bills of exchange in commercial transactions. Paper and paper products became commonplace in households, marking China's pioneering role in the Age of Paper.

According to systematic analysis of paper from this period, the thickness of hemp paper and bark paper generally ranged from 0.05 to 0.14 mm, with thicker varieties measuring 0.15–0.16 mm, and even thicker ones being rare. The dispersion of fibers increased, with a tight weave and a smooth surface, and there were significantly fewer fiber bundles. Watermark patterns had four grades: (1) coarse patterns, with each pattern being 0.2 cm wide; (2) medium patterns, 0.15 cm wide; (3) fine patterns, 0.1 cm wide; (4) extra-fine patterns, 0.05 cm wide.[57] Fine and extra-fine patterns were less common in previous periods, indicating improvements in pulping and screen-making technologies. Tang dynasty paper was typically around 30 cm in height and 36–55 cm in width, with some larger sheets measuring 76–86 cm. Late Tang dynasty manuscripts, such as the handwritten *Bore Boluomi Jing* ("Perfection of Wisdom"), had a width of 94 cm. Toward the end of the Tang dynasty, giant-sized "*pi* paper" was produced, with some sheets reaching the length of a *pi* of silk (300 cm), approximately one *zhang*. This type of paper was used for posting exam results during the imperial examination system.

The addition of plant mucilage or "paper medicine" to the pulp is an important invention, as mentioned earlier. In 1901, when Wiesner analyzed Tang dynasty documents unearthed in Xinjiang, he found viscous substances extracted from lichen in the paper, which is plant mucilage. However, commonly used plant mucilage includes those from Abelmoschus manihot L. and Chinese gooseberry vine. During the Tang dynasty, cultural paper was divided into raw paper and processed paper. In a broad sense, the latter referred to paper treated with sizing, coating, waxing, fillers, dyeing, etc. In a narrow sense, it specifically referred to sized paper, which is paper brushed with starch paste on the surface or mixed into the pulp. Paper dyed yellow, coated with wax, and polished is called "hard yellow paper". This type of paper was mostly produced in the early and middle Tang dynasty (seventh to eighth centuries). The seventh-century *Miaofa Lianhua Jing—Fashi Gongde Pin Di Shijiu* ("Lotus Sutra – No. 19 of the Chapter on the Merits of the Master") discovered in Dunhuang is an example of hard yellow paper (Fig. 3.33). In 1100, Mi Fu mentioned in his *Shu Shi* ("A History of Calligraphy") that the calligraphers Zhiyong and Chu Suiliang used powdered wax paper for their works *Qianziwen* ("Book of a Thousand Characters") and *Kumu Fu* ("Ode to the Withered Tree") respectively. During the Tang dynasty, gold powder and gold leaf were also used to decorate various colored papers, known as "gold flower paper". For writing letters or poems, sometimes specially processed paper was used. Duan Chengshi (803–863) in Jiujang, Jiangxi produced *Yunlan* (blue cloud) paper for letter writing. The paper had light blue cloud patterns on it. The manufacturing process involved dyeing a small amount of paper pulp into blue. The blue pulp was placed on one side, and during the papermaking process, the pulp flowed to create a wave-like cloud pattern. During the Tang Yuanhe period (806–820), in Chengdu, Sichuan, the female poet Xue Tao produced red short paper dyed with hibiscus flower juice, which is called *Xue Tao Jian* (Xue Tao's paper), for writing eight-line poems.

During this period, people also created *yahua zhi* (embossing paper) by pressing wooden boards engraved with raised patterns or designs onto white or colored paper. When viewed against the light, the paper revealed patterns and designs. The embossing paper was used for letters, calligraphy, or painting. In the Ming dynasty,

[57] Pan Jixing. *A History of Papermaking Technology in China* [M]. Beijing: Cultural Relics Press, 1979: 199.

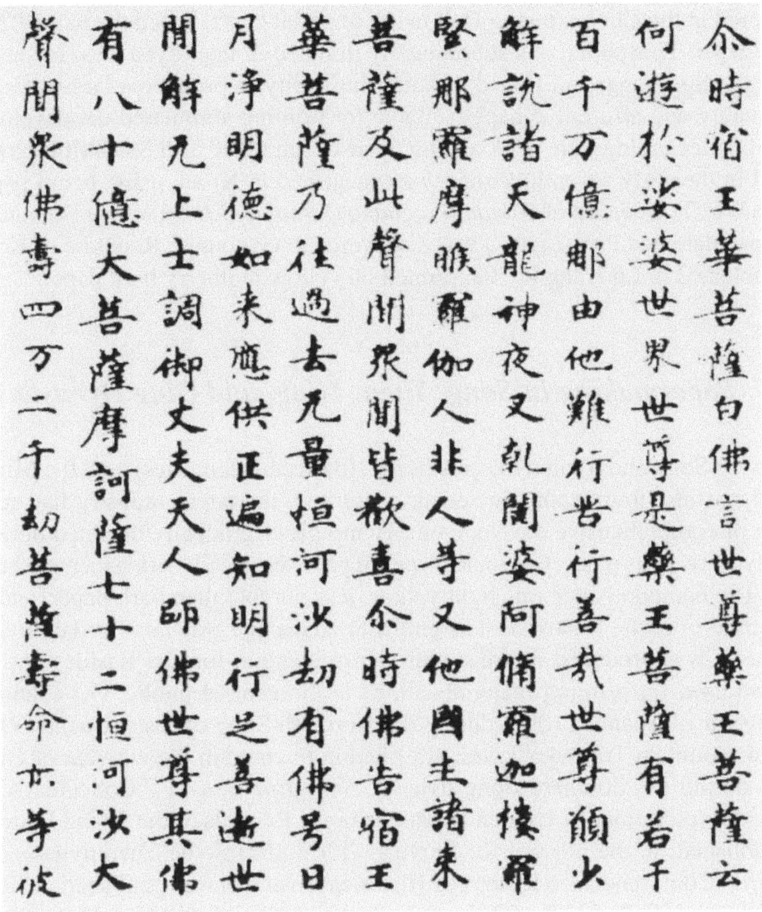
尒時宿王華菩薩白佛言世尊藥王菩薩云
何遊於娑婆世界世尊是藥王菩薩有若干
百千万億那由他難行苦行善哉世尊願少
解說諸天龍神夜叉乹闥婆阿脩羅迦樓羅
緊那羅摩睺羅伽人非人等又他國土諸來
菩薩及此聲聞衆聞皆歡喜尒時佛告宿王
華菩薩乃往過去无量恒河沙劫有佛号日
月淨明德如來應供正遍知明行足善逝世
間解无上士調御丈夫天人師佛世尊其佛
有八十億大菩薩摩訶薩七十二恒河沙大
聲聞衆佛壽四万二千劫菩薩壽命亦等彼

Fig. 3.33 In the early Tang dynasty of the seventh century, *Lotus Sutra* was written on hard yellow paper made from hemp (Pan Jixing collection)

it was referred to as *gonghua*. In the Tang dynasty, watermarks paper, also known as *yanbo jian*, emerged. Patterns and designs were woven into the sieve, and when the paper was made, these patterns became visible. Another type was *luowen jian* (cloth-pattern paper) or *yuzi jian* (fish-roe paper), where fine cloth treated with starch and glue was pressed onto the paper, revealing a cloth pattern, resembling fish roe. During the reign of Li Yu, last lord of Southern Tang, from 961 to 970, in Shezhou and Chizhou (modern-day Shexian County, Guichi District, Anhui), high-quality paper suitable for calligraphy and painting was produced using paper mulberry bark as a raw material, known as Chengxintang paper, renowned for a time as the best paper. In the last month of the lunar calendar year, the paper pulp was mixed with ice water, and after meticulous processing, the paper was polished on both sides, smooth as jade tablets. It was used by Li Yu and frequently presented to ministers. The paper was stored in the Chengxintang Hall in the imperial court, hence the name Chengxintang paper. This paper was subsequently imitated in later dynasties. In summary, during the Sui, Tang, and Five dynasties, the ability to produce a large quantity of high-quality and affordable paper suitable for printing stimulated the development of printing technology. In 1974, a Tang dynasty sutra leaf with Sanskrit inscriptions printed in the early seventh century was unearthed in Xi'an, using hemp paper. In 1966, a Wu Zhou edition of *Wugou Jingguang Datuoluoni Jing* ("Great Dharani Sutra of Immaculate and Pure Light") was discovered in Gyeongju, Republic of Korea. It was published in Luoyang in 702, printed on yellow mulberry bark paper.

3.3.3 Papermaking in Song, Yuan, Ming, and Qing Dynasties

During the Song and Yuan dynasties (tenth–fourteenth centuries), and the Ming and Qing dynasties (fourteenth–nineteenth centuries), the paper industry had reached a stage of comprehensive development. Hemp paper gradually declined during the Song dynasty. In its place, two major types of paper emerged: bark paper and bamboo paper. The bamboo paper was light yellow, less durable than bark paper, and more susceptible to pests. However, its significant advantage was its affordability, especially as it was produced in areas with dense bamboo forests. Besides writing, it was also used for printing materials aimed at the general public. For example, in the 5th year of Yuanhe (1090) during the Northern Song dynasty, *Gushan Dazang* ("Drum Mountain Tripitaka") was published in Fuzhou; in the 7th year of Qiandao (1171) during the Southern Song dynasty, *Shiji Jijie Suoyin* ("Collected Annotations and Explorations of Hidden Meanings in the Records of the Grand Historian") was published; in the 6th year of Zhiyuan (1269) during the Yuan dynasty, *Shilin Guangji* ("Comprehensive Record of Historical Events") was published in Jichengtang, Jianyang, Fujian. All of them were printed on bamboo paper. In the 2nd year of Mingdao (1033) during the Northern Song dynasty, the Minister of War Hu Ze printed *Dabeixin Tuoluoni Jing* ("Dharani Sutra of Great Compassion Heart") using high-quality bamboo paper. Literary figures such as Su Shi, Mi Fu, and Wang Anshi also used bamboo paper. For example, Mi Fu used bamboo paper for his *Shanhu Tie* ("Coral Rubbings"). Many editions in Fujian were printed on bamboo paper, and there are many surviving artifacts from this period. Bark paper, known for its high quality, became the primary paper type and was widely used. Painting on bark paper became popular, with artists like Su Shi and Emperor Huizong frequently using it for their calligraphy and paintings. The size of paper increased, compared to previous dynasties. Records suggest that the area of painted surfaces in the Tang dynasty reached 650 cm^2, in the Song dynasty, it reached 2412 cm^2, and in the Yuan dynasty, it reached 2937 cm^2. Emperor Huizong's cursive script for *Qianziwen* (Thousand-character Classic) was 3 *zhang* (approximately 10 m) long, setting a historical record. Large-sized paper required multiple people to handle during the papermaking process (Fig. 3.34).

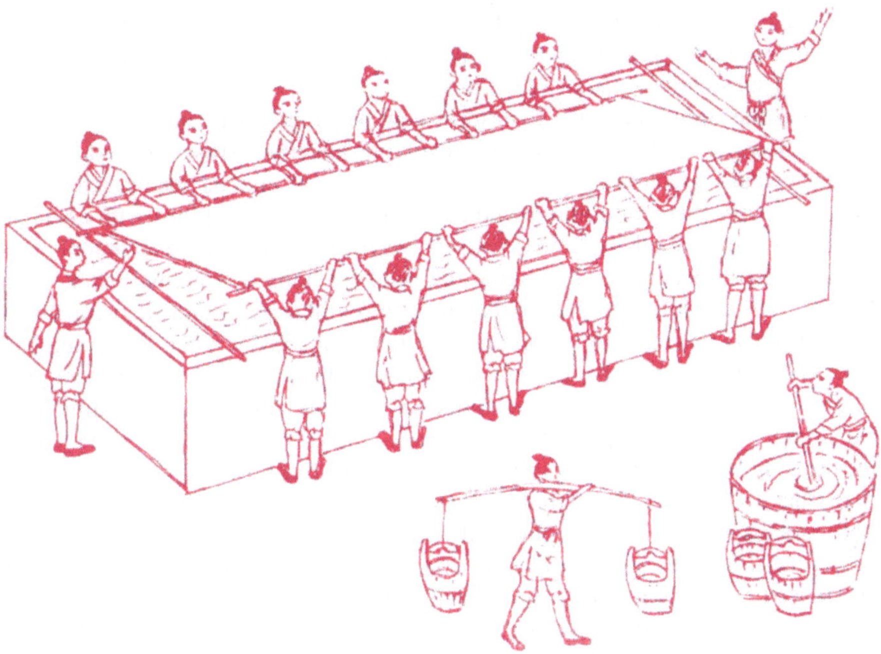

Fig. 3.34 Illustration of the production of large bark paper by multiple people in the Tang dynasty (Illustrated by Pan Jixing, 1998)

In the Song and Yuan dynasties, high-quality printed editions were still predominantly produced using bark paper. For instance, the 973 edition of *Kaibao Zang—Foshuo Awei Yuezhi Zhe Jing* ("Kaibao Tripitaka: The Buddha's Discourse on the Awei Yuezhi Zhe Sutra") utilized premium mulberry bark paper, coated on both sides with wax and dyed yellow. *Changle Xiansheng Ji* ("Changle Collection") published by the Liao family in the Southern Song dynasty used thin white mulberry bark paper. *Wenyuan Yinghua* ("Best Literary Works") published in Ji Prefecture, Jiangxi, in the first year of Jingding (1260) of the Southern Song dynasty, and *Mengxi Bitan* ("Dream Pool Essays") published in Chaling in the ninth year of the Yuan dynasty (1305), used paper made from the bark of the paper mulberry. In 1978, a copy of *Miaofa Lianhua Jing* printed in the first year of Tianxi (1017) during the Northern Song dynasty was discovered in the pagoda of Ruiguang Temple in Suzhou. This edition used a mixture of mulberry bark and bamboo fibers. According to the records of Ye Mengde in *Shilin Yanyu* ("The Swallow's Words from the Stone Forest", 1136), during that time, "printing books in Hangzhou was considered superior, followed by Sichuan, with Fujian being the least preferred." This preference for books printed in Hangzhou was due to the use of quality mulberry bark paper, precise proofreading, and skilled engraving. Many books from Fujian were privately printed using bamboo paper, aiming for cost-effectiveness rather than premium quality. Sichuan editions were positioned between those of Zhejiang and Fujian. From the Northern Song to the Southern Song, Jin, and Yuan dynasties, official government-issued banknotes were predominantly printed on specially manufactured mulberry bark paper, which consumed substantial amounts of the material.

During the Song and Yuan dynasties, new paper products emerged, including paper pillows, folded paper fans, and revolving lanterns. Artists created intricate designs by cutting colored paper into representations of famous poems, achieving a level of artistry that could mimic authentic handwriting. In regions with abundant water resources in southern China, water-powered stamp mills were commonly used to pound paper materials, eliminating the need for human or animal labor. Wang Zhen's *Nong Shu* introduced a continuous stamp mill with a main shaft capable of simultaneously driving four stamping operations (Fig. 3.35). When producing bark paper and bamboo paper, people commonly added plant mucilage to the paper pulp. There were innovations in sizing techniques as well, with the addition of animal glue and alum to the paper pulp. Paintings such as Li Gonglin's (1049–1106) meticulous line drawing *Weimo Yansan Tu* ("Illustrations of the Teachings of Vimalakirti") and Zhao Chang's (998–1022) colored *Xiesheng Xiadie Tu* ("Illustrations of Butterflies") used bark paper treated with glue and alum. In the third year of Chunxi during the Southern Song dynasty (1176), the official publication *Chunqiu Jingzhuan Jijie* ("Collected Annotations on the Spring and Autumn Annals and its Commentaries") utilized a solution made from the fruit of the Zanthoxylum bungeanum Maxim to prevent insect damage. In the Song dynasty, Changzhou (now in Jiangsu province) artisans applied white mica powder to paper surfaces, resulting in a silver-white luster, creating what was known as *yunmu jian* (mica paper). Various processed papers flourished throughout history, with the imperial court even using golden mud to paint dragon and phoenix patterns on colored waxed papers, turning paper into an artistic medium with artistic value for writing and painting.

During the Ming and Qing dynasties, bamboo paper production surged to the forefront, displaying significant improvement in quality compared to the Song and Yuan periods. The paper was whiter, thinner, and contained fewer bamboo fibers. In contrast, bark paper fell to the second place, mainly due to its high raw material consumption and increased costs, which made it less competitive compared to bamboo paper. However, as the finest paper, its status remains unchanged. Paper for writing, painting, and printing dominated paper consumption, with surviving artifacts surpassing the total from all previous dynasties. Paper production is widespread across the country, with the main production areas concentrated in the southern provinces of Jiangxi, Fujian, Zhejiang, Anhui, Guangdong, and Sichuan as well as the northern regions, Shandong, Shanxi, Shaanxi, Hebei, and Henan. Bamboo paper produced in Taiwan region has seen significant development compared to the past. Xuande paper made from paper mulberry bark in Jiangxi during the Ming dynasty, and Xuan paper made from sandalwood bark in Jingxian, Anhui during the Qing dynasty were considered top-quality paper, used for high-grade cultural purposes. These two types of paper, with various forms and processing methods, formed two

Fig. 3.35 The continuous mill from Wang Zhen's *Nong Shu* (1313) (Reproduced from the Ming Jiajing ninth-year edition)

major series that led the technological trends of their respective eras. What distinguishes the Ming and Qing dynasties from earlier periods is the emergence of systematic treatises discussing papermaking techniques, which provide detailed insights into the craft.

In the first year of the Yongle era (1403) of Emperor Chengzu of the Ming dynasty, an official paper mill was established in Xinjian County, Nanchang Prefecture, Jiangxi, to produce mulberry bark paper for imperial use. The following year, under the direction of Grand Secretary Xie Jin (1369–1415), *Yongle Dadian* ("Great Encyclopedia of the Yongle Reign", 1408), a vast encyclopedic compilation, was created using this imperial paper. During the Xuande reign (1426–1435), this paper evolved into Xuande paper. In the late sixteenth century, during the Longqing and Wanli periods, the paper mill moved to Qianshan County, Guangxin Prefecture, Jiangxi, where it continued to produce paper using the original method. The paper was pure white and thick, resembling the Chengxintang paper of the Southern Tang dynasty.[58] In the 25th year of the Wanli era (1597), the Ming scholar Lu Wangai wrote *Chu Shu* ("Paper Made from Paper Mulberry") as a supplement (Volume 8) to *Jiangxisheng Dazhi* ("Great Annals of Jiangxi Province", 1556). The technical details of the official paper mill in Qianshan reflected the manufacturing techniques of Xuande paper. A summary of the process is provided below.

1. Peel off the paper mulberry bark, bundle it, and soak it in the river for several days.
2. Use feet to remove some of the outer green skin, then bundle and retrieve the material.
3. Boil in clean water.
4. Pound off the outer shell, and tear the inner bark into fibers.
5. Cut into small pieces.
6. Soak the material in lime slurry, and pile it for over a month.
7. Boil again.
8. Place the material in cloth bags and wash it in flowing water for several days.
9. Spread out by the river to naturally bleach.
10. Pound into fine pulp.
11. Pile the material and thoroughly soak it with scalding plant ash water, then air-dry in the shade for half a month.
12. Steam again.
13. Wash in the river.
14. Spread out by the river for another round of natural bleaching.
15. Manually remove any remaining colored particles.
16. Place in cloth bags and wash clean with river water.
17. Put the material into a stone trough and mix with mountain spring water to make the paper pulp.
18. Add plant mucilage and stir.
19. Weave fine bamboo strips into a paper-making screen, then place it on a screen rack.
20. Lift the screen to make paper: six people operate the large screen, with three on each side; two people operate the small screen.
21. Filter water from the paper pulp on the screen, layering the wet paper.
22. Press out water with a wooden press, and let it sit overnight.
23. Peel off the half-dried paper and spread it on a brick fire wall to dry. The fire wall is hollow, with a smooth lime-brushed surface on both sides; wood is burnt at one side, and smoke and fire heat the wall.
24. Remove the dried paper and stack it.

[58] Pan Jixing. *History of Chinese Science and Technology: Papermaking and Printing Volume* [M]. Beijing: Science Press, 1998: 227–228.

25. Bundle and package, with each bundle containing one hundred sheets.[59]

Lu Rong's *Shuyuan Zaji* ("Jottings from the Shu Garden", 1945) volume 13 records the process of making mulberry bark paper presented to the imperial court from the two counties of Changshan and Kaihua in Quzhou, Zhejiang.[60] This process involves 18 steps. Wang Shunmin's *Huizhoufu Zhi—Wuchan Zhi* ("Records of Huizhou Prefecture – Records of Local Products") describes the technology of making mulberry bark paper in Huizhou, Anhui. The two processes are similar, representing the typical techniques in mulberry bark paper production in the region. *Tiangong Kaiwu* ("Exploitation of the Works of Nature", 1637) provides a more concise description, and later works by Qing dynasty authors show a lack of innovation, often borrowing from Ming dynasty scholars. Jiangxi, Zhejiang, and Anhui provinces are key regions for the production of bark paper, each with representative techniques. Compared to Zhejiang and Anhui, the paper mills in Jiangxi, which produce imperial paper for official use, have higher quality requirements. As a result, their process is more complex, involving eight additional steps, i.e. three rounds of boiling, two rounds of sunlight bleaching, three washes, and multiple removals of colored outer bark. The production cycle in Jiangxi was longer, requiring significant manpower and resources, resulting in high costs. The paper produced was of superior quality, only suitable for imperial use. In contrast, Zhejiang and Anhui, with local or private operations, followed a less complex process, involving only two rounds of boiling and one round of sunlight bleaching, yet still producing good-quality paper. Consequently, there is a tendency for waste in the Jiangxi official paper mills, which is not seen in private or local paper production. An ideal solution would be a compromise between Jiangxi and Zhejiang/Anhui methods, possibly refining the processes of the latter for more precise operations, as seen in the private papermaking industry producing Xuan paper in Jing County.

Xuan paper is primarily made from the blue sandalwood bark produced in Jingxian, which was historically part of Xuanzhou, hence the name "Xuan paper". During the Song and Yuan dynasties, the Cao family observed that blue sandalwood bark was similar to paper mulberry bark and began producing paper from it. Their descendants continued this craft through the Ming and Qing dynasties. After the decline of the Jiangxi official paper mills, Jingxian paper became the tribute paper. In the 47th year of the Qianlong reign (1782), *Siku Quanshu* ("Complete Library in the Four Branches of Literature") was printed using Jingxian paper. From a technical perspective, there is a clear lineage of paper-making techniques from Chengxintang paper, Xuande paper to Xuan paper. Xuan paper inherited the craft of Xuande paper but with simplified processes, omitting one round of boiling, one round of sunlight bleaching, and one wash, and using plant mucilage such as Chinese gooseberry vine and Chinese winterberry. The process is as follows:

1. Cut trees, peel the bark, and bundle them.

---

[59] Lu Wangai. *Paper Made from Paper Mulberry* [M] // Wang Zongmu. *Great Annals of Jiangxi Province: Vol. 8*. 1597.

[60] Lu Rong. *Jottings from the Shu Garden: Vol. 13* [M]. Beijing: Zhonghua Book Company, 1985: 157.

2. Boil in clean water.
3. Pound the bark, tear it into fibers, and remove the green bark.
4. Bundle the bark and soak it in a vat.
5. Soak in lime water and pile for one month.
6. Boil again.
7. Wash in water while kneading.
8. Spread out by the riverbank or hillside to bleach for 3–6 months.
9. Wash again to remove impurities.
10. Pound the material into pulp.
11. Wash in cloth bags.
12. Prepare the white pulp with mountain spring water.
13. Add plant mucilage and stir.
14. Use a screen to make paper, with two or four people holding the screen.
15. Drain water, remove the wet paper from the screen, and stack it layer by layer.
16. Press to remove water and let the paper sit overnight.
17. Dry the paper on a fire wall.
18. Remove the paper.
19. Stack the paper, trim the edges, and stamp.
20. Package the paper, with each bundle containing one hundred sheets.

In the Ming dynasty, the scientist Song Yingxing (1587–1666) provided a detailed description of the bamboo paper-making process in Fujian in his *Tiangong Kaiwu—Shaqing* ("Exploitation of the Works of Nature: Killing the Green", 1637), accompanied by six illustrations (Fig. 3.36). The process he described includes:

1. Cut bamboo in the sixth month of the lunar calendar and bundle it.
2. Soak the bamboo in a vat for one hundred days.
3. Wash and pound the bamboo in the river to make it fibrous, removing the outer bark.
4. Soak the bamboo in lime water and pile it for ten days.
5. Boil the bamboo for eight days.
6. Wash and knead the bamboo in the river.
7. Soak the material in plant ash water and boil it again for seven days.
8. Wash the material.
9. Pound the material into pulp in a water mill.
10. Prepare the white pulp with mountain spring water.
11. Add Chinese gooseberry vine mucilage to the pulp and mix well.
12. Use a screen to make paper.
13. Remove the wet paper from the screen and stack it layer by layer, up to a thousand sheets.
14. Press to remove excess water and let the paper sit overnight.
15. Use copper tweezers to peel a corner of the paper, then spread it by hand and brush on a fire wall to dry.
16. Remove the dried paper, and stack it neatly

Fig. 3.36 Flow chart of making bamboo paper in Fujian (Extracted from *Tiangong Kaiwu*)

3. Dry the paper

Fig. 3.36 (continued)

17. Trim the edges, and package it with each bundle containing one hundred sheets.[61]

During the Qing dynasty, Yan Ruyu (1759–1826) documented the bamboo paper-making techniques in Yang County and Xixiang in southern Shaanxi in his work *Sansheng Bianfang Beilan* ("A Survey of Border Defenses in the Three Provinces", 1821), Volume 10, *Shan Huo* ("Mountain Products").[62] These techniques can be compared with those described in *Tiangong Kaiwu.* While there are similarities between the two, the bamboo paper-making techniques in southern Shaanxi were not as advanced as those in the lower reaches of the Yangtze River. Huang Xingsan (1850–1910), in his work *Zaozhi Shuo* ("Treatise on Papermaking", ca. 1885), described the bamboo paper-making techniques in Changshan, Jiangsu and Zhejiang, which largely align with those in *Tiangong Kaiwu.* However, he introduced a natural bleaching process for bamboo material. Although *Tiangong Kaiwu* was written during the Ming dynasty, the described bamboo paper-making actually reflects traditional techniques dating back to the Song and Yuan dynasties. This book also influenced the development of paper-making in other countries in the East and West.

---

[61] Song Yingxing, Pan Jixing. *Exploitation of the Works of Nature: Translation and Annotation* [M]. Shanghai: Shanghai Classics Publishing House. 1990: 151–154, 292.

[62] Yan Ruyu. *A Survey of Border Defenses in the Three Provinces: Vol. 10, Mountain Products: Paper* [M] 1830: 5–7.

During the Ming and Qing dynasties, new paper products were introduced, including paper inkstones, paper cups, paper flutes, paper woven paintings, and wallpapers that were more exquisite than those of previous generations. Paper inkstones originated in Haining, Zhejiang, during the Qianlong and Jiaqing periods of the Qing dynasty. They were made by mixing paper, stone sand, and lacquer. Paper woven paintings were a new variety in folk arts and crafts, emerging during the Jiajing period of the Ming dynasty. Thin paper was dyed in different colors, cut into long strips, and then twisted into fine ropes. These ropes were used to weave replicas of paintings and calligraph. The book *Liuqing Rizha* ("Journal of Preserved Green", 1579) by Tian Yiheng mentioned that paper woven paintings were discovered during the confiscation of the property of the corrupt minister Yan Song in the 44th year of Jiajing (1565). Another source, *Shuyuan Zaji* volume 12 mentioned the use of high-quality processed paper as wallpaper during the Tian Shun period (1457–1464). In the Beijing Palace Museum, varieties of wallpaper from the seventeenth to eighteenth centuries are still preserved. During the Ming and Qing dynasties, the art of papermaking reached its peak. Jiangxi Xuande paper and Anhui Xuan paper regions became prominent centers for producing a wide array of specialty paper. These included five-color paper, powdered paper, wax paper, five-color powdered paper, sprinkled-gold paper, five-color gold-flower paper (Fig. 3.37), porcelain blue paper, sprinkled-gold five-color powered paper, and sheep-brain paper, among others, showcasing a rich variety of types. During the Northern Song dynasty, official and private papermaking workshops existed in Nanjing, Suzhou, and Hangzhou. Many of their products can still be seen today. Tu Long (1543–1605) introduced some technical methods for manufacturing processed paper in his work *Kaopan Yushi* ("Addressing the Outstanding Issues", 1600). In the Kangxi period of the Qing dynasty (1662–1722), flower-patterned paper was made using a copper mesh. According to Xu Kang's *Qianchen Mengying Lu* ("Chronicles of Past Dreams and Illusions", 1897), Zhejiang papermakers used copper wires to create patterns on a copper mesh. After papermaking, the patterns appeared on the paper. One of the significant achievements in the Qing dynasty was the production of tube-shaped paper with oblique screen marks. In *Yangjizhai Conglu* ("Collected Works of Yangji Studio", ca. 1863) volume 26, Wu Zhenyu (1792–1871) mentioned that:

During the Dingchou year of Qianlong (1757), Emperor Gaozong received two pieces of tube-shaped paper during his southern inspection tour. He kept one for himself and used one for writing. He also wrote a poem about the paper. Later, during a check of the old library, five more sheets were found. In Renyin year (1782), Zhejiang successfully produced the tube-shaped paper with oblique screen marks, which was presented to the emperor and led to various poems being written about it. This paper, with its tube shape, is often referred to as "round tube paper". It was once distributed to the court officials, and Peng Yuanrui had a poem titled "Respectfully Composed with the Emperor's Original Rhyme" written on it.[63]

[63] Wu Zhenyu. *Collected Works of Yangji Studio: Vol. 26* [M]. Beijing: Beijing Cultural Relics Publishing House, 1983: 274.

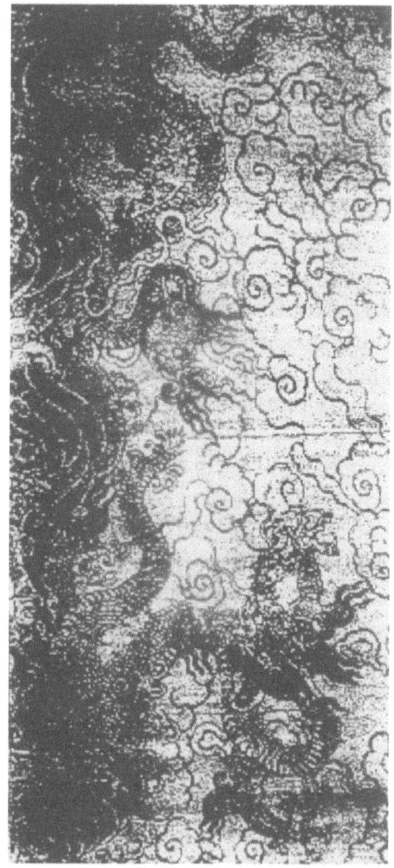

Fig. 3.37 Colored powdered paper with golden cloud-dragon- pattern made during the Xuande period of the Ming dynasty

In the 51st year of the Kangxi era (1712), Kong Shangren (1648–1718) mentioned in *Xiang Jin Bu* ("Record of the Enjoyment of Gold") that he received a type of paper with a width of about one *zhang* (around 320 cm), marked with a texture similar to that of a grinding wheel, given to him by a friend. Ruan Yuan (1764–1849) noted in *Shiqu Suibi* ("Random Notes from the Stone Canal", 1793) that during the Qianlong era, a similar tube-shaped paper was imitated. Its color resembled that of bitter rice, and it left a mark when rubbed, with lengths extending over three meters. It is evident that this paper was first manufactured in the late Kangxi period in Zhejiang. After being presented to the court, the Emperor greatly appreciated its use. In 1757, during Emperor Gaozong's southern inspection, two pieces of this paper were presented, and poems were composed to commemorate it. In 1782, Zhejiang again produced large quantities of this paper, and Emperor Gaozong wrote poems for it, which were also recorded by court officials.

In 1965, I saw remnants of this tube-shaped paper from Li Zongren's collection in the library of Sichuan University. In 1973, I saw another complete piece from 1782 in the collection of Zhang Boju (1897–1982) at the China National Museum. This piece had been removed from the Forbidden City in the early years of the Republic of China. Upon closer inspection, the paper seems to be machine-made, featuring a deep skin color, a thick and heavy texture, and an uneven surface with oblique screen marks. It is crafted from mulberry bark fibers and has been beaten to a high degree. The paper is tube-shaped, seamless, and unfolds to over a *zhang*, consistent with historical records. The term "tube-shaped paper with oblique screen marks" is quite fitting.

The production process of the tube-shaped paper with oblique screen marks during the Kangxi era is worth exploring. It appears to have been formed by using a cylindrical copper mesh. The coarse copper wires on the mesh were woven into oblique lines, protruding on the mesh surface. To form wet paper on a cylindrical copper mesh, the copper mesh must be rotated. Paper pulp flows from a high-level trough through a duckbill-shaped outlet onto the rotating mesh. As the pulp is applied, excess water is filtered out. After one complete rotation, the cylindrical wet paper is formed. The internal structure of the cylindrical copper mesh had channels to discharge the filtered water. Another rotating roller, covered with soft material, needed to be made. This roller was brought close to the paper screen but rotated in the opposite direction to squeeze out the water from the wet paper, so as to ensure uniform thickness. After drying, the paper was lifted from both ends of the screen, and a long, thin strip was used to lift the paper forward until the entire paper tube was detached from the copper mesh. When using this type of paper, it still needed to be cut into rectangular pieces and polished before writing.

Therefore, the production of the tube-shaped paper with oblique screen marks must address three technical challenges: (1) the design and manufacturing of a cylindrical copper-mesh papermaking device; (2) the concept and manufacturing of a device that allows the cylindrical paper screen to rotate for paper retrieval; (3) utilizing the principle of a sugar press (formerly known as a sugar mill) to bring the paper-retrieving cylinder close to another roller, achieving dewatering through their counter-rotational movement. These problems were already solved during the Kangxi era (1661–1722). They constitute the basic elements of the modern mono-cylinder round-mesh papermaking machine, which was only invented and patented by the Englishman John Dickinson (1782–1871) in Europe as late as 1809 and later adopted worldwide. China's tube-shaped papermaking device shares structural principles similar to Dickinson's round-mesh papermaking machine but predates it by nearly a century. It serves as an early prototype of modern round-mesh machines. It is no wonder that, at first glance, the tube-shaped paper with oblique screen marks appears similar to modern machine-made paper. The fact that China, during the Kangxi era, could develop the principles of the round-mesh papermaking machine and apply them to paper production is a technological marvel.[64]

[64] Pan Jixing. From Tube-shaped Paper with Oblique Screen Marks to the Round-mesh Paper Machine [J]. *Cultural Relics*, 1994 (7): 91–93.

Chapter 4
Origin and Development of Woodblock Printing

4.1 From Classic Reduplication Techniques to Printing Techniques

4.1.1 Role of Seals

There has been a long history of the idea and application of reproduction of texts or images. The so-called reproduction, in printing, is the process of replicating the texts or images repeatedly by using ink or other coloring agents to reveal the details that are engraved or etched on the same text or image model. For this reason, characters on the model must be in reverse form before they can be reproduced and then displayed in upright type format. Long before the advent of woodblock printing, there existed various classical reproduction techniques in China, such as affixing seals, creating rubbings from tablet inscriptions, etc., which inspired the reproduction of books by woodblock printing. Seals have been in use (Fig. 4.1) since the later period of the Shang dynasty (also known as the Yin dynasty) in the fourteenth–eleventh centuries BC and have continued to be used till the present day. Most seals are made of hard materials such as jade, wood, metal, horns, and ivories, usually square, rectangular, and cylindrical in shape, with different forms of seal knobs (Fig. 4.2). The seal inscription, which may consist of a few characters or dozens of characters, usually indicates the owner's name, official position, affiliated institution, etc., either in intaglio or in relief forms due to the difference in the concave and convex nature of those characters. Affixing seals to instruments and contracts signifies responsibility, credit, and authority, and also serves as a safeguard against forgery. A seal on a painting or a calligraphic work indicates ownership, while a seal on a letter indicates solemnity. Seals are classified into official seals and private seals. As cited in *Han Shu·Baiguan Zhi* ("Book of Han: Treatise on the Officials"), Wei Hong in the Han dynasty stated in *Han Jiuyi* ("Old Rites of the Han Dynasty") that according to the official ranking system the seal of a high-ranking official with

P. Jixing, *History of Printing Techniques in China*,
https://doi.org/10.1007/978-981-96-6119-0_4

Fig. 4.1 Seals of various periods in ancient China

an annual salary of 2000 *dan* (a volume unit) of grain was called *zhang* (the seal for a higher ranking official), the seal of an official with an annual salary of 200, 400, or 600 *dan* of grain was called *yin* (the seal for a lower ranking official), which is later collectively referred to as *yinzhang* (seal), and the imperial seal of an emperor was called *xi* (the royal seal for an emperor). After the Han dynasty, seal inscription was mostly engraved in seal script and became a unique art. Over the years, various seals from the Zhou, Qin, Han, and Tang dynasties have been unearthed in different parts of China (Fig. 4.1). Early seals were mostly affixed to silk manuscripts or clay used for sealing documents, or pressed onto pottery blanks or clay casting molds to reveal the inscriptions on potteries and bronze wares.

In the period before paper was invented or commonly used, during which bamboo slips had been used as the main writing materials, when an important official document or personal letter was written, the bamboo slips were folded or rolled up, and the exterior of the document was covered with a blank bamboo slip, with the name of the author, his official position, delivery address, and the title of the document written on it, and tied up with a string. A piece of clay was placed at the tie, and a seal was affixed to it. When the seal impression became dried and solidified, no one else could read the document, and the sealing clay was called lute (Fig. 4.3). Lute produced after the Warring States Period and the Han dynasty have been unearthed occasionally. Seals were also pressed onto lute on papyrus documents in ancient Egypt, while in Europe wax was used in place of lute. After the Jin dynasty, Chinese paper became the main writing material to replace bamboo slips, and therefore the practice of sealing with clay gave way to paper sealing, which involved stamping a seal on the seams of joined sheets of paper to prevent forgery, or affixing a seal to the sealed part of the paper envelope containing documents to prevent unauthorized access. After the disappearance of bamboo slips in the fourth century, lute was discarded, and seals were mostly affixed to paper or silk. Ink seal impressions were

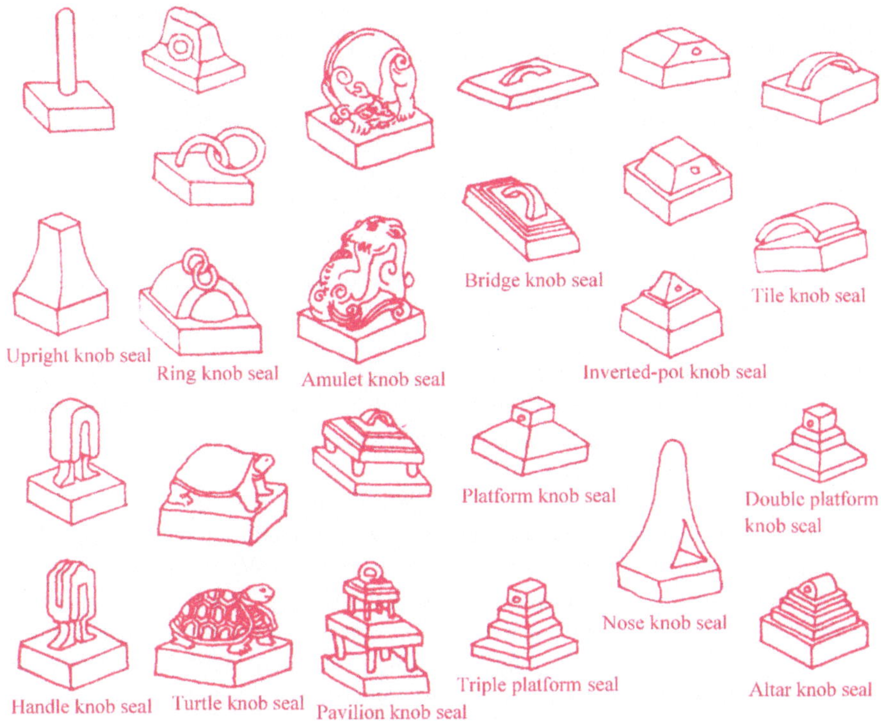

Fig. 4.2 Various forms of seals in ancient China[1]

found on paper documents of the Wei and Jin dynasties unearthed in Xinjiang. Later, people noticed that seal impression of ink might be easily confused with the writing in ink, so they made red inkpad with vermilion, which was used to affix seals for red inscriptions. Seals with vermilion inkpad have been used at the latest in the fifth-sixth centuries in the Northern and Southern dynasties. However, there was a transition period during which red vermilion seals were used in conjunction with red ink seals.[2] According to *BeiQi Shu: Lu Fahe Zhuan* ("Biography of Lu Fahe in Book of Northern Qi Dynasty"), during the reign of Emperor Yuan of Liang (Xiao Yi), Lu Fahe was appointed the governor in 553. "Although Fahe did not declare himself a vassal in the service of the emperor, all his writings were affixed with red seals, with the inscription indicating the title of *situ* (a high ranking official almost equivalent to a minister)". In *Tong Dian* ("Comprehensive Institutions") by Du You in the Tang dynasty, it was also recorded that during the Northern Qi dynasty large wooden seals were affixed to the seams of official documents,[3] and multiple reproductions of a small number

[1] Wang Zhimin. *Chinese Seals and Seal Engraving* [M]. Beijing: The Commercial Press, 1991.

[2] Carter T F, Goodrich L C. *The Invention of Printing in China and Its Spread Westward* [M]. 2nd ed. New York: Ronald, 1955. Qian Cunxun. *Papermaking and Printing* [M] // Needham J. Science and Civilisation in China [M]. Beijing: Science Publishing & Media, 1990.

[3] Du You. *Tong Dian* [M]. Shanghai: The Commercial Press, 1937: 3586.

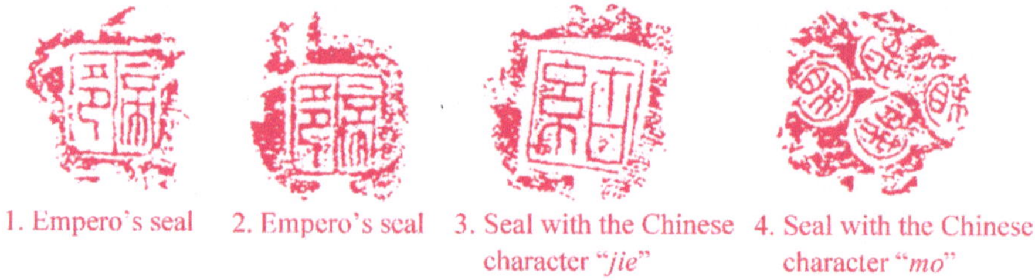

Fig. 4.3 Lute from the tomb of Nanyue king of the Western Han dynasty (122BC) in Guangzhou unearthed in 1983[5]

of characters could be realized by stamping seals on paper. *Saṃyuktābhidharma-hṛdaya-śāstra* ("Treatise on the Mixed Heart of the Dharma") found in Dunhuang stone chamber bears a large red seal inscription showing "Yongxing County Seal" of the Southern Qi dynasty (479–502).[4]

Affixing a seal to paper and woodblock printing both have the commonality of reproducing texts, but the two approaches are different in utility and operation. Seals hold less printing surface and thus less character capacity. When affixing a seal, the paper will be placed under the seal, and the seal will be pressed on the paper with the pressure of the hand, making seal text in upright type form. Woodblock printing, on the other hand, is featured by a large printing plate, more character capacity, and a greater weight. Ink is always applied to the upper side of the plate, while the paper is placed on the printing plate, and the pressure of the brush is applied to the back of the paper. The final printed characters form the contents of the reading products. By increasing the size of the printing surface, carving more characters in reverse form, and inverting the direction of affixing seals, woodblock printing is invented. Needless to say, it is easy to realize this reversal. In fact, if the seal is large and heavy, paper is sometimes put on the printing surface to save effort when pressure is applied, which is only one step away from woodblock printing. After the Wei and Jin dynasties in the fourth century, the further development of Taoism and Buddhism led to the emergence of two new directions for seal technology heading towards the evolution of woodblock printing. Firstly, the Taoists made large wooden seals engraved with many characters to print Taoist magic figures or incantations, and secondly, the Buddhists carved wooden seals with the reverse images of the Buddha, which were printed on the Buddhist scriptures.

Ge Hong (284–363), a Taoist scholar of the Eastern Jin dynasty, claimed, "those who seek the Tao and produce medicinal concoctions, usually desiring to escape from turmoil and live in seclusion, tend to live in the mountains. Ancient people who ventured into the mountains would carry the seal of *huangshen yuezhang* ("Seal of the Yellow Deity"). This seal is four inches wide (ca. 13.5 cm × 13.5 cm), inscribed

[4] Luo Fuyi. *Introduction to Ancient Chinese Seals* [M]. Beijing: Cultural Relics Press, 1981: 71–72.

[5] Guangzhou Municipal Cultural Relics Management Committee, Institute of Archaeology, Chinese Academy of Social Sciences, Guangdong Museum. *The Tomb of Nanyue King of Western Han Dynasty: Vol. 1* [M]. Beijing: Cultural Relics Press, 1991.

with one hundred and twenty characters. By placing the sealed clay at the distance of a hundred paces in each of the four directions around their dwellings, people could prevent tigers and wolves from approaching them."[6] With regard to the common practice of wearing amulets for those people entering the mountains, Ge explains that "in ancient times, people would carry a special kind of seal to ward off various ghosts, snakes, tigers, wolves, and other beasts when they entered the mountains and forests. This seal was made from the heartwood of the jujube tree, measuring two inches wide". In his description, "the seal of *huangshen yuezhang*" may be interpreted as follows. "The Taoists would carve jujube wood to make seals, measuring four inches wide", as quoted in *Huang Jun Zhi Shi Hubao Fa* ("The Way to Tame Tigers and Leopards") in Volume 26 of *Chuxue Ji* ("References for Primary Learning", 700). Jujube wood is used to make a four-inch square (about 182.25 cm^2) wooden board, inscribed with 120 characters, which is equivalent to a small woodblock plate. "Ancient people", in Ge Hong's terms, refer to the people from the Wei dynasty to the early Jin dynasty (third to fourth centuries), during which the Taoists had already used large wooden seals and lutes. When paper became common during the fourth and fifth centuries, the Taoists affixed wooden seals to paper for seal script as amulets, marking a step forward to woodblock printing. In 1959, amulets and motifs written or painted on paper in the sixth century were unearthed in Astana, Turpan, Xinjiang, while seal script printed on paper have yet to be discovered.[7] However, we cannot deny that the above-mentioned seal script printed on paper did exist in history. Amulets are known as religious signs or mystical writings, evolving from Chinese characters and yet unreadable to the public, several examples of which have been included in *Baopuzi* ("Master Embracing Simplicity"), treated as a special type of reading material.

In addition, to make the Buddhist scriptures they recite illustrated with images accompanied by descriptive captions, Buddhists often print Buddha figures and relevant images in ink on the scrolls with woodcut models, which can also add sacred and solemn features to the scriptures. According to T.F. Carter, "Reproduction of molded small Buddha figures on the scrolls marks the transition from inscription of seals to woodblock printing. Thousands of such kind of small Buddha figures have been found in Dunhuang, Turpan, and Xinjiang, sometimes at the beginning of a line in a hand-copied book, and sometimes throughout a scroll. The British Museum exhibits a scripture scroll that is 17 feet (ca. 518.16 cm) long with 468 Buddha figures printed on it."[8] Although the images of those Buddha figures were actually inscribed on the scroll by affixing the seals engraved with those figures repeatedly one by one, this process was still more convenient than freehand sketching. Printing Buddha figures on written sutras began in the Northern and Southern dynasties and was still prevalent in the Tang dynasty. The practice of seal-engraving has shown that in order to

[6] Ge Hong. *Baopuzi: Inner Chapter Vol. 17, Dengshe* [M]. Shanghai: The Commercial Press, 1936: 311, 346, 352.

[7] Xinjiang Uygur Autonomous Region Museum. Briefing on Burial Excavations in Astana North District, Turpan, Xinjiang [J]. *Cultural Relics*, 1960 (6): 13–21.

[8] Carter T F. *The Invention of Printing in China and Its Spread Westward* [M]. Translated by Wu Zeyan. Beijing: The Commercial Press, 1957: 43–44.

accommodate more characters or images on a seal, thereby increasing the seal face, the shape of the seal must be changed to a flat plate, which then becomes the shape of a woodblock for printing. In other words, it was religious believers who contributed to this change.

4.1.2 Tablet Inscription Rubbing Technique

The classic reproduction technique associated with the development of Confucianism is the rubbing of classic inscriptions on stone tablets. Stone inscriptions enjoy a long history. After the Han dynasty, most of the inscribed stones were rectangular, thick slabs used to record the deeds of dead people or important events. The use of stone tablets to inscribe Confucian classics for reading or copying by scholars was a major innovation in the Eastern Han dynasty. Due to the copying of Confucian classics based on different master copies, the contents might vary occasionally. In order to assure scholars of access to standardized versions of those classics, the Empress Dowager Deng, who was in power during the fourth year of Yongchu Period of Emperor An (Liu Hu) of Han's reign in the Eastern Han dynasty, ordered Liu Zhen (67–127) and over fifty learned scholars and ceremonial officers to proofread the Five Classics, the biographies of various scholars, and hundreds of books in *dongguan* (a place to proofread and store documents in the royal palace of the Eastern Han dynasty), and then to compile them to be stored in *mifu* (a place where books and documents are kept in secret).[9] In the fourth year of Xiping Period (175) of Emperor Ling (Liu Hong) of Han's reign, Cai Yong (132–192) submitted a petition to the emperor, asking for the permission of ordering craftsmen to inscribe the classics stored in *mifu* (a place where books and documents are kept in secret) on stone tablets to be shared by the public. The suggestion was accepted by the imperial court. Cai wrote down those classics in clerical script, ordering craftsmen to engrave them on stones (Fig. 4.4). Those stone tablets were placed outside the gate of *taixue* (imperial college) in Luoyang, known as standardized versions for scholars,[10] marking the beginning of engraving Confucian classics on stones. As quoted by *Houhan Shu* ("Book of Later Han") from Lu Ji's (261–303) *Luoyang Ji* ("Bibliography of Luoyang") those stone tablets of classics include seven Confucian classics: *Shang Shu* ("Book of Documents"), *Yi Jing* ("I Ching"), *Shi Jing* ("Classic of Poetry"), *Yi Li* ("Book of Etiquette and Ceremonial"), *Chun Qiu* ("Spring and Autumn Annals"), *Gongyang Zhuan* ("Gongyang's Commentary on Spring and Autumn Annals"), and *Lun Yu* ("The Analects"), which were placed outside Kaiyang Gate in the south of Luoyang City. The seven classics contain a total of 209,000 characters, engraved on 46 tablets, each of which is 175 cm high, 90 cm wide, and 20 cm thick, with a capacity of about

[9] Fan Ye. *Book of Later Han: Vol. 10, Biography of Empress Deng* [M] // *Twenty-five Histories: Vol. 2*. Compact Edition. Shanghai: Shanghai Classics Publishing House, 1986: 35.

[10] Fan Ye. *Book of Later Han: Vol. 90, Biography of Cai Yong* [M] // *Twenty-five Histories: Vol. 2*. Compact Edition. Shanghai: Shanghai Classics Publishing House, 1986: 216.

Fig. 4.4 Residue of the Eastern Han dynasty's Xiping stone classics

5000 characters each. The tablets were originally engraved on both sides in the fourth year of Xiping Period (175), so they were known as "Xiping Stone Classics". Then engraving was completed in the sixth year of Guanghe Period (183). The tablets were arranged in a U-shape with the opening to the south, protected by a cover, surrounded by wooden fences, and guarded by designated personnel.[11] Since then, scholars from all directions had gathered here to read these stone classics. Every day, there were up to a thousand carriages parked outside the gate of *taixue* (imperial college), causing traffic jams in nearby areas and making it a grand academic event. Firstly, the craftsmen chose suitable stones before carrying out the engraving process. Then they carved the stones into tablet shapes, smoothed the surface, added wax, marked the ink lines of the character grid, and wrote down the inscriptions in cinnabar and glue. Finally, those tablets were engraved into standardized-form characters cut in intaglio.

During the Zhengshi Period (240–249) of the Wei dynasty in the Three Kingdoms Period, three stone classics were again completed in Luoyang, namely, *Guwen Shangshu* ("Ancient Book of Documents"), *Chun Qiu* ("Spring and Autumn Annals"), and *Zuo Zhuan* ("Zuo Qiming's Commentary on Spring and Autumn Annals"), which were later called "Zhengshi Stone Classics", marking the second time of emergence of stone classics in the history of China. "Zhengshi Stone Classics" consisted of 35 tablets, each 192 cm in height and 96 cm in width, engraved in three scripts, namely, the Chinese script before the Qin dynasty (large seal script),

[11] Qian Cunxun. *History of Ancient Chinese Books* [M]. Hong Kong: The Chinese University of Hong Kong Press, 1975: 66–71.

small seal script, and clerical script (Fig. 4.5), also known as "three-style stone classics" or "three-script stone classics". Each tablet contained about 4000 characters, with a total of 147,000 characters, erected to the east of the lecture hall of *taixue* in Luoyang, arranged in L shape. However, during the period between the third and sixth centuries, the capitals of the Wei, Jin, and Northern and Southern dynasties were sometimes located in Luoyang and sometimes moved to other places. At that time, the custody of stone classics began to slacken off, and the stone tablets even became unattended. In the relocation process, some stone classics got lost, so no piece of a complete stone classic has been preserved so far. What we can see now is only some fragments or partial rubbings. In the Northern and Southern dynasties, some people took advantage of the lax care of the tablets and made rubbings of these classics for their own use or for sale. The so-called rubbing is to moisten thin and tough hemp paper or kozo paper, attach it to the surface of the tablet with a brush and beat the paper with a bat so that the paper tightly fits into the concave part of the tablet inscriptions. After the paper is dried, dip the ink into a small bag wrapped with silk cotton, pat it evenly on the paper, and peel it off to make a copy of the inscriptions in white characters on a black background, which is called a rubbing. If the paper is not big enough, several pieces of paper can be used for rubbing different parts of the tablet surface. Then the rubbings are spliced together and bound into a rubbing booklet.

Tablet inscription rubbing is another way to make multiple copies of written materials, quite similar to woodblock printing, highlighting the function of the final product for reading use. The common feature shared by the two mentioned above is that the characters or pictures engraved on the large hard surface are brushed with ink to be pressed onto the paper. The difference between the two is described as follows. The inscribed characters on stone tablets are in intaglio regular script. When making a rubbing, the paper is placed on the surface of the tablet, and the finished product exhibits white characters on a black background. On the contrary, the characters on the woodblock are in relief mirror script, where the ink is applied on the woodblock plate and then paper is overlaid to be brushed with ink. The finished product is black characters on a white background. Rubbing is also a unique invention in China that converts the inscriptions, patterns, and parts of complex objects from three-dimensional form to two-dimensional form, keeping the original appearance of inscriptions on ancient bronzes and stone tablets. It is equivalent to the role of a camera before the invention of photographic technology, and the rubbing technology has been flourishing for a long time. The earliest surviving inscription rubbings were found in the early twentieth century in Dunhuang stone chamber: *Huadu Si Ta Ming* ("Huadu Temple Pagoda Inscription") engraved in the sixth year of Zhenguan Period of the Tang dynasty (632), and *Wenquan Ming* ("Hot Spring Inscription") (Fig. 4.6) engraved in the fifth year of Yonghui Period (654), which was written by Emperor Taizong of Tang dynasty (Li Shimin). The popularity of rubbing in the Tang dynasty was clearly developed based on the experience from previous generations.

Suishu Jingji Zhi ("Treatise on Classics in Book of Sui") records that in the third year of Kaihuang period (583), the Secretary Supervisor Niu Hong (545–610) wrote, "Envoys were sent to search for different copies of books. Each book was

Fig. 4.5 Three-style stone classics engraved during the Zhengshi period (240–248) in the Wei dynasty, from Zheng Songxian

rewarded with a piece of silk. When proofreading was finished, the original book was immediately returned to the owner. Encouraged by this policy, various rare and exotic books from the folk often emerged from time to time. After the collapse of the Chen dynasty in 589, collection of the classics were gradually completed."[12] *Sui Zhi* lists 33 volumes of seven classics, including *Yi Jing*, *Shang Shu*, *Shi Jing*, *Chun Qiu*, *Gongyang Zhuan*, *Yi Li*, and *Lun Yu*, which belong to the "one-script stone classics", and it also lists nine volumes of *Shang Shu*, five volumes of *Shang Shu*, and three volumes of *Chun Qiu*, which belong to "three-script stone classics".[13] The second item, "five volumes of *Shang Shu*", should be a clerical error of "five volumes of *Zuo Zhuan*". *Sui Zhi* also records that, "The 'seven classics' were engraved on stone tablets in the later Han dynasty, all written by Cai Yong. In the middle of the Zhengshi Period of the Wei dynasty, the tablets of the 'three-script stone classics' were erected, which were inherited as the standardized characters of the 'seven classics'." Today's

---

12 Wei Zheng, *Book of Sui: Vol. 32, Treatise on Classics* [M] // *Twenty-five Histories: Vol. 2.* Compact Edition. Shanghai: Shanghai Classics Publishing House, 1986: 115.

13 Wei Zheng, *Book of Sui: Vol. 32, Treatise on Classics* [M] // *Twenty-five Histories: Vol. 2.* Compact Edition. Shanghai: Shanghai Classics Publishing House, 1986: 119.

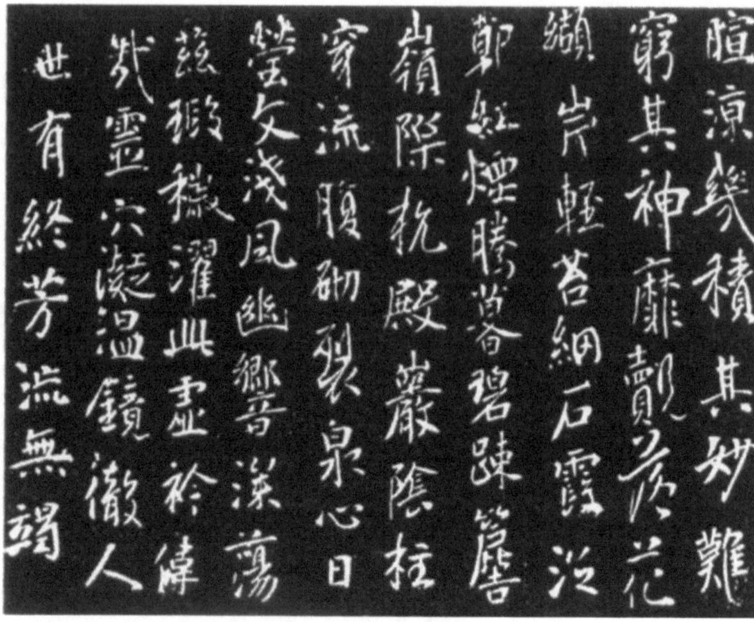

Fig. 4.6 *Wenquan Ming* written by Emperor Taizong of Tang dynasty in the fifth year of Yonghui period (654) (Collected by Bibliothèque Nationale de France in Paris)

edition was wrongly written as "one-script stone classics", so we revised it to "three-script stone classics". Therefore, the "one-script stone classics" mentioned by Wei Zheng, the author of *Sui Zhi*, refer to the rubbing version of the "Xiping stone classics" of the Eastern Han dynasty, and the "three-script stone classics" refer to the rubbing version of the "Zhengshi stone classics" or "three-script stone classics" of the Wei dynasty. Since he clearly stated that these incomplete Confucian classics were "the foundation of inheritance and extension", Wei Zheng made them counted in the volumes. Wei also told us that these rubbings were obtained from the folk under the command of Emperor Wen (Yang Jian) of the Sui dynasty from 583 to 589. The rubbings must have been made before that time. Based on this evidence, the invention of the Chinese rubbing technique was no later than the Southern and Northern dynasties in the fifth century. At that time, rubbings of stone inscriptions of *Qinshihuang Xun Kuaiji KeshiWen* ("Stone Inscriptions to Memorize Emperor Qinshihuang's Visit to Kuaiji") was made, which was also stored in *mifu* in the early Tang dynasty.

Just as reversing the common way a seal is affixed to paper can produce woodblock printing, the same effect can be achieved if certain procedures concerning the techniques of stone engraving and rubbing are reversed. One of the most important reversals would be inscribing the characters on the tablet surface in reverse form and changing the way of rubbing. This kind of reversal was made in the Northern and Southern dynasties, which can be found in the Holy Way stele (ca. 556) in front of the mausoleum of Xiao Gang, the Emperor Jianwen of Liang (503–551), in the suburbs of Nanjing. The inscriptions on the front side are in intaglio regular script, while those on the back side are in relief mirror script (the reverse form) (Fig. 4.7).[14] According to the usual way of rubbing, the rubbing of the front tablet should be in white standardized-form characters (regular script) on a black background, and that of the back side should be in white reverse-form characters (mirror script) on a black background. However, if ink is applied to the back of the tablet with paper placed on the surface for constant pounding, the white standardized-form characters on a black background will be revealed, quite similar to the practice of woodblock printing. *Shiping Gong Zaoxiang Ji* ("Record of the Statue of Duke Shiping") included in *Longmen Shiku* ("Longmen Grottoes", 1961) in Longmen Grottoes in the south of Luoyang, Henan Province, was inscribed in the 22nd year of Taihe Period in the Northern Wei dynasty (498). The inscriptions on the back of the tablet were standardized-form characters in relief, which could produce black characters on a white background, in contrast to the black background of the traditional rubbing technique. Since people were able to engrave intaglio mirror-script and relief regular script inscriptions during the Northern and Southern dynasties, they could also engrave relief mirror-script inscriptions, which essentially became woodblock printing. Therefore, as far as tablet inscriptions and rubbings are concerned, there already existed ways to use printing techniques as early as the Southern and Northern dynasties.

4.1.3 Pattern Printing Technique

Another classical reproduction technique that led to woodblock printing was the pattern printing technique. In ancient China, there were two types of pattern printing plates: the relief plate and the stencil plate. The technique of carving embossed patterns on wooden boards and printing them on fabrics with dyes was first practiced as late as the Warring States period. In 1979, fragments of printed linen with plate patterns were unearthed from a tomb in the early to middle period of the Warring States (fifth–fourth centuries BC) in Xianshuiyan Cliff, Guixi County, Jiangxi Province.[15] Since the Qin and Han dynasties, plate printing has developed rapidly. In 1972, two pieces of cloud-pattern printed gauze were unearthed from the Mawangdui No. 1 tomb (165 BC), a Western Han tomb in Changsha, Hunan Province, one of which was a gold and silver printed gauze with a length of 64 cm and a width of 47 cm. The gauze was printed with silver-white and silver-gray cloud

[14] Central Antiquities Preservation Commission. *Report on the Investigation of the Tombs of the Six Dynasties* [M]. Nanjing: 1935: Plate 11, Plate 20.

[15] Cheng Yinglin, Liu Shizhong. Briefing on the Excavation of Cliff Tomb in Guixi, Jiangxi [J]. *Cultural Relics*, 1980 (11): 29.

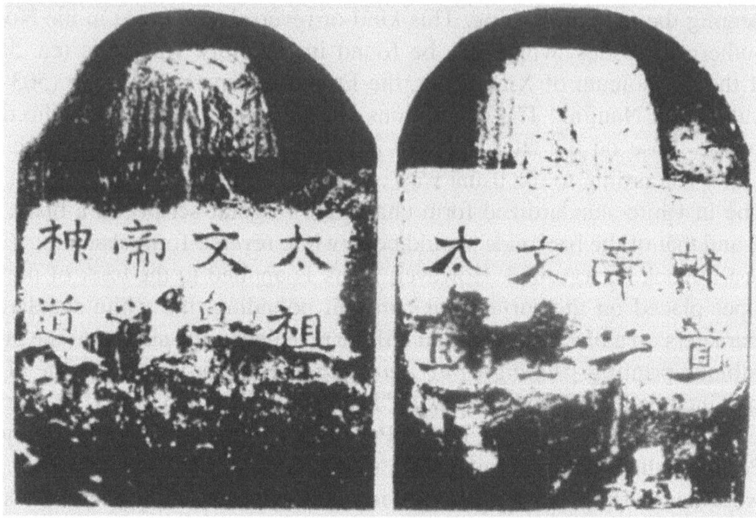

Fig. 4.7 Standardized and reverse-form inscriptions on the tombstone of the Emperor Jianwen of Liang (556)[16]

patterns of various forms, accompanied by gold-yellow dots printed on the surface. Each cloud-pattern unit is 10.4 cm long and 7.6 cm wide, slightly diamond-shaped, interlaced up and down, arranged horizontally with a total of 13 units of patterns (Fig. 4.8). Experts believe that it was printed using three pieces of relief printing plates for color overprinting.[17] In 1983, in Yuexiu Mountain, Guangzhou, similar cloud-pattern printed gauze was unearthed in the tomb of Nanyue King Zhao Mo, with different colors. The cloud patterns were golden yellow or silver-white, and the dots were red, which indicated that this kind of plate printing technique had been commonly used all over the country in the Western Han dynasty in the second century BC.

It is worth noting that in the west chamber of the tomb of Nanyue King of the Western Han dynasty in Guangzhou, where a large number of silk fabrics were unearthed, two bronze printing plates were found, the larger of which was 5.7 cm long, 4.1 cm wide, and 0.1 cm thick, with a perforated button at the back (Fig. 4.9) and embossed cloud patterns on the plate.[18] The other plate is smaller, with traces

---

[16] Central Antiquities Preservation Commission. *Report on the Investigation of the Tombs of the Six Dynasties* [M]. Nanjing: 1935: Plate 11, Plate 20.

[17] Shanghai Textile Science Research Institute. *Research on Textiles Unearthed from Mawangdui No.1 Han Dynasty Tomb in Changsha* [M]. Beijing: Cultural Relics Press, 1980: 110–111.

[18] Guangzhou Municipal Cultural Relics Management Committee, Institute of Archaeology, Chinese Academy of Social Sciences, Guangdong Museum. *Tomb of the Nanyue King of the Western Han Dynasty: Vol. 1* [M]. Beijing: Cultural Relics Press, 1991: 485–487. Guangzhou Municipal Cultural Relics Management Committee, Institute of Archaeology, Chinese Academy of Social Sciences, Guangdong Museum. *Tomb of the Nanyue King of the Western Han Dynasty: Vol. 2* [M]. Beijing: Cultural Relics Press, 1991: Plate 48-2.

Fig. 4.8 Schematic diagram of three-color overprinting plates unearthed in Changsha in 1972 from Mawangdui No. 1 Tomb in the Western Han dynasty (165 BC)

of wear and tear on the plate, indicating that it has been used. The method of use is the same as that used for the printed gauze unearthed in the Mawangdui Tomb in Changsha. That is, the first small printing plate is used to print and dye the positioning patterns, and then a large printing plate is used to print and dye on a larger scale. Finally, print and dye by using a plate with embossed dots. In this way, patterns from those three types of printing plates constitute a unit. Then, another unit is printed and dyed accordingly, following the same procedures, and the pattern lines of each unit are closely connected. In the west chamber of this tomb, a fragment of printed gauze was also unearthed, with a length of 8 cm and a width of 2.2 cm, and the cloud patterns on it were the same as those on the bronze printing plate. Thus, it can be concluded that the bronze printing plate was actually the tool for printing and dyeing this piece of gauze. Pattern plate printing has a long history, which refers to printing and dyeing embossed patterns on the plate with dye and pigment on silk, linen, and cotton textiles. Strictly speaking, it is not a printing process. Only when the patterns are printed on paper does the product become printed, which is the printed wallpaper later invented in China. When the patterns to be printed on paper are Buddha figures

Fig. 4.9 Convex copper printing plate unearthed in 1983 in Guangzhou from the tomb of the Nanyue King of the Western Han dynasty (122 BC)

or characters, the products turned out to be printed reading materials. Therefore, it can be summarized that the three classical reproduction techniques of affixing seals, tablet inscription rubbing, and pattern plate printing all eventually evolved to woodblock printing after their wide application in the Northern and Southern dynasties.

4.2 Origin and Early Development of Woodblock Printing

4.2.1 Factors Contributing to Invention of Printing Technology

Woodblock printing emerged after the advent of classical reproduction techniques such as affixing seals, tablet inscription rubbing, and pattern plate printing. It is not simply the outcome of improvement or innovation of the former, but a completely new mechanical invention, for it differs from previous methods of reproduction in terms of the materials, tools, processes, operations, and usages. In place of handwritten copies, direct production of cultural readings by mechanical reproduction methods formed the basic idea that served as the starting point for printing activities. Nevertheless, this idea was absent with regard to the classical reproduction techniques as mentioned earlier. The emergence of printing technology in China resulted from a long period of historical evolution. Inspired by previous classical reproduction techniques, it was also necessary to cultivate a comprehensive environment with suitable social, economic, cultural, technological, and material foundations, which would be even more closely related to historical traditions, language and writing, religious beliefs, and other factors. Only when these conditions were mature, could the printing technology emerge. Printing technology was invented in China because China was the first country to fulfill all the conditions that might lead to its emergence, while other countries or regions only acquired some of the conditions, yet inadequate for the invention of this technology. By the time other countries had all the conditions ready to develop printing, Chinese printing technology had already been invented and spread overseas. Although affixing seals, tablet inscription rubbing, and pattern plate printing techniques appeared early in ancient times in both Eastern and Western countries, the evolution of the above-mentioned techniques to printing technology was hindered because paper, the material necessary for the development of printing, was not available in other countries. Printing technology could not have been developed without paper. China was the first country to make paper and also the first to have a material carrier for printing tools. Only after the invention of paper and the phase of hand-written copying of reading materials on paper did the need for new reproduction technologies arise, as well as the material prerequisites for the development of printing technology.

Paper was used in China hundreds of years before other countries. During that gap period, there had been plenty of time to develop printing technology. Taking seals as an example, the decisive steps that led to the evolution of woodblock printing were first found in China, such as the printing of characters inscribed in reverse form on paper, which took place at the latest in the Jin dynasty from the fourth to fifth centuries, a time when paper was still unknown or just starting to be used in certain countries. The Taoists affixed wooden seals engraved with many characters to paper as amulets, especially during the period from the Jin dynasty to the Northern and Southern dynasties. China realized the evolution from seals to woodblock plates earlier than other countries. Although stone inscriptions are found in both the East and the West, the topography of inscriptions on paper is a unique phenomenon that has never been seen in the West. Rubbing appeared very late in Japan and Korea. Inscriptions of characters in reverse form are also rare in other countries. Rubbing of Confucian classics on paper began in the Northern and Southern dynasties, and the evolution from rubbing to printing technology could only have occurred in China. Fabric printing was practiced in other countries in ancient times, but the printing of engraved plates on paper came much later and was not used for religious or cultural purposes. In terms of cultural and technological traditions, the evolution of various classical reproduction techniques towards printing technology took place first in China rather than in other countries or regions.

As a driving force for cultural development, printing technology is the product for an era of social stability, economic prosperity, and cultural and educational prosperity, and it can only be invented under such circumstances, regardless of any country or region. In the case of China, the establishment of the great unified feudal empire by Qin Shi Huang (Ying Zheng, the first emperor of the Qin dynasty) in the third century BC put an end to the separation and chaos of the Warring States period, which was further consolidated in the Han and Jin dynasties. Thereafter, after a brief split between the Northern and Southern dynasties, China returned to unity in the Sui and Tang dynasties (sixth–tenth centuries), which brought about relative social stability, unprecedented economic, cultural, educational, and academic development, as well as the establishment of a unified imperial examination system and a rapid increase in the number of schools and scholars. Successive rulers emphasized civilized governance, making China a state advocating civilization and education. During that time, like Confucianism, Buddhism developed rapidly on a large scale, with numerous monasteries, monks, nuns, and followers all over the country. *HanShu·Yiwen Zhi* ("Literature Records in Book of Han") contains nearly 15,000 volumes of various types of books in history, while books included in *Suishu·Jingji Zhi* ("Treatise on Classics in Book of Sui") have increased to more than 50,000 volumes, and the collection of books in Sui dynasty's inner palace, *Jiazedian*, reaches 370,000 volumes. Hand copying of this vast amount of literature on paper was very laborious, and hence extremely time-consuming. Therefore, there was an urgent need for a new reproduction technology to replace hand-copying labor, and woodblock printing could precisely meet this requirement. Since the unification of characters in China during the Qin dynasty, the development of Chinese characters has entered a new stage. Since the Wei and Jin dynasties, clerical script and regular script have become prevalent. In the Sui dynasty, a stable regular script was formed, easy to recognize and engrave, which served as a suitable script for typesetting in block printing. Since the Northern and Southern dynasties, papermaking and ink-making technologies both have undergone new developments, providing raw materials and substantial basis for the printing industry. Thus, in the later years of the Northern and Southern dynasties (532–589), all the necessary conditions were ready for the emergence of printing technology.

On the contrary, Europe had a long period of slavery, and it entered feudal society 1000 years later than China. Although the fall of the Western Roman Empire in 476 marked the collapse of slavery, early feudal societies in the West were still characterized by serfdom. In the medieval ages, Europe was in the so-called Dark Age for a long time, with stagnant social development, slow economic progress, sluggish culture and education, and very few people were literate. Some books were copied on parchment boards or papyrus slices, adequate enough to satisfy the needs of the society, which incurred no urgent need for a new type of reproduction technology. While the Sui and Tang empires stood in the East as highly civilized, rich, and united powers, the West was still amid constant conflicts. The struggles for power between the monarchs and popes within each country led to social unrest, creating a social environment in sharp contrast to China.[19] As for other areas of ancient civilization, for instance, culture development in Egypt and also in the region between the Tigris and Euphrates rivers was later interrupted, with their written languages even extinct,

[19] Wells. *The Outline of History: A Brief History of Life and Mankind* [M]. Translated by Wu Wenzao, Xie Bingxin, Fei Xiaotong, et al. Beijing: People's Publishing House, 1982. Bernal. *Science in History* [M]. Translated by Wu Guangfu, et al. Beijing: Science Press, 1959.

let alone the development of printing technology. The Indian slavery society lasted until the sixth century before it was gradually dismantled. When India started making paper, the Chinese were already using printed books. Arab civilization shone for a time, but it came too late. In East Asia, social, cultural, scientific, and technological achievements of the Sui and Tang dynasties were higher than those of Japan and Silla in the same period. Therefore, it is impossible for those two countries like Japan and Silla to stay ahead of China in the Sui and Tang dynasties in terms of printing culture.

4.2.2 *On the Origin of Printing Technology in Sui Dynasty*

The main issue to be explored in quest of the origin of woodblock printing is the exact time of its emergence in China. This problem has not been properly resolved for a long time in the past because people mainly drew on fragmented records in ancient books, with very few comprehensive studies on archaeological discoveries, regular patterns of the development of printing technology, and the socio-economic and cultural background. In addition, there are also different understandings of the meaning of ancient books. All these factors resulted in the divergence of opinions at one time. With intensive research and accumulation of information in recent decades, it is time to put an end to this disputation. Woodblock printing did not suddenly appear within one day. Its origin should be studied as the outcome from a gradual process, that is, the process of all sectors of society exploring new reproduction techniques to replace hand-copying labor, or the process of evolution and transformation of classical reproduction techniques into mechanical reproduction techniques. After a certain stage of maturity, printing technology naturally emerged. With this regard, the time of the origin of printing cannot be designated in a specific year but should be delineated over an appropriate period of time to find out the upper and lower time limits. Then, according to the literature, unearthed objects, and technical reasoning, we can set a time of origin close to the actual situation between the upper and lower limits. This is probably more sensible than the earlier studies of the time of origin based solely on a single sentence recorded in a certain book.

Based on the foregoing discussions, from a technical point of view, we can take the Northern and Southern dynasties (420–589) as the upper time limit for the origin of printing technology. That is because, at that time, paper-making and ink-making technologies were sufficient to provide the paper and ink needed to meet the needs of printing, and the technological preparations for the transition from affixing seals and tablet inscription rubbing techniques to printing technology were all mature at that time. The thriving of Buddhism during that period, which acted as a stimulus to printing technology, also gained momentum in the Northern and Southern dynasties. Printing technology originally came from the folk and was related to the religious activities among a vast number of Buddhist believers. Almost all the Buddhist scriptures conveyed the Buddha's teachings. Repeated recitation of mantras, copying of Buddhist scriptures, and worshiping of Buddhist statues would accumulate blessings, eliminate calamities, and save them from suffering in hell after death. Since believers were tired of copying mantras, the reproduction of sutras by the woodblock printing technology could provide a large number of inexpensive copies. Thus, believers could buy ready-made scrolls of sutras and printed images of Buddha by filling in their names and wishes, so they were happy to do so, In this way, printing technology flourished among the folk. Many Chinese and foreign scholars, therefore, tended to hold the view that there had already existed printing activities in the Northern and Southern dynasties. Although the historical evidences they resorted to might be debatable, their conclusions seem to be sensible to some extent, which still need to be verified by new evidences or future archaeological discoveries. The chance of printing technology appearing before the Northern and Southern dynasties is rather slim. Although there already existed paper in the Han and Wei dynasties, it was mainly used for writing and was not yet separated from the stage of concurrent use of paper and bamboo slips. Only in the Eastern Jin dynasty were bamboo slips completely eliminated. The transition from affixing seals and tablet inscription rubbing to printing technology occurred after the Jin dynasty, and there was a lack of technological accumulation to invent printing prior to the Northern and Southern dynasties. Literary records and archaeological discoveries indicate that there were printing activities in the early years of the Tang dynasty (618–713), which should be the lower time limit for the origin of printing technology. Therefore, as far as the current evidences are concerned, the Sui dynasty (581–618) should be the decisive period for the emergence of printing technology.

Let's start with the documentary records. As early as in the Ming dynasty, Lu Shen (1477–1544) proposed in *Hefen Yanxian Lu* ("Collection of Hefen in Spare Time") that woodblock printing began in the Sui dynasty. "On December 8 of the Kaihuang Period (January 5, 594 according to the solar calendar) of the Emperor Wen (Yang Jian) of Sui's reign, the emperor ordered to inscribe and rewrite all of the ruined statues and the remains of scriptures, which signified the beginning of the printing of books, and this view was put forward before Feng Dao." Hu Yinglin, a Ming dynasty bibliographer and cataloger, also argued in *Shaoshi Shanfang Bicong* ("Collection of Hu Yinglin, with the courtesy name of Shaoshi Shanren") and *Jiabu Jingji Huitong Si* ("The First Part in Comprehensive Study of Catalog of Ancient Books"):

Lu Shen (also called Lu Ziyuan) proposed in *Hefen YanXian Lu* that, on December 8 of the Kaihuang Period of the Emperor Wen of Sui's reign, the emperor ordered to rebuild all of the ruined statues and the remains of scriptures, which marked the beginning of the printing of books. According to Lu, the printing of books began in the Sui dynasty. This view was put forward before Liu Pin (848–898), Feng Dao (882–954), and Mu Zhaoyi (about 902–967)… I think that the carvings in the Sui dynasty were mostly Buddhist statues and scriptures, owing to the fact that Buddhism was widely worshiped in the Six Dynasties and thus there was not enough time to engrave other books as well. Since the middle of the Tang dynasty, this engraving method began to be used gradually for printing books, which was practiced in the Five Dynasties, flourished in the Song dynasty, and is extremely popular in the present day (the Ming dynasty according to the author)… In summary, the woodblock printing technique originated in the Sui dynasty, prevailed in the Tang dynasty, expanded in the Five Dynasties, and were refined in the Song dynasty. I have considered all sorts of opinions to come up with the conclusion above, and I am sure that they are credible.

When Lu and Hu presented those opinions, they apparently relied on what was stated in Volume 12 of Fei Zhangfang's (557–610) *Lidai Sanbao Ji* ("Records of the Three Treasures Throughout the Successive Dynasties", 597).

On December 8 of the 13th year of the Kaihuang period, Emperor Wen (541–604) of the Sui dynasty, a disciple of Buddhism, paid homage to the Buddha... Due to the chaos of the Zhou dynasty, the holy relics have been insulted, the pagodas have been destroyed and ruined, and the scriptures and statues have fallen into ruin... The people should be rescued, the veneration should be restored, and the divinity should be honored again. The situation is even more serious with the destruction of the base. The emperor ordered to rebuild all the ruined statues and the remains of scriptures... Then, vegetarian diets were provided, and the celebration of scriptures and statues was consecrated. Hundreds of thousands of people came here to a ceremony of washing Buddha statues.[20]

In the above record, what Fei Zhangfang called "chaos of the Zhou dynasty" refers to the fact that in the third year of the Jiande period (574) of Emperor Wu (Yuwen Yong)'s reign in the Northern Zhou dynasty (557–581), the emperor ordered the banning of Buddhism and Taoism, destroying the scriptures and statues of the temples and forcing *shamen* (Sramana, Buddhist monks) to resume secular life, thus destroying a lot of religious artifacts. Yang Jian (541–604), Emperor Wen of Sui, who overthrew the Northern Zhou regime and established the unified Sui dynasty, was a great believer in Buddhism. After the country's economic situation had improved and social culture and Buddhism had further developed, he honored the Buddha on 5 January, 594, with the hope of restoring the pagodas, statues, and scriptures that had been destroyed during Emperor Wu's reign in the Northern Zhou dynasty and those that had been lost, to revitalize Buddhism. He and the Empress promised to "re-carve all of the ruined statues and rewrite the remains of scriptures", and gave alms to the public. After the bronze statues were cast, people gathered for the ceremony of collecting scented water from the congregation to bathe Buddha statues, which is a significant ritual in Buddhist festivals, symbolizing the cleansing of sins and the attainment of purity. The "statues" mentioned here mainly refer to the bronze casting of Buddha statues, and the "scriptures" mainly refer to the paper Buddhist scriptures. "Re-carve" is engraving, and "rewrite" is making, so "re-carve and rewrite" should be interpreted as "re-carve and rebuild" here. "Rewrite" here should not be interpreted as "rewriting", so "re-carve and rewrite all of the ruined statues and the remains of scriptures" is actually "re-carve and rebuild all of the ruined statues and the remains of scriptures", where "rebuild" refers to the casting of statues and the carving or printing of paper Buddhist scriptures and statues. In Fei Zhangfang's terminology, "re-carve and rewrite" or "re-carve and rebuild" are two

[20] Fei Zhangfang. *Li Dai San Bao Ji: Vol. 12* [M]. *Taisho-pitaka: Vol. 49*. Tokyo: Daizo Shuppansha, 1924: 108.

transitive verbs, the complements of which are Buddha statues and Buddhist scriptures. In the case of Buddha statues, it means that their dignified presence reappears through casting; in the case of Buddhist scriptures, it means that they are re-divinized through printing and manufacturing. Therefore, what is recorded in *Lidai Sanbao Ji* ("Records of Three Treasures Throughout the Successive Dynasties") can be taken as an account of printing in the Sui dynasty and should not be questioned or criticized due to its abbreviated wording.

When studying this book, it should be analyzed in the broader context of the Sui dynasty at that time, while also considering the close connection with the printing literature that appeared shortly thereafter in the early Tang dynasty and the dating of unearthed prints. At the time of Emperor Wen of Sui, the whole country was unified, wealthy, and prosperous, and the rulers vigorously fostered Buddhism. To re-circulate the lost Buddhist scriptures of the previous dynasties as soon as possible, printing was the most convenient way. However, Wang Shizhen (1634–1711), a scholar in the Qing dynasty, commented on Lu Shen's viewpoint in Volume 25 of *Juyi Lu* ("Collected Records of Calligraphies and Paintings"): "I have delved into the meaning of the text; what were carved are statues, and what were written are scriptures, and Lu Shen's interpretation of the two notions linked together has caused misunderstandings." In recent time, some people have also held the view that "re-carve and rewrite" refers to the carving of Buddha statues and the compilation of Buddhist scriptures. It is suspected that the carving of Buddha statues and printing of Buddhist scriptures appeared in the Sui dynasty.[21] Such an understanding of the original words of *Lidai Sanbao Ji* may not be convincing but rather creates new misunderstandings. A problem arises if the word "carve" in the phrase "re-carve and rewrite" refers exclusively to carving Buddha statues. Zan Ning (919–1001) explained in the first volume of *Seng Shilve* ("Topical Compendium of the Buddhist Clergy") that during the ceremony of bathing statues of Buddha, bronze statues were used. However, bronze statues were all cast and forged, so how could they be carved? The interpretation of "what were carved are statues" does not correspond to the meaning of the original text. The original text of Fei Zhangfang should be understood in its entirety, not in a compartmentalized way. Another historical material is also worth noting. In Volume 78 of *Sui Shu* ("Book of Sui"), *Lu Taiyi Zhuan* ("Biography of Lu Taiyi") noted that Lu Taiyi (548–618) was a scholarly man, appreciated by Emperor Wen of Sui, "who then became blind and knew his words by touching books with his hands". He died in Luoyang several years later after traveling with Emperor Yang to Liaodong in the ninth year of the Daye Period (613). Wang Renjun (1866–1914) explained as follows, "since he could touch books by hand to know the characters, the books must have been woodblock books… It is very clear that the books at that time had their own woodblock plates, so we know that what he touched were printing plates."[22] Characters engraved on woodblock plates are in reverse form. Lu knows the standardized-form characters by touching the

[21] Zhang Xiumin. *The Invention of Printing in China and Its Influence* [M]. Beijing: People's Publishing House, 1958.

[22] Wang Renjun. *Ge Zhi Jing Hua Lu: Vol. 4* [M]. Shanghai: Stone print, 1896.

reverse-form characters, which reveals Lu's ingenuity. Some people say that what he touched were the inscriptions on stone tablets, which may not be appropriate. Historical recordings clearly show that what Lu Taiyi touched were books, and stone tablets cannot be called books. Moreover, the inscriptions are mostly in standardized forms of regular script. Touching the standardized-form characters in regular script to understand them cannot reveal Lu Taiyi's ingenuity. It is therefore still justified to interpret the books he touched as book plates, which is another record of printing in the Sui dynasty.

Records of printing appeared successively since the beginning of the Tang dynasty. When the monk Yancong (625–690) wrote a biography of his mentor Xuanzang (602–664), he pointed out that in his later years, between the third year of Xianqing period of Emperor Gaozong of Tang dynasty (Li Zhi) and the third year of Longshuo period (658–663), Xuanzang "vowed to print a million Buddha statues and finally managed to print them".[23] The word *juzhi* is the transliteration of the Sanskrit word "koti" for quantity, which in this context means one hundred thousand. The character *zao* (meaning "making" nowadays) in the Tang dynasty referred to "printing", as in the inscription on *Jingang Jing* ("Diamond Sutra") published in the ninth year of Xiantong period (868), which reads, "Wang Jie respectfully made it for his parents". Therefore, Yancong cited that in 658–663, Xuanzang vowed to print a million Buddha statues and finally managed to print them. In the tenth century, Feng Zhi, a native of Jincheng (now Lanzhou), also mentioned in volume 5 of *Yunxian Sanlu* ("Miscellaneous Notes of Unzen", 926), "Xuanzang printed the figures of Samantabhadra Bodhisattva on paper and gave them to other people. Five loads of Buddha figures were distributed each year." The two mentioned above referred to the same thing.[24] In fact, the two records can complement and corroborate each other. In the past, some people suspected that *Yunxian Sanlu* might be a "forgery" written by Wang Zhi (1090–1161) of the Northern Song dynasty. However, this book has a version engraved by Guo Yingxiang in the first year of Kaixi period (1205). At the beginning of the volume, there is a preface written by the author, Feng Zhi, in the first year of Tiancheng period in the later Tang dynasty. This book is also cited in Kong Chuan's *Kongshi Liutie* ("Six Copybooks for Calligraphy of Kong"). The record of Xuanzang's printing of Buddha statues is still credible and was supplemented by the records of his disciple, Yancong.

It is a pity that very few early prints survived due to the ban on Buddhism in the fifth year of the Huichang period (845) in the reign of Emperor Wuzong of the Tang dynasty, when he ordered to burn Buddhist scriptures and figures. Fortunately, early prints buried in tombs prior to this time have been unearthed in recent years, allowing us to have a glimpse of them. As reported by Mr. Han Baoquan, an archaeologist in Shaanxi Province, the archaeologists of the Xi'an Municipal Commission of Cultural Relics Management in 1974 obtained a single-page printed *Dharani Sutra Mantra*

[23] Huili, Huicong. *The Biography of Master Sanzang: Vol. 10* [M] // *Taisho-pitaka*: *Vol. 50*. Tokyo: Daizo Shuppansha, 1927: 275.

[24] Feng Zhi. Y*un Xian San Lu: Vol. 5* [M]. *Complete Library in the Four Branches of Literature Vol. 1355*. Photocopy. Taipei: Commercial Press, 1983: 666.

(Fig. 4.10) from a Tang tomb at the Xi'an Diesel Machinery Factory. When the mantra was unearthed, it was contained in a copper armlet (bracelet). The artifacts unearthed together include a round-shaped bronze mirror with mysterious patterns of four deities, while other artifacts were basically lost. The printed copy is square, 27 cm in length and 26 cm in width, with a blank box 7 cm in width and 6 cm in height in the center, and the four characters of "Wu De □ Fu" (□ representing the character missing) written on the upper right, with one character peeled off, which means that Wu De should be the name of the owner of the tomb. Around the outside of the square frame are mantra texts, with 13 rows of characters both on the top and at the bottom. The left and right corners were fragmented, so it is assumed that there were originally 13 rows of mantra as well. The mantra is surrounded on all sides by a triple double-column border with a 3 cm spacing between the inner and outer borders, with patterns of lotus flowers, flower buds, Dharma vessels, mudras, and constellations. It is difficult to determine whether the mantra is in Sanskrit or in Pāli (Pali Bhasa), which is spelt in the Sinhalese alphabet, so Mr. Han treated it as a foreign language inscription tentatively.

The bronze armlet containing the mantra is in the form of a curved circle, 4.5 cm in length and 4.2 cm in width, tapering to 12 cm on each side. Underneath

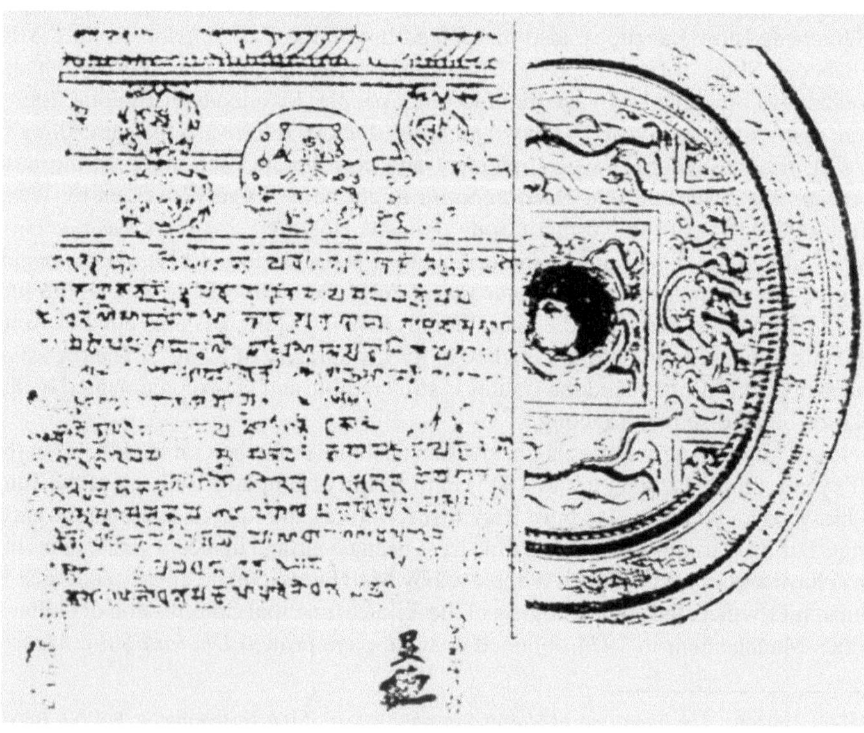

Fig. 4.10 A Single-page print of Sanskrit Dharani Mantras from the early seventh century, unearthed at the Xi'an Diesel Machinery Factory in 1974

is a hollow semicircular cylinder with a diameter of 2.8 cm. The two ends of the cylinder are sealed with copper sheets, and the mantra is stacked in the cylinder, which has been damaged after unfolding. The round-shaped bronze mirror with four deities (mysterious creatures, also known as the four symbols of the Chinese constellations) is 19.5 cm in diameter, 0.3 cm in thickness, and 0.8 cm in height along the rim. The mirror body is silver-white, with a rounded dragon square seat. Within the inner area are the four deities, namely, Blue Dragon, White Tiger, Rose Finch, and Tortoise, adorned with four rims of geometric patterns that hold cultural and cosmological significance, on each of which is positioned a head of those four mysterious creatures, and there is an inscription within the outer circle of the inner zone, which is separated from the outer area by a saw tooth (Fig. 4.10). From the style and form, it can be identified as an object from the Sui dynasty to the early Tang dynasty, the same as other bronze mirrors of the same period unearthed in Xi'an. The Institute of Archaeology at the Chinese Academy of Sciences conducted research on typology and periodization of mirrors, in the compilation of *Xi'an Jiaoqu Suitang Mu* ("Tomb of Sui and Tang Dynasties in the Suburb of Xi'an", 1966). By comparison, the inscriptions and decorations of this mirror can be categorized as Type I Style II and Type II Style IV respectively, so it belongs to the period between the Sui dynasty and the early Tang dynasty. Compared with other Tang Dynasty Dharani prints found in Xi'an, the engraving technique is more rudimentary, with the corners and the ink color of the mantras being blurred and unclear, indicating an earlier dating.

Mr. Han, taking into account the aforementioned circumstances and integrating the context of the development of Esoteric Buddhism during the Tang dynasty, has attributed the sutra spell in question to the early Tang dynasty, positing that "it should be the oldest known printed item in existence today."[25] Following the publication of this assertion, a decade passed without contestation until 1997 when Zhao Yonghui challenged this dating in his scholarly article. He proposed that the sutra spell was more likely a product of the mid-to-late Tang dynasty (756–845), citing three primary reasons: Firstly, the printed text and the bronze mirror were retrieved from a site that had been ravaged, with no clear archaeological context or documentation, making it impossible to ascertain their relationship; Secondly, the practice of wearing talismans inscribed with spells, arranged in a square formation akin to mandalas and encircled by hand seals, ritual instruments, and patterns, only became prevalent during the mid-to-late Tang dynasty; Thirdly, the inscription "Wu De □ Fu" lacks the calligraphic characteristics of the early Tang dynasty.[26] However, proponents of this counterargument have yet to decipher the sutra's text. To elucidate whether Zhao's rationale suffices to overturn the original chronological conclusion that the Dharani Sutra printed edition discovered in Xi'an in 1974 belongs to the early Tang dynasty, it is imperative to conduct a fresh investigation into this manuscript. In an effort to

---

[25] Han Baoquan. *The World's Earliest Printed Mantra: A Dharani Mantra Print Unearthed from a Tang Tomb in Xi'an* [M] // *Shi Xingbang. The Collected Works of Chinese Archaeology: In Commemoration of the Fiftieth Anniversary of Mr. Xia Nai's Archaeology*. Xi'an: Sanqin Publishing House, 1987: 404–410.

[26] Zhao Yonghui. On the Origin of Printing Technology [N]. *Weekly of China's Cultural Relics*, 1997–02–16 (03).

elucidate the circumstances surrounding the discovery of this specific manuscript, we initially sought the expertise of Han Baoquan, the author of the original report. Han recounted that in 1974, the Xi'an Cultural Relics Management Committee had conducted an investigation at the site where the artifact was unearthed, verifying that despite the unfortunate loss of numerous items, "luckily, there was a serendipitous preservation of the armlet adorned with the foreign script printed spells and the Four Deities Ruled Bronze Mirror, both of which were extracted from the same tomb".[27] The 1987 report had also been unambiguous in its assertion that the printed volume and the bronze mirror shared a common origin in the same Tang dynasty tomb. Furthermore, the "Cultural Relics Collection Receipt" issued by the Xi'an Cultural Relics Management Committee in 1974 corroborates this account, documenting that the bronze mirror, the wooden block printed spells, and the copper armlet—presumably the very armlet mentioned—were unearthed from a single location, sequentially numbered, and officially entered into storage on the 2nd of August, 1974. It is imperative to confirm this foundational truth; the connection between the printed volume and the bronze mirror is not open to question, with evidence accessible for scrutiny. Given that this was not the result of a scheduled archaeological excavation and considering that research institutions were not operational during the tumultuous period of the "Cultural Revolution", it would be inaccurate to suggest that the origin of these cultural relics is shrouded in ambiguity. It is, in fact, a testament to their achievement that they were able to secure a cultural relic of great national significance during that era.

In order to discern the script on the printed pages, the author sought the expertise of the renowned Sanskrit scholar Mr. Jiang Zhongxin from the Institute of Asian and Pacific Studies, Chinese Academy of Social Sciences, in 1997. The text was identified as a Sanskrit script that had been in common use since at least the sixth century.[28] At that time, during the reign of King Śīlāditya (589–647), who unified northern India and moved the capital to Kanyākubja, Buddhism was greatly developed. When Xuanzang resided in India (628–643), he was received with honor, and thus the Buddhist scriptures he brought back were also written in this type of Sanskrit script. The script of this spell is arranged around a central blank square, forming a square array, indeed presenting the shape of a mandala altar, with 13 lines above and below, and 13 lines left and right, totaling 52 lines. If we regard the part inscribed with the name of the tomb owner as the front, it can be observed that the lines of characters below the center of the spell are all correctly oriented, while the characters at the top are inverted. Rotating the spell text by 90° in any direction also reveals the same phenomenon. It is believed that the text begins at the innermost line from the lower left corner of the spell center, continues to the lower right corner, then turns to the upper right corner and the upper left corner, circling the spell center once, completing the first loop. Subsequently, the same method is used for arrangement until the 13th loop is completed (Fig. 4.11).

[27] Letter from Han Baoquan to Pan Jixing, December 5, 1996, from Xi'an.

[28] Jiang Zhongxin's conversation with Pan Jixing, April 15, 1997, at Jiang's residence in Beijing.

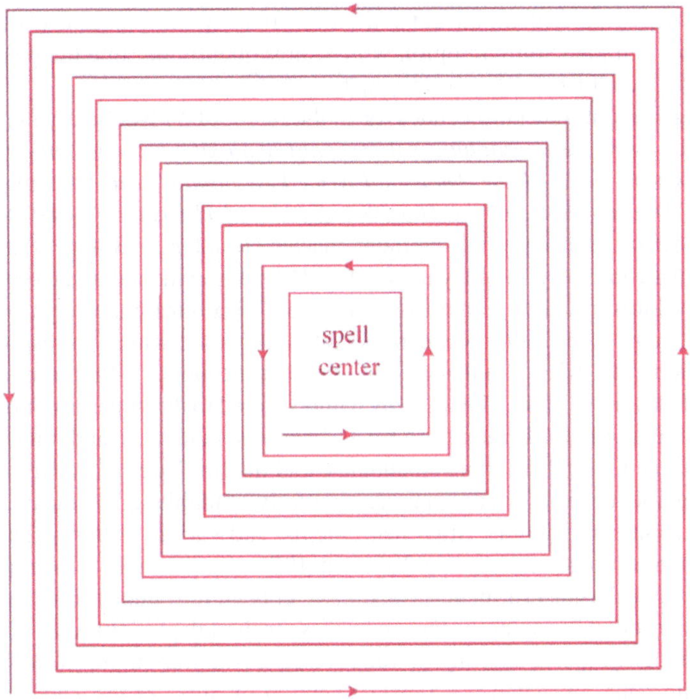

Fig. 4.11 The spell's text arrangement and circular reading direction of the Sanskrit Dharani Sutra unearthed from a Tang Tomb in Xi'an in 1974 (Restored by Pan Jixing)

Sanskrit, a language of the Indo-European linguistic family, is transcribed horizontally from left to right. To fully comprehend this printed incantation, one must read it by continuously altering the angle of perception, allowing for a seamless reading experience. Historical evidence from Chinese Esoteric Buddhism reveals that the custom of wearing talismans and arranging the characters of these spells and their accompanying images into the geometric patterns reminiscent of mandalas was already in vogue by the early Tang dynasty, rather than emerging in the later periods. Vidya-dhārani-yāna Buddhism, which was introduced to China during the latter half of the sixth century, during the waning years of the Liang and the nascent years of the Sui dynasty (550–600), had reached a zenith by the early Tang dynasty. This religious movement was not only propagated by foreign monks but also embraced by native Tang monks who established altars for its promotion, leading to the eastern and western capitals, namely Chang'an and Luoyang, becoming the epicenters of Esoteric Buddhism. From these hubs, the teachings radiated outwards to various regions and neighboring East Asian nations. Among the pioneers in the Tang dynasty for translating and disseminating these esoteric texts and establishing altars for their propagation was the venerable Zhitong (580–645), who during the Sui dynasty's Daye period (605–618), resided in what was then known as Chang'an Great Zen Buddhist Site. His particular focus was on the Dharani (Sojiji) School, a branch of

Esoteric Buddhism. With the advent of the Tang dynasty, this monastery was renamed the Dharani Temple (Sojiji Temple), which subsequently became the pivotal center for Esoteric Buddhism in the western metropolis.[29]

Zhitong's initial translation endeavor in the realm of Vidyadhara-yana scriptures commenced with *Sabacrabhuja-sahasraneta-avalokitevara-dharani-sutra* otherwise known as *Sutra of the Thousand-Armed Avalokitesvara.* Drawing from the preface of *Dharani Sutra of the Thousand-Eyed and Thousand-Armed Bodhisattva Avalokitesvara,* penned by the eminent master (Bhadanta) Bolun of the Fushou Temple during the early Tang Dynasty (660–714), it is evident that the esoteric practices of Buddhism had migrated from India to China during the reign of Emperor Gaozu of Tang, specifically between the years 606–647, which corresponds to the Wude period (618–626). "In that epoch", the text recounts, "a Brahmin monk from the central regions of India, Gotādeva by name, offered to the imperial court a tapestry depicting the divine form, alongside ritual diagrams, mandalas, and mudras."[30] The aforementioned "ritual diagrams, mandalas, and mudras" are indicative of the intricate arrangement of Sanskrit dharani in a circularly-readable square matrix, encircled by the bodhisattva's mudras, ritual instruments, and various patterns. The Sanskrit dharani manuscript discovered in Xi'an in 1974 was a single-page reproduction crafted from such Sanskrit manuscripts that had been introduced from India during the Wude period. Advancing further in history, around the twelfth year of the Zhenguan era under Emperor Taizong of Tang (638), a Brahmin monk from North India named Vajnryana presented the Sanskrit version of *Sutra of the Thousand-Armed Avalokitesvara.* Emperor Taizong decreed that Zhitong collaborate with this monk on the translation, which, upon its completion, was dutifully presented to the imperial court. Zhitong, however, propagated a different version of the translation, one that he had undertaken with the siblings Jati and Bhaga, and this version was subsequently passed down to his disciple Xuanmu. The fact that hundreds of monks dedicated to the chanting of dharani sought instruction from Zhitong is a testament to the extensive circulation and influence of this text.

In the Wuzhou period (690–705), a Brahmin monk from the kingdom of Uddiyana, named Dharma-Candā, disseminated the practice of the thousand-arm technique at the Buddha's Prophecy Temple in Luoyang. As recounted by Bolun, this monk was "an expert of the esoteric dharani spell verses and was frequently commissioned by the imperial court to translate these texts, presenting them along with depictions of a thousand-armed bodhisattva on exquisite felt cloth, as well as the original sutra spells." The divine Empress Wu Zetian, desiring their widespread veneration, decreed that the palace maids embroider these images or have artisans render them in paint, thus ensuring their proliferation across the realm, preserving their sanctity

[29] Lv Jianfu. *History of Esoteric Buddhism in China* [M]. Beijing: China Social Sciences Press, 1995: 162–163.

[30] Bolun. *Preface to Dharani Sutra of the Thousand-Eyed and Thousand-Armed Bodhisattva Avalokitesvara* [M] // *Complete Literature Works of Tang Dynasty: Vol. 913, Book 10.* Beijing: Zhonghua Book Company, 1983: 9511–9522.

for posterity. The iconography and incantations he referred to align with those introduced by an Indian monk during the reign of Emperor Gaozu of Tang in the Wu De period and translated by Zhitong during the Zhenguan period of Emperor Taizong's reign. Zhitong's rendition of the scripture was a two-volume work: the first volume encompassed ninety-four verses of the great body spell, instructions for its reception and application, a dozen types of seal spells, the ritual for constructing a ten-elbow mandala altar, and the method for creating icons; the second volume presented a further thirteen to twenty-five seal spell techniques and the original Sanskrit text of the great body spell, also known as the great dharma body spell. Given the extensive circulation of this scripture in Chang'an during the early Tang period and the substantial demand for it, the decision to engrave it on woodblocks was a natural progression. Consequently, the Sanskrit dharani unearthed in Xi'an is believed to have been extracted from the original Sanskrit text of the great body spell found in the Vidyadhārai-yāna Buddhist scriptures, specifically *Sutra of the Thousand-Armed Avalokitesvara.*

During the second year of the Changshou period in the Wuzhou dynasty (693), the Kashmiri monk Ratnacina (619–721), translated *Mahashvara Dharani Mantra* ("Great Compassionate and Response-Fulfilling Vairocana Dharani Sutra") at the Tiangong Temple in Luoyang. The sacred text delineates that those who aspire to inscribe or bear this incantation must adhere to the Dharma's mandate to construct a mandala, circumscribing the spell around an unadorned square—avoiding the heart of the spell—so that it radiates outward in the form of a mandala, with the inner sanctum of the spell's heart reserved for the transient etchings of one's desires. Once inscribed, these talismans, when suspended from the person, are said to harness the dharani's potent force. For instance, to invoke rainfall, one would depict a nine-headed dragon at the core; for a woman's wish to conceive a male heir, a youthful boy would be sketched therein, and so forth. Subsequently, the periphery of the script is encircled with boundary lines, demarcating the formation of a mandala as a ritual altar, within which are rendered lotus blossoms, vajra clubs, swords, axes, and hand seals. These amulets were to be worn by women around their necks and by men on their right arms, concealed within the coils of a bracelet. The scripture culminates with the promise that "those who, in accordance with the Dharma, inscribe and carry this spell shall know perpetual peace, see their endeavors crowned with success, revel in the joys of this world, and upon death, ascend to the heavens, their karmic impediments dissolved."[31] Hence, it was deemed fitting for both the living and the deceased to carry these talismans. Zhao's discourse singularly alludes to the rendition of *Mahashvara Dharani Mantra* by Amoghavajra (the Chinese name Bukong) in the sixth year of the Qianyuan period (758), seemingly uninformed that this scripture had already been translated during the Wuzhou era, and that analogous esoteric practices had been in circulation long before the advent of the Wuzhou period.

In the initial half of the seventh century, as the Sui dynasty gave way to the Tang, the essence of Esoteric Buddhism was encapsulated in the intricate rituals of Dharma, mandala, and offerings, each governed by a set of stringent protocols. The

[31] *Mahashvara Dharani Mantra* [Z]. Translated by Manicinta. The National Library of China: 7444.

iconography of bodhisattvas during this epoch was characterized by a proliferation of heads, arms, and hands, each bearing an array of ritual instruments, precious beads, lotus blossoms, vajra clubs, tridents, Dharma wheels, and lassos, interwoven with hands sealed in a multitude of mudras. These figures were customarily presented in mandala formations, with the altar practices often entailing the construction of square mandalas. The physical manifestations of these rituals are vividly captured in the Sui dynasty's grottoes in Dunhuang.[32] The Dharma, mandala, and offering practices are not only directly correlated with the Sanskrit incantations unearthed in Xi'an but also serve as the foundational precursors to their emergence. The iconography of bodhisattvas represents a distillation of the Vidyadharani-yana Buddhist scriptures, transposed into the visual realm of art. Similarly, the mantras and mandalas represent a condensation of the scriptural content, elegantly displayed on a singular plane replete with both script and imagery. In these depictions, the presence of the bodhisattva may be implied rather than explicit, with the various ritual instruments they wield suggesting the enduring presence of their dharmakaya. The act of wearing these incantations upon one's person symbolizes a perpetual communion with the bodhisattva and the safeguard of their blessings. This underscores that the discovery of Sanskrit spell prints in Xi'an at the dawn of the Tang dynasty was not a serendipitous event but rather the inevitable fruition of the evolution of Esoteric Buddhism during the transition from the Sui to the Tang era.

The custom of using armlets placed with such mantras as talismans, as chronicled in the Buddhist scriptures from the early Tang dynasty and as immortalized in the sculptural representations of that period, is exemplified in the Grottoes of Guangyuan, Sichuan. Particularly, within Cave 13 of the Lotus Flower Grotto, one can observe a depiction of a bodhisattva embellished with a necklace, armlets, and bracelets, which dates back to the first year of the Wansui Tongtian period of the Wuzhou dynasty (696) or possibly earlier.[33] *Mahashvara Dharani Mantra*, which emerged during the early Tang era, articulates that a bodhisattva would encase the dharani spell within an armlet, thereby safeguarding against all manner of perils and adversities. "He who bears this spell shall recognize that the Tathagata shields him with his divine might, shall discern that such an individual is the very manifestation of the Tathagata, and shall comprehend that this person is endowed with an adamantine physique." As elucidated by the sutra, the act of wearing the spell upon one's person bestows upon the wearer the sanctity of the bodhisattva. This suggests that the tradition of talisman adornment was not a practice that emerged in the mid-to-late Tang dynasty but was, in fact, already in vogue by the early Tang dynasty. Spanning across the echelons of society, from imperial figures and military leaders to the monastic and lay communities, there existed a profound conviction in the potency of the dharani, a belief that engendered a substantial demand for these mystical inscriptions and, consequently, catalyzed their engraving and dissemination.

[32] Lv Jianfu. *History of Esoteric Buddhism in China* [M]. Beijing: China Social Sciences Press, 1995: 189–191.

[33] Lv Jianfu. *History of Esoteric Buddhism in China* [M]. Beijing: China Social Sciences Press, 1995: 196.

The Sanskrit dharani printed paper unearthed from a Tang dynasty tomb in Xi'an, upon meticulous examination by our team, was identified as high-quality hemp paper, distinguished by its pure white color and uniformly fine fibers. The tripartite double-lined border encircling the mantra is indicative of a specific type of mandala altar. As delineated in *The Collection of Dharani Sutras*, translated by the esteemed Atikuta (617–672) during the fourth year of the Yonghui period of Emperor Gaozong of Tang (653), and specifically in Volume 12, various mandala practices were described, detailing altars that ranged from four, eight, twelve to sixteen cubits in length, each surrounded by a graduated number of courts, from single, double, to triple, and even extending to sevenfold courts, with each court gridded. Accordingly, the mantra mandala found in Xi'an adheres to the altar method of a four-cubit altar with triple courts. The term "court" is a translation of the Sanskrit word "ārāma", which traditionally refers to a dwelling enclosed by earthen walls, a concept represented by lines on the manuscript. The selection of the altar method is at the discretion of the dharani master, hence the diversity in the depiction of dharani altars on Tang dynasty manuscripts and printed texts. The quantity of border lines does not indicate a chronological progression. Some, unfamiliar with this nuance, have incorrectly dated the age of the printed dharani based on the presence of single or triple double lines and hand-drawn patterns between them, erroneously pushing back the timeline of the Xi'an discovery.[34] This approach is misguided, as these differences do not reflect the complexity of the technique; the capability to engrave a single set of double lines implies the skill to engrave triple lines, and the ability to engrave text also extends to intricate patterns. These variations are a result of the different altar methods chosen by the dharani masters. Hand-drawn sections are subject to change based on the specific requirements of the dharani bearers, and printed texts can still be embellished with hand-drawn elements within the blank spaces of the mantra heart, without any chronological implication.

As for the inscription "Wu De □ Fu" on the dharani, initially believed by some to resemble the "running and cursive script popular in the early Tang dynasty, akin to the style of the renowned calligrapher Wang Xizhi", Zhao's discourse disputes this, suggesting it lacks the calligraphic essence of the early Tang period. After a comparative analysis with the works of Wang and other scripts from the early Tang dynasty, it became evident that the original scribe was not a virtuoso of calligraphy but rather wrote in haste, without any discernible artistic finesse. From the perspective of printing technology, upon comparing this manuscript with other existing Sanskrit dharani editions from the Tang dynasty, it revealed a more archaic quality in both engraving and ink application, and its Sanskrit script appears to be of an earlier period. On November 20, 1996, the Shaanxi Provincial Cultural Relics Appraisal Committee convened a panel of experts for a collective appraisal of this manuscript, reaffirming its classification as a first-class cultural relic from the early Tang dynasty. Subsequent to this, our team conducted a comprehensive re-examination of this manuscript from multiple perspectives, arriving at the following insights:

---

34 Su Bai. *Woodblock Printing in the Tang and Song Dynasties* [M]. Beijing: Cultural Relics Press, 1999: 8–9.

1. The Sanskrit dharani manuscript on paper, the copper armlet, and Four Deities Ruled Bronze Mirror, all three cultural artifacts were excavated from a single Tang dynasty tomb, with the mirror exhibiting features that date it to the period from the Sui to the early Tang dynasty.
2. The text of the dharani spell is extracted from the original Sanskrit of the Vidyadharani-yana Buddhist scriptures, which were introduced from India during the Wu De and Zhenguan periods in the early Tang dynasty, and the printed characters are in an ancient Sanskrit script that has been in vogue since at least the sixth century.
3. Documentation from Buddhist scriptures and the physical evidence provided by Buddhist iconography demonstrate that the custom of wearing dharani spells and arranging the text in the shape of a mandala square altar, complete with illustrations of hand mudras and ritual instruments, was already in vogue by the early Tang dynasty, rather than emerging in the mid-to-late Tang period.
4. As delineated in *The Collection of Dharani Sutras*, compiled in the fourth year of the Yonghui era during the reign of Emperor Gaozong of Tang (653), the tripartite double-lined border encircling the printed dharani spell confers a specific religious meaning, signifying the selection of a particular altar method.[35]

Therefore, the initial report in 1987, which identified the Sanskrit dharani printed manuscript unearthed from a Tang tomb in Xi'an Diesel Engine Factory in 1974 as a seventh century Tang dynasty print, and the subsequent validation by the Shaanxi Provincial Cultural Relics Appraisal Committee in 1996, are correct. Counterarguments that suggest a mid-to-late Tang dynasty origin for this manuscript are at odds with a wealth of historical facts and thus cannot be substantiated. This manuscript is recognized as the earliest known printed work, with its period of engraving and circulation likely situated between the latter part of Emperor Taizong's Zhenguan era and the Xianqing era of Emperor Gaozong (640–660). The aforementioned activity of Xuanzang in printing the images of the Thousand-Armed Avalokitesvara between the third year of Emperor Gaozong's Xianqing period and the third year of Longshuo period (658–663) aligns temporally with the aforementioned Sanskrit dharani print from 640 to 660, offering a genuine glimpse into the printing activities of the early Tang period. Tracing back to the thirteenth year of Emperor Wen of Sui's Kaihuang period (594), when there was an imperial decree for "discarded statues and left-behind scriptures to be carved", we find ourselves less than half a century removed from the inception of this manuscript. The advent of printing in the early Tang dynasty must have been preceded by an embryonic phase, a period of gestation for the technology, which aligns with the principles of technological evolution, as evidenced by the aforementioned analysis. Therefore, the Ming dynasty scholar Hu Yinglin's theory that "woodblock printing began in the Sui dynasty and flourished in the Tang dynasty" is today reinforced with additional context. It is more prudent

[35] Pan Jixing. A Study of the Sanskrit Dharani Print of the Early Tang Dynasty Discovered in Xi'an in 1974 [J]. *Guangdong Print*, 2000 (6): 56–58. Pan Jixing. A Study of the Sanskrit Dharani Print of the Early Tang Dynasty Discovered in Xi'an in 1974: Continued [J]. *Guangdong Print*, 2001 (1): 63–64.

to attribute the inception of woodblock printing to the Sui dynasty. Other scholars[36] concur with this perspective. Following the early Tang dynasty, a succession of documentary evidence and archaeological findings related to printing emerges, indicating that the craft had indeed entered its nascent phase of development, a fact that need not be reiterated when discussing its origins.

4.2.3 Early Development of Woodblock Printing in Tang Dynasty

In the early Tang dynasty, succession passed from Emperor Taizong to Emperor Gaozong, and eventually to Wu Zetian, who ascended to the throne and proclaimed herself empress, thus revolutionizing the Tang mandate by renaming the dynasty to Zhou, a period historically known as the Wuzhou era (690–705). Wu Zetian, a fervent adherent of Buddhist teachings, adeptly utilized Buddhist parables to legitimize her imperial claim. During the summer of the second year of the Tianshou period (691), she decreed an order that exalted Buddhism over Taoism, effectively designating it as the preeminent state religion. Her reign witnessed the invitation of erudite monks from regions as diverse as Central Asia, West Asia, and India, who were summoned to the Tang courts to partake in the translation of sacred Buddhist texts in the imperial translation bureaus established in the dual capitals. This initiative not only fostered the propagation of Buddhist teachings but also catalyzed the advancement of printing technologies within the realm. Among those who enjoyed Wu Zetian's patronage was the esteemed Dharma Master Fazang, a native of Chang'an and a lay follower of the surname Kang. Fazang, who demonstrated a profound understanding of printing techniques, left behind detailed chronicles of the craft. His academic journey began in the fourth year of the Xianqing era (659) under the tutelage of Zhi Yan at the Yunhua Temple, where he studied *Avatamsaka Sutra.* In the first year of the Xianheng era (670), Wu Zetian converted her residence into the Taiyuan Temple, ordaining a new generation of monks and commissioning Fazang to translate Buddhist scriptures. As the architect of the theoretical framework for the Huayan School of Buddhism, Fazang earned the reverential title of "Xianshou Jieshi" from the empress and was posthumously awarded the honorific "Honglu Qing", a distinction reserved for the fourth rank of nobility. Between the first year of the Zhengsheng era and the second year of the Shengli era (695–699) during the Wuzhou period, Fazang dedicated himself to the compilation of the 80-roll *Avatamsaka Sutra* and immersed himself in the profound study of its esoteric teachings.

Before the advent of the new translation of *Avatamsaka Sutra*, the 60-roll version translated in the year of 421 was utilized. This scripture is divided into eight assemblies, each recounting the Buddha's discourses to his disciples during eight distinct

[36] Xiao Dongfa. *History of Book Publishing and Printing in China* [M]. Beijing: Peking University Press, 2001: 41.

dharma gatherings following his attainment of enlightenment. There exists a diversity of interpretations regarding the formation of these assemblies. The Tiantai sect suggests that the assemblies are bifurcated, with the initial seven encompassing the teachings delivered over the first three seven-day periods post-enlightenment, and the eighth assembly presenting the teachings beyond that timeframe. In contrast, the Huayan theorist Fazang dissented from this perspective, asserting that the Buddha remained silent during the initial seven days after his enlightenment, only to articulate the complete Buddhist doctrine in the second seven-day period. He argued that despite the sequential arrangement of the sutras, all of the Buddha's teachings were comprehended instantaneously at the moment of enlightenment and were fully articulated during the second seven-day period, with the eighth assembly's teachings not being expounded subsequently. To substantiate this perspective, Fazang elaborated extensively in his text *Essay on the Five Teachings of Hua-yen* (ca. 677). As for the compilation of this work, the Japanese Huayan scholar Shōei Yuki posits in his *Lectures on the Five Teachings of Hua-yen* (part of *Lectures on Tripitaka*) that the manuscript was composed by Fazang around the age of thirty, coinciding with the second year of the Yifeng era under Emperor Gaozong (677). After elucidating his stance, Fazang employed the metaphor of printing to demonstrate:

Hence, drawing from this universal auditory comprehension, it is proclaimed that the entirety of the Buddha's teachings were imparted at a single moment, encompassing both sequential exposition and simultaneous revelation, akin to the earthly technique of printing, where the text's meaning unfolds progressively when read, yet all characters emerge in an instant upon the press. The concepts of simultaneity and sequence are not in conflict; one must apprehend that the underlying truth in this scenario is, indeed, harmonious.[37]

In Fazang's perspective, *Avatamsaka Sutra* contains eight assemblies, the order of which is sequential, yet all the teachings of the Buddha were realized simultaneously upon his enlightenment, much like the characters on a printed page that, despite their ordered sequence, are all imprinted at once. Thus, the concepts of "sequence" and "simultaneity" are not contradictory but dialectically unified. Fazang used the metaphor of printing to explain the mechanism behind the formation of the eight assemblies in *Avatamsaka Sutra* in 677, indicating that the printing technique was quite widespread in China by this time. After the new 80-roll translation of *Avatamsaka Sutra* was completed, Fazang revisited this issue in his *Exploration of the Profound Meaning of the Avatasaka Sutra*, where the Japanese scholar Matsubara Kyūryō, in his *Explanatory Dictionary of Buddhist Books,* suggests that Fazang wrote this work between the ages of 45 and 50, around the third to the first year of the Chuigong era of the Wuzhou dynasty (687–692). In volume 2 of this text, he stated:

The concept of capturing the sequence has three levels of depth: firstly, at the time of these two seven-day periods, it encompasses the eight assemblies, spoken all at

---

[37] Fazang. *Essay on the Five Teachings of Hua-yen: Vol. 1* [M] // *Taisho-pitaka: Vol. 42*. Tokyo: Daizo Shuppansha, 1924: 482.

once. If that's the case, why is there a sense of sequence? The answer is: like printed text, which is read in sequence but printed on the paper all at once.[38]

In an article initially referenced by the Japanese printing historian, Mr. Kōichi Goda, titled *On the Origin of Printing in China* and published in the *Proceedings of the Japanese Academy* in Volume 34, Issue 2, dating back to 1975, there exists a historical account that unveils the implementation of printing techniques for the issuance of official documents during the early years of the Wuzhou dynasty. According to the chronicles of Liu Su, a notable figure from the Tang dynasty who lived between 770 and 830, as documented in his work *New Anecdotes of Tang Dynasty*, completed in the year of 807, it was during the nascent phase of Empress Wu Zetian's reign, specifically in September of the Tianshou period's second year (691), that Zhang Jiafu, a golden figure within the Phoenix Pavilion, orchestrated a movement. This movement saw the involvement of Wang Qingzhi, a native of Luoyang, and a collective of others who submitted a joint petition advocating for the appointment of Wu Zetian's nephew, Wu Chengsi, to the esteemed role of Crown Prince. Their aim was to dethrone Wu Zetian's own offspring, Li Dan, in a bid to establish the Wu family's dominion over the political landscape. Despite the fervor of their pleas, Empress Wu Zetian chose not to entertain their proposal. Wang Qingzhi, in his desperation, is said to have prostrated himself before the Empress, offering his life in his fervent request. However, Wu Zetian, determined to dismiss him, resorted to employing an inner palace seal to imprint a message upon a sheet of paper, remarking, "Take this and present it to the gatekeepers when you wish to see me, and they shall grant you passage." With this, Wang Qingzhi, now in possession of the paper, moved with impunity, his comings and goings becoming a common and irritating sight to Empress Wu, eventually incurring her displeasure. This episode, along with others, was similarly chronicled in *Zizhi Tongjian*, specifically within its 204th volume, under the "Tang Records, Chapter Twenty". These records are all derived from eyewitness accounts of the happenings within the imperial palace. The term "inner seal printing paper", as used during the Tang dynasty, pertains to the passes that were printed on Assortedpaper by the imperial court for the purpose of granting access to and from the palace. The practice of utilizing "printed paper" within the officialdom of the Tang dynasty was indicative of the specialized printed materials used for governmental affairs. For instance, *Old Book of Tang,* under the section *Records of Food and Goods*, documents the use of printed paper as tax vouchers as early as the fourth year of the Jianzhong period (783). This system of employing printed paper continued through the subsequent Song dynasty.

Historical records indicate that during Empress Wu Zetian's reign (690–705), she created and promulgated 18 unique characters, collectively known as the "Wuzhou System of Characters". Subsequently, upon his ascension to the throne, Emperor Zhongzong of the Tang dynasty, Li Xian, in the inaugural year of the Shenlong period (705), decreed the abolition of these characters, alongside other reforms imposed by Wu Zetian, reinstating the systemic structure that had been in place during the

---

[38] Fazang. *Explanatory Notes on the Mahāvaipulya Buddhāvataṃsaka Sūtra: Vol. 2* [M] // *Taisho-pitaka: Vol. 35*. Tokyo: Daizo Shuppansha, 1926: 127.

Yongchun period of Emperor Gaozong's reign (682). Thus, the identification of Wuzhou's characters within the manuscripts and printed editions of the Tang dynasty has become a pivotal criterion for discerning the authenticity of artifacts from the Wuzhou period.

In October 1966, a small scroll containing the esoteric scripture *Aryarasm-vimalvi-suddha-prabha-nama-dharani-sūtra* was discovered in the Shakyamuni Pagoda of Bulguksa Buddhist Temple in Gyeongju, South Korea. It is printed on 12 sheets of yellow parchment paper; each sheet is 6.5 cm in height and 52.5–54.7 cm in width, with a total length of 643 cm. The frame height of each plate is 5.4 cm, with varying horizontal lengths (35.2–57.3 cm). Each plate has 55–63 lines, 7–9 characters per line, most of which are 8 characters. They were engraved in Tang dynasty regular script, with a diameter of 4–5 mm per character, equivalent to the size of today's 15.6-point Song typeface. The strokes of each character are vigorous, with a neat cutting technique, and there are cutting marks on the characters (Fig. 4.12). There is no inscription, and the scripture contains a mixture of characters from the Wu and Zhou dynasties, as well as a large number of variant characters used before the Song dynasty. There exist differences between the scripture and the Song and Goryeo (936–1391) editions, indicating that it is a carved edition from the Wu and Zhou dynasties. But the specific publication year and location need to be verified. It is worth noting that Gyeongju, once the capital of Silla (688–936) and historically referred to as Geumseong, was inextricably linked to the Tang Dynasty. The Bulguksa Temple was erected in the 10th year of the Tianbao era of Emperor Xuanzong of Tang and the 50th year of the reign of King Gyeongdeok of Silla (751), through the initiative of the Silla chancellor Kim Daeseong (700–774), who sought craftsmen and technicians from the Tang dynasty.

As chronicled in *Kaiyuan Catalog of Buddhist Scriptures* (730) by the eminent monk Zhi Sheng (695–750) in the Tang dynasty,

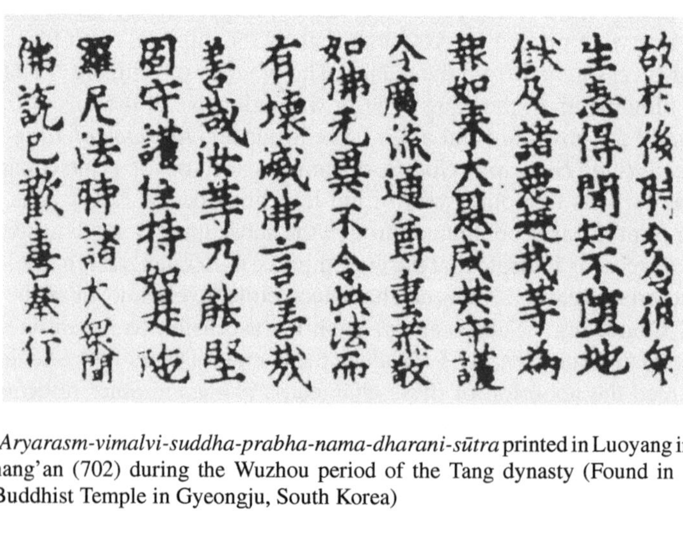

Fig. 4.12 *Aryarasm-vimalvi-suddha-prabha-nama-dharani-sūtra* printed in Luoyang in the second year of Chang'an (702) during the Wuzhou period of the Tang dynasty (Found in 1966 in the Bulguksa Buddhist Temple in Gyeongju, South Korea)

Aryarasm-vimalvi-suddha-prabha-nama-dharani-sūtra (hereinafter referred to as *Wugou Jing*) was translated by Fazang, a Dharma master of the Tang dynasty, and the Central Asian monk Mitrasanda from the region of Tukhāra during the waning years of the Wu Zhou period, a period denoted as the "last years of the Heavenly Empress".[39] This era lasted for 15 years (690–705), and its "last years" is obviously the final five years, i.e. the Chang'an period (700–705). During the interlude from the second year of the Sheng Li period to the third year of the Chang'an period (699–703), Empress Wu, beset by the infirmities of old age and frequently confined to her bed, sought fervently to alleviate her maladies and extend her lifespan. It was in this context that she decreed the translation of a corpus of esoteric Buddhist scriptures. In the years 699–700, the monk Siksananda (652–712) from Khotan initially rendered *The Great Dharani Sutra of Immaculate and Pure Light* into Chinese, a text that proclaimed the merits of reciting its mantras or sutras repeatedly, or of constructing Buddhist pagodas in veneration of the text, as means to augment fortune and lifespan. This was precisely what the empress was in quest of, yet she remained dissatisfied, suspecting that the translation was derived from an antiquated Sanskrit manuscript that had been lost to time. Consequently, she commanded Fazang to undertake a new translation, which came to be known as *Wugou Jing*. The timing of this translation was constrained to the period subsequent to Siksananda's version and prior to Mitrasanda's departure from China, that is, sometime after 700 and before 702, discounting any other years. *Kaiyuan Catalog of Buddhist Scriptures* further elucidates that Mitrasanda, upon the completion of *Wugou Jing*, took his leave of the imperial court to return to his homeland. Given that his return to Tukhāra occurred in 702, it allows us to infer that the scripture was translated in the year 701, with the translation work being carried out at the Scripture Translation Academy of the Buddha's Prophecy Monastery in Luoyang. Empress Wu, upon the scripture's presentation, was reportedly delighted and bestowed her commendation upon the translators. It can be inferred that the scripture was published in the year subsequent to the bestowment of this honor, which corresponds to 702. We have delved into a meticulous scholarly investigation,[40] and thus further discussion is unnecessary.

Some have misunderstood the term "the last years of the Heavenly Empress" as the final year of the Wuzhou period, suggesting that the publication date of the text was between 706 and 751. This interpretation is erroneous. By that time, Mitrasanda had already returned to his homeland of Tukhāra, making it impossible for him to have collaborated with Fazang in translating the sutra in Luoyang. Moreover, Empress Wu passed away in 705, and by 706–751, the Wuzhou system of characters had been abolished in the Tang empire. It is also unlikely that Silla would have continued to use

[39] Zhisheng. *Kaiyuan Catalog of Buddhist Scriptures: Vol. 9* [M] // *Taisho-pitaka: Vol. 55*. Tokyo: Daizo Shuppansha, 1928: 567.

[40] Pan Jixing. On the Discovery of Aryarasm-vimalvi-suddha-prabha-nama-dharani-sūtra in Korea [J]. *Chinese Science Bulletin*, 1997, 42(10): 1009–1028. Pan Jixing. On Chinese Paper Making in the Century and Publication of Aryarasm-vimalvi-suddha-prabha-nama-dharani-sūtra [J]. *Yellow River Culture Forum*, 2001 (6): 43–66. PAN J X. On the Origin of Printing in the Light of New Archaeological Discoveries [J]. *Chinese Science Bulletin*, 1997, 42(12): 976–981.

Wugou Jing at that time, so how could the Wuzhou system of characters still exist? The initial translation and publication of this sutra were intended to accumulate merit and blessings for Empress Wu, harnessing the power of the dharani to alleviate her illnesses and extend her life. After her death, such efforts were no longer necessary; it only made sense to do so in 702 when she was alive and suffering from illness. A comprehensive comparison of the printing style of this sutra with that of manuscripts and editions from the Tang dynasty discovered in the Mogao Grottoes of Dunhuang reveals identical fonts, all being the popular regular script of the Tang period, with the same variant characters appearing in both. The format of printing blocks and binding style of this edition are also consistent with those of Tang manuscripts and printed editions. The Japanese edition of *Million Pagodas Dharani* (770) was based on a Tang edition, which is similar to the one found in Gyeongju. The intermingling of Wuzhou characters with standard characters in this edition is also observed in other Buddhist scriptures from the Tang period. The sutra was printed on paper dyed yellow with berberine, which was fashionable in the Tang dynasty, regardless of whether it was hemp or mulberry paper. All of these indicate that *Wugou Jing* was first engraved in the Tang dynasty of China in 702 and later transmitted to Silla. There are no records of printing in Silla, nor are there any other surviving printed editions; it was not until the Goryeo dynasty that printing activities took place in the Korean Peninsula. The discovery of *Wugou Jing* in South Korea reflects one aspect of the Buddhist cultural exchange between the Tang dynasty and Silla.

In 1975, a single-page printed edition of *Buddha Speaks Mahashvara Dharani Mantra* (Fig. 4.13) in Chinese was discovered from a Tang tomb (Fig. 4.13) in the Metallurgical Machinery Factory in the western suburbs of Xi'an. When unearthed, it was placed in a small bronze box, agglutinated into a ball. When unfolded, it revealed itself to be rectangular, measuring 35 cm × 35 cm, printed on hemp paper. There is a square frame in the center (spell heart, 5.3 cm × 4.6 cm) with a colored image of two figures, one standing and one kneeling. The outer perimeter of the frame is surrounded by dharani scripts, with 18 lines on each side for a total of 72 lines, and there are boundaries between the lines. This sutra was translated by the Buddhist master Ratnacina in the second year of the Changshou era (693) in the Tiangong Temple in Luoyang. Its typographical layout bears a striking resemblance to another printed Sanskrit dharani unearthed in Xi'an in 1974, a fact that testifies to the enduring legacy of esoteric Buddhist practices, despite the transition of the text from its original Sanskrit to Chinese.

In 1944, an archeological discovery was made in a Tang dynasty tomb near the Wangjianglou outside the eastern gate of Chengdu, Sichuan, where a single printed Sanskrit dharani page was unearthed. This artifact, measuring 31 cm by 34 cm and housed within a silver armlet that adorned the deceased, was found in a square shape. At the heart of the page, within a square frame, an engraving depicted the six-armed Avalokitesvara seated on a lotus throne, each arm wielding a distinct religious instrument. Surrounding this central imagery, the dharani text was inscribed in a circular pattern, with 17 lines on each side. Beyond the sacred text, figures of various bodhisattvas adorned the periphery, though the print had fallen into disrepair upon being flattened for examination (Fig. 4.14). The presence of Avalokitesvara within

Fig. 4.13 A single-page print of *Buddha Speaks Mahashvara Dharani Mantra* in the flourishing period of the Tang dynasty (713–766), unearthed in Xi'an in 1975[41]

the heart of the dharani indicated the fulfillment of all devotees' prayers, rendering further detailed illustrations unnecessary. This shift suggests that this printed edition postdates the aforementioned Sanskrit dharani scriptures and *Great Dharani Sutra of the Greatly Victorious*. The deceased in the tome was found with two Kaiyuan copper coins in mouth, with the character "Yi" on the reverse, indicating minting in the Yizhou Prefecture. Additionally, the deceased held a Kaiyuan coin and a jade rod in each hand, and four pieces of pottery were also discovered within the tomb.

[41] Han Baoquan. *The World's Earliest Printed Mantra: A Dharani Mantra Print Unearthed from a Tang Tomb in Xi'an* [M] // Shi Xingbang. *The Collected Works of Chinese Archaeology: in Commemoration of the Fiftieth Anniversary of Mr. Xia Nai's Archaeology*. Xi'an: Sanqin Publishing House, 1987: 404–410.

Fig. 4.14 A sanskrit dharani print from the late flourishing period of Tang dynasty (757–766), unearthed in Chengdu in 1944

Within the border of the dharani's frame on the right side, there was a line of Chinese characters that read "□□□ Chengdu County □ Dragon Pond Lane… near Bian □□ dharani scripts sold in …", with some characters missing, and no date inscribed. The first three characters before "Chengdu County" were fragmented, leading the original report's author to speculate "Chengdu Prefecture", suggesting that the copper coins were minted during the Huichang years of Emperor Wu Zong (841–848), thereby designating this tomb and its contents as relics from the post-Huichang era (ninth century).[42]

The above-mentioned opinions on the dating of Sanskrit scriptures and mantras unearthed in Chengdu had been widely cited by many people before the discovery of scriptures and mantras in Xi'an, becoming the standard opinion for dating such printed materials. All other Tang dynasty prints with similar shapes were pushed back to the mid to late Tang dynasty. However, upon closer examination, these dating assumptions require a more nuanced evaluation. The tomb's brickwork, characterized by a "three horizontal and one vertical" pattern, was a construction technique prevalent in the middle and lower reaches of the Yangtze River during the Northern and Southern dynasties. This tomb belongs to the early brick tomb in Sichuan. The

[42] Feng Hanji. Records of the Discovery of the Tang Dynasty Print of Dharani Mantra [J]. *Cultural Relics Reference Materials*, 1957 (5): 48–51.

pottery was fired in the Tang kiln of Chengdu Liuli Factory and can be traced back to the prosperous Tang dynasty. The printing paper has been tested to be semi-transparent and tough leather paper, which was mostly produced in the first half of the Tang dynasty and was rare in the late Tang dynasty. The arrangement of the Sanskrit dharani in a square mandala form, worn within an armlet, is documented in early Tang dynasty Buddhist scriptures and is consistent with the Sanskrit dharani printed editions found in Xi'an tombs, indicating that it is too late to date this tomb and its printing to the end of the Tang dynasty in the ninth century.

The original dating primarily relied on copper coins. As recorded in *The Chronicles of Food, Silks Clothing or Currency* in *Old Book of Tang*, Emperor Gaozu initiated the casting of Kaiyuan Tongbao in Luo, Bing, You, and Yi states during the fourth year of the Wude era (621), with subsequent remoldings taking place on multiple occasions. The same source, in its "Treatise on Geography", notes that Chengdu was established as a county in the second year of the Chuigong era (686), under the jurisdiction of Yizhou Prefecture. In the twenty-first year of the Kaiyuan era (733), the nation was divided into 50 provinces, with Sichuan belonging to the Jiannan Province, governed from Yizhou. During the first year of the Tianbao era under Emperor Xuanzong (742), Yizhou was renamed Shu Prefecture, and during the fifteenth year (756), to evade the An Lushan Rebellion, the emperor resided in Shu, making it a secondary capital. In 756, Emperor Suzong ascended to the throne, initiating the Zhide era, and Emperor Xuanzong abdicated. In the second year of the Zhide era (757), Suzong renamed Shu Prefecture to Chengdu Prefecture. Excavations of various tombs in Sichuan have indicated that the copper coins found were mostly in common use at the time, suggesting that the coins found in this tomb, minted between 713 and 741, must predate 848 at the latest. Considering the architectural style of the brick tomb, the shape of the pottery, the paper used for printing, the form of the scriptures, the timing of the coinage, and the administrative divisions, it can be deduced that the tomb and the printed artifacts within date to the first half of the eighth century, no later than the late prosperous Tang period (757–766). The previous opinion that dated it to the late Tang in the ninth century requires revision.

The mid-Tang dynasty, a period of opulence often referred to as the Kaiyuan era, should have been a prolific time for the emergence of numerous printed works. However, the widespread destruction of such materials, particularly Buddhist scriptures, ensued following the anti-Buddhist persecutions instigated by Emperor Wu Zong in 845. The majority of early printed works were, in fact, religious in nature. During the 1960s and 1970s, additional single-sheet printed texts in both Sanskrit and Chinese from the Tang dynasty were discovered, but they have not been the subject of rigorous academic scrutiny and thus are not discussed in detail here. It is important to highlight that, alongside religious publications, the Tang dynasty also saw the proliferation of secular printed materials serving a variety of purposes. According to *The Chronicles of Food, Silks Clothing or Currency* in *Old Book of Tang*, in the fourth year of the Jianzhong era under Emperor Dezong (783), the Ministry of Revenue, in an effort to ease the financial strain on the nation, petitioned and received approval to levy two additional forms of taxation: a housing tax and a business profit tax. The document states, "The market offices were tasked with providing printed paper", referring to tax receipts that were pre-printed with categories for tax items, revenue amounts, tax amounts, and the names of taxpayers, to be filled out by officials and stamped with an official seal as proof of tax compliance. Individuals who defaulted on their taxes faced hefty fines and physical punishment. The implementation of this policy was met with considerable discontent and was eventually abolished after a mere two years.[43] This episode illustrates the utilization of printed materials in the state's fiscal management during that era. Later, during the Yuanhe era of Emperor Xianzong (806–820), promissory notes known as *feiqian*[44] should have also been printed, given their widespread use across various provinces with the stipulation that they were "redeemed upon presentation of the voucher". The necessity for a standardized format, coupled with the substantial volume in circulation, necessitated the use of "printed paper" for such transactions.

In the late Tang dynasty, there was a diversification of printed materials, with Chinese language reference books such as Buddhist scriptures, dictionaries, phonetics, fortune telling books, and almanacs being published successively for the general public. However, the almanacs, sanctioned by the Ministry of Rites and circulated across the realm, were often privately printed for profit. This led to an imperial edict in the ninth year of the Taihe era (835) during Emperor Wenzong's reign, which proclaimed that "no local governments should privately establish plates for the printing of almanacs".[45] Cities like Chengdu, Huainan, and Yangzhou became epicenters for private printing of these calendars, which were then peddled by merchants far and wide, such that by the time the official calendars were distributed in the twelfth month, the privately printed versions had already saturated the market. The British Library possesses a fragmentary page of an almanac from the fourth year of the Qianfu era (877) during the reign of Emperor Xizong of Tang, a piece that exemplifies the exquisite printing technique at the time (Fig. 4.15). The layout is intricate, with a comprehensive array of charts and diagrams. Delicate lines of text are separated by both vertical and horizontal rulings. The extant portion of the almanac spans from April to August, and aside from the calendar dates and solar terms, it also includes fortune-telling *Shier Xiangshu Zaiji Fa* ("The Twelve Zodiacal Afflictions Method"), illustrations of the twelve zodiac animals, and geomantic diagrams for *Wuxing Anzhi Menhu Jingzao Tu* ("The Placement of Doors, Wells, and Kitchens According to the Five Surnames"). Additionally, it features *Gongnan Gongnv Tui Younian Bagua Fa* ("The Method of the Court Ladies and Eunuchs for Promoting the Annual Excursion of the Bagua"), which also concerns fortune-telling. The depth and breadth of information contained are such that they nearly rival the comprehensiveness of almanacs from the Qing dynasty.

---

[43] Liu Xu. *Old Book of Tang: Vol. 49, Chronicle of Geography* II [M] // *Twenty-five Histories: Book 5*. Compact edition. Shanghai: Shanghai Classics Publishing House, 1986: 254–255.

[44] Liu Xu: *Old Book of Tang: Vol. 19 The Chronicles of Food, Silks Clothing or Currency* [M] / / *Twenty-five Histories: Book 6*. Compact edition. Shanghai: Shanghai Classics Publishing House, 1986: 152.

[45] Liu Xu: *Old Tang Book: Vol. 17, Zong Ji* [M] // *Twenty-five Histories: Book 5*. Compact edition. Shanghai: Shanghai Classics Publishing House, 1986: 76.

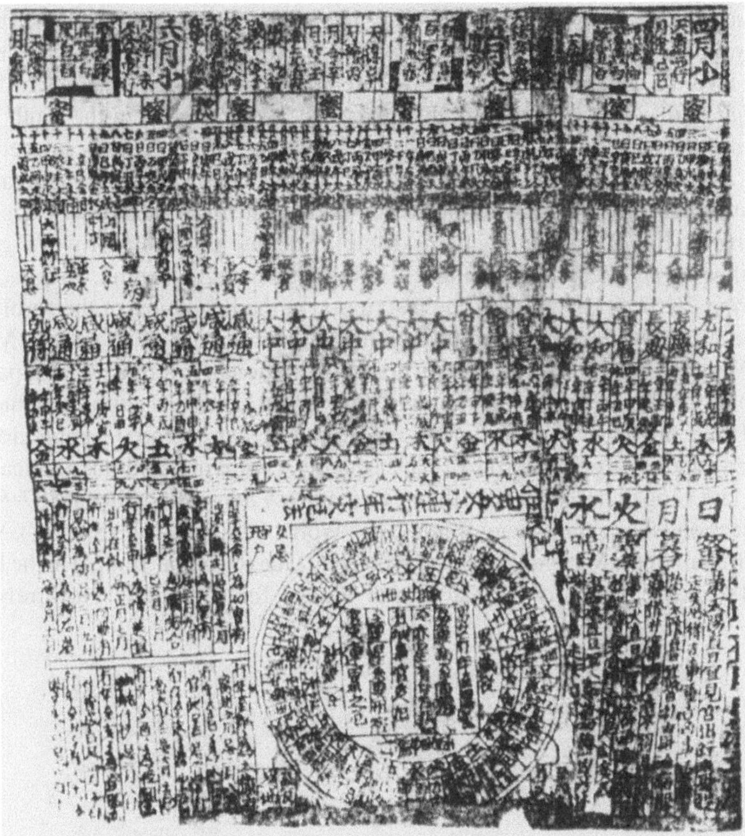

Fig. 4.15 A calendar published in the fourth year of the Qianfu period (877) of the Tang dynasty, found in a stone chamber at Dunhuang

Wang Dang (1075–1145) remarked: "When Emperor Xizong took refuge in Shu, the official calendar from the Imperial Astronomical Bureau did not reach Jiangdong, yet there were privately printed versions for sale in the market. Each version had discrepancies in the dates of the new and full moons, leading to disputes among the sellers."[46] This indicates that in January of the first year of the Zhonghe era (881), as Emperor Xizong fled from Chang'an to Chengdu, private individuals took the opportunity to print calendars. There were private calendars in both Sichuan and Jiangdong (the eastern road of Jiangnan), which were beyond the reach of the official calendars. Jiangdong refers to the areas of present-day Jiangsu, Zhejiang, Fujian, and Taiwan. The printing of calendars was concentrated in Yangzhou, Suzhou, Hangzhou, and Yuezhou (present-day Shaoxing). The methods of calculating the private calendars varied, leading to differences in the dates of the new and full moons,

[46] Wang Dang. *Tangyulin* ("Collection of Historical and Literary Works from the Tang Dynasty"): *Vol. 7* [M]. Shanghai: Shanghai Cultural Relics Press, 1978: 256.

and consequently, disputes arose. During the reign of Emperor Yi of the previous dynasty, in the Xiantong era (860–873), the printing technique had already reached a high level of sophistication. In 1907, Stein discovered a complete scroll of the *Diamond Sutra* (Vajracchedikā-prajña-pā-ramita sūtra) printed in the ninth year of Xiantong (868) in the Mogao Grottoes of Dunhuang, which is now housed in the British Museum. This sutra, measuring 525 cm in length and composed of seven sheets of paper connected end-to-end, features exquisite illustrations at the beginning, depicting the Buddha preaching to his disciple Subhuti in the Jeta Grove (Fig. 4.16). The following six sheets contain the text of the sutra in regular script, with each sheet measuring 26.67 cm by 75 cm. The inscription at the beginning of the scroll reads, "On the fifteenth day of the fourth month of the ninth year of Xiantong (May 11, 868), Wang Jie made this sutra for universal distribution in reverence for his parents." In October 1982, upon my observation of this sutra in London, I confirmed that it was printed on hemp paper, with a smooth surface and a cream-colored complexion. This sutra is a superior work in terms of paper quality, engraving, and ink application.

The late Tang dynasty literati Sikong Tu (courtesy name Biao Sheng, 837–908), in his collection *Sikong Biaosheng Wenji* ("The Works of Sikong Biao Sheng"), Volume 9, "On Behalf of the Vinaya Preaching Monk Hui Que of Jing'ai Temple in the Eastern Capital to Raise Funds for the Carving of Vinaya Texts", stated: "Now, relying on

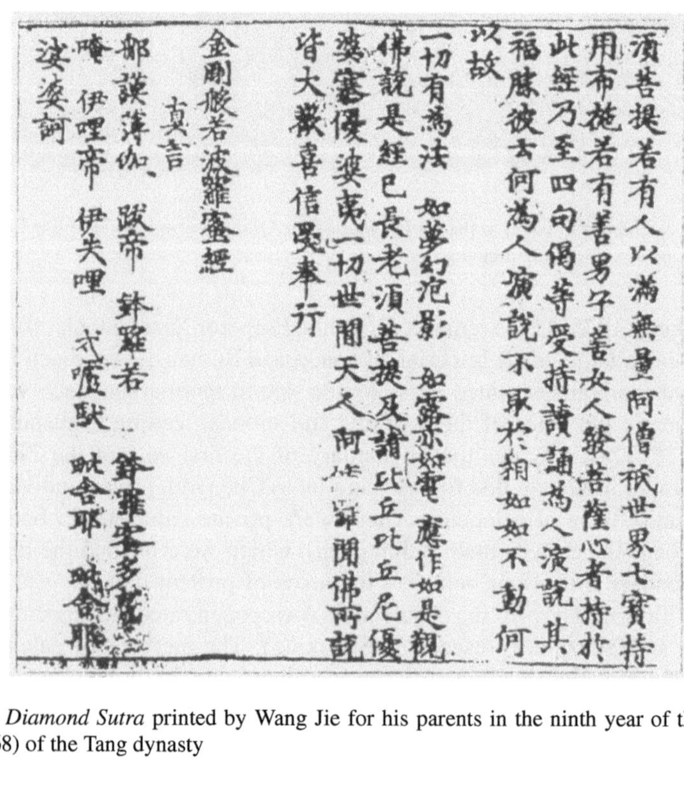
須菩提若有人以滿無量阿僧祇世界七寶持
用布施若有善男子善女人發菩薩心者持於
此經乃至四句偈等受持讀誦為人演說其
福勝彼云何為人演說不取於相如如不動何
以故
一切有為法 如夢幻泡影 如露亦如電 應作如是觀
佛說是經已長老須菩提及諸比丘比丘尼優
婆塞優婆夷一切世間天人阿修羅聞佛所說
皆大歡喜信受奉行
金剛般若波羅蜜經
真言
那謨薄伽跋帝 鉢羅若 鉢羅蜜多曳
唵 伊哩帝 伊失哩 式嚧馱 毗舍耶 毗舍耶
娑婆訶

Fig. 4.16 *Diamond Sutra* printed by Wang Jie for his parents in the ninth year of the Xiantong period (868) of the Tang dynasty the old 'Sunlight' texts, the grand upholding of the dragon elephants, and the grand feasts of the capital temples... ever since Luoyang was deprived of timely exchanges, the Chu editions gradually feared loss, and wished to engrave again." Here, "old 'Sunlight' texts" refers to *Sifen Lvshu* ("Commentary on the Four-Part Vinaya") by the monk Fale of the Sunlight Temple in Xiangzhou (including Anyang, Henan) from the early Tang dynasty (569–635). From Sikong Tu's account, it is understood that before the anti-Buddhist movement in 845, this *Sifen Lvshu* had already been printed, and was taught by the monk Hui Que of Jing'ai Temple in Luoyang. In 845, this edition was burned, and the temple was destroyed. After Emperor Xuanzong of Tang (846–859), the ban on Buddhism was lifted, and Hui Que entrusted Sikong Tu to write a fundraising letter to raise funds for re-engraving. The fundraising letter was probably written in the first year of the Qianfu era under Emperor Xizong (874), noting that "the printed edition has a total of eight hundred sheets", which means that 800 copies of the leaflet were printed and widely distributed. A late 9th-century Tang edition of *All Tathagatas Supreme and Victorious Buddha Crown Dharani* is now housed in the National Library of France (Fig. 4.17). At the same time, He Ganzhong (817–884), the Observer of Jiangxi, had studied the art of alchemy for many years and published thousands of copies of *Liu Hong Zhuan* ("Biography of Liu Hong") during the Dazhong era (847–859), sending them to fellow practitioners, as recorded in Fan Shu's (840–912) *Yunxi Youyi* (Yunxi's Discourse with Literati Companions, ca. 876), Volume Two.

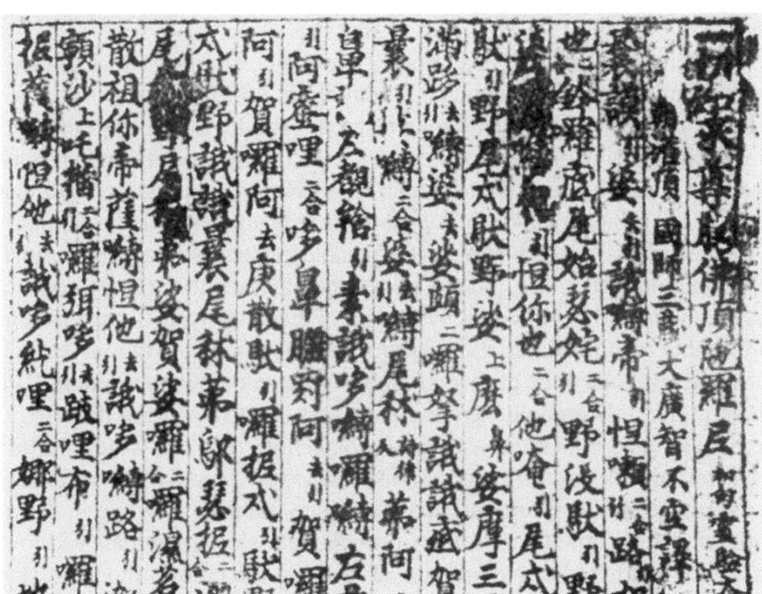

Fig. 4.17 A nineth-century Tang inscription of *All Tathagatas Supreme and Victorious Buddha Crown Dharani* found in Dunhuang

In the Xiantong period, Chinese language reference books such as *Tangyun* and *Yupian* (China's first regular script dictionary) were also published in Chengdu, Sichuan, and found their way to Japan. In the third year of Xiantong (862), the Japanese scholarly monk Soei came to China, and after a three-year sojourn (865), he returned with the ship of Tang merchant Li Yanxiao and brought back many Chinese books. He listed the following books in his *Catalog of Writing Invitations to Buddhist Schools*:

...Xichuan Yinzi (The Western Sichuan imprint) features *Tang Yun*, a comprehensive work with five volumes, and *Yu Pian* with thirty volumes. Such a diverse array of texts, while not belonging to the sacred Buddhist canon, are nevertheless deemed essential by the scholars and literati of the secular realm.[47]

"Xichuan Yinzi" refers to the Sichuan imprints, and these works were procured by the Japanese scholarly monk Enryaku during his sojourn in Chang'an in the year 865, who entrusted the monk Soei to take them back. In the same year, they were stored in the Todai-ji Temple in Nara.

4.3 Woodblock Printing After Five Dynasties and Song Dynasty

4.3.1 Printing in Five Dynasties and Ten Kingdoms

In the era spanning from the inception of woodblock printing technology in the Sui dynasty during the sixth century to the twilight of the Tang dynasty, the established supremacy of handwritten books faced a formidable challenge posed by the emerging technology of printed books. Despite the dominance of handwritten books, printed books gradually carved out a niche in the market, gaining momentum among consumers. This growing accessibility was primarily attributed to the economic efficiency of mechanical reproduction, which made printed books a more financially viable option for the populace, thereby positioning them as the primary consumers of this burgeoning medium. However, by the end of the Tang dynasty, the enticement of these affordable reading materials had proven irresistible to the literati and scholars, who subsequently joined the growing number of printed book consumers. This shift in the market not only emboldened the publishers but also spurred them to produce a range of reference books that catered to the scholarly class, which, in turn, met with considerable commercial success. The monopoly held by the authorities over almanac printing was further contested by private publishers, while in the domain of traditional, hand-copied books, the Confucian classics alone managed to uphold their exclusive and hereditary dominion, at least for the present moment. However, the wave of innovation could not be halted, and by the ninth century, China's book

[47] Kimiya Thai-yan. *History of Japan-China Cultural Exchange*. Translated by Anthony Woo (Hu Xinian). The Commercial Press, 1980: 202.

industry had embarked on a path leading to the adoption of woodblock printing, where arduous task of hand-copying was increasingly overshadowed by the advent of mechanical reproduction. In the era of the Five Dynasties, the state-sponsored publication of the Confucian *Jiu Jing* ("Nine Classics"), which was indispensable to the scholars, represented a continuation of the trajectory set in the late Tang dynasty. This initiative not only bolstered the printing industry but also signified the moment when printed books finally gained entry into the hallowed halls of high culture. The impetus for this initiative was driven by Feng Dao, the chancellor of the Later Tang dynasty (923–936). As recorded in *Cefu Yuangui* ("Prime Tortoise of the Record Bureau", 1013), a compilation ordered by the imperial court and authored by Wang Qinruo (962–1025) and others, specifically in Volume 608, Feng Dao, in the third year of the Changxing period (932), submitted a memorial to Li Siyuan, the Emperor Mingzong of the Later Tang dynasty (867–933), stating,

We, your humble servants, have taken note that merchants of Wu and Shu regions are engaged in the commerce of a diverse array of printed materials, yet the esteemed Confucian classics stand unmatched in their significance. Should the esteemed classics undergo a meticulous process of collation and be subsequently disseminated through the meticulous craft of engraving, it would serve as a pivotal catalyst for the enrichment of our literary heritage and the nurturing of ethical enlightenment.

In his endeavour to rejuvenate the cultural and educational landscape of the northern regions, Feng Dao proposed using the "Kaicheng Stone Classics" of Chang'an as the definitive template, having the various sutra scholars of *Guozijian* (Imperial College) to meticulously collate and refine the text, subsequently commissioning the engraving printing to disseminate the Confucian *Jiu Jing* for publication. Emperor Mingzong of the Later Tang dynasty granted Feng Dao's petition and decreed that Ma Gao (854-ca. 938), a follower of the Crown Prince, would preside over the project. Despite the changes of the four dynasties in the north, Feng Dao consistently held his position as a chancellor, and the work of engraving the classics, which he advocated, continued without interruption. Among the officials who were initially part of this affair, some had retired due to the infirmities of old age while others had passed away; however, Tian Min (ca. 881–972) remained steadfast in his position throughout these changes. From the third year of the Changxing period in the Later Tang dynasty (932) to the third year of the Guangshun period in the Later Zhou dynasty (953), the entire project was completed, spanning a period of 21 years. A total of 130 volumes were printed, including *Yi Jing* ("I Ching"), *Shang Shu* ("Book of Documents"), *Shi Jing* ("Classic of Poetry"), *Zuo Zhuan* ("Zuo Qiming's Commentary on Spring and Autumn Anuuals"), *Gongyang Zhuan* ("Gongyang's Commentary on Spring and Autumn Annals"), *Chunqiu Guliang Zhuan* ("Guliang's Commentary on Spring and Autumn Annals"), *Yi Li* ("Book of Etiquette and Ceremonial"), *Li Ji* ("Book of Rites"), and *Zhou Li* ("Rites of Zhou"). Aiming to achieve consistency in the layout and typographical style of the classics, in the third year of the Kaiyun period of the Later Jin dynasty (946), *Guozijian* meticulously published a solitary volume of *Wu Jing Wen Zi* ("The Characters of the Five Classics") and *Jiu Jing Zi Yang* ("The Forms of the Characters in the Nine Classics") respectively. These authoritative texts laid down the procedural guidelines and typographical standards for the various stages of the printing process. These two books, from the standpoint of the annals of printing technology, are of monumental importance. Regrettably, they had been lost following the Song dynasty. The official dissemination of *Jiu Jing* in the northern dynasties during the Five dynasties period represents a groundbreaking and historically significant milestone, exerting far-reaching influence on the evolution of printing techniques in subsequent generations.

At this time, private book publishing was experiencing an expansion. The National Library of France in Paris houses a large bundle of printed Buddha statues unearthed from Dunhuang (No. 4514), which includes: (1) Five copies of the single-page printed edition of *Avalokitesvara Bodhisattva Statue* printed by Cao Yuanzhong (ca. 905–980), the governor of Guiyi Army in the fourth year of the Kaiyun period (947), engraved by the craftsman named Lei Yanmei; (2) Eleven copies of *The Great Sage Vaisramana Statue* printed by Cao Yuanzhong in the fourth year of the Kaiyun period; (3) Eleven single-page printed copies of *Manjusri Bodhisattva Statue*, each with a size of 31 cm × 20 cm; (4) Five copies of *Amitabha Bodhisattva Statue*; (5) One printed page of *Ksitigarbha Bodhisattva Statue*. All these artifacts feature an image superimposed above the text (Fig. 4.18), published in Dunhuang. In the fifteenth year of the Tianfu period (950), Cao Yuanzhong also published *Vájra-Cchedikā-Prajñā-Pāramitā-Sūtra*, engraved by Lei Yanmei, in the form of a booklet, which is also housed in the National Library of France (No. 4515). The printed editions identified as No. 4516 and No. 4515 are, respectively, the upper and lower sections of a single Buddhist scripture, together constituting a complete volume. Cao Yuanzhong, serving as a local official in Longyou, made significant contributions to the advancement of local papermaking and printing industries, as well as to the preservation of the Dunhuang grottoes.

As chronicled in *Hening Zhuan* ("Biography of He Ning") in Volume 127 of *Jiu Wudai Shi* ("Old History of Five Dynasties"), it is documented that He Ning (898–955), a celebrated lyricist and composer, was appointed as chancellor (an equivalent of prime minister) in the fifth year of the Tianfu period of the Later Jin dynasty (940). His compositions were popular in the two capitals of Kaifeng and Luoyang. "He had a collection of a hundred volumes, which he personally inscribed on the printing blocks, and printed hundreds of copies, distributing them as gifts to others." Shi Jingtang, the Emperor Gaozu of the Later Jin dynasty, was an ardent admirer of Taoism. According to *Gaozu Jing* ("Annals of Emperor Gaozu") in Volume 79 of *Jiu Wudai Shi*, the emperor ordered the Taoist Zhang Jianming to carve *Daode Jing* ("Classic of the Way and Virtue") onto printing blocks. Furthermore, he commissioned He Ning to write a new preface for the volume, with the intention of disseminating this sacred text throughout the realm.

The era of the Five Dynasties was marked by a succession of five distinct dynasties that governed the northern regions of the Yellow River basin in a continuous alternation. Simultaneously, in the southern territories, ten separate regimes were established, collectively referred to as the Ten Kingdoms. These polities, which lacked temporal continuity and spanned in size and scope, were spared from extensive warfare, allowing for the development of their economies and cultures, which in turn created favorable conditions for the expansion of printing technology. During

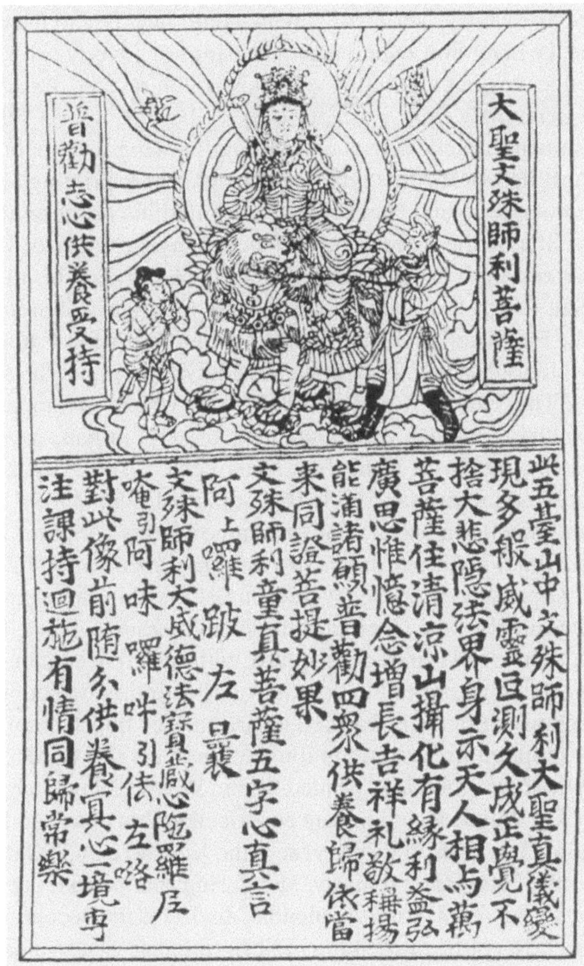

Fig. 4.18 A single-page figure of *Manjusri Bodhisattva* printed in the Five Dynasties (950) and found in Dunhuang (collection of the National Library of China, Beijing)

this period, the region of Sichuan saw the establishment of the Former Shu (907–925) and Later Shu (934–965), both of which published a variety of texts disseminating Confucianism, Buddhism, and Taoism. During the period from 909 to 913, Lin Guangting (850–933), a Taoist printer of Former Shu, undertook the monumental task of producing 30 volumes of *Daode Jing Guang Sheng Yi* ("Broad and Sacred Meaning of *Daode Jing*") employing a staggering 460 printing blocks for this endeavor. Subsequently, in the fifth year of the Qiande period of the Former Shu (923), the eminent monk Guanxiu (832–912) composed *Chanyue Ji* ("Zen Moon Collection"), a compilation of a thousand poetic pieces, which was subsequently published in Chengdu by his disciple, Tan Yu. The Former Shu also published almanacs. As delineated in the annals of *Huizhulu·Houluyuhua* ("Records of Shaking the Flywhisk·Supplementary Records"), Volume 2, penned by the scholar Wang Mingqing (1127-ca. 1216) of the

Song dynasty, Mu Zhaoyi, who would later rise to the exalted position of chancellor in the Later Shu, experienced the pangs of poverty in his youth. In his quest for knowledge, he sought to borrow a copy of the revered *Wen Xuan* ("Selections of Refined Literature") from a friend, only to be met with a refusal. He subsequently grew immensely industrious on learning, vowing that should he ever attain a lofty position, he would insist on inscribing texts onto printing blocks and disseminating them among scholars. He eventually fulfilled his promise after serving as the chancellor of Shu. In the seventh year of the Guangzheng period (944), following Mu Zhaoyi's ascension to the chancellorship of the Later Shu, his disciples Gou Zhongzheng and Sun Fengji took it upon themselves to engrave *Wen Xuan*, *Chu Xue Ji* ("Record of the First Learners") and *Baishi Liutie* ("Bai's Six Categories") onto printing blocks, which subsequently circulated widely throughout the scholarly community.

The Wuyue Kingdom (907–978), which encompassed regions in what is now southern Jiangsu, Zhejiang, and eastern Fujian, saw the flourishing of printing technology with Hangzhou at its center. The Wuyue King, Qian Chu (929–988) ordered the printing of 84,000 copies of *Dhtū Krand Dhrani Sūtra* on oparchment and bamboo papers. The full name of this scripture is *Sarvatthikāra Mahāguhya Śarīra-śūnyāśaya-ratnakūṭa Dharani Sūtra*, which has three printed copies[48]: The first printed copy, discovered in 1917 within the pagoda of Tianning Temple situated in Huzhou, Zhejiang Province, measures 7.5 × 60 cm in length per sheet. It comprises 341 lines of scripture, with each line containing a range of 8–9 characters. At the forefront of the volume, an illustration depicting the Buddha, his attendants, and worshippers is featured, characterized by its simplicity in lines and shapes. The inscription preceding the illustration bears the words, "The 84,000 copies of *Dhtū Krand Dhrani Sūtra*, printed by the Wuyue King Qian Chu, the Grand Commander of the Wuyue Kingdom, are enshrined within this pagoda. This is recorded in the year Bingchen (956), the third year of the Xiande reign." Xiande was a reign title belonging to the Later Zhou dynasty, and during that period, the Wuyue Kingdom adhered to the Later Zhou's reign calendar. In 1971, the second printed copy was discovered inside a gold-coated stupa in Shaoxing, encased in a bamboo tube measuring 10 cm in length and featuring 11–12 characters per line. This edition bears illustrations and inscriptions that are akin to those found in the first printed copy, albeit with a refinement in the quality of its engraving. Crafted on pristine white bark paper, it boasts an inscription that, interestingly, omits any mention of a reign title, instead solely referencing the year Yichou (965), which corresponds to the third year of Emperor Taizu's Qiande reign during the Song dynasty. The third printed copy was discovered within the Leifeng Pagoda in Hangzhou in 1925. It was crafted on bamboo paper, boasting dimensions of 3.6 cm in height and an impressive 190.5 cm in length. This edition featured 27 lines, with each line containing 10 or 11 characters. The inaugural printed edition features an illustration at the commencement of the sutra that portrays the queen and her attendants paying homage to the Buddha. The inscription on this artifact declares, "Wuyue King Qian Chu, the Grand Commander of the Realm, had

[48] Needham J. *Science and Civilisation in China*: *Vol. 5*. Cambridge: Cambridge University Press, 1985: 157–158.

Fig. 4.19 *Dhtū Krand Dhrani Sūtra* printed in 975 and found in Hangzhou in 1925 (Collection of the National Library of China, Beijing)

84,000 volumes of this sutra printed and donated them to the brick pagoda in Xiguan for perpetual veneration. This record is inscribed in the month of August during the year Yihai (975)" (Fig. 4.19).

Throughout the span of 19 years from 956 to 975, Qian Chu sponsored the printing of *Dhtū Krand Dhrani Sūtra* and commanded their distribution for enshrinement among the numerous pagodas scattered across Wuyue. During this period, the monk Yanshou (904–975) from the Lingde Temple in Hangzhou embarked on a prolific printing endeavor, producing a diverse array of scriptures, mantras, and Buddhist images, amassing an impressive 400,000 copies. Notably, among these, a sum of 160,000 were meticulously rendered onto silk.[49] Wuyue, a flourishing kingdom situated to the east of the Southern Tang dynasty (937–957), held under its dominion a vast expanse of territory that comprises the majority of modern-day Jiangsu, Anhui, Jiangxi, and the western part of Fujian. Establishing itself as a formidable state, Wuyue boasted its capital at Jiangning (present-day Nanjing). It was within this very kingdom that the celebrated Chengxintang paper was crafted, earning Wuyue international acclaim for its papermaking expertise. Feng Fang (1510–1567), a renowned book collector of the Ming dynasty, penned in his *Zhenshangzhai Fu* ("Ode of Zhengshangzhai Studio"): "Liu's (Liu Zhiji) *Shi Tong* ("Theory of Historiography and Historical Criticism") and *Yutai Xinyong* ("New Songs from a Jade Terrace") hold the distinction of being the inaugural printed works originating from the Southern Tang dynasty." The note states that these books bear the mark "Seal of Jianye Wenfang". Southern Tang was established after Wu (919–936). In the aforementioned memorial of Feng Dao, it was said that "engraved Confucian classics with abundant styles mostly came from Wu and Shu", which indicates that Wu was not only engaged in the production of a multitude of books but also in their distribution, having dispatched them to Luoyang and various other regions within the dominion of the Later Tang dynasty.

Taking a panoramic view of the Five dynasties and Ten Kingdoms period, it becomes clear that even amidst the disunity that characterized China at the time,

[49] Zhang Xiumin. Printing of Wuyue Kingdom in the Five Dynasties [J]. *Cultural Relics*, 1978 (12): 74.

there existed an extraordinary level of eagerness among the various regional powers, both in the north and the south, to foster the growth of printing technology. This spirit of emulation between rival states played a pivotal role in the broader propagation of these techniques throughout the land. Especially under the northern dynasties of the Later Tang, Later Jin, Later Han, and Later Zhou, the impetus provided by Chancellor Feng Dao played a pivotal role in the advancement of printing technology. His advocacy led to the esteemed Imperial Academy undertaking the monumental task of publishing the Confucian *Jiu Jing* published by *Guozijian*, a development that significantly propelled the proliferation of printing techniques. During the Tang dynasty, the nascent streams of printing that had begun to flow in disparate regions culminated, by the epoch of the Five Dynasties, into an irresistible deluge within the span of nearly half a century. This burgeoning wave crashed against the long-standing dominion of handwritten books with the force of a tempest. In conclusion, the tenth century stands as a transformative threshold, fulfilling a role of critical succession in the narrative of printing history. It forged a link between the pioneering efforts of bygone ages and the innovations that were to follow, thus providing the impetus for the Song dynasty to step into a golden era of printing. The groundwork established by the Five dynasties for the monumental leap in printing technology during the Song dynasty is largely due to the highly evolved papermaking technology and the thriving cultural and educational pursuits of the Tang dynasty. These enhancements provided the printing art with the necessary lift to ascend to unprecedented levels of achievement.

4.3.2 Woodblock Printing in Song, Liao, Western Xia, Jin, Uighur, Mongolia and Yuan Dynasties

The Northern Song dynasty (960–1127), born from the resolution of the discord that characterized the era of the Five Dynasties and Ten Kingdoms, stood as a nascent imperial power that achieved a long-sought unification. It was under this dynasty that the economies and the scientific and cultural intellect of the northern and southern realms harmoniously merged, fostering an era of unified progress and development. The Song dynasty's sovereigns accorded significant importance to the printing industry, considering the publication of books as a strategic initiative aimed at invigorating literary and educational pursuits, fortifying their governance, and extolling the nation's prowess. Following his ascension to the imperial throne, Emperor Taizong of the Song dynasty (976–997), promptly commissioned the esteemed Hanlin scholar Li Fang (925–996) to establish a library during the second year of the Taipingxingguo period (977). Li Fang was entrusted with the stewardship of a monumental scholarly endeavor: the compilation and engraving of three major books, *Taiping Yulan* ("Taiping Imperial Encyclopedia"), *Taiping Guangji* ("Extensive Records Compiled in the Taiping Years"), and *Wenyuan Yinghua* ("Best Literary Works"), which together constituted a vast collection of 2500 volumes. The issuance of these scholarly works marked a prodigious accomplishment in the annals of academic research. The magnitude of their printing was such that the volume of these three books eclipsed 19-fold the entirety of *Nine Classics* produced throughout the Five Dynasties. In 986, Emperor Taizong bestowed his imperial sanction upon the publication of rarefied texts, most notably *Shuowen Jiezi* ("Analytical Dictionary of Characters"), with the compilation and printing endeavors being orchestrated under the aegis of *Guozijian*. Under the successive reigns of Emperor Taizong and Emperor Zhenzong of the Song dynasty, *Shier Jing Zhushu* ("*Annotations and Commentaries on the Twelve Classics*") were meticulously printed between the years 988 and 996. Subsequently, in 1006, a reissue of the antiquated edition of *Shier Jing* ("The Twelve Classics") from the Five Dynasties was undertaken. In conjunction with *Mengzi* ("Mencius"), which was published in 1011, these texts collectively constituted *Shisan Jing* ("The Thirteen Classics"), mandatory books that became indispensable for scholars in subsequent eras. Furthermore, the monumental *Shiqi Shi* ("The Seventeen Histories") commenced its printing in 994, and the endeavor was ultimately concluded in 1061. These printed books were meticulously edited and exquisitely printed, marking the first imperially commissioned editions of the orthodox historical records. With these books, both the classics and historical texts were now fully available. In the chronicles detailed within the *Xingbing Zhuan* ("Biography of Xing Bing") in Volume 431 of *Song Shi* ("History of Song"), it is recorded that in May of the second year of the Jingde period (1005), Emperor Zhenzong made a sojourn to *Guozijian* to review the library's collection. There, he inquired of the officer Xing Bing (932–1010), who was also a librarian, about the number of classics currently housed within the institution. Bing replied,

In the fledgling days of the realm's inception at the dawn of the Song dynasty, the nascent library boasted a mere 4000 books. However, the current holdings have swelled to 100,000 volumes, with the entirety of classics, commentaries, and annotations fully at hand. In my youth, while I delved into the study of Confucianism under the tutelage of a master, it was a rarity to encounter annotated editions; perhaps one or two among a hundred books bore such scholarly marks, a scarcity likely due to the onerous task of transcription. However, the current collection is not only extensive but also accessible to every household, be they scholars or laymen. This indeed signifies a boon for the purveyors of Confucian knowledge. The emperor, Zhenzong, was pleased with this discourse and remarked, "Though the state champions Confucianism, without the tranquility that pervades our lands, such an achievement would be unattainable."

The discourse that transpired between the sovereign and his officials reveals that, propelled by societal tranquility, economic flourishing, and renaissance of scholarship, the realm had arrived at the peak of its publishing endeavors by the reign of Emperor Zhenzong (997–1022). Considering the collection of *Guozijian* as a paradigm, it exemplifies the remarkable expansion of the publishing industry, having multiplied 25-fold within the span of 45 years since the dynasty's inception. This growth ensured a comprehensive array of scriptures and historical texts, accessible to scholars and the populace alike. In the second year of the Jingde period (1005), the imperial court commanded the scholar Wang Qinruo and the edict official Yang Yi

(974–1020) to commence the editing of the 1000-volume *Lidai Junchen Shiji* ("The Deeds of the Emperors and Ministers of Successive Dynasties", 1013). Emperor Zhenzong personally penned the preface and entitled the work *Cefu Yuangui*, a text of such magnitude that it stands shoulder to shoulder with *Taiping Yulan*. Following these scholarly footsteps, Emperor Shenzong (1067–1085) also championed the compilation and dissemination of literature. He honored Sima Guang's (1019–1086) opus, *Zizhi Tongjian* ("Comprehensive Mirror to Aid in Government", 1084), with an imperial title and graced it with a preface written under the imperial seal.

Owing to the impetus provided by the sovereigns for the collation and dissemination of literary works, the evolution of woodblock printing technology experienced a significant boon. The publications issued by *Guozijian* distinguished by their exceptional quality and stringent oversight, emerged as a paragon for the publishing houses throughout the realm. Local governments also exerted their utmost efforts to contribute to this endeavor, propelling the art of printing in the Song dynasty to new heights. Printed books had ascended to a position of preeminence, encompassing a vast array of scholarly domains, from the venerable teachings of Confucianism, Buddhism, and Taoism to the diverse philosophies of the Hundred Schools of Thought. The scope of printed materials had even broadened to encompass the spheres of economics and the intricacies of everyday life. Printing technology has also flourished in regions inhabited by various ethnic minorities. Consequently, a comprehensive network of printing has emerged across the country, encompassing government-sponsored, commercial, and private printing houses. Building upon the established tradition of woodblock printing, the innovation of movable type printing was introduced, and alongside the existing practice of monochrome printing, the technique of multicolor printing was developed. These advancements catapulted China to the status of a preeminent printing power on the global stage. In Chaps. 5 and 6, we will undertake an in-depth analysis of multicolor printing and movable type printing techniques. In response to the distinctive requirements of the burgeoning printing industry, the Song dynasty crafted a standardized typographical style that has come to be recognized as the Song typeface today. This development precipitated a revolution in the binding of books, as the traditional scroll format was supplanted by innovative binding techniques. The design of the page layout and the framework of the printing blocks have set a precedent for future ages. Numerous volumes from the Song dynasty have survived to this day, and they are esteemed by subsequent eras for their meticulous collation, exquisite engraving, and superior printing quality.

In the Song dynasty, government presses were actively engaged in the publication of a vast array of Confucian classics, official histories, and academic works. Concurrently, they undertook the printing of a comprehensive series of Buddhist texts, with the most significant being the extensive collection known as *Tripitaka*. During the fourth year of Emperor Taizu's Kaibao reign (971), the imperial court dispatched Gao Pin and Zhang Congxin to Yizhou, now recognized as Chengdu, with the mandate to supervise the intricate process of engraving Buddhist sutras. Upon the culmination of this endeavor, which spanned a dozen years (971–983), the compilation was completed, comprising an extensive collection of 5048 volumes, titled *Kaibao Great Tripitaka* (abbreviated as *Kaibao Tripitaka*, as depicted in Fig. 4.20). This grand

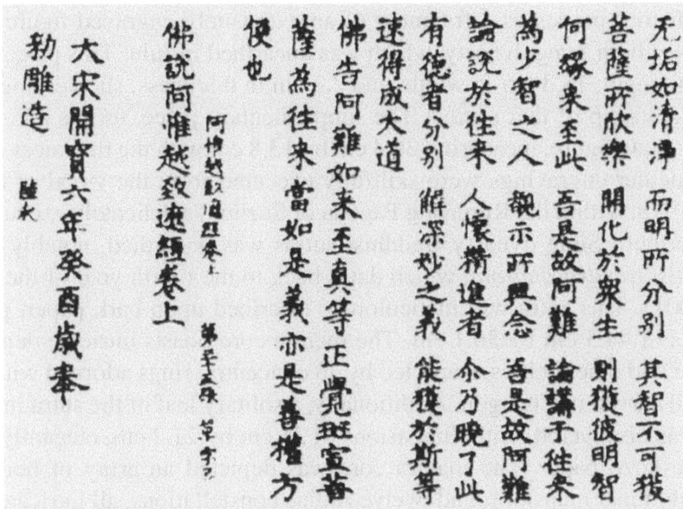

无垢如清淨 而明所分別 其智不可獲
菩薩所欣樂 開化於衆生 則獲彼明智
何緣來至此 吾是故阿難 諭講于往來
為少智之人 觀示所興念 吾是故阿難
諭說於往來 人懷精進者 尒乃曉了此
有德者分別 解深妙之義 能獲於斯等
速得成大道
佛告阿難如來至真等正覺班宣菩
薩為往來當如是義亦是善權方
便也
佛說阿惟越致遮經卷上
大宋開寶六年癸酉歲奉
勅雕造

Fig. 4.20 *Fo Shuo A Wei Yue Zhi Zhe Jing* (partial, collection of the National Library of China) from *Kaibao Tripitak*a, a Buddhist sutra printed in Chengdu in the fourth year of the Kaibao period of the Northern Song dynasty (971)

endeavor marked an unparalleled achievement in the global annals of engraving extensive Buddhist collections, establishing a pioneering model for subsequent eras and foreign realms to follow in the compilation and publication of their respective scriptural repositories. Subsequently, the period of the Northern and Southern Song dynasties witnessed the successive publication of six additional Buddhist scripture collections across various regions, including *Chongning Tripitaka* (1080, Fuzhou), *Vairocana Tripitaka* (1112, Fuzhou), *Yuanjue Tripitaka* (1132, Huzhou), *Zifu Tripitaka* (1175, Anji), and *Qisha Tripitaka* (1231, Suzhou), with each collection encompassing a vast range of 5000–7000 volumes. Among them, *Kaibao Tripitaka* required the use of 130,000 wooden printing blocks, but was bound in scroll form, while the others were bound in the folded-sutra format. The publication of the scriptures represented a large-scale printing activity; prior to the Song dynasty, most were manuscripts.

Archeological findings from the Song dynasty have repeatedly unveiled the intricately carved wooden blocks and printed editions of Buddhist scriptures. For instance, in 1920, fragments of a block for Buddhist prints were excavated from a Song dynasty tomb in Julu, Hebei Province. The block measured 43.1 cm × 12.5 cm and was engraved on both sides. One side depicted the Dharani mantra and eight Buddha statues, while the other side portrayed the Amitabha Buddha. Engraved approximately in the second year of the Daguan period (1108), the artifact is currently part of the collection of the New York Public Library. Additionally,[50] the Museum of

[50] Qian Cunxun. *Collection of Papers on the History of Books, Paper, Ink and Printing in China.* [M]. Hong Kong: Chinese University of Hong Kong Press, 1992: 127–138.

Chinese History possesses a fragment of an exquisitely engraved figure of a lady from the Northern Song dynasty, which was unearthed in Julu. This piece measures 59.1 cm in length, 15.3 cm in width, and 2.3 cm in thickness, showcasing the intricate craftsmanship of that period. The supplementary piece, too, is an exquisitely rendered female figure, measuring 26.4 cm by 13.8 cm with the thickness of 2.5 cm. These particular engravings were skillfully executed from the wood of the jujube tree.[51] In 1978, within the Ruiguang Pagoda of Suzhou, a delicately executed exemplar of Northern Song dynasty Buddhist sutras was unearthed, notably including *Mahā-pratisāravidya-dhāraṇi*, which dates back to the fourth year of the Xianping period (1001). This sutra was meticulously inscribed upon bark paper, presenting dimensions of 44.5 cm by 36.1 cm. The mantra core boasts intricate depictions of Sakyamuni and other deities, encircled by 26 concentric rings adorned with Chinese mantras, all carved in Chengdu. Additionally, a solitary leaf of the sutra inscribed in Sanskrit was discovered, with dimensions of 25 cm by 21.2 cm, elegantly rendered on the fine kozo paper. The mantra core was depicted an array of bodhisattvas, twenty-eight lunar mansions, and twelve zodiac constellations, all intricately carved in the second year of the Jingde period (1005). Simultaneously unearthed were ten volumes of *Saddharma Pundarika Sutra*, engraved in the inaugural year of the Tianxi period (1017), and printed on paper crafted from a blend of mulberry bark and bamboo, with each sheet measuring (16.9–17.1 cm) × (51.5–55.5 cm), in a hue of yellow.[52] The Taoist collection *Dao Zang* ("Taoist Canon") also debuted its publication during the Song dynasty. For instance, Fujian's *Wanshou Dao Zang* ("Taoist Canon, Wanshou Edition") (1116–1117) comprised 540 cases and 5481 volumes, an impressive assemblage. This extensive compilation of texts was later reprinted.

During the Song dynasty, scientific, technological, and medical publications were of significant importance. Works such as the mathematics series *Suanjing Shishu* ("Ten Mathematical Classics", 1074), *Qimin Yaoshu* ("Important Arts for the People's Welfare"), *Sishi Xueyao* ("Essentials of Learning During the Four Seasons", 1018), *Shanghan Lun* ("Treatise on Cold Pathogenic Disease", 1061), *Mai Jing* ("Pulse Classic", 1068), *Zhubing Yuanhou Lun* ("General Treatise on Causes and Manifestation of All Diseases", 1027), *Huangdi Neijing Suwen* ("Emperor Huang's Inner Canon: Basic Questions"), were all exceptional ancient texts that were previously circulated only in handwritten copies. These publications greatly advanced the dissemination and application of knowledge in various fields. A number of new works completed during the Song dynasty were published in a timely manner, such as *Wujing Zongyao* ("Complete Essentials for the Military Classics", 1044), *Yingzao Fashi* ("Treatise on Architectural Methods", 1103), *Xin Yixiang Fayao* ("Standard Essentials of the New Armillary Clock"), *Kaibao Bencao* ("Materia Medica of the Kaibao Era", 973), *Bencao Tujing* ("Illustrated Classic of Materia Medica", 1062),

---

51 Shi Zhilian. Two Examples of Northern Song Dynasty Portrait Engravings [J]. *Cultural Relics*, 1981 (3): 70–71.

52 Le Jin, Liao Zhihao. A batch of Five Dynasties and Northern Song Dynasty Cultural Relics Discovered in Ruiguang Pagoda, Suzhou [J]. *Cultural Relics*, 1979 (11): 21–26.

Zhenglei Bencao ("Classified Materia Medica from Historical Classics for Emergency", 1108), *Taiping Shenghui Fang* ("Taiping Holy Prescriptions for Universal Relief", 1100), and *Tongren Zhenjiu Tujing* ("Acupuncture and Moxibustion Classic of the Bronze Man", 1026). These scientific volumes often incorporated detailed illustrations, and intended to aid the readers in grasping the intricacies of the subject matter presented within the narratives. Moreover, the spectrum of publications was staggeringly comprehensive, encompassing literature, history, philosophy, geography, poetry, fiction, drama, music, fine arts, archaeology, and even the esoteric fields of divination and astrology. It was a cornucopia of knowledge that could leave one quite bedazzled upon perusing the bibliographic records of the era.

The official publishing entities encompassed *Guozijian*, Chongwen Academy, Palace Library, Imperial Astronomer, Imperial Medical Service, as well as various other departments within the central government apparatus, with *Guozijian* located in the capital being the most prominent among them (Fig. 4.21). These institutions were dedicated to publishing scholarly works with a mission of service, not profit. For those desiring to acquire such books, merely the expenses associated with the paper, ink, and the very act of production were levied.

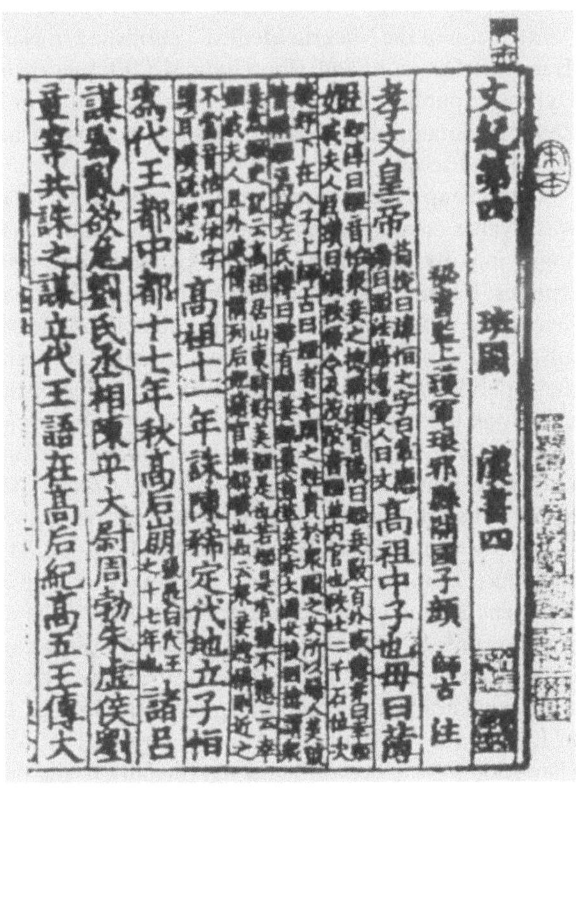
文紀第四 班固 漢書四
孝文皇帝 高祖中子也母曰薄
姬
高祖十一年誅陳豨定代地立子恒
爲代王都中都十七年秋高后崩 諸呂
謀爲亂欲危劉氏丞相陳平大尉周勃朱虚侯劉
章等共誅之謀立代王語在高后紀高五王傳大

Fig. 4.21 Northern Song edition of *Hanshu Zhu* printed by *Guozijian* at the beginning of the Southern Song dynasty (Collection of the National Library of China)

Manuscripts were expected to exhibit a high degree of scholarly merit, and they underwent rigorous review and correction. Utilizing superior paper and ink for printing, these publications were emblematic of the nation's utmost level of expertise. Books published by *Guozijian* include official classics, official history books, the aforementioned major books such as *Taiping Yulan*, *Taiping Guangji*, *Wenyuan Yinghua* and *Cefu Yuangui* and medical books such as *Taiping Shenghui Fang* ("Taiping Holy Prescriptions for Universal Relief") and *Qianjin Yifang* (" A Supplement to Recipes Worth a Thousand Gold"). The Chongwen Academy published the codes promulgated by the government, in addition to issuing a variety of significant books such as *Wu Zhi* ("Records of Wu"), *Guang Yun* ("Extensive Rhymes"), *Qimin Yaoshu.* The Imperial Medical Service was principally dedicated to the publication of medical literature. Local administrative bodies, ranging from the provincial level down to the state, county, and further, along with an array of specialized institutions and educational establishments at various levels, exhibited considerable zeal for the publication of books. They were supported either by direct financial allocations from the central government or through the diversion of local tax revenues for this purpose. The following books, which remain in circulation to this day, encompass a diverse range of historical and scholarly works. Among them are *Zizhi Tongjian* (published in Yuyao in 1185) by the Tea and Salt Division of Eastern Zhejiang, *Bencao Yanyi* ("Extension of the Materia Medica", published in Nanchang in 1185) by the Jiangxi Transport Division, and *Han Guan Yi* ("Etiquettes of the Officialdom of the Han Dynasty", published in 1139 by Lin'an Prefecture in Hangzhou, Fig. 4.22), and *Liji Zhu* ("Annotation on *Li Ji*", published in Linshui in 1177) by Fuzhou Minister's Office, and so on.

In the Song dynasty, the private sector's involvement in the book printing industry was diverse, encompassing a range of methods such as familial engravings (Fig. 4.23), engravings by specialized workshops, and those produced within the confines of temples. Engravings undertaken by private individuals were often financed by the literati and scholar-officials, who would collaborate to produce books that were then gifted to their friends and relatives. Despite the limited quantities, the standard of these publications was exceptionally high. For instance, Zhou Bida (1126–1204) undertook the engraving of 153 volumes of *Ouyangwenzhong Gong Ji* ("Collected Works of Ouyang Xiu") in 1191, *Shicaitang,* the ancestral hall of Fujian scholar Liao Yingzhong (ca. 1200–1275) engraved *Changli Xiansheng Ji* ("Collected Works of Mr. Changli") during the Xianchun period (1265–1274), and *Zhijingtang,* the engraving workshop of Huang Shanfu (1155–1225) from Jian'an, engraved *Qian Han Shu* ("Book of Former Han") in 1195. Workshop engravings were produced with the intent of generating profit, catering to the popular and necessary books among the common people. These books encompassed a vast array of types and were produced in large quantities, striving for affordability. They were often printed on bamboo paper with small fonts and compact layouts. Bookshops like those of the Chen family in Lin'an, the Yu family in Jian'an, Fujian (Fig. 4.24), and the Zhong family in Hangzhou have all contributed significantly to the publication of numerous books. The temple engravings handed down include *Gushan Cang* ("Collection of Gushan"), which was carved by the Dongchan Dengjue Zen Temple in Fuzhou. According to

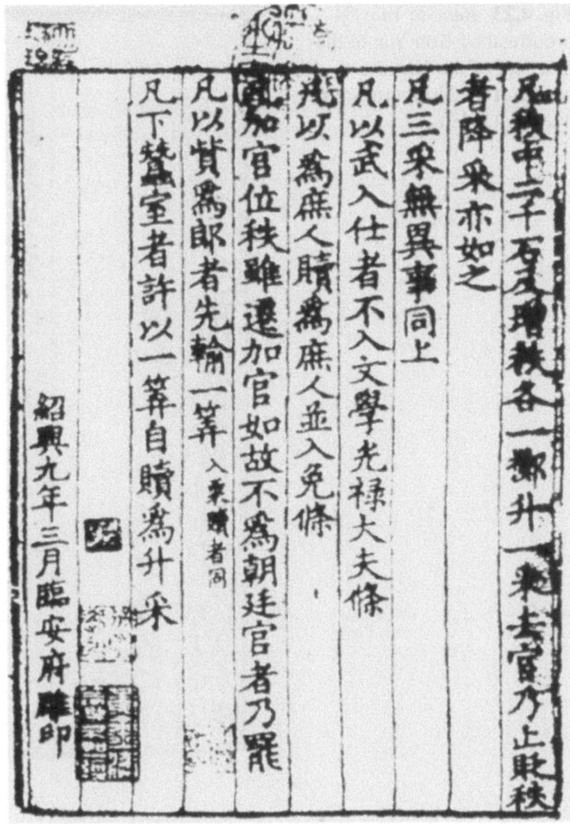
凡秩中二千石及增秩各一等升一等去官乃止貶秩
者降等亦如之
凡三等無異事同上
凡以武入仕者不入文學光祿大夫條
凡以爲庶人贖爲庶人並入免條
凡加官位秩雖還加官如故不爲朝廷官者乃罷
凡以貲爲郎者先輸一算 入粟贖者同
凡下蠶室者許以一算自贖爲升等
紹興九年三月臨安府雕印

Fig. 4.22 *Han Guan Yi* engraved in Lin'an Prefecture in the ninth year of the Shaoxing period of the Southern Song dynasty (1139)

Meng Yuanlao's *Dongjing Menghua Lu* ("Reminiscences of the Eastern Capital", 1187), as recorded in Volume 3, in front of the Buddha Hall of Xiangguo Temple, "there are all books, antiques, and paintings in front of the Zisheng Temple Gate", which can be described as a cultural street with numerous bookstores, offering a vast array of literature for acquisition. Notably, Hangzhou and Jianyang in Fujian were recognized as epicenters where these private publishing houses were particularly concentrated.

Given the surge of publications by private publishing houses and the intermittent emergence of pirated editions, the Song dynasty administration enacted a suite of regulations to oversee and govern the realm of book publishing. The unauthorized reproduction and sale of documents concerning state secrets and official records were categorically forbidden. Furthermore, the private printing by the populace of government edicts, the emperor's scripts, almanacs, and paper currency were also strictly interdicted. Numerous imperial decrees and statutory regulations regarding this domain can be found in *Song Huiyao Jigao·Xingfa Jinyue* ("Draft for the Collected Statutes of Song Dynasty") and *Qingyuan Tiaofa Shishi* ("Legal and Economic Historic Materials during Qingyuan Period"). The state's legislative edicts

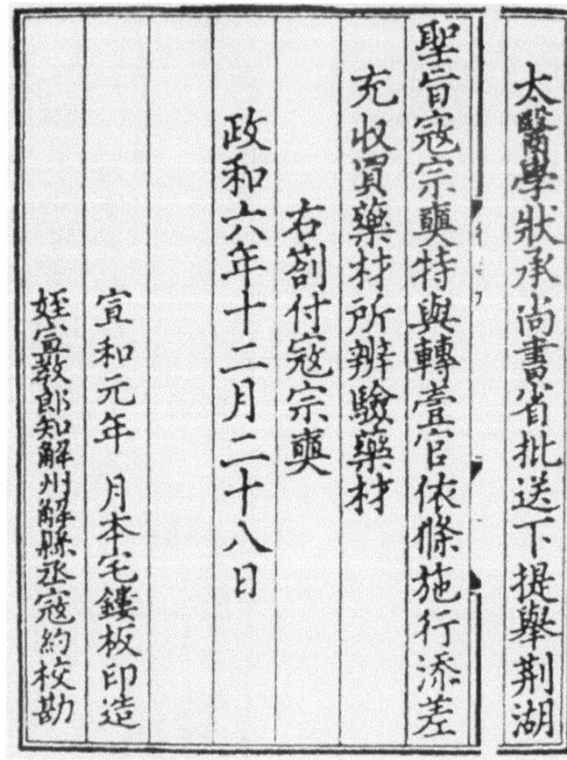
太醫學狀承尚書省批送下提舉荊湖
聖旨寇宗奭特與轉壹官依條施行添差
充收買藥材所辨驗藥材
右劄付寇宗奭
政和六年十二月二十八日
宣和元年　月本宅鏤板印造
姪宣教郎知解州解縣丞寇約校勘

Fig. 4.23 *Bencao Yanyi*[53] proofread by Kou Yue in the first year of the Xuanhe period of the Northern Song dynasty (1119)

were granted publication authority by the Dali Temple and Chongwen Academy. In a bid to safeguard the lawful rights and interests of the book engravers, the government proclaimed that the reproduction of copyrighted books was strictly prohibited without permission. Despite the absence of a complete codified law regarding publishing, the Song dynasty populace had cultivated a sense of copyright protection, and there were numerous records of individuals being held accountable for the infringement of these rights. However, there existed explicit legislation concerning the unauthorized printing of paper currency; counterfeiters were subject to the death penalty, while informants were entitled to a reward. For an exhaustive discussion on the printing of paper currency, refer to Chap. 7.

The Song dynasty's well-developed and comprehensive printing system exerted a significant impact on the governance structures established by various ethnic minorities within the geographical expanse of China during that era. Contemporaneous with the Northern Song dynasty, the Liao dynasty (916–1125) and the Uighurs coexisted in the northern regions, while the Western Xia (1038–1227) established their

[53] Nakayama Kyushiro. *A General History of World Printing* [M]. Tokyo: Sanshusha Publishing, 1930.

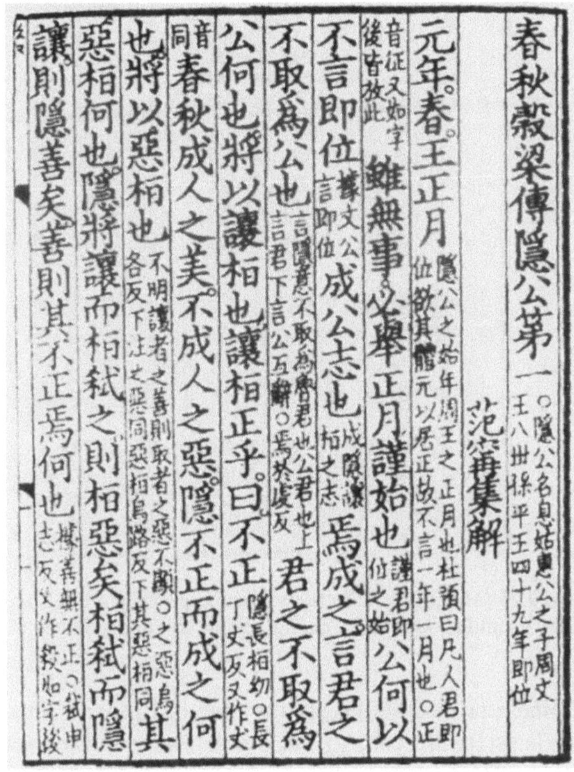

春秋穀梁傳隱公第一 ○隱公名息姑惠公之子周文王八世孫平王四十九年即位
范甯集解
元年。春。王正月 隱公之始年周王之正月也杜預曰凡人君即位欲其體元以居正故不言一年一月也○正
音征又如字後皆放此 雖無事。必舉正月謹始也 謹君即位之始 公何以
不言即位 據文公言即位 成公志也 成隱讓桓之志 焉成之言君之
不取爲公也 言隱意不取爲魯君也公君也上言君下言公互辭○焉於虔反 君之不取爲
公何也將以讓桓也讓桓正乎。曰不正 隱長桓幼○長丁丈反又作丈
音同 春秋成人之美。不成人之惡。隱不正而成之何
也將以惡桓也 不明讓者之善則取者之惡不顯○之惡烏各反下注之惡同惡桓烏路反下其惡桓同 其
惡桓何也隱將讓而桓弑之則桓惡矣桓弑而隱
讓則隱善矣善則其不正焉何也 據善無不正○弑申志反又作殺如字殺

Fig. 4.24 *Chunqiu Guliang Zhuan* ("Guliang's Commentary on Spring and Autumn Annals"),[54] published by Yu Renzhong from Jian'an in the second year of the Shaoxi period of the Southern Song dynasty (1191)

dominion in the northwest, sandwiched between the two Song dynasties. Concurrently, the Jin dynasty (1115–1234) and the Mongol Empire (1206–1279) also rose to power in the northern territories. The Liao dynasty, established by the Khitan people, was the largest minority regime in the north. In 920, Emperor Taizu of Liao developed an ideographic writing system based on Chinese characters, which was called the Khitan Large Script. In 926, under the influence of the Uighur script, the younger brother of Emperor Taizu of Liao ingeniously devised the Khitan Small Script, a unique script that, while preserving a select number of ideographic characters, predominantly employs phonetic characters. However, while the Liao dynasty's block-printed books did incorporate the Khitan script, a significant portion of these texts were preserved in the traditional Chinese characters, and it is noteworthy that the majority of the Khitan-scripted volumes were, in fact, rendered translations of the original Chinese literary works. The Liao rulers held Confucianism in high esteem and embraced Buddhism. Despite military conflicts with the Song dynasty, there was a continuous exchange of economy, culture, technology, and personnel between the Liao and Song, which allowed for the timely introduction of a large number of

[54] Nakayama Kyushiro. *A General History of World Printing* [M]. Tokyo: Sanshusha Publishing, 1930.

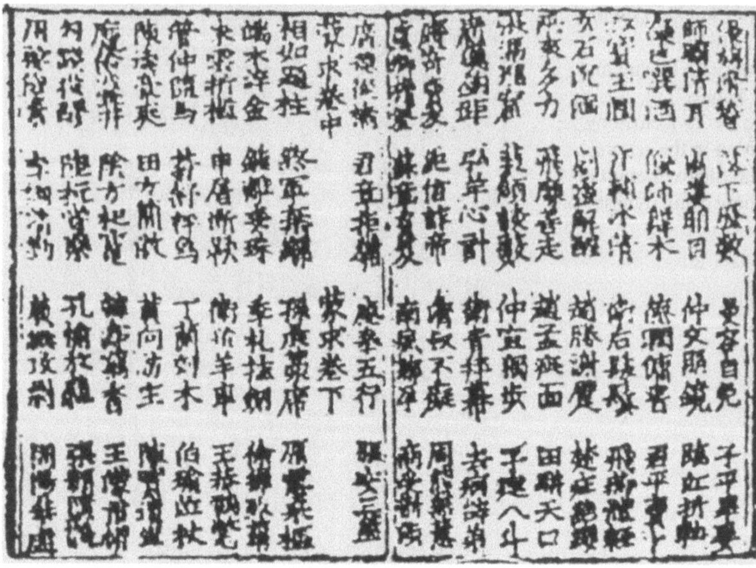

Fig. 4.25 *Meng Qiu* in butterfly binding printed in the Liao dynasty and found in 1974 in the Sakyamuni Pagoda of the Fogong Temple in Yingxi, Shanxi (Collection of the Shanxi Museum)

Chinese printed books. By the time of Emperor Muzong and Emperor Jingzong at the latest (tenth century), the Liao had already mastered the art of printing (Fig. 4.25).

In 1974, a large number of printed materials from the Liao dynasty was unearthed in the wooden pagoda of Ying County, Shanxi Province, which was established by the Liao regime. This collection comprised a total of 61 items, with the earliest being *Shangsheng Jingshu Kewen* (a Buddhist text) from the eighth year of the Tonghe period under the reign of Emperor Shengzong (990), inscribed with the words "Made by the Yang's Printing in front of Yangshan Temple in Yanjing", where Yanjing refers to present-day Beijing. Within the pagoda, there also resided a collection of 12 fragmented scrolls from the Liao dynasty's engraved *Tripitaka,* crafted in the form of traditional scrolls. These relics were printed on sheets of yellow paper, adorned with the grandeur of regular script characters, and each scroll bore a numerical designation derived from *Qianziwen.* Among these precious fragments was *Chengzan Dasheng Gongde Jing* ("Praise of the Mahayana Merit Sutra"), which was inscribed in the year 1003. Notably, *Chishengguang Jiuyao Tu* ("Chart of the Nine Planets with the Surpassing Brightness") was an extraordinary full-page woodblock carving, boasting dimensions of 120 cm in height and 48 cm in width, thereby earning it the distinction of being the largest known woodblock print to date. Beyond the engraving of Buddhist sutras and figures, the Liao dynasty also engraved Confucian classics and works of history and philosophy. As chronicled in Volume 22, *Daozong Ji* ("Annals of Emperor Daozong") of *Liao Shi* ("History of Liao"), within the annals of Emperor Daozong, it is documented that during the winter of the ninth year of the Xianyong period (1073), "the emperor ordered the officials to promulgate *Shi Ji* ("Records of the

Grand Historian") and *Han Shu* ("Book of Han")". Similarly, *Shengzong Ji* (Annals of Emperor Shengzong), Volume 15 of *Liao Shi,* states that in the first year of the Kaitai period (1012), Emperor Shengzong of Liao ordered the granting of a Buddha figure of Karuṇikarāja Rāṣṭrapāla Prajñāpāramitā, along with copies of *Yi Jing*, *Shi Jing*, *Shang Shu*, *Chun Qiu*, and *Li Ji*, to Tieli (modern-day Tieli, Heilongjiang). These esteemed texts were the official editions sanctioned by the Liao *Guozijian*. *Longkan Shoujing* (997), a Chinese character dictionary compiled by the Liao dynasty monk Xingjun (957–1017), contains over 26,000 characters with phonetic annotations and explanations. The Liao block-printed edition of this book was transmitted to the Northern Song dynasty, where Shen Kuo, in his esteemed collection *Mengxi Bitan* ("Dream Pool Essays", 1088), Volume 15, articulated his admiration by stating, "Upon examining the systematic arrangement of its phonetics, one can perceive a coherent and logical structure." This book underwent reprinting by the Song dynasty, Goryeo, and Japan, with the extant editions from the Liao dynasty predominantly featuring Chinese text, while the Khitan version remains a rare find.

The Jin dynasty, which originated in Northeast China and was established by the Jurchen ethnic group, embarked on a southern expedition and successfully invaded Kaifeng, the imperial city of the Northern Song dynasty, in the year 1126. During this conquest, they seized a considerable quantity of books, printing plates, and skilled printing artisans, subsequently transporting them all northward. This significant event also marked the relocation of their capital to Yanjing, now known as Beijing. Many prominent figures within the Jin dynasty possessed a deep understanding of Chinese literature and were well-versed in Chinese classics. According to Volume 51 of *Jin Shi* ("History of Jin"), *Xuanju Zhi* ("Treatise on the Imperial Examinations"), all books such as *Yi Jing*, *Shi Jing*, *Shang Shu*, *Zuo Zhuan*, *Yi Li*, *Zhou Li*, *Lun Yu*, *Mengzi*, *Xiao Jing*, *Shiqi Shi*, *Daode Jing*, and *Xunzi* were "all printed by *Guozijian* and distributed to numerous educational institutions throughout the realm". In the year 1130, an official printing house dedicated to scriptures and classics, the Jingji Institute, was founded in Pingyang Prefecture (present-day Linfen, Shanxi Province). Additionally, this era witnessed the growth and expansion of private bookstores. For instance, Huimingxuan, owned by Zhang Cunhui in Pingyang Prefecture, reprinted *Zhenglei Bencao* ("Classified Materia Medica from Historical Classics for Emergency"), drawing upon the Song dynasty edition as its basis. Meanwhile, Jing's Bookstore, located in Ningjin, Hebei Province, published books such as *Chongqing Xindiao Gaibing Wuyin Jiyun* (a work of phonology). The Jin dynasty's publishing centers were predominantly situated in Yanjing (present-day Beijing), Shanxi Province, and Kaifeng, with the majority of their publications being grounded in the editions of the Song dynasty. Consequently, these publications have garnered significant attention from bibliographers. Among the books that have been preserved to this day, we find *Chongjiao Shengji Zonglu* (a medical book) published in 1164, *Shanghan Zhige* ("Straightforward Rules for Treating Cold"), *Huangdi Neijing Suwen* ("The Emperor Huang's Inner Canon: Basic Questions"), and *Liu Zhiyuan Zhugongdiao* ("The Stories of Liu Zhiyuan in Zhugongdiao-Style Songs"), which were published in 1186 by Liu Wansu. Between 1140 and 1178, the renowned *Zhaocheng Tripitaka*, based on the Northern Song dynasty's *Kaibao-Edition Tripitaka*, was printed in Shanxi in a total of 7000 volumes, bound in scroll format (Fig. 4.26). The publication of this Buddhist collection also reflects the robust publishing prowess of the Jin dynasty. Figure 4.27 presents an edition of *Xiaoxian Laoren Mingxiu Jizhu* ("Annotations on the Collected Works of the Leisurely Elder Mingxiu"), which is the Pingshui edition from the Jin dynasty.

The Jurchen people, initially devoid of a national writing system, subsequently adopted and adapted Chinese characters and Khitan scripts to devise their own writing system, termed the Large Script, which was promulgated in 1119. A further development occurred in 1138 with the creation of the Small Script. The Jurchen script occupies an intermediate position, drawing influences from both Chinese characters, Khitan script, and Western Xia script. Upon the creation of their script, the Jin dynasty took the initiative to establish the Jurchen National Imperial Academy, thereby affording the noble offspring the opportunity to enroll and acquire education. From 1164 onwards, a selection of Chinese literary works such as *Yi Jing*, *Shang Shu*, *Zuo Zhuan*, *Lun Yu*, *Xiao Jing*, *Laozi*, *Shi Ji*, *Han Shu*, and *Zhenguan Zhengyao* ("the Political Program of Zhenguan Period") were translated into the Jurchen script for the study of scholars, and these translations were all published. These translations were carried out under the direction of the distinguished Jurchen scholar Tudan Yi (ca. 1144–1214), at the translation institute he presided over. As recorded in *Shizong Ji* ("Annals of Emperor Shizong"), Volume 8 of *Jin Shi* ("History of Jin"), Emperor Shizong stated to the chancellor, "The rationale behind my directive to translate *Five Classics* stems from my desire for the Jurchens to comprehend the essence of benevolence, righteousness, and morality." Upon Emperor Zhangzong's ascension to the throne in 1190, all the classic translations were completed. To cater to the requirements of both the Han and Jurchens in the imperial examinations, the Jin dynasty also printed books on language and writing, including *Shuowen Jiezi* ("Analytical

Fig. 4.26 Scroll-bound *Zhaocheng Tripitaka*, published in Shanxi in 1140–1178

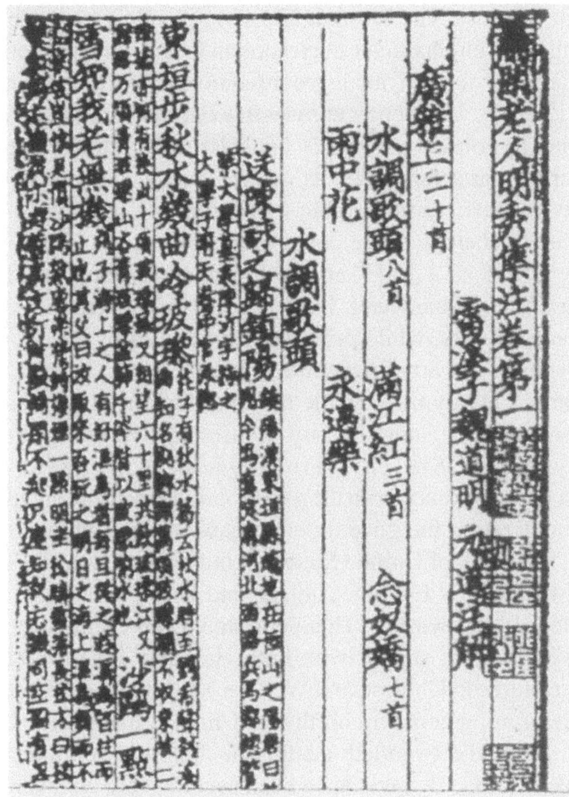

Fig. 4.27 Pingshui edition of *Xiaoxian Laoren Mingxiu Jizhu* in the Jin dynasty

Dictionary of Characters"), *Yupian* (China's first regular script dictionary), and *Er Ya* (China's earliest collation and interpretation of Chinese basic vocabulary and encyclopedia words).

The Western Xia dynasty, a sovereign state erected by the Tangut ethnic group, was nestled within the territories of present-day Ningxia and Gansu provinces, with its imperial capital firmly established in Xingqing Prefecture (present-day Yinchuan, Ningxia). This flourishing realm fostered intricate and vibrant economic and cultural exchanges with the neighboring Northern Song dynasty, and within its flourishing borders, a substantial community of Han Chinese coexisted. Emperor Jingzong of the Western Xia, Li Yuanhao, in concert with his chancellor Yeli Renrong, embarked upon a cultural synthesis by assimilating the refined elements of Han Chinese civilization while vigilantly safeguarding the indigenous traditions of the Tangut ethnicity. They crafted a dual governance framework that harmoniously integrated the established Han bureaucratic apparatus with the intrinsic Tangut tribal administrative structures. The Tangut language, also known as the Western Xia language, belongs to the Tibeto-Burman language family and is more closely related to Tibetan. In the third year of the Guangyun period (1036), Yeli Renrong standardized the Western Xia script and promulgated it, which circulated alongside the Chinese script within the territory. The Western Xia script, analogous to Chinese, is a logographic system where each character corresponds to a single phonetic unit. As documented in the Tangut lyrics and music recorded in *Yin Tong* (a dictionary of Western Xia characters, 1132), the script encompasses a lexicon of more than 6000 characters, a figure which may be conservative when considering the full extent of its vocabulary. The Tangut script, characterized by its square-shaped characters, is categorized into four distinct styles: seal script, regular script, running script, and cursive script. The strokes of Tangut characters are constructed from a combination of elements such as "一", "丨", "丿", "㇏", "㇇", and "㇕", rendering them more intricate in structure compared to Chinese characters. Papermaking in Western Xia can be traced back to the early Tang dynasty, while printing technology was introduced during the Northern Song dynasty. The art of papermaking in Western Xia has a rich history, dating back to the early Tang dynasty, while the introduction of printing technology occurred during the Northern Song dynasty. Consequently, during the Tiansheng reign of Emperor Renzong of Xia (1149–1169), *Tiansheng Lvling* (Revised edition of the Tiansheng Code) mentioned both the paper-making institute and the engraving department while enumerating the government organizations and factories. The postscript of *Yintong* ("The Book of Unified Sounds") indicates that the "engraving department was established, led by foreign scholars, and so on, to engrave and print books and distribute them to the world." This department represented the largest printing house in the Western Xia, staffed with both Han and Tangut craftsmen. It published Buddhist scriptures in Chinese and Western Xia language, as well as secular works, becoming the printing center in northwest China at that time.

Since the twentieth century, a significant number of printed materials from the Western Xia dynasty have been successively uncovered. Between 1907 and 1909, a Mongolian and Sichuan expedition, headed by the Russian explorer Peter Kuzmich Kozlov (1863–1935), unearthed over 2000 documents, manuscripts, and printed materials at the Kharahota site (present-day Ejin Banner, Inner Mongolia), situated along the Ejina River in Inner Mongolia.[55] These materials were written in Western Xia, Chinese, and Mongolian languages. These documents possess immense historical and academic significance, serving as invaluable physical materials for the study of Western Xia printing. Currently, they are housed at the Institut Aziatskich Natsiey, where L. N. Menshikov and E. N. Gerbacheva have catalogued the Chinese and Western Xia scripts.[56] This collection of documents has been photocopied and subsequently published by the Shanghai Classics Publishing House, with the 29th volume seeing its release in 2019. In summary, the Western Xia documents that have been unearthed encompass a wide array of materials, including Buddhist scriptures, Confucian and Taoist texts, language dictionaries and reference works, historical chronicles, military treatises, political and legal documents, literary

[55] Kozlov P K. *Mongolia, Amdo and the Dead City of Qara Qota* [M]. Translated by Wang Xilong and Ding Shuqin. Lanzhou: Lanzhou University Press, 2011.

[56] L. N. Menshikov. *Narrative of the Chinese Manuscripts Unearthed in Heicheng* [M]. Translated by Wang Kexiao. Yinchuan: Ningxia People's Publishing House, 1994. Gerbacheva E N, Kochekov E I. *Katalog Tangutskikh Manuskriptov I Pechatnykh Proizvedeniey Sobrannykh V Leningradskom Otdelenie Institut Aziatskikh* [M]. Moscow: AN SSSR, 1963.

compositions, works on astronomy and almanac, as well as medical manuscripts. This collection boasts original Chinese works, translations into Western Xia, and Tangut creations, with Buddhist sutras constituting an overwhelming 80% of the entire content. The printed copies were presented in various formats such as scroll-bound, butterfly-bound, and folded-sutra styles. The printing locations were either at Xingqing Temple in the capital or within government agencies. Initially printed by the private sector, these publications were soon integrated into the official publishing system. Notably, the earliest Chinese edition with a year mark is *Prajñāpāramitā Hṛdaya Sūtra* (Fig. 4.28), printed by Lu Wenzheng between 1073 and 1074. This edition features poems as annotations, adopts the folded-sutra format, and is printed on hemp paper with a page size of 8 cm × 13.5 cm. The colophon reads, "On the first day of the eighth month of the fifth year of Tianci Lisheng Guoqing period (1074), namely, the year of Guichou (1073), by Lu Wenzheng." This dating aligns with the reign of Emperor Huizong of Xia (1068–1086), albeit with a one-year discrepancy when compared using the reign year versus the heavenly stems and earthly branches. If the latter were used as the standard, the printing date would be September 5, 1073.

In the tenth year of the Da'an period (1084), the Great Yanshou Temple in Xingqing Prefecture, under the patronage of the Buddhist monk Shouqiong, published the scroll-bound edition of *Avatamsaka Sutra: The Chapter of the Practices of Samantabhadra*. Subsequently, *General Preservation of the Function as Basis in the Sutra Record* was crafted with the intricate butterfly binding during the reign of the esteemed Xia Emperor Renzong (1140–1193). In the imperial aspirational text, Emperor Renzong proclaimed his imperial decree that commanded the artisans to inscribe this sacred text in both Western Xia and Han scripts, an endeavour that resulted in a prodigious output of 15,000 volumes, which were then distributed among officials and civilians, thereby summoning up a chorus of recitations that echoed throughout the expanse of the Dharma realm. He vowed to engrave seven volumes of *Saddharma Pundarika Sutra*, which were in concertina-binding form,

Fig. 4.28 Folded-sutra form of *Prajñāpāramitā Hṛdaya Sūtra*, published in the fifth year of Tianci Lisheng Guoqing period of Western Xia (1073), collected by Russia

with Buddha figures engraved at the commencement of each volume. This monumental task was undertaken in the third year of the Renqing period (1146), resulting in a total of 70,000 engraved Chinese characters. The craftsmen responsible for this endeavour were Wang Shanhui, Wang Shanyuan, He Shanhai, and Guo Goumai. Western Xia Buddhist scriptures in print, including *The Buddha Speaks of Amitabha Sutra* published in the eleventh year of the Da'an period (1085), were among the religious texts disseminated during that period. Among the non-religious texts excavated from the ruins of Heishuicheng, *Fanhan Heshi Zhangzhongzhu* ("Dictionary of Tangut and Chinese") (Fig. 4.29) penned by the Western Xia scholar Gule Maocai, stands out as a paramount work, printed in the twenty-first year of the Qianyou period (1190) during the reign of Emperor Renzong of Xia. It was a bilingual dictionary that correlates the Western Xia and Han scripts in terms of phonetics and semantics. It has become an essential language tool, indispensable for the Tangut people in their pursuit of proficiency in Han characters and for Han people endeavouring to comprehend Tangut, the Western Xia language.

Between 1902 and 1907, the Prussian expedition team, led by Albert von Le Coq (1860–1930) and Albert Grünwedel (1856–1935), unearthed a trove of printed fragments in Turpan, Xinjiang, encompassing multiple languages, including Tangut and Chinese (Fig. 4.30). These fragments are now part of the Ethnologisches Museum

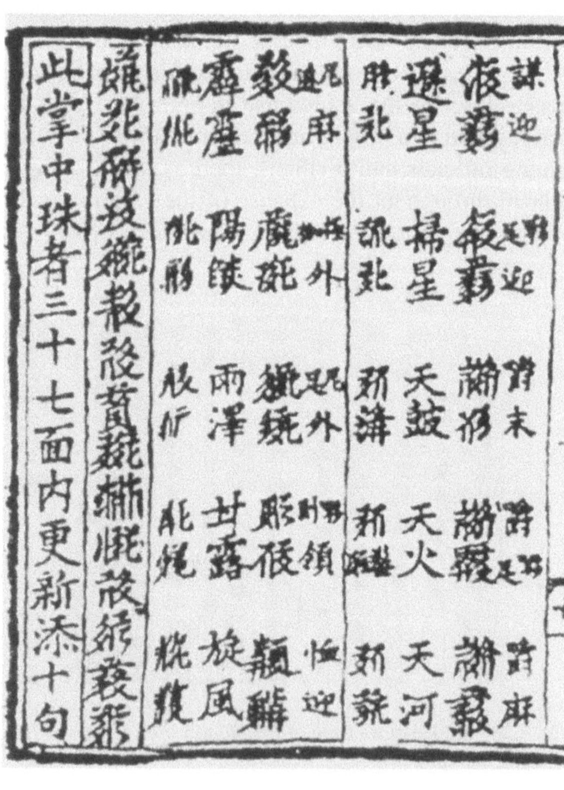
此掌中珠者三十七面内更新添十句

Fig. 4.29 *Fanhan Heshi Zhangzhongzhu* published in 1190 in the Western Xia dynasty

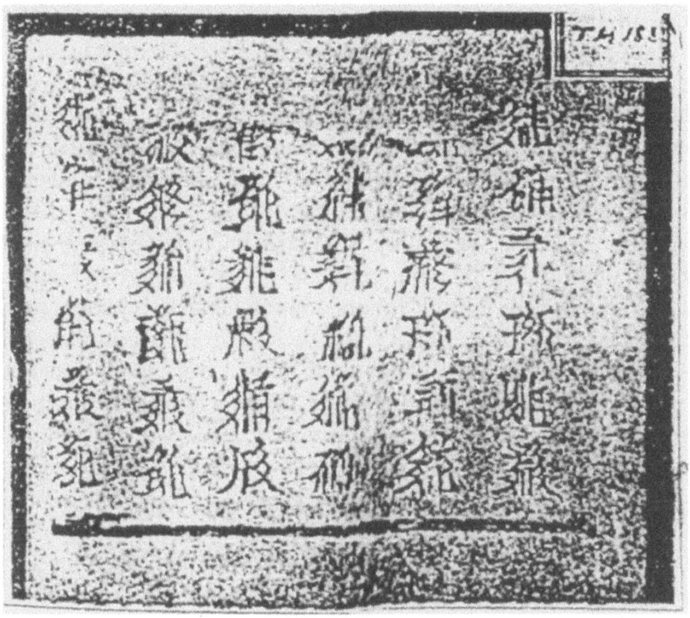

Fig. 4.30 Fragmented pages from a concertina-bound Buddhist scripture in Tangut published in the Western Xia dynasty and unearthed in Xinjiang in 1907 (Formerly in the collection of the Ethnologisches Museum Berlin)

Berlin's collection. For details, refer to Le Coq's *Auf Hellas Spuren in Ostturkestan* (literally translated as "Buried Treasures of Chinese Turkestan", meaning "Treasure Hunt in Xinjiang", 1926).

In the subsequent years following the 1950s, cultural relics from the Western Xia period have been intermittently unearthed, such as the instance in 1959 when an archaeological squad from the Dunhuang Research Institute of Cultural Relics discovered in a Yuan dynasty lamaist temple, situated on the eastern bank of the Dangquan River, three volumes of concertina-bound *Avalokiteshvara Sutra.* These texts, exemplifying the intricate folding binding technique, were inscribed in Tangut script and were embellished with initial illustrations, utilizing hemp paper for the printing process.[57] In 1972, in a cave in Wuwei, Gansu Province, two pages of the Tangut-script edition of *Siyan Zazi* ("Four-Character Miscellaneous Characters") and fragments of the calendar in Chinese script from the second year of the Renqing period of Emperor Renzong of Western Xia (1145) were discovered, along with butterfly-bound Tangut-script Buddhist scriptures.[58] In 1989, within the confines of the Hongfo Tower in Helan County, Ningxia Hui Autonomous Region, a discovery

[57] Zhang Zhong. *A Brief History of Dunhuang* [M]. Lanzhou: Dunhuang Foreign Cultural Exchange Association, 1990: 120.

[58] Gansu Provincial Museum. A Batch of Western Xia Relics Discovered in Wuwei, Gansu [J]. *Cultural Relics*, 1974 (3): 200–204, 216.

was unveiled of fragmentary remnants of printing blocks used for Tangut-script Buddhist scriptures, which had been scorched by fire.[59] Despite its brief existence of 189 years, the Western Xia dynasty managed to produce a vast array of printed works in both Han and Tangut scripts, a testament to the highly developed culture of this minority region in the west. The region not only fostered the development of woodblock printing but also pioneered movable-type printing, a subject that will be delved into with greater depth in Chap. 6.

Adjacent to the Western Xia Empire, a Uyghur-inhabited territory established its headquarters in Ganzhou, a location today known as Zhangye in Gansu Province, and came to be known as the Hexi Uyghurs. Simultaneously, another Uyghur faction migrated westward to Xizhou, establishing their central base in Gaochang, which in present-day terms is Turpan in Xinjiang, thereby acquiring the appellation of the Xizhou Uyghurs. The Uyghur language is a member of the Turkic branch in the Altaic language family. During the eighth century, the Uyghur script was devised, utilizing the Sogdian script of Central Asia as its phonetic foundation, and this script remained in use until the fifteenth century. The Uyghur region was situated at the western terminus of the ancient Silk Road in China. During the Han dynasty, significant administrative entities such as *xiyu duhufu* (Western Regions Protectorate) and *xiyu zhangshibu* (Western Regions Prefecture Office) were established there. Further, in the Tang dynasty, *longyoudao* (Longyou administrative division) was founded within the same region. The region maintained close economic and cultural interactions with the Central Plains, with a significant population of Han Chinese residing within its borders. From the Five Dynasties period through to the Song dynasty, the Uighur people, adherents of Buddhism, found themselves severed from direct interaction with the Central Plains, as the Western Xia Empire had effectively erected a barrier between the two regions. Beginning in 1030, the Western Xia dynasty claimed dominion over the Hexi Corridor, leading to a dependency of the Uighurs residing there upon the Western Xia, which in turn fostered more intimate cultural and intellectual exchanges between the two entities. Following the year 1209, the Uighur territories were entirely subsumed under the dominion of the Mongols, who referred to these regions as "Uyghur", a phonetic adaptation of the Uyghur people's own collective designation signifying "unity and union" in their own language. Extensive documentary evidence, corroborated by the unearthing of subterranean archaeological findings, confirms that by the latest count, during the period of Uighur governance under the aegis of the Mongol Yuan dynasty (1209–1324), both woodblock and movable-type printing techniques had been cultivated in the region, with the methods seemingly having been directly transplanted from the Western Xia.

In his *Histoire Generale Des Huns* (1758), the eighteenth-century French sinologist Joseph de Guignes (1721–1800) observed that "on the ancient Mongolian Plateau, there existed a tribe known as the Uyghurs, who held considerable fame

[59] Ningxia Hui Autonomous Region Cultural Relics Management Committee Office, Helan County Cultural Bureau. A Brief Report on Cleaning of Hongfo Pagoda in Helan County, Ningxia [J]. *Cultural Relics*, 1991 (8): 1–13, 26, 97–101.

within the Mongol Empire. They fostered advancements in science and technology....Their writing style mirrored the vertical flow of Chinese characters, and they were the first to use woodblock printing techniques."[60] The use of the word "first" implies that the Uyghurs developed woodblock printing prior to the Mongolians, a claim that was substantiated by archaeological discoveries in the twentieth century. Between 1902 and 1907, Albert von Le Coq uncovered a significant quantity of printed book fragments at the ruins of Xizhou Uyghurs in Turpan. A substantial portion of the artifacts uncovered comprised Buddhist scriptures and statues, inscribed in six languages—Uyghur, Chinese, Sanskrit, Tangut, Mongolian, and Tibetan—and these were eventually preserved in the Ethnologisches Museum in Berlin. Given that the majority of Central Asian monks were fluent in Sanskrit during that era, the printing of Buddhist scriptures in Sanskrit was tailored to accommodate the export requirements, thereby ensuring their widespread dissemination. Certain non-Chinese printed materials were frequently interspersed with Chinese characters, such as the names of the sutras, page numbers, and the names of the engravers, suggesting a collaborative effort between Han craftsmen and the Uyghurs. These fragments, however, lack publication marks and dates, necessitating further verification of their publication time. Experts have discovered that the name of Genghis Khan (1206–1227) appears in Mongolian and Lantsa script of the printed Sanskrit sutras, indicating a publication date no earlier than the thirteenth century. Yet, post the thirteenth century, Turpan boasted a relatively mature woodblock printing industry that continued to evolve for a considerable duration.[61]

There exist two types of Sanskrit printed fonts. Ancient Sanskrit was utilized more predominantly, whereas a minority of prints employed the Lantsa script, which was prevalent during the thirteenth century. Notably, the Lantsa script printed copies exhibited exceptional craftsmanship. Figure 4.31 depicts the Vájra-Cchedikā-Pajñā-Pāramitā-Sūtra in the concertina-binding format, with each page measuring 15.5 cm in height and 64 cm in length. The volume name and page number were alternately printed in Sanskrit and Chinese on alternating pages to facilitate binding. Figure 4.32 features the Chinese title *The One Hundred Thousand Verse Perfection of Wisdom Sutra*, accompanied by the Chinese character *sa* (meaning the number of 30) and others. Notably, the character "万" (meaning the number of ten thousand) appears in its current simplified form, while the character *sa* signifies the 30th page. This assortment of multilingual printed materials was available in scroll-bound, concertina-bound, and palm-leaf forms, with the Lantsa script Sanskrit sutras specifically presented in the palm-leaf form. In 1930, Dr. Huang Wenbi acquired a printed Buddhist sutra in Uyghur from Turpan. This sutra, in the concertina-bound form, measured 24.5 cm in height and 53.5 cm in width. It featured a Chinese page number *shi* (meaning ten) and the three characters *chen ning kan* (meaning "engraved by

[60] De Guignes J. *Histoire générale des Huns, des Turcs, des Mongols, et des Autres Tartares Occidentaux* [M]. Paris: Chez Desaint & Saillant, 1758.

[61] Von Le Coq A. *Buried Treasures of Chinese Turkestan: an Account of the Activities and Adventures of the 2nd and 3rd German Turfan Expedition* [M]. London: Allen & Unwin, 1928: 52. Carter T F, Goodrich L C. *The Invention of Printing in China and its Spread Westward* [M]. 2nd ed. New York: Ronald, 1955: 146–147, 218.

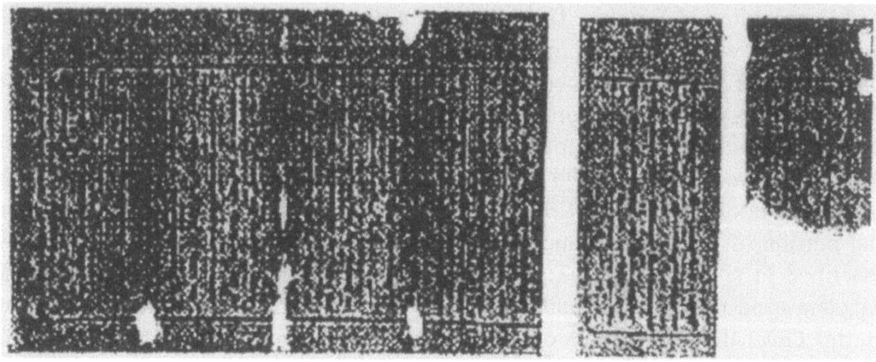

Fig. 4.31 A concertina-bound Buddhist scripture printed in the Uighur language from the Yuan dynasty (thirteenth century) unearthed in Xinjiang

Chen Ning"), denoting the name of the engraver. Upon examination by Mr. Feng Jiasheng (1904–1971), an authority on the Uyghur language and culture, the sutra was identified as *Buddha Speaks the Scripture of the Eight Buddhas Spell.* Additionally, the volume began with *Illustration of the Buddha's Teachings.*[62] Between 1908 and 1909, Ōtani Kōzui (1876–1948), a Japanese explorer, acquired a collection of printed books in Uyghur, Sanskrit, and other languages during an expedition to Turpan. Among these was a fragment of a Buddhist sutra, now housed in a museum in Kyoto, which bore a four-line Uyghur inscription. The general meaning of this inscription is, "…May it be accomplished. Printed in Ganzhou on March 1st, in the 21st year of the Zhizheng period. How righteous it is!"[63] This inscription reveals that the sutra was printed on April 6, 1361, in Ganzhou (present-day Zhangye, Gansu Province). It can be asserted that Uyghur woodblock printing flourished prominently during the thirteenth and fourteenth centuries, with Turpan and Ganzhou serving as pivotal printing centers. The topic of moveable-type printing will be elaborated on in Chap. 6.

The Mongolians, who were nomads residing in the northern steppes of the desert, were initially governed by the Liao empire. In 1125, the Jin empire supplanted the Liao empire and assumed control over the Mongolian steppes. During this period, an exceptional individual emerged: Temüjin. In 1206, he was elected as the Ikh Mongol Ulsyn Khaan, famously known as Genghis Khan, and subsequently founded the Mongol empire. Initially, the Mongolians adhered to Shamanism, but they later converted to the Gelug sect of Lamaism, which is a branch of Tibetan Buddhism. Mongolian belongs to the Mongolian language family in the Altaic language family,

[62] Huang Wenbi. *Archaeological Records of Turpan* [M]. Beijing: Chinese Academy of Sciences Press, 1954: 64. Li Guangbi, Qian Junye. *Collection of Essays on Chinese Scientific Inventions and Scientific Figures* [M]. Beijing: Joint Publishing, 1955: 225–235.

[63] Kōgi Kudara. *Amitayurdhyana Sutra in the Uyghur Language* [M]. Kyoto: Bunkoudo, 1985: 31. Shi Jinbo, Wu Shou'er. *A Study on the Movable-Type Printing Technology of Western Xia and Uighur* [M]. Beijing: Social Sciences Academic Press, 2000: 83.

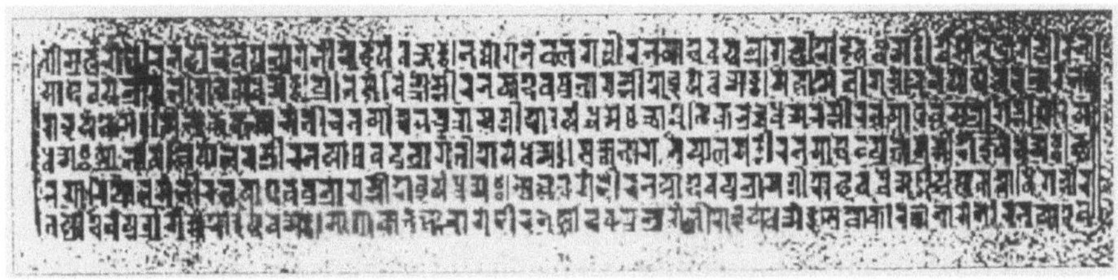

Fig. 4.32 *Diamond Sutra* in Sanskrit published in the Yuan dynasty at the end of the thirteenth century, discovered in Xinjiang

and initially lacked a written form. In 1204, after Temüjin defeated the culturally advanced Naiman tribe, he captured their seal holder, Tatatungya (1169–1234), and instructed him to devise a writing system for the Mongolian language. Utilizing the Uyghur alphabet as a foundation, Tatatungya created 19 letters to document the Mongolian language, thereby establishing a writing system for the Mongolians. The shape of the Mongolian characters closely resembled those of the Uighur characters. After the establishment of the Mongol empire, it swiftly expanded westward and southward, unifying northern China and extending its influence across Central Asia and Eastern Europe. Kublai Khan (1215–1294) ascended to the Khanate and founded the Yuan dynasty, relocating the capital to Yanjing, renamed Dadu (now Beijing), and became the emperor of a unified China, known as Emperor Shizu of Yuan. The great unification accomplished by the Yuan dynasty brought to an end the prolonged coexistence of various regimes that had persisted since the fall of the Tang dynasty, marking a significant event in Chinese history. In 1269, Emperor Shizu commissioned Drogön Chögyal Phagpa (1235–1280), also known as Phags-pa, the Tibetan state preceptor, to devise new Mongolian characters based on the Tibetan alphabet. These characters, introduced in 1270, were later named Phags-pa script, and they were written vertically from right to left, paralleling the old Mongolian characters inspired by the Uyghur style. Following the collapse of the Yuan dynasty, the Phags-pa script fell into disuse, while the old Mongolian characters remained prevalent during the Ming and Qing dynasties. During the Mongolia and Yuan dynasties, Chinese printing technology had undergone further advancements, and publications in various ethnic languages continued to be produced. The diverse printed materials discovered by the German expedition led by Albert von Le Coq in Xinjiang serve as a testament to the development of China's publishing and printing industry in the western regions during this period.

Among the printed materials unearthed by Albert von Le Coq, there were four fragments of Buddhist sutras printed in Phags-pa script Mongolian (Fig. 4.33), which serve as evidence that these sutras were published in the late thirteenth century. The center of the page featured Chinese characters such as *bashi* and *shi*, and the fragment measured 14.2 cm by 20 cm, suggesting that it was likely still printed in Turpan. Similar to Chinese and Uyghur Buddhist sutras, the Mongolian Buddhist sutras were read from top to bottom and from right to left. The layout of this sutra resembled Han books from the Central Plains, featuring a black margin, double fishtail designs, and two pages per board. The binding form was neither a folded sutra nor a scrolled binding, but rather likely a thread binding. This represents one of the earliest surviving printed materials in the Mongolian language. During Mongolia's conquest of Western Xia, Jin, and Southern Song dynasties, it invariably transported the books housed in imperial libraries and the printers from the capitals back to Mongolia for its own use. The printing centers of Mongolia were located in Karakorum (in present-day Mongolia), Shangdu (present-day Zhenglanqi in Inner Mongolia, also known as Plain Blue Banner or Shuluun Höh khoshuu in Xilingol League), Yanjing, and Turpan. According to *Yuan Shi·Xuanju Zhi* ("History of Yuan Dynasty: Treatise on the Imperial Examinations") and *Bai Guan Zhi* ("Treatise on the Officials"), in the sixth year of the Zhiyuan period (1269), Emperor Shizu mandated the establishment of Mongolian schools in each province. Two years later, in 1271, the Mongolian Imperial Academy was established in Dadu, with the aim of recruiting Mongolian and other ethnic children to study the Mongolian language. These schools employed Mongolian translations of texts such as *Tongjian Jieyao* ("Essential Summary of the Comprehensive Mirror") and others as their teaching materials. Subsequently, in 1275, the Mongolian Hanlin Academy was founded to oversee tasks including book translation. The Mongolian rulers issued directives for various Chinese books, including *Tongjian Jieyao*, *Xiao Jing* ("The Classic of Filial Piety"), *Daxue Yanyi* ("Extended Meaning of Great Learning"), to be translated into Mongolian and subsequently printed by the government. Regrettably, these editions have not survived to the present day. Currently, libraries, museums, and large temples in the Inner Mongolia Autonomous Region house Mongolian handwritten and printed manuscripts from various periods. Nevertheless, to ascertain their historical significance and integrate them into the scholarly discourse of printing history, a rigorous and time-consuming process of dating and authentication is imperative. This endeavor, while essential, is not one that yields its fruits hastily.

The Mongolian Khans who established the Yuan dynasty, notably Emperor Shizu, Kublai Khan, grasped the fundamental principle that "while one may conquer the world on horseback, governing it on horseback is impossible." They adhered to the counsel of advisors such as Yelv Chucai, Xu Heng, Wu Cheng, and Yao Shu, and embarked on a mission to propagate Confucianism across the country. Instead of abandoning the Han system, they opted to build schools, conduct imperial examinations, and permit Han Chinese to engage in politics, thereby fortifying the reign of this minority ethnic group in the Central Plains. As far back as the eighth year of Emperor Taizong's reign, in 1236, an editing office was established in Yanjing, accompanied by the construction of a classics office in Pingyang dedicated to book publishing. Following the downfall of the Southern Song dynasty, numerous publishing entities emerged across various provinces in the south. Throughout the Yuan dynasty, government agencies involved in book publication at the central level encompassed *Guozijian* (Imperial College), Imperial Library (founded in 1273), Xingwen Office (established in 1290), Imperial Hospital, and Administration of Heavenly Observatory. At the local level, Confucian schools and academies existed in various provinces, prefectures, states, and counties. Folk bookstores and private schools spread throughout

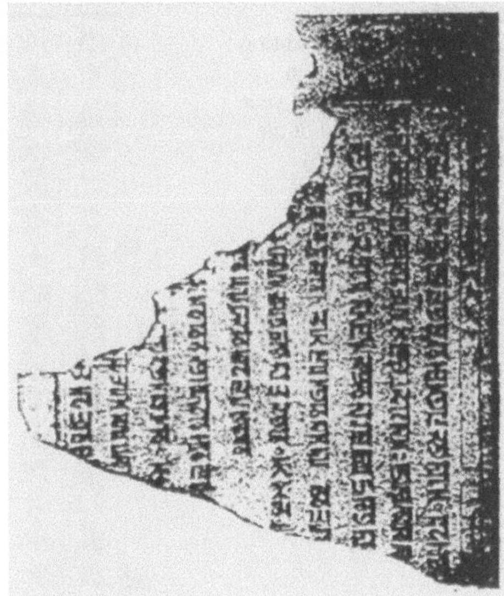

Fig. 4.33 A thirteenth-century printed copy of the Buddhist sutra in Mongolian Phags-pa script unearthed in Turpan, Xinjiang (Formerly in the collection of the Ethnologisches Museum in Berlin)

the landscape of the entire country. Printing in the Yuan dynasty carried on the tradition of the Song dynasty, with a regulation stipulating that books could only be published after submitting an application to the local government of the province where they were to be printed. In certain instances, these applications were required to be submitted to the Ministry of Rites or Jixian Academy of the Central Government, or to the Central Secretariat or local officers, for review prior to publication. However, over time, the enforcement of this regulation became lax, rendering it impractical for officials to individually review such a vast number of books.

Books published by the central bodies of the Yuan dynasty are scarce in circulation, yet the Imperial Secretariat alone employed over a thousand engravers, indicating a prolific production of books. Numerous copies of books commissioned by the central government and published by local officials have survived to this day, such as *Shiqi Shi* ("Seventeen Histories") published during the Dade period, with the printing site in what is now Anhui and Jiangxi provinces. In the sixth year of the Zhizheng period (1346), the Central Secretariat set up a regional office in Hangzhou, Zhejiang, to publish *Song Shi* ("History of Song"), *Jin Shi* ("History of Jin", Fig. 4.34), and *Liao Shi* ("History of Liao"), which were regarded as exquisite "fine editions", featuring ten lines per half-page and twenty-two characters per line. Notable among the academy editions is *Jiaxuan Changduanju* ("Works of Xin Qiji", Fig. 4.35) by the Song dynasty poet Xin Qiji (with the courtesy name of Jiaxuan, 1140–1207), published in 1299 by the Guangxin Academy in Guangxin (present-day Qianshan, Jiangxi), a representative publication printed in the calligraphic style of Zhao Mengfu (1254–1322). Privately published editions have also left many surviving copies, such as *Wangshi Maijing* ("Wang's Pulse Classic") published in the third year of the Tianli period (1330) by Ye Rizeng's Guangqin Hall in Jian'an, Fujian Province (Fig. 4.36),

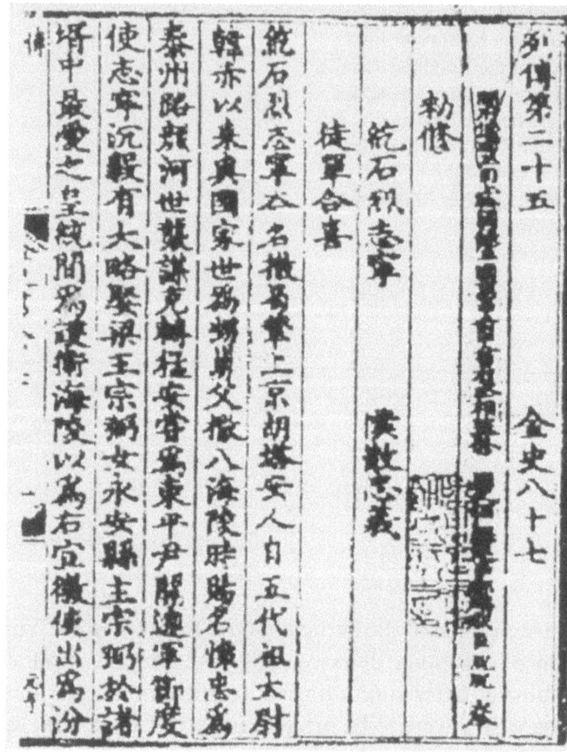

列傳第二十五 金史八十七
敕修
紇石烈志寧 僕散忠義
徒單合喜
紇石烈志寧本名撒曷輦上京胡塔安人自五代祖太尉
韓赤以來與國家世爲甥舅父撒八海陵時賜名懷忠爲
泰州路頗河世襲謀克轄猛安睿宗爲東平尹闢邊軍節度
使志寧沉毅有大略娶梁王宗弼女永安縣主宗弼於諸
將中最愛之皇統間爲護衛海陵以爲右宣徽使出爲汾

Fig. 4.34 Government-issued edition of *Jin Shi* published in Hangzhou in the sixth year of Zhizheng period of the Yuan dynasty (1346)

and *Guochao Mingchen Shilue* ("A Brief Account for Famous Officials in Yuan Dynasty") published in the third year of the Yuantong period (1335) by Yu Zhi'an's Qinyou Studious Hall in Chong'an (Fig. 4.37).

A detailed list of publications from the Yuan dynasty will not be enumerated here. Overall, due to the relatively short duration of this dynasty, its publications paled in comparison to the prosperity of the Song dynasty, and therefore it did not surpass the Song dynasty in terms of printing technology.

4.3.3 Woodblock Printing in Ming and Qing Dynasties

The Ming and Qing dynasties marked the zenith of traditional Chinese printing technology, surpassing previous eras in terms of the quantity, variety, quality, thematic diversity, and geographical distribution of printed materials. Woodblock printing technology reached new heights of sophistication, with the artistry and functionality of prints at that time unmatched by those of earlier generations. Technological advancements, particularly in multicolor printing techniques, allowed for the reproduction of intricate and large-scale artistic works, with the number of illustrated

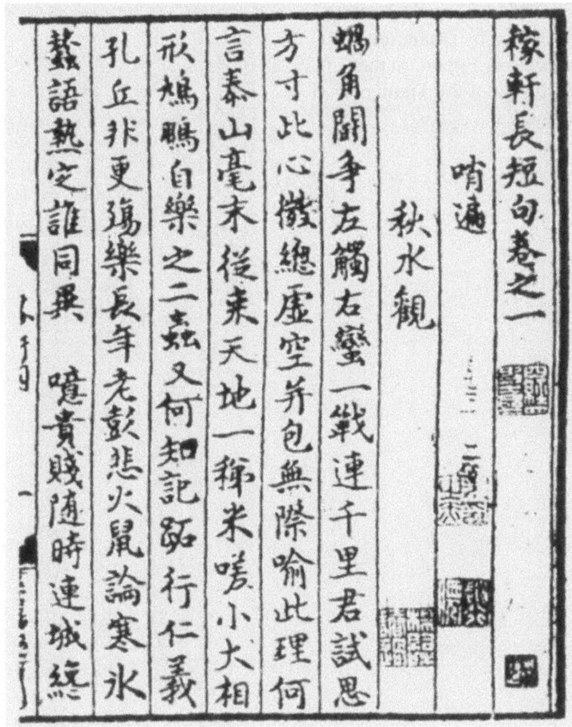
稼軒長短句卷之一
哨遍
秋水觀
蝸角鬪爭左觸右蠻一戰連千里君試思
方寸此心微總虛空并包無際喻此理何
言泰山毫末從來天地一稊米嗟小大相
形鳩鵬自樂之二蟲又何知記跖行仁義
孔丘非更殤樂長年老彭悲火鼠論寒氷
蟻語熱定誰同異 噫貴賤隨時連城纔

Fig. 4.35 *JiaXuan Changduanju* published by Jiangxi Guangxin Academy in the third year of the Dade period of the Yuan dynasty (1299)

editions surging dramatically, bringing high-quality prints into the homes of the masses and elevating printed materials to the status of art. Various forms of movable-type technology had undergone comprehensive development, with successive waves of innovation in typography, especially in terms of movable-type printing. Previously, printing fonts were predominantly handwritten in regular script, imitating the style of famous calligraphers, leading to a lack of standardization. However, during the Ming and Qing dynasties, there emerged a vogue for a square-shaped standard typeface in printing, characterized by its horizontal strokes that were light and vertical strokes that were bold, which not only adhered more rigorously to the intrinsic attributes of the printing industry but also epitomized the emblem of technological progress. In terms of bookbinding, methods such as wrapped-back binding and thread binding became prevalent, offering greater ease of use and storage. The range of subjects covered in publications was vast, catering to the interests of the general public. Illustrated editions of operas, novels, and picture books were produced in large quantities. Additionally, works on Confucian classics, history, philosophy, religion, science and technology, as well as local chronicles, genealogies, collected works, books of categories, texts in various ethnic minority languages, and translations of Western literature, were all extensively published, with some subjects not seen in previous eras.

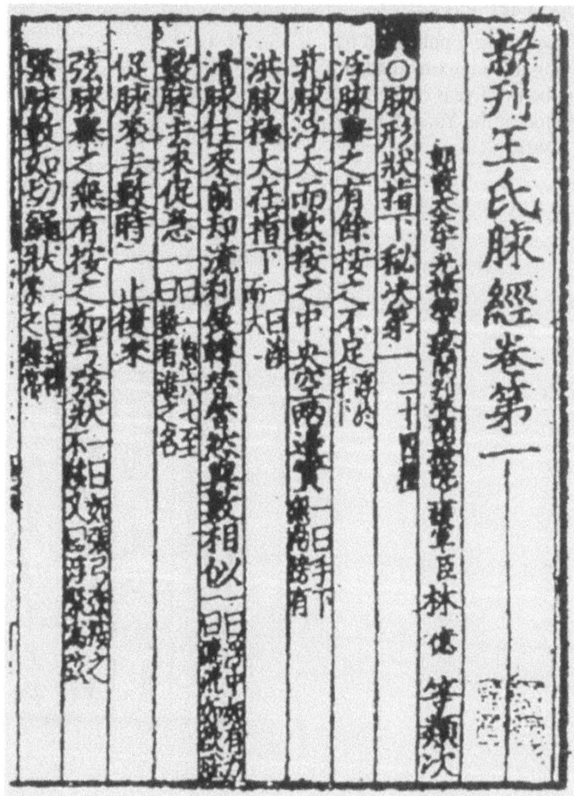

Fig. 4.36 *Wangshi Maijing* published by Guangqin Hall in Jian'an, Fujian, in the third year of the Tianli period of the Yuan dynasty (1330)

The Ming and Qing dynasties were unified, with a well-established imperial examination system, highly developed school education system, and fruitful academic research achievements, providing a rich source of manuscripts for printing and a vast readership as well. These dynasties, particularly during the reigns of eminent rulers like Emperors Hongwu and Yongle of the Ming dynasty and Emperors Kangxi and Qianlong of the Qing dynasty, were characterized by relative social stability, economic prosperity, and strong national power—often referred to as "golden ages". These periods saw a surge in the compilation and publication of books, laying the foundation for the development of the printing industry. Hence, papermaking and printing during the Ming and Qing dynasties represented the culmination of historical advancements. However, from a global perspective, after the eighteenth century, traditional manual production methods gradually yielded to modern, advanced machine printing technologies. As part of the world, China could not resist this general trend of technological development. Therefore, while the printing technology of the Ming and Qing dynasties reached its historical peak, it also signaled its impending exit from the historical stage, making way for modern printing technology. Towards the end of the Qing dynasty, the introduction of Western lithographic and letterpress printing technologies revolutionized the Chinese print industry. As the

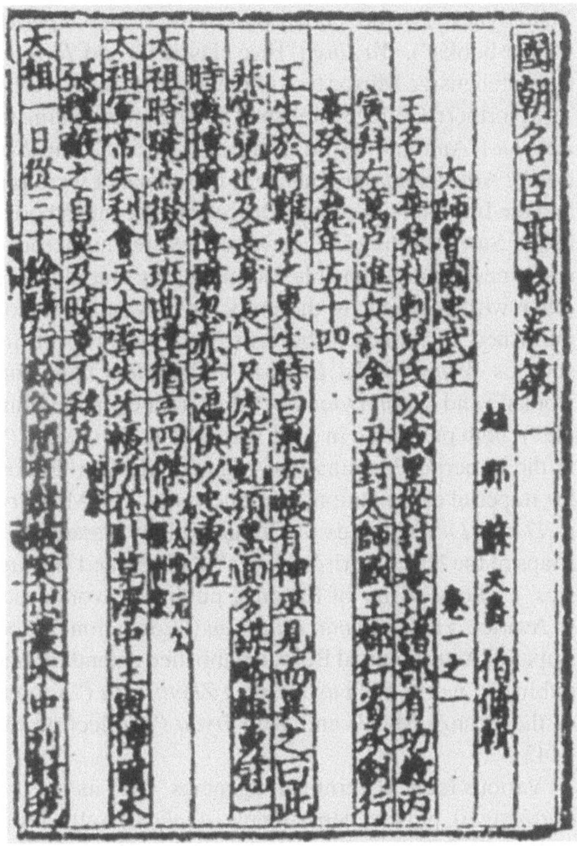
國朝名臣事略卷第一

趙郡蘇天爵伯脩輯

卷一之一

太師魯國忠武王

Fig. 4.37 *Guochao Mingchen Shilue* published by Yu Zhi'an's Studious Hall in Chong'an in the third year of the Yuantong period in the Yuan dynasty (1335)

1920s unfolded, modern movable-type lead typesetting technology had usurped the traditional methods, solidifying their position as the predominant means of printing in China.

In the early Ming dynasty, Nanjing served as the capital for more than 50 years (1368–1421), becoming both the political center and the largest printing center. The *Guozijian* (Imperial College) and government departments all engaged in book printing. The *Guozijian* provided standard textbooks for schools at various levels and other important works approved by the government (Fig. 4.38). Emperor Chengzu (1403–1424) moved the capital to Beijing in the 19th year of the Yongle period (1421) and established *Guozijian* there, while retaining the one in Nanjing, hence the term "Southern and Northern Colleges." Other ministries and agencies in Nanjing were not abolished, and a new set was established in Beijing. The Beijing ministries and agencies had great power, and orders were issued from there. During the reign of Emperor Taizu (1368–1398), the Southern College took over the books and printing plates from the West Lake Academy in Hangzhou, the former Imperial College of the Southern Song dynasty, as well as *Kuizhangge* and *Chongwenyuan* in Dadu of the

Yuan dynasty. In the second year of the Hongwu period (1369), books such as *Si Shu* ("Four books"), *Wu Jing* ("Five classics") and *Tongjian Gangmu* ("Compendium of Comprehensive Mirror to Aid in Government") were published, totaling 271 books. The Northern College published 85 books, including the Wanli edition of *Shisan Jing Zhushu* ("Annotations on Thirteen Classics") and *Ershiyi Shi* ("Twenty-one Histories"). After Emperor Chengzu, the eunuchs were promoted to positions of power, and the Imperial Ceremony Office also became a book publishing agency. The established Sutra Printing Office, which boasted numerous skilled carvers and substantial resources, was responsible for publishing Buddhist scriptures, Taoist scriptures, and other works issued by the emperor to various regions. The Sutra Printing Office published 172 kinds of books (Fig. 4.39). The Ministry of Rites published works such as *Wujing Sishu Daquan* ("Complete Collection of Five Classics and Four Books") and *Xingli Daquan* ("The Great Compendium of Human Nature and Principle") (both published in 1415), as well as *Dengke Lu* ("List of Successful Candidates in the Imperial Examination") and *Huishi Lu* ("Records of Imperial Examination") for imperial examinations over the years. The Ministry of War published works such as *Dayue Lu* ("Records of Grand Military Parade"), *Jiubian Tushuo* ("Records and Maps of the Nine Garrison Zones") (1538), and *Wujing Qishu* ("Seven Military Classics"). The Ministry of Revenue published works such as *Xingtan Lu* ("Awakening to Avarice") and *Jiaomin Bangwen* ("Educational Proclamation to the People"). The Imperial Astronomical Bureau published calendars, and the Imperial Medical Bureau published works such as *Tongren Zhenjiu Tu* ("Acupuncture and Moxibustion Atlas of the Bronze Man") and *Yilin Jiyao* ("Collection of Essential Medical Texts") in 1443.

Various local government agencies, such as the provincial financial departments (*buzhengsi*), judicial departments (*anchasi*), salt administration offices (*yanyunsi*), as well as prefectures, states, and counties, also occasionally published books, particularly local chronicles, and reprinted editions issued by the central government. To consolidate his rule, the founding emperor of the Ming dynasty, Emperor Taizu, granted his sons the titles of princes and stationed them in various regions. The publication of books by princely courts was a distinctive feature of the Ming dynasty, exemplified by Zhu Su (1361–1425), Prince of Zhou, who, in the fourth year of the Yongle reign (1406), published his *Jiuhuang Bencao* ("Edible Herbs and Other Plants Needed in Disaster Relief") and the 168-volume *Puji Fang* ("Prescriptions for Universal Relief"). Zhu Zaiyu (1536–1610), the son of Prince Zhenggong, published *Yuelv Quanshu* ("Collection of Tone-system"). The topic of currency printing will be discussed in Chap. 7. Meanwhile, private publishing and book printing by commercial presses were also prevalent, mainly concentrated in the northern and southern capitals, Suzhou, Hangzhou, Changzhou, Yangzhou, Jianning, Fuzhou, Nanchang, and Huizhou. Commercial presses published popular books such as reference materials for imperial examinations, recreational readings, daily household books, reference books for industrial and agricultural production, and medical books. One of the most renowned private publishers was Mao Jin (1599–1659) of Changshu (now in Jiangsu Province). As a bibliophile, he established Jigu Pavilion, where he housed a collection of 80,000 books. He set up a printing workshop at home, invited scholars to

叙重刻三國志

Fig. 4.38 *Sanguo Zhi Zhu* ("Notes to History of the Three Kingdoms")[64] published by the *Guozijian* of Nanjing in the 24th year of the Wanli period of the Ming dynasty (1592)

proofread, and published 600 types of books, including *Shisan Jing* ("Thirteen Classics"), *Shiqi Shi* ("Seventeen Histories"), *Liushi Zhong Qu* ("Sixty Plays"), and *Jindai Mishu* ("Collection of Rare Editions", 751 volumes). Another publisher renowned alongside Mao Jin was Hu Wenhuan (1561–1630) of Hangzhou, who published 450 types of books. The Ming dynasty witnessed the publication of more scientific and technical books in terms of variety and quantity than any previous dynasty. Various government agencies, commercial presses, and private individuals produced books in this field, such as *Bencao Gangmu* ("Compendium of Materia Medica", 1596), *Wubei Zhi* ("Treatise on Military Preparedness", 1621) (Fig. 4.40, with a format size of 14 cm × 20.5 cm), *Nongzheng Quanshu* ("Complete Treatise on Agriculture", 1639), and *Tiangong Kaiwu* ("Exploitation of the Works of Nature", 1637). Additionally, some Western scientific books were also published, such as *Jihe Yuanben* ("Elements of Geometry").

The official printing system of the Qing dynasty differed significantly from its predecessors. In that period, *Guozijian* and central ministries seldom published

64 There are three Chinese characters *dai ming ke* (engraved by Dai Ming) under the book mouth.

叙重刻三國志

Fig. 4.39 *Zhouyi Zhuanyi* published by the Sutra Factory of the Imperial Ceremony Office in Beijing in the 12th year of the Zhengtong period of the Ming dynasty (1447)

books; instead, the responsibility fell to the Wuyingdian Scriptorium, established during Emperor Kangxi's reign. In the 29th year of the Kangxi period (1690), it was renamed the Wuyingdian Book Revision Office, located inside the Xihua Gate of the Forbidden City and supervised by the Imperial Household Department, under the oversight of the Chief Minister of the Imperial Household Department. Books intended for nationwide circulation were supplied from this office, known as Wuyingdian editions or imperial editions. After their distribution to various regions, local governments were permitted to reprint them. During the reigns of Kangxi, Yongzheng, and Qianlong, approximately 500 titles were published,[65] most of which were contemporary works, such as *Daqing Huidian* ("The Code of Qing Dynasty"), *Daqing Lvli* ("Statutes and Regulations of Qing Dynasty"), *Daqing Yitongzhi* (book about historical geography of Qing Dynasty), *Huangchao Liqi Tushi* ("Imperial Qing Ceremonial Paraphernalia"), *Wanshou Shengdian* ("Grand Celebration for the Emperor's Birthday"), *Baqi Tongzhi* ("General Records of Eight Banners"), and *Rixia Jiuwenkao* ("The Beijing

[65] Lu Xiuju. *The Royal Printing Industry in the Prosperous Qing Dynasty* [M] // Xu Zhuoyun, et al. *Collection of Essays on Chinese Books, Literature and History*. Beijing: Modern Press, 1992: 33–74.

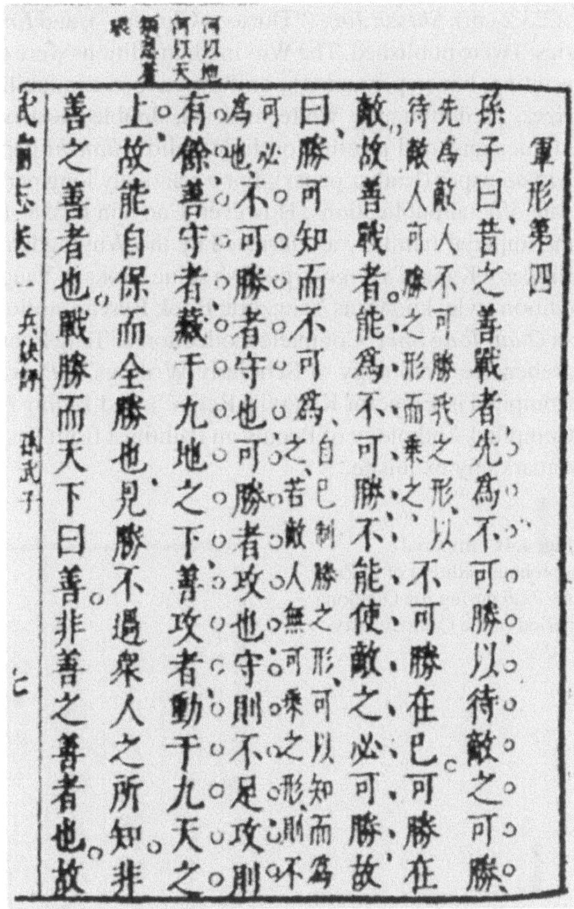
軍形第四
孫子曰昔之善戰者先為不可勝以待敵之可勝不可勝在己可勝在敵故善戰者能為不可勝不能使敵之必可勝故曰勝可知而不可為不可勝者守也可勝者攻也守則不足攻則有餘善守者藏於九地之下善攻者動於九天之上故能自保而全勝也見勝不過衆人之所知非善之善者也戰勝而天下曰善非善之善者也故
武備志卷一 兵訣評 孫武子 七

Fig. 4.40 *Wubei Zhi* published in Nanjing in the first year of the Tianqi period of the Ming dynasty (1621)

News & the Study of Beijing News"). During Emperor Kangxi's reign, a vast number of "imperially commissioned editions" were published, including *Shuli Jingyun* ("Essence of Mathematical Principles"), *Lixiang Kaocheng* ("Astronomical Observations and Calculations"), *Xingchen Kaoyuan* ("Origins of the Stars"), *Kangxi Zidian* ("Kangxi Dictionary"), *Peiwen Yunfu* ("Rhyme Repository Compiled by Emperor Kangxi"), *Yuanjian Leihan* ("Comprehensive Anthology of Scholarly Writings"), *Xingli Jingyi* ("Essence of Neo-Confucian Philosophy"), *Gengzhi Tu* ("Illustrated Treatise on Farming and Textile Production"), *Quan Tang Shu* ("Complete Tang Dynasty History"), and *Zishi Jinghua* ("Essence of Historical and Philosophical Writings"). During Emperor Qianlong's reign, works such as *Shiqu Baoji* ("Treasures in the Stone Channel"), *Xiqing Gujian* ("Ancient Mirrors of the Western Studio"), *Shoushi Tongkao* ("A General Survey of Agriculture"), *Yizong Jinjian* ("Golden Mirror of Medical Science", Fig. 4.41, with a format size of 15.2 cm

× 23.2 cm), *Shisan Jing* ("Thirteen Classics"), and *Ershisi Shi* ("Twenty-Four Histories") were published. The Wuyingdian editions were distinguished by their exquisite printing, fine paper and ink, uniform layout, availability in both large and small font sizes, predominantly white margins, double borders around the text, single fish-tail designs, and printing on light yellow *lianshi* paper (bamboo paper) and white *kaihua* paper (leather paper). For a relatively long period, imperial editions were the sole official publications. However, Cao Yin (1658–1712), who had close ties with the imperial family, was entrusted by the Wuyingdian Book Revision Office during Emperor Kangxi's reign to publish some books in Yangzhou according to the imperial edition style, known as Yangzhou Book Bureau editions. These included works such as *Quan Tang Shi* ("Complete Collection of Tang Poems"), *Yuanjian Leihan* ("Comprehensive Anthology of Scholarly Writings"), *Peiwen Yunfu* ("Rhyme Repository Compiled in Emperor Kangxi's Reign"), and *Yuding Lidai Tihuashi Lei* ("Imperially Compiled Anthology of Poems on Paintings from Past Dynasties"), which were also remarkably exquisite.

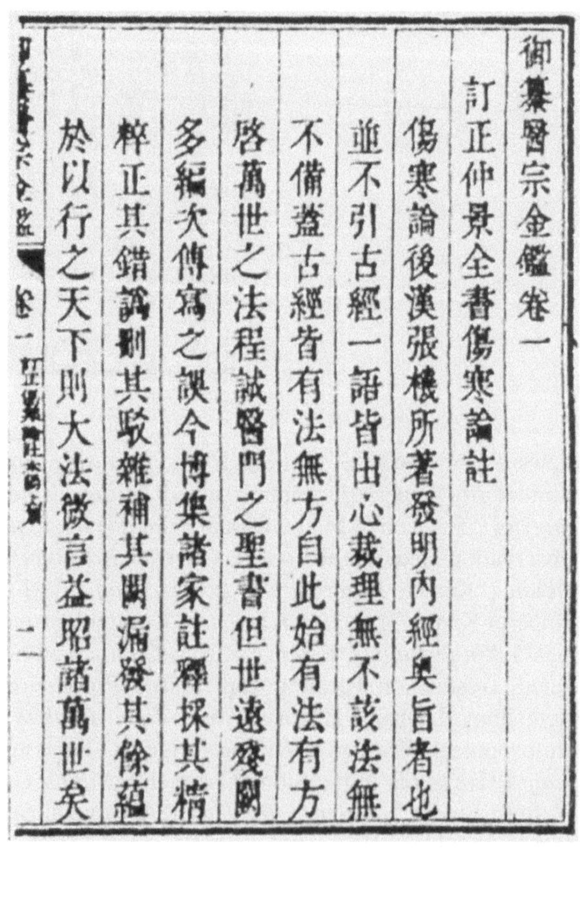
御纂醫宗金鑑卷一
訂正仲景全書傷寒論註
傷寒論後漢張機所著發明內經奧旨者也
並不引古經一語皆出心裁理無不該法無
不備蓋古經皆有法無方自此始有法有方
啓萬世之法程誠醫門之聖書但世遠殘闕
多編次傳寫之誤今博集諸家註釋採其精
粹正其錯譌刪其駁雜補其闕漏發其餘蘊
於以行之天下則大法微言益昭諸萬世矣

Fig. 4.41 Imperial storehouse edition of *Yi Zong Jin Jian* during the Qianlong period of the Qing dynasty

The Wuyingdian Book Revision Office also published works from previous dynasties, such as *Tong Dian* ("Comprehensive Institutions"), *Tong Zhi* ("General Treatises"), *Wenxian Tongkao* ("Comprehensive Textual Research of Historical Documents"), *Tang Huiyao* ("Economic and Political Structure of Tang Dynasty"), *Lunyu Jijie Yishu* ("Collected Interpretations and Commentaries on *The Analects*"), *Shisan Jing* ("Thirteen Classics"), *Ershisi Shi* ("Twenty-Four Histories"), and *Tripitaka*, among others. After the reign of Jiaqing (1796–1820), imperial editions continued to be published, but the quantity dwindled sharply. Following Emperor Tongzhi's reign (1862–1874), official book bureaus were established in provinces such as Jiangsu, Zhejiang, Jiangxi, Hubei, Sichuan, Hunan, Shanxi, Shandong, Fujian, Guangdong, Yunnan, Guizhou, Hebei, and Anhui, where they reprinted imperial editions and published other books, totaling over 1000 titles. Private bookstores were ubiquitous across the Qing dynasty, producing a wide range of content in large quantities. The variety of printed materials in the Qing dynasty was extensive, including books, newspapers, governmental proclamations, deeds for houses and land, advertisements, bills, folk new year paintings, and paper money. Among these, proclamations were the most extensive in length. For instance, the "Announcement for Reassuring the People" issued by Regent Dorgon after the Qing troops entered Beijing on July 8 in the first year of Emperor Shunzhi's reign (August 9, 1644), is now housed in the China Printing Museum. It was carved on large wooden blocks, surrounded by cloud and dragon designs. Similarly, a proclamation from the Shanxi Provincial Court of Justice on the use of old and new paper money, dated September 15 of the twenty-second year of the Yongle period in the Ming dynasty (September 30, 1422), was discovered in 1974 at the Shaka Tower of Fogong Temple in Ying County, Shanxi. This proclamation measures 94.5 cm in height and 276 cm in length, printed as a single sheet. These two proclamations may be the largest single-page printed materials to date.

During the Ming and Qing dynasties, the printing of ethnic minority scripts also flourished, with particular attention to the printing in Tibetan regions and the publication of Tibetan classics. Historical records indicate that Tsampa Gade Kongpo (ca. 1259–1319) visited the interior of China seven times during the reign of Kublai Khan (1260–1294) of the Yuan dynasty, introducing printing technology to Tibet and establishing a printing press at the Tsampa Monastery in the eastern suburbs of Lhasa. Meanwhile, Phags-pa, the National Preceptor favored by Kublai Khan, who assisted the central government in managing Tibetan affairs, also established a printing press at his base, the Sakya Monastery, located in Xigaze Prefecture's Sakya County, which is divided into a Northern Temple and a Southern Temple. The Southern Temple was built by Xiaka Sambhu under Phags-pa's commission in the sixth year of the Zhiyuan period (1269). Inside the main hall of the temple, there are woodblock prints and Tibetan Buddhist scriptures, including *Sajia Lidai Shilue* ("A Brief History of the Sakya Lineage") and *Sajia Ge Jiaozhu Fawang Zhuan* ("Biographies of the Dharma Kings of the Sakya School") and Phags-pa and his disciples introduced mainland printing technology to Tibet between the second and fifteenth years of the Zhiyuan period (1265–1278). Therefore, printing presses were already established in Lhasa and Xigaze in the thirteenth century.[66] From 1902 to 1907, the Le Coq Expedition discovered fragments of Tibetan Buddhist scriptures carved on woodblocks in Turpan, which also dated back to the thirteenth century.

Early Tibetan printed works mostly consisted of Buddhist scriptures, Buddhist images, or individual Buddhist texts. At the end of the Yuan dynasty, Tsampa Kunga Dorje (1309–1364) compiled *Kangyur*, which included over 1000 Tibetan Buddhist scriptures. Meanwhile, Buton Rinchen Gyalpo (1290–1364) compiled *Tāngyur*, which contained 3461 works (Dege edition). These two collections formed the basis of the Tibetan *Tripitaka*. Due to their immense volume and late completion, they were not published during the Yuan dynasty. In the early Ming dynasty, Emperor Yongle, in response to requests from Tibetan monks, issued an imperial command in the ninth year of his reign (1411) to establish a Tibetan scriptures printing factory in Nanjing. He dispatched the eunuch Hou Xian as an imperial envoy to oversee the printing process, along with the Great Precious Dharma King Chogyal Sangpo (1384–1415). With the cooperation of Han and Tibetan workers, *Kangyur* was printed (Fig. 4.42, with a plate frame size of 8.1 cm × 5.8 cm), consisting of 108 volumes. The original Yongle edition was once housed in the Sakya Monastery and later moved to the Potala Palace in Lhasa. It was discovered during a cultural relics inventory in 1985. In the 22nd year of the Wanli period (1594) of Emperor Shenzong in the Ming dynasty, the imperial court issued a command to establish a scripture printing factory in Beijing to publish *Tāngyur*. By this time, the Tibetan *Tripitaka* had been fully printed. This process accumulated experience and cultivated a group of Tibetan printers, promoting the development of printing in Tibet. During the late Ming and early Qing dynasties (seventeenth century), the Litang Monastery in western Tibet published new editions of the *Kangyur* and *Tāngyur*. In the seventh year of Emperor Yongzheng's reign (1729) of the Qing dynasty, Dege Chief Qoji Dengba Tsering (1678–1738) established the famous Dege Printing House in Dege County, Ganzi Tibetan Autonomous Prefecture, Sichuan Province, which is close to Tibet and became the largest printing center in Tibetan region.[67] It has published a large number of Tibetan works and currently possesses 217,500 book blocks from various dynasties. Each block contains two pages, and with about 600 words per page, the total number of words reaches 260 million.

The Yi people, an ethnic minority in southwestern China, believing in polytheism, are distributed across Yunnan, Sichuan, Guizhou, and Guangxi, with Yunnan having the largest population. The Yi language belongs to the Sino-Tibetan language family's Tibeto-Burman branch. The Yi script, modeled on Chinese characters, is a syllabic ideogram that transcends dialects. Each character of the Yi language has a unique meaning, and there are over 10,000 characters in total. By the Yuan dynasty, the Yi script had reached its mature form. Countless works written in the Yi script by

---

[66] Pan Jixing. *History of Chinese Science and Technology: Papermaking and Printing Volume* [M]. Beijing: Science Press, 1998: 482–485.

[67] Dege Scripture Printing House: *Tibetan Edition* [M]. Chengdu: Sichuan Nationalities Publishing House, 1981: 1–7. Yang Jiaming. Dege Scripture Printing House: *Chinese, Tibetan, and English Commentary Editions* [M]. Chengdu: Sichuan People's Press, 2000.

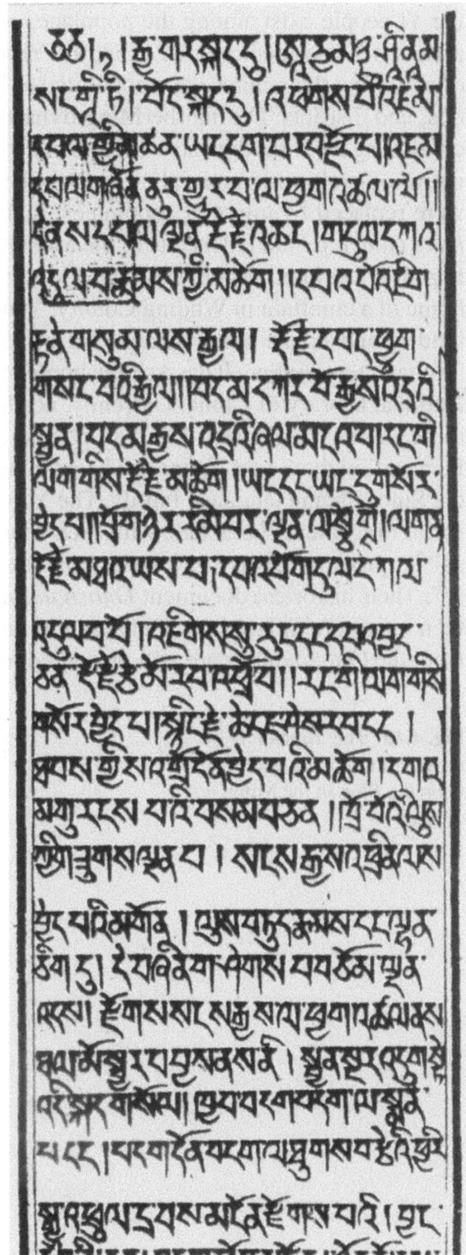

Fig. 4.42 *The Sutra of the True Names of Manjushri* from *Kangyur*, a Tibetan Buddhist Tripitaka published by the Nanjing Buddhist Tripitaka Factory in the ninth year of the Yongle period of the Ming dynasty (1411) (Collection of the National Library of China)

the Yi people exist among the populace, covering topics such as history, literature, astronomy, and medicine. In 1253, the Yi regions were unified in the Mongol-Yuan dynasty, and their chiefs were appointed as local officials in prefectures, cities, counties, and districts. During the Ming dynasty, garrison commands were established, continuing the Yuan dynasty's system of native officials. After the reform of replacing native officials with imperial officials in the Qing dynasty, hereditary native officials were replaced by imperial appointees. In 1940, an old copy of the Taoist text *Taishang Ganying Pian* ("Tractate of the Most High One on Actions and Consequences", Fig. 4.43), printed in the Yi script during the Ming dynasty, was discovered in the home of a chieftain in Wuding County, Yunnan. The book features a black-mouthed binding and a plate frame size of 13.9 cm × 22.3 cm, with 10 lines per half-page and 25 characters per line. It has a single border on all sides and was later transferred to the National Library of China. Currently, academic research on Yi literature still needs to be further conducted. The Yao people are distributed in Guangdong, Guangxi, Hunan, Guizhou, and Yunnan. Their language belongs to the Miao-Yao branch of the Sino-Tibetan language family. The Yao script was also created by borrowing the forms of Chinese characters, with each character representing a single syllable, such as "I", pronounced as "yā", written in the form similar to the Chinese character "仡". Their historical document *Guoshan Bang* records the origin of the Yao people, their ancestors' migrations, and agricultural practices. Wooden block-printed copies from the Qing dynasty are still preserved today.

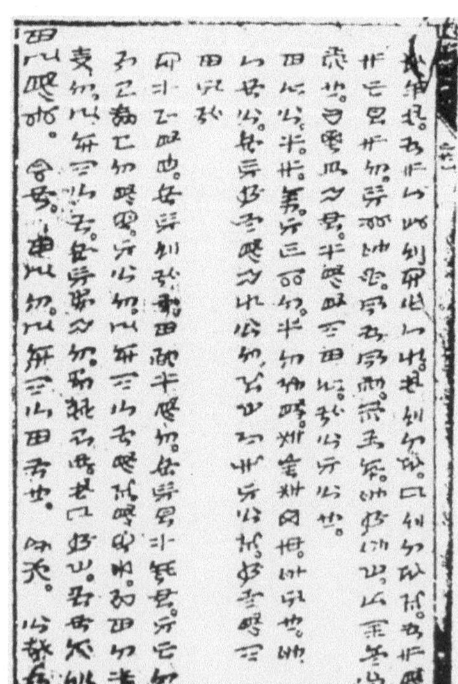

Fig. 4.43 The Yunnan Yi script edition of *Taishang Ganying Pian* in the Ming dynasty

During the Yuan dynasty, the cultural heritage of the Mongol ethnicity reached its zenith by the Ming and Qing periods. In the reign of Emperor Taizu of Ming, the Mongolian scholar Huo Yuanjie (1342–1402), who served as a lecturer in the Imperial Academy, translated Mongolian into Chinese characters and compiled *Huayi Yiyu* ("Chinese-Foreign Language Translation Dictionary") in the 15th year of the Hongwu period (1382). The edition published by the Imperial Household in the 22nd year of the Hongwu period (1389) is still preserved today and serves as a reference tool for studying the Mongolian language. Lobsang Danjin (1686–1758) wrote *Altan Tobchi* ("Golden History of the Mongols") in Mongolian, which chronologically records the history of the Mongolian people from ancient times to the Ming and Qing dynasties and was published during the Qing dynasty. *Mengwen Qimeng* ("Mongolian Primer"), compiled by Danzandageba during the Yongzheng reign, has various editions and represents a comprehensive collection of Mongolian grammar. *Menggu Wenzi Xiyi* ("Explanation of Mongolian Characters"), written by Jing Hui, a successful candidate in the imperial examinations from the Mongolian people, and published during Emperor Daoguang's reign, is a reference book for studying the relationship between Chinese, Mongolian, and Manchu characters. Mijur translated the Tibetan medical text*Sibu Yidian* ("The Four Medical Tantras", ca.750) by Yutuo Yundangongbu into Mongolian, which was also published during the Qing Dynasty. Early Qing Dynasty publications also included Buddhist scriptures, such as the Mongolian-language *Qifo Rulai Gongyang Yigui Jing* ("Rituals for Offering to the Seven Buddhas and Tathagatas"), published in Beijing in the 21st year of the Kangxi reign (1682) (Fig. 4.44, with a plate frame size of 11.8 cm × 41.6 cm).

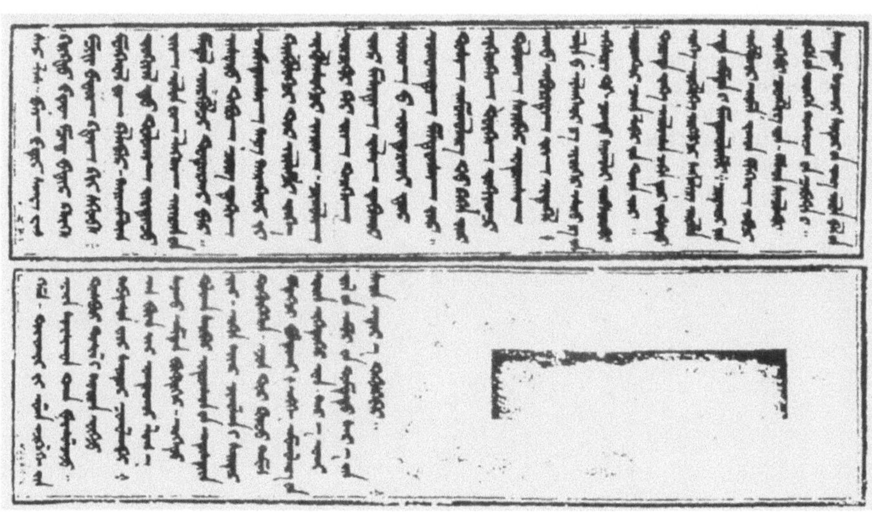

Fig. 4.44 Mongolian edition of *Qi Fo Ru Lai Gong Yang Yi Gui Jing* published in Beijing in the 21st year of Emperor Kangxi's reign in the Qing dynasty (1682) (Collection of the National Library of China)

The Qing dynasty was established by Manchu rulers, leading to a substantial number of Manchu-language block-printed books during the Qing period, with Beijing serving as the printing hub. Xifu (1588–1652), a Manchu scholar who served as an academician in the Hongwen Guan in the first year of Chongde (1636) during the reign of Emperor Taizong, had already translated excerpts of *Liao Shi* (*"History of Liao"*), *Jin Shi* (*"History of Jin"*), and *Yuan Shi* (*"History of Yuan"*) into Manchu during his time in Shengjing (present-day Shenyang, Liaoning Province). After entering the Shanhaiguan Pass, he presented these translations in the first year of Emperor Shunzhi's reign (1644) and they were subsequently published in Beijing for the study by the youth of the Eight Banners. In the seventh year of Emperor Shunzhi's reign (1650), Manchu-Chinese bilingual edition of *Sanguo Yanyi* ("Romance of the Three Kingdoms") translated by Dahai (1595–1632) in Shengjing was reprinted, titled *Manhan Hebi Sanguo Zhi Yanyi* ("Romance of the Three Kingdoms: A Bilingual Version in Manchu and Chinese", Fig. 4.45, with a plate frame size of 28.1 cm × 20 cm), which facilitated mutual learning of the two languages by both Manchus and Han Chinese. According to Volume 1 of *Xiaoting Xulu* by Zhao Lian (1776–1830): After the establishment of the capital in Beijing in 1644, a translation and reproduction office was set up in the west corridor of the Taihe Gate [in the Forbidden City], staffed by Manchu bannermen proficient in Manchu language and literature, with no fixed number of personnel. Books such as *Zizhi Tongjian*, *Xingli Jingyi*, and *Guwen Yuanjian* were all translated into Manchu for circulation.

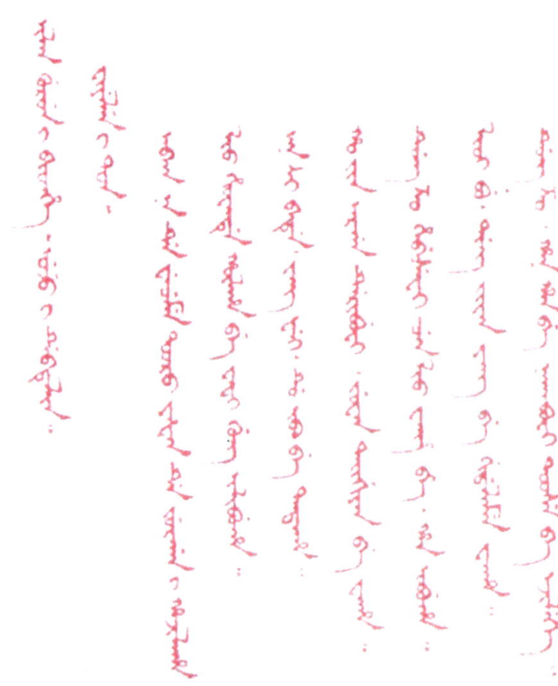

Fig. 4.45 The Manchu section of the Manchu-Chinese version of *Romance of Three Kingdoms* published by the imperial storehouse in Beijing in the seventh year of Emperor Shunzhi's reign of the Qing dynasty (1650) (Collection of the National Library of China)

It is evident that *fankefang* (Translation and Reproduction Office) was the earliest institution established within the imperial palace during the early Qing dynasty to publish works in Manchu. The aforementioned histories of Liao, Jin, and Yuan dynasties, as well as *Sanguo Yanyi*, were all published here in the early years of Emperor Shunzhi's reign. According to Zhao Lian's account, *Xingli Jingyi* and *Guwen Yuanjian* were completed during Emperor Kangxi's reign, and their Manchu translations should have been published during the same period. In the third year of Emperor Shunzhi's reign (1646), *Hongwu Baoxun* ("Precepts of Emperor Hongwu") was translated and published in both Manchu and Chinese, and distributed nationwide. Emperors Kangxi, Yongzheng, and Qianlong's reigns marked a period of economic and cultural prosperity in the Qing dynasty, and also a peak in the publication of Manchu books. In the twenty-second year of Emperor Kangxi's reign (1683), Shen Qiliang completed his 14-volume *Daqing Quanshu* ("Complete Book of the Great Qing"), which is the first large-scale Manchu-Chinese dictionary in China, containing 12,000 entries arranged according to Manchu characters with 12 initial consonants. When it was reprinted in 1713, *Qingshu Zhinan* ("Guide to Qing Literature") was added as a supplement. In the forty-ninth year of Emperor Kangxi's reign (1710), *Futu Manhan Xixiangji* ("Illustrated Manchu-Chinese Romance of the Western Chamber") was published. To accommodate Manchu and Mongolian students participating in the imperial examinations, *Si Shu* ("Four Books") and *Wu Jing* ("Five Classics") were already published in official Manchu editions, along with various privately printed editions. For example, in the third year of Emperor Qianlong's reign (1738), Hongyun Bookstore in Beijing published an illustrated edition of *Manhan Zishujing* ("Dictionary in Manchu and Chinese") (Fig. 4.46). In the thirteenth year of Emperor Qianlong's reign (1748), the emperor ordered the Grand Secretary Fu Heng (1720–1769) to design Manchu seal script inspired by Chinese seal script. In the same year, Emperor Gaozong authored *Shengjing Fu* ("Ode to Shengjing"), which was published by Wuyingdian. This book was available in three versions: a Manchu edition, a Manchu-Chinese bilingual edition, and a Manchu seal script edition (Fig. 4.47, with a plate frame size of 21.1 cm × 15.3 cm). This type of script posed great difficulties in block printing, as the Manchu-Chinese bilingual edition required the corresponding characters of the two languages to be aligned in position. Printing one line of one language and then another line of the other language was also challenging.

4.4 Technical Process of Woodblock Printing

4.4.1 Technique of Engraving Printing Plates

Historically, scholarly works pertaining to the evolution of printing have often centered on the publication of books, neglecting the intricacies of the printing techniques per se. To be more precise, these studies are more accurately characterized

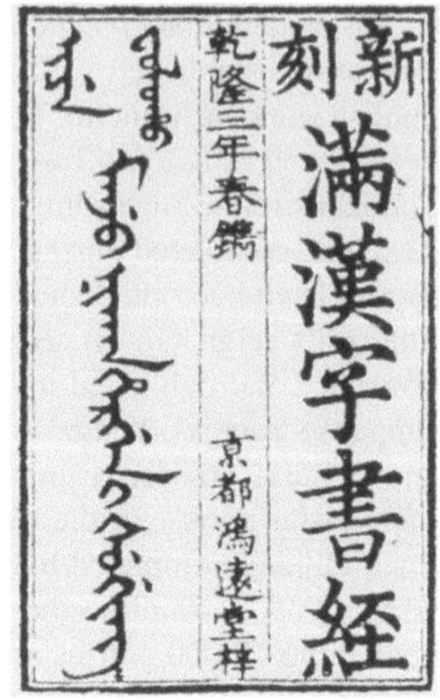

Fig. 4.46 *Manhan Zishujing* printed by Beijing Hongyun Bookstore in the third year of Emperor Qianlong's reign in the Qing dynasty (1738)

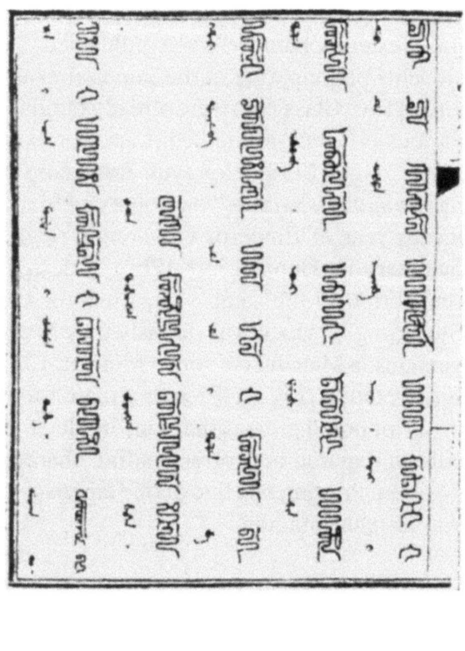

Fig. 4.47 The Manchu edition of *Shengjing Fu* written by Emperor Gaozong of the Qing dynasty, printed by Wuyingdian in the 13th year of Emperor Qianlong's reign (1748) (Collection of the National Library of China)

as historical accounts of publication rather than the history of the printing craft. The narrative of printing should be woven into the fabric of technological history, standing apart from the chronicles of publication. Our curiosity lies not in the mere enumeration of the bibliographic outputs of successive dynasties; instead, it lies in deconstructing and examining the technical mechanisms and methods by which these literary works were brought to life. The discourse presented herein on the technology of woodblock printing delineates a traditional paradigm that was universally employed by artisans across various dynasties and embraced by both the Central Plains and the regions inhabited by numerous ethnic minorities. Regrettably, extant documentation pertaining to this specialized craft from ancient times is exceedingly rare. The posthumous work of Lu Qian (courtesy name: Yi Ye, 1905–1951), the former director of the Nanjing Gazetteer Office, *Shulin Biehua* ("Collection of Reviews of Books", 1947), stands as an exception; through his meticulous survey of the traditional woodblock printing techniques that were preserved in the workshops of the common folk, he has bequeathed to us an invaluable record.[68] Furthermore, the scholarly contributions of Qian Cunxun, who bestowed considerable focus on the intricacies of printing technology within his treatises on the history of printing, have also shed considerable light on this subject.[69] Inspired by their pioneering efforts, our own investigative journey in the 1990s led us to visit workshops in both the northern and southern realms where traditional techniques continue to be practiced. By juxtaposing modern manual prints with their ancient counterparts and employing analytical methods from the annals of technological history, we have been able to delineate, in essence, the overarching contours of the printing technology as it existed in antiquity.

The initial discussion pertains to the discerning selection of woodblock materials, a process that demands the timber be derived from sturdy and upright trees to ensure a sufficiently expansive surface for engraving. The wood's density must be poised between pliability, which facilitates the carving of characters, and durability, providing the necessary strength for the craft. The texture of the wood should exhibit a fine, even grain, with a regular pattern that is devoid of knots and irregularities. Given the substantial quantities required for such endeavors, the trees selected should be not only widely distributed but also economically viable; thus, the use of rare and precious trees is categorically excluded. Softwoods like pine and poplar, though abundant and cost-effective, are unsuitable for engraving due to their flimsy nature, and the resin present in pine is also incompatible with the printing process. Having weighed these technical and economic factors, our ancestors typically selected woods such as catalpa, pear, and jujube, which meet these criteria, for their engraving boards. In the inscriptions of woodblock printed editions from the Song and Yuan dynasties, the selection of wood for the blocks is sometimes discussed. For instance, in the Southern Song dynasty, in the seventh year of the Qiandao period (1171), Cai Mengbi published *Shiji Jijie Suoyin* ("Historical Records of Collected Commentaries and Hidden Traces") Volume 2, which recorded "engraved on catalpa in the eastern study". In the early Jin dynasty (1140–1178), *Zhaocheng Collection* included a record in *Abhidharma Vibhasha Shastra* Volume 31, stating that Yang Chang and others from Jing Village, Wanquan County, donated 50 pear trees for engraving blocks. The Chinese National Museum houses a Northern Song dynasty woodblock print carved from jujube wood. These three types of wood were commonly used in

---

[68] Lu Qian. *Shulin Biehua* [M] //Zhang Jinglu. *Historical Materials on Modern Chinese Publishing: Ding Edition.* Beijing: Zhonghua Book Company, 1959: 627–654.

[69] Needham J. *Science and Civilisation in China: Vol. 5.* Cambridge: Cambridge University Press, 1985: 195–201.

printing, and thus terms like *zixing* (engraving on catalpa), *fuzi* (delivering to the engraver), and *fuzhilizao* (literally meaning committing to pear or jujube) became synonymous with "publishing".

The catalpa wood, scientifically denoted as Catalpa ovata G. Don, is a deciduous arboreal species belonging to the Bignoniaceae family, which can soar to heights of approximately 6 m (Fig. 4.48). It is indigenous to the southern reaches of Northeast China, extending its range to the Yangtze River basin. This species is celebrated for its rapid growth and wood that is firm yet moderate in hardness, featuring a straight grain and notable durability against decay. Its utility extends beyond the realm of engraving printing blocks; it is also favored for the construction of coffins. The Chinese pear, a deciduous tree from the Rosaceae family specifically known as Pyrus sinensis, is a fruit-bearing variety that has been cultivated in China for over two millennia, standing as a testament to the country's rich horticultural history. With a broad distribution across various regions, this tree is available in multiple cultivars. The jujube, scientifically known as Ziziphus vulgaris, is a deciduous tree that originates from China (Fig. 4.49). It is predominantly found in provinces such as Hebei, Shandong, Shanxi, Henan, Shaanxi, and Gansu, where it thrives in abundance, and there are numerous varieties available. Whether a woodblock is made of jujube wood or pear wood, it meets the technical requirements. Jujube wood is slightly harder, making it suitable for engraving illustrations. Other trees with similar qualities can also be used, such as the apricot tree, which is also native to China and distributed in the northwest and north regions. Depending on the availability of resources, different areas can select their materials accordingly, such as Ficus microcarpa of the Moraceae family, and in the Dege Tibetan area, the red birch (Betula albo-sinensis) is used for making woodblocks.

Upon felling the trees, the outer bark is meticulously stripped away, and the logs are subsequently sawed into sections of appropriate length. Each segment is crafted along the grain into rectangular planks, the dimensions of which are contingent upon the spatial requirements of the print design. A solitary plank is akin to a single printing plate or equates to the collective expanse of two pages within a book. The planks are typically fashioned to a thickness of 2.5 cm, with a vertical dimension

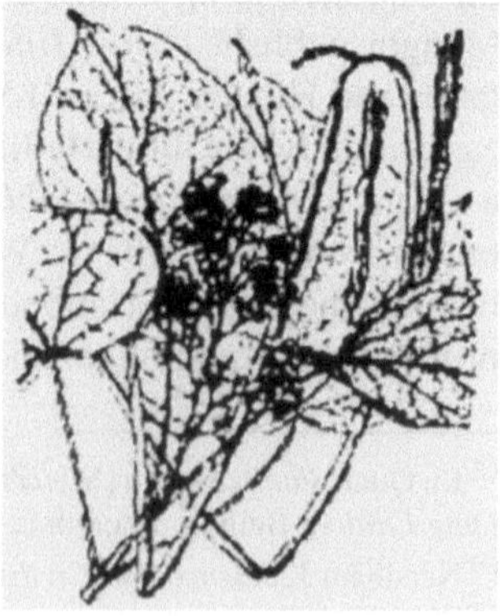

Fig. 4.48 Catalpa ovata

Fig. 4.49 Jujube tree

of 20 cm, and an approximate horizontal extent of 30 cm. The surfaces and edges of these planks are then planed to an impeccably smooth finish, with the stringent proviso that no knots or blemishes are permissible. To avert the risk of fissures within the printing plates due to temperature fluctuations, they are subjected to a period of immersion in water. This process facilitates a fermentation that expels pectin and other impurities from the wood's cellular structure. The duration of this soaking phase is contingent upon the seasonality; in the heat of summer, the immersion is abbreviated to approximately one month. In instances deemed necessary, the timber undergoes a process of steaming or boiling, which, while serving the same purpose as water immersion, offers the advantage of conserving time. Once the planks have been sufficiently soaked, they are subjected to a drying process in a shaded environment, with direct sunlight or exposure to fire strictly eschewed. Post-drying, the planks are meticulously planed to achieve a uniformly flat surface, which is then coated with a layer of soybean oil or rapeseed oil. The act of polishing the surface to a state of refined smoothness is executed using the stem of the Alternanthera sessilis plant (Fig. 4.50). While alternative grasses may be employed for this purpose, they must be devoid of rigid spines. Should any section of a plank reveal a stubborn knot, it is imperative that it be excised and supplanted with a piece of wood of commensurate size, which is then planed to a smooth uniformity.

When the wooden boards have been meticulously prepared, the subsequent phase involves the meticulous transposition of the manuscript onto their surfaces. Skilled artisans of calligraphy are tasked with the precise replication of the text upon paper adorned with a grid, with the dimensions and typographical style of the characters having been meticulously predetermined. The grid is imprinted in vermilion upon the paper, guiding the scribe to write along the central axis of each square, thereby ensuring that each character is rendered with symmetrical balance and uniformity in size. Annotations are inscribed in a dual row of minuscule script (Fig. 4.51). The paper employed for the manuscript was traditionally of hemp or bark before the Song dynasty; subsequently, it was exclusively bark paper. Upon the completion of

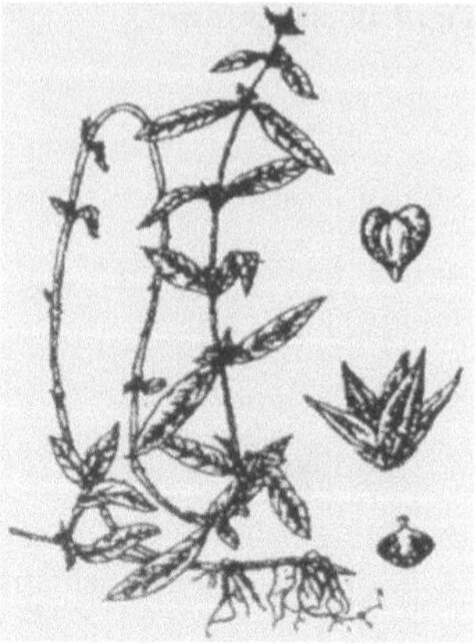

Fig. 4.50 Alternanthera sessilis

the manuscript, an initial round of proofing is conducted against the original text, with any errant characters being meticulously corrected and affixed, followed by a thorough secondary review. Thereafter, a uniform layer of thin, well-cooked rice paste is applied to the wooden board with a brush. The manuscript is then positioned on the board in reverse, with the character-bearing side in contact with the board. Utilizing a fine brush crafted from palm fibers, the back of the paper is gently rubbed, facilitating the transfer of the inked characters onto the board in mirror image (as depicted in Fig. 4.52). Once the characters have been allowed a brief period to set firmly onto the board, a coarse brush is employed to rub the back of the paper until it forms a soft nap, which is then meticulously brushed away. In instances where adherence is too strong, the paper may be manually peeled away. The board is subsequently air-dried, with any lingering paper debris being abraded with the stem of the Alternanthera sessilis plant. In this manner, the board is left adorned with reversed characters or illustrations, giving the impression that they have been inscribed or painted directly onto the board. These meticulous preparations are the essential precursors to the engraving process.

Upon the transference of the manuscript onto the wooden blocks, the intricate art of engraving begins. It is within the hands of seasoned engravers, masters of their craft, that a diverse array of chisels and knives (Fig. 4.53) are employed to meticulously etch the reversed inked characters into the surface of the wood, transforming them into characters cut in relief known as *yangwen* (as depicted in Fig. 4.54). Concurrently, the artisans deftly remove the surplus blank areas of the block, creating a concave recess of about 2 mm deep in relation to the plane of the block. These craftsmen utilize slanted-edge knives and a flat-edge knives to delineate the perimeter of each character with precision, initiating with straight lines followed by horizontal ones, thus encapsulating each character within a quadrangular framework etched by the blades. The trajectory of the knife is oriented from the periphery towards the interior, commencing the engraving from the left side of the character. Each stroke, be it a "horizontal", "slanting," "dot", or "vertical", is carved with careful attention to detail. The knife should not be held perpendicularly but at a slight angle, initially

Fig. 4.51 Writing the printed samples

Fig. 4.52 Transferring the reverse characters to the wooden board

to carve the "vertical" and "slanting" strokes, and subsequently rotated by a right angle (90°) to execute the "horizontal" ones. Accounting for the diverse contours and thicknesses of the lines to be etched, a variety of engraving tools with distinct shapes and dimensions are meticulously selected. Some of these chisels boast broad and flat or narrow and flat cutting edges, while others are characterized by beveled

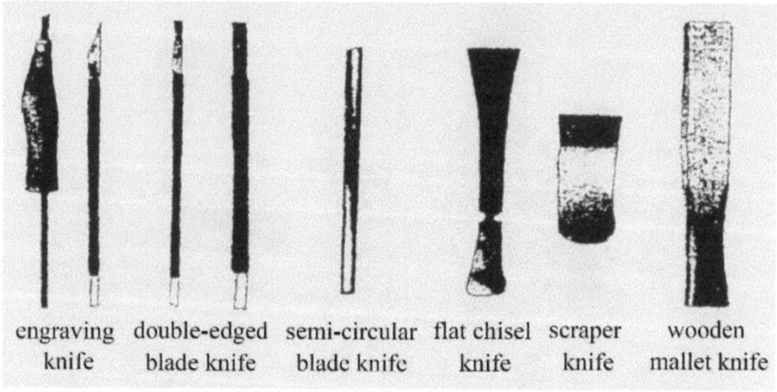

Fig. 4.53 Cutting tools for engraving

cutting edges that bear different angles. There exists a particular type of engraving tool that features cutting edges at both ends, adeptly crafted from steel and adorned with wooden handles. The sections devoid of characters are meticulously excised using chisels of varying sizes, thereby elevating the text in a manner that it protrudes from the surface. In instances where there is a substantial expanse to be cleared, a round-mouthed chisel is employed; by striking the rear of the chisel with a small wooden mallet and advancing it diagonally, any surplus wood is adeptly removed. The characters that are carved stand elevated on the plane of the board in a trapezoidal relief, a technique that ensures their durability against damage. When delineating the boundaries between rows of text, one must firmly press a ruler with the left hand while wielding the engraving tool with the right, executing two incisions for those with a heavier touch or three for those with a lighter one, thus ensuring that the lines remain impeccably straight. Ultimately, in accordance with the design of the print area, a peripheral border is preserved. Upon the completion of the engraving, the surface of the board is rinsed with hot water to cleanse it of any residual wood chips.

4.4.2 Printing Technology

Before the commencement of the official printing process, a series of proof impressions are meticulously produced to facilitate a final review of the text. In the event that any typographical errors are detected, they are meticulously removed and subsequently replaced with the accurate characters. Should there be an absence of certain strokes within the engravings, these omissions are remedied by additional engraving work. If substantial alterations are deemed necessary, necessitating the rearrangement of lines, the defective plate is consequently discarded, and the process begins anew with a fresh plate. However, such occurrences are infrequent, as thorough

Fig. 4.54 Carving characters with knives

proofing is conducted prior to the engraving stage, and engravers are typically characterized by their meticulous attention to detail. In preparation for the printing, paper and ink must be readied with precision. The ink is formulated with a binder content that is less substantial than that found in solid ink sticks, adhering to a weight ratio of carbon black to binder that fluctuates between 100:20 and 100:25. By blending carbon black with cowhide glue, or gelatin, and stirring the mixture until it resembles a thick porridge, the ink is then stored in a container. A modest quantity of alcohol is introduced to the mixture, which is left to mature for approximately half a month. Over time, the mixture transforms into a thin, black gruel, emitting an odor that is initially pungent. After being subjected to a period of maturation through three to four days of mold exposure, the odor eventually subsides, signaling that the ink is ready for application. The fermentation of the gelatin ensures that the carbon black is evenly dispersed throughout the mixture. It is crucial that the ink is not hastily prepared at the last minute; instead, it must be allowed to rest for a sufficient duration to ensure optimal performance. The engraver meticulously numbers each plate in accordance with the sequence of the original manuscript before they are transferred to the printing chamber, where they are joined by sheets of paper cut to the precise required dimensions.

The printing plates are then affixed to a wooden table that stands at a comfortable waist height, with the paper positioned on an adjacent table, ensuring a narrow gap exists between them. On the surface where the plates rest, one finds ink and a variety of ink brushes. The printer initiates the process by using a cylindrical brush with a flat bottom to blend the ink on an ink pad; subsequently, with another rounded brush, the ink is carefully picked up and evenly distributed across the surface of the plate. Once the ink is applied, the obverse side of the paper is delicately laid over the inked plate. Employing a clean, flat brush, the printer then methodically strokes the paper from top to bottom along the orientation of the type, ensuring that the paper

Fig. 4.55 Inking and printing

conforms to the plate and captures the obverse image of the ink. Upon completing a single impression, the sheet is flipped and suspended in the interstice between the two tables, allowing it to dry undisturbed. This cycle of inking and impressing is repeated until the desired quantity of prints is achieved. The sheets produced from each plate are amassed into individual stacks, creating numerous piles of printed sheets that are subsequently arranged in sequence and bound into volumes (Fig. 4.55). The application of ink requires uniformity of color across each sheet, a process that is pivotal to the quality of the final product and is executed by seasoned craftsmen. Typically, a single individual can produce between 1500 to 2000 prints in a day, with each plate exhibiting the durability to withstand tens of thousands of impressions before it requires cleansing. The plates are meticulously cleaned after printing and stored for potential future reprints. The multicolor printing process adheres to the same protocol, with the distinction of employing pigments of varying hues.

4.4.3 Evolution of Binding Techniques and Binding Forms

Judging from historical records and existing artifacts, it can be observed that the form and binding style of printed editions have undergone a continuous evolution, reaching a final stable form by the Ming and Qing dynasties. Mr. Jiang Yuanqing has provided a comprehensive discussion on traditional binding techniques,[70] which is a

[70] Jiang Yuanqing. The Development of Chinese Bookbinding Technique [J]. *Library Science* Newsletter, 1957 (6): 20–25.

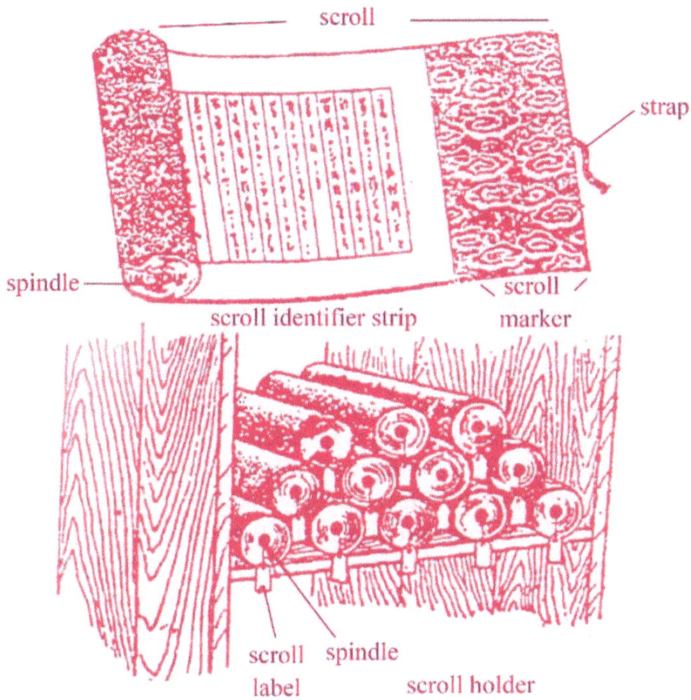

Fig. 4.56 Scroll binding (from Liu Tujun)

rare record in literature. The earliest form of printed books in history was the scroll binding, which was commonly used in the Tang dynasty (Fig. 4.56), such as *Diamond Sutra* published in the ninth year of the Xiantong period of the Tang dynasty (868), discovered in the Dunhuang stone chamber in 1907, and *Kaibao-Edition Tripitaka* carved in the fourth year of the Kaibao period of the Northern Song dynasty (971).

This format, which emerged from the manuscript traditions predating the Sui and Tang dynasties, involved the meticulous cutting of each printed sheet to uniform height, subsequently joining them to create an extensive horizontal scroll. At the conclusion of the scroll, a blank space was allocated for the attachment of a scroll shaft, which was commonly crafted from wood, occasionally adorned with vermilion lacquer, and for the most esteemed volumes, embellished with materials such as jade or ivory. The scroll was intended to be rolled up from left to right, with the initial section bearing the title, chapter name, and author's name, followed by the main body of text that varied in length from a dozen to several dozen characters per line, and culminating in an inscription. These scrolls, ranging in length from several meters to over a decade of meters, presented considerable challenges for reading and were typically suited for horizontal display only.

In the waning years of the Tang dynasty, circa the mid-ninth century, a revolutionary form of bookbinding emerged, characterized by the intricate process of folding a lengthy sheet of printed paper into a series of uniform widths, approximately 10 cm, and securing the exposed extremities with robust paper. These ends were then adorned with a protective layer of silk, serving as a cover that not only shielded the text but also added an element of elegance. Subsequently, the evolution of this technique saw the adoption of thin wooden boards as covers, inscribed with the book's title and the names of its various sections, thus transforming the book's physical presence from a cylindrical form to a more slender, rectangular prism. This innovative method, which persisted from the Song and Yuan dynasties through to the Ming and Qing dynasties for the binding of Buddhist scriptures, came to be known as "concertina binding" or "fold binding" (Buddhist-classic binding). This type of binding style may have been influenced by the Indian palm-leaf manuscripts, either as a direct emulation of their form or as a result of cultural osmosis. The Buddhist scriptures inscribed on palm leaves, commonly referred to as *Beiye Jing* (Indian palm-leaf manuscripts), were distinct in their method of preservation. Each leaf was punctured with two holes and bound together, with the content of each leaf forming a separate page, which contrasted with concertina binding (or fold binding). A collection of such scriptures was assembled from multiple volumes, encased in a protective covering of silk or cloth, and secured within a thick paperboard wrapper, enabling the book to be positioned either horizontally on a shelf or in an upright stance. The act of reading involved a progressive turning of pages, as opposed to the unwieldy rolling required by scroll-bound texts. Various editions of Buddhist scriptures, such as *Vairocana Tripitaka*, printed in 1112 at Kaiyuan Temple in Fuzhou in the Song dynasty, and those from the subsequent Yuan and Ming dynasties, often adopted this format, a practice also mirrored in the preservation of certain rubbings from stone inscriptions. The books, with their paper's creases exposed on both sides, were susceptible to wear and tear over time, particularly at the folds, necessitating the application of thin paper for reinforcement at the sites of potential rupture. Conversely, the scroll format, though less convenient for reading, had the advantage of protection; its rolled pages, shielded by a cover, were not exposed outside, thereby reducing the likelihood of damage and extending their "life span".

In order to facilitate the ease of page-turning and to prevent the pages from creasing and tearing, a bookbinding method known as "Whirlwind Binding" or "Cyclone Binding" emerged during the mid-Tang dynasty in the ninth century. This method, due to the scarcity of physical specimens, is imagined to be similar to the "concertina binding" (or fold binding), but with the initial and final pages connected by a single sheet of thick paper, thereby protecting the other side of the page's fold, leaving only one side exposed. Should the pages tear, they would be difficult to repair, and thus the understanding of the "Whirlwind Binding" lacks substantial documentary and physical evidence to support it. In the annals of the Song dynasty, Zhang Bangji (1090–1166) meticulously documented in the third volume of his literary work *Mozhuang Manlu* ("Rambles from Ink Cottage") written in 1131, an account of Wu Cailuan, a female calligrapher of repute. "Wu Cailuan was renowned for her finesse in inscribing minuscule characters, and had once transcribed *Tang Yun* ("Tang Rhymes"), a piece of work she subsequently sold. The manuscripts of *Tang Yun* that have descended through the ages are characterized by their Cyclone Binding, pages that resemble the whorls of a wind, with script so delicately etched that they exude a vigorous clarity, and these tomes are not uncommon in the abodes of the scholarly". Wu Cailuan, a lady of the Tang dynasty, was as well-known for her pursuit of alchemical transmutation and Taoist asceticism as for her calligraphic prowess. Upon her marriage to Wen Xiao, she sustained their livelihood by offering her expertly penned *Tang Yun* to the marketplace. The Palace Museum in Beijing houses a manuscript, traditionally ascribed to Wu Cailuan, titled *Kanmiu Buque Qieyun* ("Amendments to *Cut Rhymes*"), which is believed to be the very artifact referenced by Zhang Bangji. This exemplar, crafted on sturdy hemp paper that had been treated with wax, exhibits inscriptions on both surfaces. Measuring 26 cm in height and 48 cm in length, it is presented in the form of a scroll. A singular, lengthy sheet of thick paper constitutes the foundation, onto which four to five sheets of inscribed paper, with their right-hand margins left blank, are affixed in a sequential manner to the left side of the base sheet. They are arranged in a staggered pattern, akin to the scales of a fish, before being rolled up from the leading edge towards the tail, with a wooden spindle secured at the commencement of the scroll. Upon unfurling, the reader is permitted to navigate through the pages, engaging with the text on both sides—a feature that may indeed characterize the whirlwind format (Figs. 4.57 and 4.58).

The whirlwind binding, while seemingly akin to the scroll binding, features pages that can be flipped back and forth akin to concertina binding (or fold binding), representing a hybrid of the two forms. Due to the protection for the pages, they are

Fig. 4.57 Whirlwind binding imagined by modern people

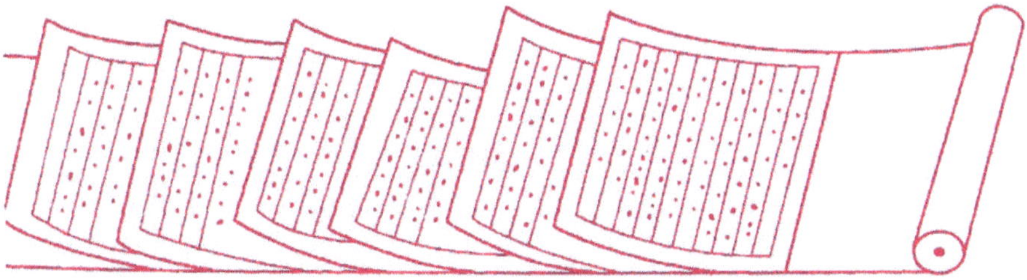

Fig. 4.58 Whirlwind building (Drawn by Pan Jixing according to collection from the National Library of China)

less susceptible to creasing, yet their curled state necessitates the use of paperweights upon unfurling, an inconvenience that is not without its drawbacks. Only manuscript copies have been preserved to this day, and there is skepticism regarding whether printed editions were bound in this manner. This format appears to have been in use for a very brief period, as the evolution of book shapes was trending from a cylindrical to a flat form, facilitating both reading and storage. Consequently, during the Song dynasty, another novel binding method emerged, known as the butterfly binding. This method involved binding each printed page into a book form, which was well-suited to the characteristics of printed material. It began with meticulous page layout design for each sheet, effectively a technical design for each printing plate. The layout was rectangular with four thicker lines down the center, creating a border area, which was the section inked and printed. In the traditional Chinese printing technique, a generous margin was intentionally left around the block frame, not only to safeguard the integrity of the text it enclosed but also to serve as a protective measure against potential damage. The frame itself was demarcated by a network of fine, straight lines, known as boundary lines, which compartmentalized the space for the text. At the heart of this layout, running perpendicular to these boundaries, was the central line, commonly referred to as the "book mouth"—a term that reflects the orientation of the text in the traditional Chinese bookbinding process, where the paper would be folded along this line, creating two pages from one sheet, with each page accommodating 5–12 lines. Each line was meticulously filled with 10–30 characters, the main text appearing in a larger font, juxtaposed with the annotations that were set in a smaller, dual-column format, nestled within each line. This meticulous arrangement allowed for an approximate character count to be discerned on each page, akin to the structured count found on modern manuscript paper (as depicted in Fig. 4.59).

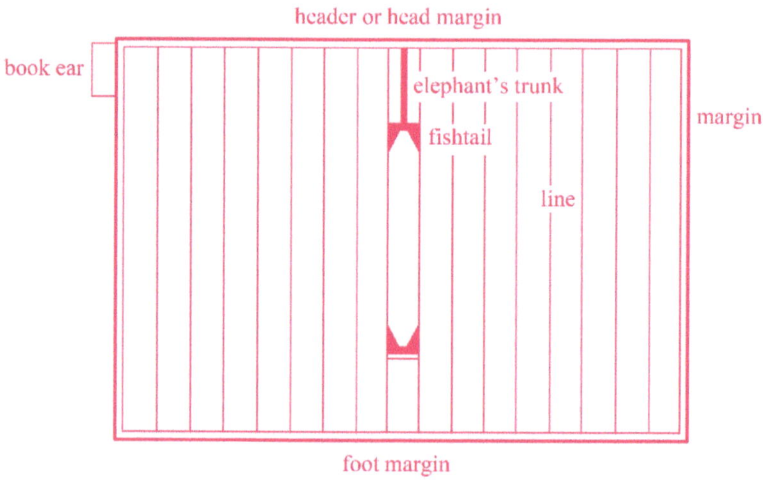

Fig. 4.59 Typical layout of printed book pages

To enhance the ease of paper folding along the central crease of the book mouth, the layout featured a distinctive element known as the "fishtail", accompanied by separatory lines. The fishtail, delineated by a pair of black triangular shapes, existed in both single and double variations, and it was at the convergence of these two shapes that the paper was intended to be folded. Occasionally, a robust central line, referred to as the "elephant's trunk", was superimposed above the fishtail, further indicating the fold line. The upper region of the book mouth typically bore the book's title, with the volume number positioned beneath the fishtail, and the page number anchored at the base. This sophisticated layout, the result of deliberate and thoughtful design, subsequently became an exemplar for the printing industry across numerous nations. The technique known as butterfly binding entails the meticulous process of taking each individually printed sheet and folding it along the central seam, allowing the two sides bearing text to face each other while the unprinted reverse sides face outward. These well folded sheets are then meticulously aligned along the crease, bound with an adhesive paste, and encompassed by a covering sheet that adorns the book's extremities and central portion (as depicted in Fig. 4.60). The title is inscribed on the cover's label, which is strategically positioned at the upper left corner of the cover. This design ensures that the crease, being nestled within the book's interior, is safeguarded from damage. Upon opening, the reader is greeted with an unobstructed view of the entire printed surface, and as the pages are flipped, they expand and retract in a manner reminiscent of a butterfly unfurling its wings (as illustrated in Fig. 4.60). The butterfly binding, a method of bookbinding that had already been observed in manuscripts by the tenth century during the Tang dynasty and was subsequently employed for printed works post the Song dynasty, exemplified by *Wenyuan Yinghua* ("Best Literary Works") which was published in 1260 during the first year of the Jingding period, allowed for the vertical placement of books when constructed with a thick paper cover. However, this technique presented a minor inconvenience for readers, who were required to navigate through two pages devoid of text before reaching a page bearing the written words, thus hindering expeditious consultation. With the passage of time and the advent of the Yuan dynasty, this style was supplanted by the more durable and user-friendly back-wrapped binding. During the subsequent Ming and Qing dynasties, the butterfly binding experienced a revival, albeit sporadic, as a means of paying homage to the aesthetic principles of antiquity, a practice that warrants separate consideration.

Back-wrapped binding, a technique that stands in opposition to butterfly binding, involves the intricate process of folding each leaf of paper along its central crease in such a manner that the text-bearing surfaces are presented outwardly, while the reverse, unadorned sides are nestled inward. This method results in the exposure of the central fold, which serves as the book mouth, with the peripheral margin transforming into the book's backbone. The edges of these meticulously aligned folds are then stacked in concert to create a cohesive volume, with the unruly pages along the back being trimmed to a uniform edge. The compilation is subsequently united with a layer of adhesive paste, and the entire book is encased within a robust paper covering, as depicted in Fig. 4.61. This design ensures that each page is imbued with text, allowing for an uninterrupted reading experience. The practice of situating

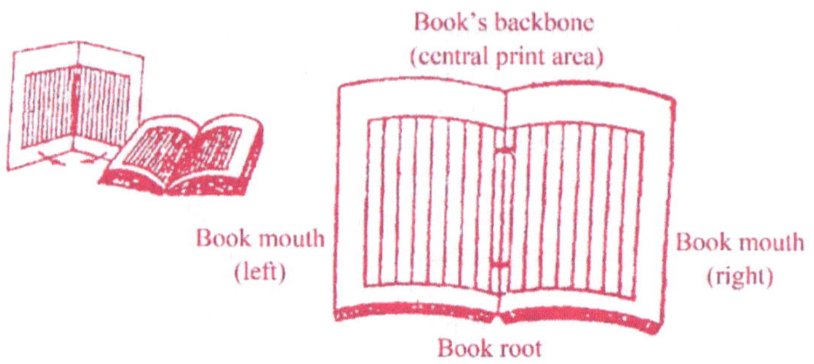

Fig. 4.60 Butterfly binding

numerous tomes within a sturdy folding case serves the dual purpose of preserving the integrity of the books while permitting their display in both horizontal and vertical orientations on bookshelves. The back-wrapped binding technique, which garnered widespread popularity during the Yuan dynasty and persisted through to the mid-Ming dynasty, that is, the mid-sixteenth century, eventually succumbed to the thread-binding method in the subsequent Qing dynasty. Notable exemplars of the back-wrapped binding include *Qianjiazhu Jinian Dugongbu Shishi* ("A Chronological History of Du Gongbu's Poetry Annotated by a Thousand Scholars"), a work published by Zhan Guangzu's Yueya Book Hall in Wuyi in the twenty-fourth year of the Zhiyuan period of the Yuan dynasty (1287), and *Sishu Ji Sansheng* ("The Four Books and the Three Sages"), issued by the Shen family's Shangde Hall in Hangzhou during the twenty-second year of the Zhizheng period, in 1362. These volumes stand as testaments to the craftsmanship of their times.

The diverse techniques of scroll binding, whirlwind binding, butterfly binding, and back-wrapped binding, which were integral to the bookbinding processes of the era, all necessitated the employment of a uniquely concocted adhesive substance. This substance was not merely a commonplace starch-based paste but an enhanced variant possessing robust adhesive qualities, meticulously made to withstand the ravages of

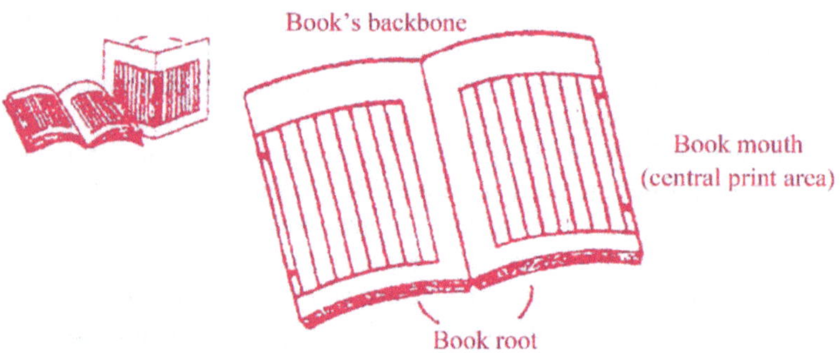

Fig. 4.61 Back wrapped binding

mold and insect infestation. Wang Shidian, a luminary of the Yuan dynasty who bore the title of Imperial Secretary and was the author of *Mishujian Zhi* ("System of the Palace Library", 1131), meticulously chronicled in the sixth volume the esoteric formulation employed by the binder Jiao Qing'an (1218–1287) during the second lunar month of the 14th year of the Zhiyuan period (1277), for the creation of this exceptional paste.

The craftsman Jiao Qing'an, who was meticulous in his selection of materials for bookbinding, employed a unique paste that was not the usual starch-based adhesive but a specially prepared compound. This compound, consisting of yellow wax, gelatin, alum, Bletila striata, Chenopodium album, soapwort, and Hierochloe odorata at a rate of one *qian*'s worth (3.73 g) each, along with half a *qian*'s worth of Pogostemon cablin and five *qians*' worth of wheat flour, was complemented by half a *jin* (500 g) of hard firewood and two *liangs* (50 g) of charcoal. This precise and intricate formula was recorded by the Yuan dynasty scholar Tao Zongyi (1316–1396) in his work *Chuogeng Lu* ("Notes Recorded During Respites from Plow", 1366), specifically in volume 29, which details a conversation between Wang Guxing and Yongguang, the abbot of Longping Temple (1288-?), regarding the binding methods of Buddhist scriptures from previous dynasties.

The ancient scriptures' bindings, with seams as slender as threads, defy the ravages of time without unraveling. What arcane knowledge preserves them so? Yongguang revealed the secret: It is an ancient technique that employs the sap of the mulberry tree, combined with flour and the ground powder of Bletilla striata. When these three ingredients are amalgamated into a paste, they serve as an adhesive for the paper edges, creating a bond that is eternally steadfast, surpassing the strength of the most resilient adhesives known to man, such as glue and lacquer.

In the concoction of this adhesive substance, the principal components are extracted from a selection of raw materials, including the protein-rich gluten, the collagen derived from cowhide gelatin, the sap harvested from the paper mulberry tree, and the tubers of Bletilla striata, which are renowned for their robust adhesive qualities and their natural resistance to insects. To this foundation, a medley of preservatives and antimicrobial agents is introduced, such as alum, Chenopodium album, Hierochloe odorata and Pogostemon cablin. The amalgamation of these ingredients culminates in a paste that exhibits not only formidable adhesive strength and insect-deterring attributes but also exudes a delightful fragrance.

Books bound in the package-back style, though frequently perused, are not immune to the pages becoming detached. To augment the binding's durability, a more permanent solution was adopted: puncturing the pages and sewing them together with silk threads or fine linen cords, thus evolving into the thread-binding method. This technique, which emerged in the manuscripts of the Five Dynasties period found in the Mogao Grottoes at Dunhuang, is generally believed to have become prevalent for printed editions only after the middle of the Ming dynasty. It eschews the use of paste, simplifies the procedure, facilitates reading, seldom leads to page loss, and is also very easy to rebind. Typically laid flat, when protected by a rigid case, they can also stand upright. Commonly, there are four-eye and six-eye binding methods (Fig. 4.62), and to prevent wear and tear on the upper and lower corners of the binding side, they are often wrapped with silk. The following is an introduction to Jiang Yuanqing's records on the thread-binding technique.

The traditional Chinese bookbinding process, a meticulous craft, involves the following intricate steps:

1. Folding: Each page is individually folded along the central crease, ensuring that the text-bearing sides are outwardly exposed without any skewness.
2. Sorting: After folding, the pages are collated in their respective order, forming a stack for each of the intended copies, and then compressed to ensure a tight formation. For instance, if there are to be 100 copies, they are arranged into 100 stacks, which are then pressed tightly together.
3. Aligning: Each page is then meticulously aligned along the crease, ensuring uniformity, before being clamped for pressure consolidation.
4. Adding Endpapers: Blank auxiliary pages, two to three in number, are added to each volume for structural support.
5. Provisional Binding: A provisional binding is performed using twisted leather paper to prevent the pages from slipping out of place.
6. Adding a Cover: A cover page is added, typically in yellow or blue, complemented by a layer of white paper underneath.
7. Trimming: The book is then trimmed with a sharp blade to align the top, bottom, and spine.
8. Sandpapering: Post-cutting, the pages are sanded with fine-grit paper to smooth out the marks left by the blade.
9. Corner Wrapping: For books that demand a touch of elegance, the spine is adorned with silk or satin corners.
10. Punching Holes: Holes are then carefully punched along the spine, determined by the height of the book, either four or six in number, using an iron awl.
11. Threading: The final step involves threading silk or linen thread through these holes, ensuring a secure and tight binding.
12. Labeling: Lastly, a label is affixed to the upper left corner of the cover, with its color chosen to complement the hue of the book's exterior.

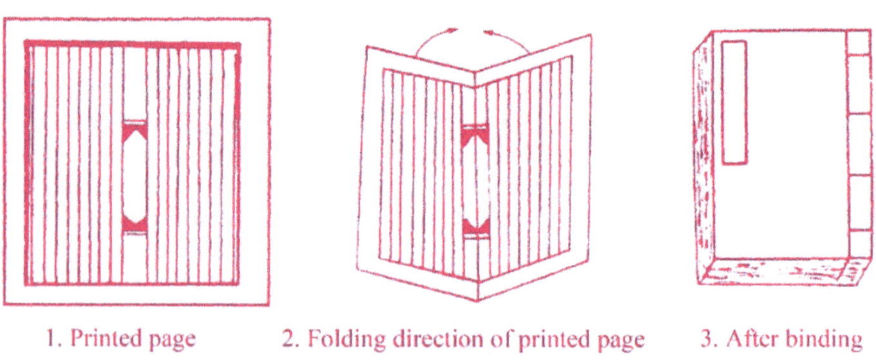

Fig. 4.62 Thread binding

The first five steps of the thread-binding process are also applicable to the back wrapped binding, after which steps such as pasting the spine and wrapping the book cover are included. Once the book is made, it can be placed in a sleeve for shelving, or it can be sandwiched between two wooden boards of the same size as the book, with cloth straps threaded through to hold the book between the boards, and the title of the book carved onto the wood. The book can also bc placed in a specially made camphor wood box. Chinese books are lighter than Western books because Chinese paper is thinner and the book covers are made of soft paper, while Western book covers are made of leather. Chinese books have ruled lines with an equal number of characters per line, which makes it difficult to skip lines when reading. Western books have smaller font sizes, no ruled lines, and often have hyphenation of words across lines. Chinese books have text on a single side of the page, while Western books have text on both sides. Thread binding is considered the best binding method for traditional Chinese books. In general, printing a book involves the following processes: (1) board preparation → (2) writing the manuscript → (3) proofreading → (4) transferring the reversed manuscript text to the board → (5) engraving → (6) cleaning the board → (7) printing a proof → (8) re-proofreading → (9) correcting errors → (10) applying ink → printing → washing the board → binding (Fig. 4.63). Each process includes several steps, and as the ancients said, "A single sheet of paper is not easily produced, it passes through seventy-two hands", in fact, producing a book is also not easy, and it also involves dozens of steps.

Fig. 4.63 Book binding operation diagram (Designed by Pan Jixing and drawn by Guo Defu)

Chapter 5
Origin and Development of Copperplate Printing, Mono-printing and Color Printing

5.1 Origin and Development of Copperplate Printing

5.1.1 A Brief History of Copperplate Printing

Subsequent to the evolution of woodblock printing, copperplate printing emerged rapidly in China. Despite the fact that both woodblock and copper-plate printing belong to mono-block printing, there exists a discernible divergence in the fabrication processes of these respective printing plates. The texts or images on a woodblock are carved with a knife, whereas those on a copper plate are created through casting. Copper plates are more durable and robust than woodblocks. The intricate designs, patterns, and texts on the copper plates remain intact through successive printing, and when not in use, the copper material can be recycled and re-melted to create new plates, a feature that woodblocks cannot replicate. Compared to woodblocks, copper plates are more costly and their manufacturing process is more intricate, typically used for printing single sheets or works of modest length. The material used for the plates is not pure copper but an alloy of copper, tin, and lead, also known as bronze. Given the significant proportion of copper in this alloy, the printing plate is often simply called a copper plate. China boasts an extensive history of crafting bronze artifacts, a legacy that dates back more than 4000 years. Following the advent of printing technology, this age-old casting expertise was adeptly repurposed to produce bronze printing plates.

Copperplate printing can be traced back to the Tang dynasty (618–907). The Qing dynasty scholar Ye Changchi (1847–1917) spent 20 years collecting over 8000 rubbings of stone carvings from previous dynasties and conducted in-depth research on them, eventually compiling the ten-volume work entitled *Yu Shi* ("A Monograph on Inscriptions on Stones", 1909). In the ninth volume of his work, after having discussed examples of inscriptions with reversed characters, Ye Changchi wrote, "Furthermore, there is also the iron pot from Junshan during the eighth year of the Xining period of the Song dynasty (1075), the copper mold of *Xin Jing* ("Heart Sutra") from the

P. Jixing, *History of Printing Techniques in China*,
https://doi.org/10.1007/978-981-96-6119-0_5

Kaiyuan period of the Tang dynasty (713–741), and the Sichuan-engraved document mold of Han [Yu], all of which also employed reversed characters." The text referred to here as *Xin Jing* ("Heart Sutra") is an abbreviated form of *Prajñāpāramitā Hṛdaya Sūtra*, translated by the eminent monk Xuanzang during the early Tang dynasty. This scripture is concise, with only one volume containing a modest amount of text, and essentially serves as a summary of *Prajñāpāramitā Hṛdaya Sūtra.* The *Xin Jing* copper matrix with reversed characters cut in relief, which Ye Changchi documented as having been left over from the Kaiyuan period in Emperor Xuanzong's reign of the Tang dynasty, and was originally thought to be a 'bronze mold' for writing templates, has been studied and found to be actually a copper printing plate used to print the sutra directly.[1] This is perhaps the earliest documented evidence of copperplate printing in China from the early eighth century, with this printing plate primarily featuring text.

Over three decades ago, a copper printing plate primarily used for printing Buddha statues was unearthed in Baoji, Shaanxi province. This relic dates back to the eighth year of the Dahe period during the reign of Emperor Wenzong of the Tang dynasty (834). The Baoji City Museum now preserves this rectangular artifact, which measures 14.8 cm in height, 11.5 cm in width, and 0.7 cm in depth, with a weight of 455.8 g. On the obverse, there are three main Buddha statues at the center, surrounded by one hundred and five smaller Buddha statues arranged in nine tiers, totaling one hundred and eight statues. On the reverse, there is a hand-carved inscription of *Jingang Diming Zhenyan* ("Vajra Life-Extending Mantra"), reading "Om, Vajra, Protect, Svaha." Below the mantra is a wish-fulfilling text: "On the eighteenth day of the fourth month in the eighth year of the Dahe period (May 30, 834), a Buddha seal was cast for the Ren family to venerate and support in perpetuity."[2] (Fig. 5.1) The dimensions of the page layout are akin to those of a contemporary 36-page booklet size, yet it boasts a vertical height that surpasses that of the Tang dynasty's woodblock-printed *The Great Dharani Sutra* (Mugu Jeonggwang Dae Darani Gyeong) (702 AD) in the Wu Zhou period and *The Sutra of Casket Seal Dharani of the Hearts of All Buddhas* printed by the Wu Yue Kingdom during the tumultuous era of the Five dynasties and Ten Kingdoms (956 AD) by a significant margin of 9.4 cm. This artifact represents the earliest surviving example of bronze printing materials discovered to this day.

The copper printing plate from the early eighth century features numerous clear and well-defined images of Buddha statues that are still in excellent condition for printing. This particular Buddha image was cast based on the story described in *Vajrasekhara Sutra* translated by the Tang dynasty monk Bukong. The sutra states that the Esoteric Buddhism's main deity, Mahā Vairocana, or the Great Sun Tathagata, has 108 Dharmakayas, hence the representation here with 108 Buddha statues. In the center of the plate, the three main Buddha statues are positioned with the Great Sun

---

[1] Zhuang Wei, Copper Mold of Kaiyuan's Heart Meridian in the Tang Dynasty to Be Copper Plate: Discussion on Copperplate Printing in the Tang Dynasty [J]. *Social Science*, 1979 (4): 151–153.

[2] Gao Ciruo, Introduction to Copperplate Prints Collected by Baoji Museum [J]. *Archaeology and Cultural Relics*, 1986 (4): 71–73.

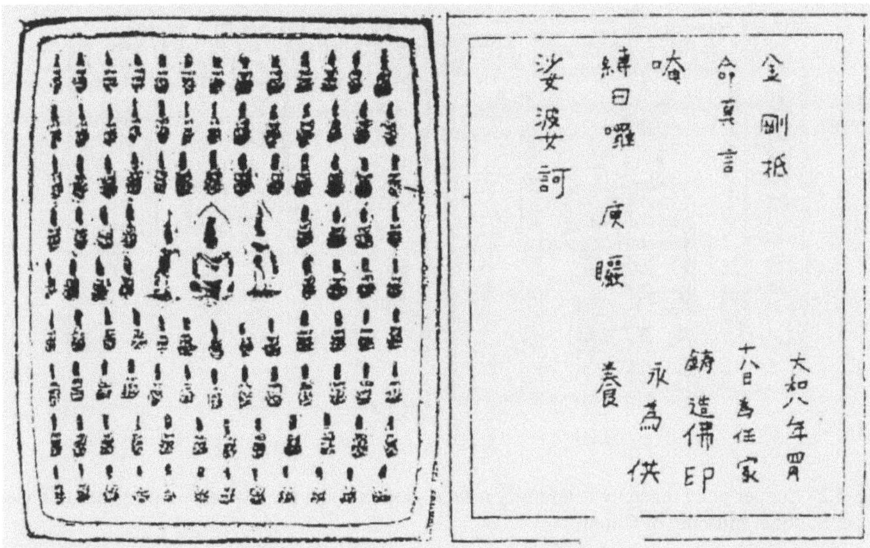

1. the front side of a copper plate 2. intaglio inscription on the reverse

Fig. 5.1 Copper plate of Buddhist images dated back to the eighth year of Dahe period in the Tang dynasty (834) (Housed by Baoji Museum of Shaanxi Province)

Tathagata in the middle, flanked by Samantabhadra Bodhisattva on the left and Avalokitesvara Bodhisattva on the right. The *Jingang Diming Zhenyan* consists of nine Sanskrit syllables translated into Chinese characters, which are the phonetic equivalents of the Sanskrit Dharani Sutras. "Vajra" means "diamond", while "Svaha" signifies "accomplishment", and "Om" denotes "devotion" or "homage". This printing plate holds significant value as a cultural artifact in the study of the history of printing and Buddhism.

Since the Song dynasty, with the development of the commodity economy, copperplate printing had spread to the economic field as well. The copper plate from the Northern Song dynasty (960–1127), which was commissioned by the Liu's Needle Shop in Jinan and housed by the National Museum of China, is slightly rectangular, measuring 12.4 cm in height and 13.2 cm in width. Upon examining the text and images on the plate, it appears to be a commercial advertisement printing plate, intended for use as product packaging for the shop's merchandise. The central part of the plate features the Liu Family Needle Shop's trademark, a white rabbit illustration. To the left and right of the image are the eight Chinese characters *Ren Men Qian Bai Tu Er Wei Ji*, meaning "recognize the white rabbit in front of the door as the sign". Above the image, in a banner, are the characters meaning "Jinan Liu Family's Expertise Needle Shop" in intaglio. Below the image, there are seven lines of 28 characters: "Purchase top-grade steel // to make fine needles, not to be delayed // for household use. For customers who sell and trade, // there will be additional benefits, // please remember this" (Fig. 5.2). The printing plate, aside from the illustration, bears

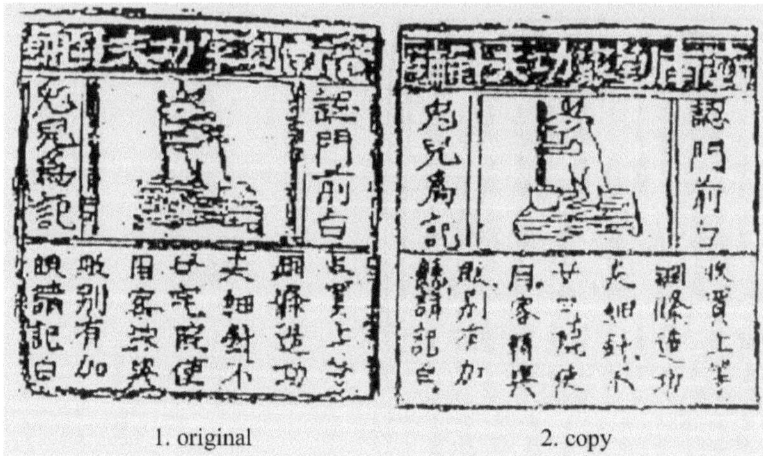

Fig. 5.2 Copper plate made by Liu family needle shop in Jinan for advertising purposes (Housed in the National Museum of China)

a total of 44 characters, eight of which are in intaglio. The combination of intaglio and relief characters on a single copper plate poses a certain level of difficulty. This is the earliest known copper casting printing plate used for advertising purposes.

From the eleventh century to the beginning of the twelfth century, the Northern Song dynasty also applied copperplate printing for issuance of banknotes, and utilized copper moveable types extracted from *Qianziwen* ("Thousand-Character Classic") to indicate the serial numbers of the notes. Therefore, banknotes were a product of combining copperplate printing and copper movable-type printing techniques. The earliest banknotes called *jiaozi* were issued during the reign of Emperor Renzong of the Northern Song dynasty (1023–1063) in Sichuan and circulated in the Sichuan-Shaanxi region. In the fourth year of the Chongning period in the reign of Emperor Huizong of the Song dynasty (1105), newly-issued banknotes called *qianyin* were in circulation in areas such as Henan, Shandong, and Anhui. These banknotes lasted for a long time (1105–1234) in the Song dynasty. From 1160 to 1279, there was another form of banknotes called *huizi* which circulated in Zhejiang, Hubei, and Anhui. The plate usually features the name of the currency, denomination, circulation area, issuing authority, printing time, reward and punishment notices, and decorative patterns or images, all of which are cast onto the copper plate. The *qianyin* was even printed in three colors: red, blue, and black. Fei Zhu, a scholar from the Yuan dynasty (ca. 1303–1363), discussed the establishment of the money-issuing institution in his work *Chubi Pu* ("Study on Collection of Banknotes", ca. 1360) in May of the first year of the Daguan period (1107). He noted that the institution had two supervisory officials, ten in charge of the classics, sixty-nine assistant scribes, eighty-one printers, six carvers, two casters, twelve miscellaneous laborers, with each receiving different rations. Among them, the printers were responsible for printing books, the carvers for carving and casting models, and the casters for making the copperplates. The

Jin dynasty (1115–1234) that coexisted with the Northern Song dynasty followed a similar banknote-issuing system to that of the Northern Song dynasty and began using banknotes called *jiaochao* since the second year of the Zhenyuan period (1154). This continued until the end of the Jin dynasty, although the name of the banknote changed over time. Subsequently, the Yuan and Ming dynasties also continued to print and circulate the *baochao*, employing the technique of copperplate printing. Therefore, copperplate printing used for banknotes in China has a history of over 700 years since the eleventh century, supported by a large number of unearthed artifacts. See Chap. 7 for details.

5.1.2 Copperplate Printing Technology

In the process of copperplate casting, the first step is to use wood or tin materials to make casting models according to pre-designed layouts, of which there are characters cut in relief on the convex side. Then, rectangular frames are used, and every two frames form a set. Ultra-fine clay and sand powder, used for shaping the models, are mixed with a small amount of water, crunched up to a plastic state, placed into a frame and compacted to create half of a casting model. Then, both the casting model and a runner system are simultaneously placed in a sand frame, and half of them are pressed into the mold sand. Another frame is placed on top of the first one, and filled with clay and fine sand to bury the casting model and the runner system before being compacted. The two frames are flipped 180 degrees so that the second frame is at the bottom and the first one is on top, causing the casting model and the runner system to fall onto the second frame. The first frame is then removed, and the casting model and the runner system exposed are carefully removed. Then, the first frame is placed on top of the second one and tied tightly with ropes. This creates a hollow casting cavity in the two sand frames, where the casting model bearing the upright letters cut in intaglio is connected to the runner cavity, and the gate lies at the junction of the two frames.[3]

Copper, tin and lead blocks are put into the crucible in certain proportion, and melted at the temperature of 800–1000 °C into alloy liquid in the melting furnace fueled by the wind from the piston bellows. The crucible containing liquid alloy is aligned at and flows into the gate to form the printing plate bearing reversed characters cut in relief after the process of cooling. Then break up the casting model to take out the castings which can be used to print banknotes after being trimmed. In order to facilitate the implantation of copper moveable types on the plate, there should be a number of grooves left in the corresponding part of the casting model, the size of which is equivalent to that of a moveable type. Finally, the plate also contains a number of grooves. The copper plates in the Ming dynasty are of excellent craftsmanship, and much improved than before, among which the masterpieces include

[3] Pan Jixing. *History of Metal Movable Type Printing Technology in China* [M]. Shenyang: Liaoning Science and Technology Publishing House, 2001: 103–126.

the Daming *baochao* banknote copperplate manufactured in Nanjing in the eighth year of the Hongwu period (1375).

Sometimes, in order to reduce the consumption of copper, the content of tin and lead is increased, leading to emergence of the tin plate, which can also be used for printing. This type of printing appeared in the Southern Song dynasty in the twelfth-thirteenth centuries, supported by unearthed cultural relics, and continued even in the Ming and Qing dynasties. In the early Ming dynasty, people who used tin plates to print counterfeit banknotes were sentenced to death. In the 52nd year of Emperor Qianlong's reign in the Qing dynasty (1787), Cheng Dun from She County, Anhui Province vividly printed a volume of *Qinhan Wadang Wenzi* ("Characters on Eave Tiles of Qin and Han Dynasties"). "Initially, jujube wood was used for carving. However, this failed as there were always discrepancies compared with the original text. Later, he adopted the ancient Han method of casting seals and creating molds, using the original tiles as models and melting tin to replicate them", achieving a lifelike precision.[4] It's not appropriate to describe this process as "creating molds, using the original tiles as models"; it might work like this: an eave tile was used as the casting model which was then filled with fine clay and sand and injected with tin alloy liquid to form the printing plate. Therefore, without damaging the original tile, a vivid plate bearing the same texts was made. Of course, the rubbing method could also have the same effect, but only one rubbing could be produced each time. This could be solved using the metal printing plate, as hundreds of thousands of sheets could be printed at the same time.

Copperplate printing failed to become the mainstream, although it lasted from the Tang dynasty to the Qing dynasty. This was mainly because the investment of funds and manpower was huge, and the copperplate printing prevailed only in governmental issuance of banknotes. Since the Northern Song dynasty, copper printing and movable type printing were combined to produce copper movable types. From this point of view, copperplate printing has its historical values. The origin of copper plate printing in Europe dates back to 1430–1440, when the plates were engraved with knives, mainly for printing individual images. The manufacturing method was relatively backward and lagged behind China by as much as 700 years. However, in the seventeenth century, Europeans invented a copperplate-manufacturing method based on chemical agents. Such an advanced method was introduced to China in the eighteenth century. In the 48th year of Emperor Qianlong's reign of the Qing dynasty (1783), the emperor ordered Italian Jesuit Joseph Castiglione (1688–1766) and two or three Chinese craftsmen to etch the copperplate with strong acid (nitric acid) and create the architectural drawing of *Yuan Ming Yuan* ("Old Summer Palace"). This move was the early attempt of the Chinese to etch the copperplate with chemical agents. Prior to this, during the Kangxi period when the geography of the whole

[4] Zhang Xiumin. *History of Printing in China* [M]. Shanghai: Shanghai People's Publishing House, 1989: 566

country was being surveyed, China had also employed new methods of copperplate printing technology.[5]

5.2 Mono-printing in Past Dynasties

5.2.1 Mono-printing in Tang and Five Dynasties

The term "mono-printing", as commonly understood, encompasses the artistic technique of inscribing or carving designs onto various substrates such as wood, stone, or copper plates, followed by the process of transferring these images onto paper through the act of printing. It is important to note, however, that the practice of utilizing chemical substances for etching on stone and copper plates originated in Europe between the seventeenth and eighteenth centuries. This particular method was not introduced to China until the eighteenth century and, therefore, does not constitute a traditional technology of ancient Chinese origin. Therefore, it lies beyond our current discussion on the subject. In this context, the primary focus is on the technique of engraving a wooden block to form a printing plate, which is then used to produce images. This represents the earliest form of mono-printing. In China, the art of woodblock mono-prints have been inextricably linked to the craft of woodblock printing since its inception. The principal purpose of this method is to create printed materials that are richly illustrated with text, achieving a harmonious integration of visual and textual elements. The inclusion of illustrations serves to enhance the comprehension of textual narratives, providing artistic imagery that words alone cannot convey, and aiding in the retention of information by the reader. Early mono-printing involved a collaborative process during which the artist would provide the design, the engraver would carve the woodblock, and the printer would then print the image, rather than the artist undertaking the carving and printing tasks themselves, which is in contrast to the modern practice of mono-printing. In early mono-printing, the images were often paired with textual narratives, serving as an illustrative supplement to the text. It wasn't until later that albums primarily focused on images began to emerge. The artist's brushwork had to be translated through the engraver's knife and finally brought to life by the printer's brush, making mono-printing a highly sophisticated fusion of art and technology.

Archaeological excavations have revealed that the earliest publications were inherently religious in nature, a fact reflected in the dominance of devotional imagery among the earliest mono-prints. This pattern is observed with remarkable uniformity across both Eastern and Western traditions. Historical records indicate that during the reign of the Tang dynasty, specifically between the years 658 and 663 AD, the revered monk Xuanzang embarked upon a sacred mission to produce an extraordinary collection of one million single-sheet Bodhisattva images through the nascent

[5] Pelliotp. *Qianlong's Western Regions Military Campaigns* [M]. Translated by Feng Chengjun. Shanghai: China Book Co., Ltd., 1956.

art of printing, a feat that was ultimately accomplished. Archaeological findings from the seventh century, such as those discovered in a single-paged *Dharani Sutra* written in Sanskrit in Xi'an in 1974, reveal that the earliest publications were religious texts often depicted with illustrations. The most renowned mono-print from the Tang dynasty was unearthed in 1907 in the Mogao Grottoes of Dunhuang, specifically the print from the ninth year of the Xiantong period (868) known as *Diamond Sutra* (*Vajracchedikā-prajñāpāramitā Sūtra*). Among its pages is a frontispiece measuring 23.5 cm in height and 28.3 cm in width. This *Sermon Scene* illustration portrays Buddha Shakyamuni delivering a discourse while seated on an altar. Elder Subhuti is depicted kneeling before him in reverence, listening intently. Surrounding Shakyamuni are deities, disciples, and 18 patrons. The backdrop of the scene is the "Jeta's Grove", and above, celestial beings known as devas are depicted in flying poses, with a total of 21 distinct characters present in the artwork. In front of the altar, there are also two lions. The composition is intricate, with a multitude of figures, props, and a well-arranged background. The relationships between the figures and the setting are skillfully managed, and the lines are fluid, reflecting the exquisite craftsmanship of the engraving. This piece is a testament to the artistic achievements of mono-printing at the time. A considerable number of such prints were produced during the Tang dynasty, but unfortunately, very few have survived to this day (Fig. 5.3).

Mono-prints from the Five dynasties saw significant advancement building upon the legacy of the Tang dynasty. A quintessential example of this era is the solitary-sheet *Manjushri Bodhisattva Image* housed in Beijing Library, measuring 31 cm by

Fig. 5.3 *Diamond Sutra* in the ninth year of the Xiantong period of the Tang dynasty (868), discovered in the Mogao Grottoes of Dunhuang in 1907 (Housed in the British Museum)

20 cm, with the image above and text below, carved by the Dunhuang craftsman Lei Yanmei between 947 and 950. Similarly, *Great Holy Pishachadvara Heaven King Image* carved in 947 and housed in the National Library of Paris (La Bibliothèque Nationale de France) also follows this style. Both prints are well-crafted, but the skills of the artists and engravers are not fully displayed due to the small size of the images.

5.2.2 Mono-printing in Song, Liao, Jin, and Yuan Dynasties

The Song dynasty, bringing an end to the era of division and warfare that characterized the Five dynasties, experienced significant advancements in agriculture, craftsmanship, commerce, and scientific technology, resulting in a period of social prosperity. The proliferation of cultural education and the imperial examination system ushered in a golden age for the papermaking and printing industries. Concurrently, the art of woodblock printing flourished like never before, riding the wave of the widespread adoption of printed books. First and foremost, the diversification of mono-printing subjects was a remarkable characteristic of this period, with both the artistic level of the drafts and the technical proficiency of the engravers surpassing those of previous generations. Beyond the ongoing refinement of religious imagery, mono-printing made its way into a vast array of scholarly works, spanning the realms of literary art, historical chronicles, medical texts, and scientific studies. The illustrations in these books were masterfully crafted to demystify intricate, specialized concepts, thereby not only fostering a broader dissemination of knowledge but also expanding the purview and application of mono-printing as an art form. The Song dynasty saw the proliferation of woodblock printing, a cultural phenomenon that extended to the Liao (916–1125), Jin (1115–1234), and Western Xia (1038–1227) dynasties, as well as the later Yuan dynasty, all of which were influenced by the Song dynasty. The legacy of printed works from these periods has been more abundant than that from the Tang and Five Dynasties. Yet, only a selection of these exemplary pieces can be highlighted here.

There was a significant and unprecedented increase in the number of illustrations appearing in medical, natural science, and engineering works. For example, *Wujing Zongyao* ("Complete Essentials for the Military Classics") (Fig. 5.4), published in the fourth year of the Qingli period in the reign of Emperor Renzong of the Song dynasty (1044), is an encyclopedia about military technology containing 239 illustrations about warships and gunpowder weapons, etc. The illustrations which provide insights into the structures and forms of weapons have attracted high attention from scholars at home and abroad. *Yingzao Fashi* ("Treatise On Architectural Methods"), published in the second year of the Chongning period in the reign of Emperor Huizong of the Song dynasty (1103), is a treatise on architectural technology containing 193 architectural engineering diagrams. *Jiayou Bencao* ("Materia Medica of the Jiayou Period"), compiled in the fifth year of the Jiayou period (1060), is a monograph on pharmacology in which there are no illustrations. Later, the author

Zhang Yuxi (990–1066) and others completed *Bencao Tujing* ("Illustrated Classic of Materia Medica"), based on samples of medicinal herbs collected from various regions. The book, published in the seventh year of the Jiayou period (1062), is the earliest illustrated pharmacopoeia printed in China. Although the original book was lost, there were surviving editions such as *Zhenglei Bencao* ("Classified Materia Medica from Historical Classics for Emergency", published in 1108; see Fig. 5.5) and *Zhenghe Bencao* ("Materia Medica of the Zhenghe Period", published in 1116, and reprinted in 1249). *Yinshan Zhengyao* ("Essentials of Diet and Nourishment") (Fig. 5.6), published in the second year of the Zhishun period of the Yuan dynasty (1331), was authored by the Mongolian scholar Husihui (1289–1355), and distributed in the northern regions. It is a treatise on nutrition and food therapy containing 185 exquisite illustrations. These highly specialized illustrations accurately express the contents described in the texts, setting a model for future scientific and technological works.

During the Song dynasty, the proliferation of illustrations in works on literature, history, archaeology, and the humanities was unprecedented, surpassing those of previous dynasties. For instance, *Lienv Zhuan* ("Biographies of Exemplary Women"), published in Jian'an, Fujian, by Yu Jing'an's *qinyoutang* (a private publishing house) in the eighth year of the Jiayou period of the Northern Song dynasty (1063), authored by Liu Xiang (77–6 BC), features 123 illustrations depicting women of various types, such as the wise, the benevolent and intelligent, and the virtuous, all following the format of image above text. Another example is *Xuanhe Bogutu* ("Xuanhe Illustrated Catalogue of Antiquities"), published in the third year of the Xuanhe period under

Fig. 5.4 The level gauge diagram in the woodblock-printed edition of *Wujing Zongyao* from the Northern Song dynasty (1044)

Fig. 5.5 Sea salt illustration in the woodblock-printed edition of *Zhenglei Bencao* (1108) from the Northern Song dynasty (Quoted from the reprint edition in 1249)

the reign of Emperor Huizong of Song (1121), which describes 839 ancient bronze vessels and 113 bronze mirrors housed in the Xuanhe Hall of the palace. It represents a comprehensive collection of ancient artifacts of the time. Each illustration was meticulously drawn by court painters based on the original items, accompanied by descriptive text, making it an essential archaeological reference book. *San Li Tu* ("Illustrations to the Ritual Classics"), created by Nie Chongyi (910–975) of the Song dynasty in 962, provided visual data on ancient costumes, customs, daily life, and objects, with an existing first edition printed in the second year of the Chunxi period (1175), containing 377 large and small illustrations.

In the twelfth century, during the Jin dynasty, a work titled *Simei Tu* ("Illustrations of Four Beauties") was engraved and printed by the Ji family in Pingyang Prefecture (now Linfen, Shanxi). It depicted four female figures from the Han and Jin dynasties: Wang Zhaojun, Ban Jieyu, Zhao Feiyan, and Lv Zhu. Made of yellow paper, it measured 77.8 cm in height and 31 cm in width (Fig. 5.7). As the earliest extant New Year paintings, the work contains the figures who were depicted in meticulous brushwork similar to the line drawing style of *Bashiqi Shenxian Juan* ("Eighty-Seven Immortals") from the Northern Song dynasty. Discovered by the Russian named Kozlov in 1909 and brought back to Russia, the work is a masterpiece famous for the vividly portrayed figures, fluid and accurate lines, and well-arranged background elements such as railings, flowers, phoenixes, and vines, which marks a shift from religious themes to folk New Year themes. Published in the second year of the Jiaxi

Fig. 5.6 Illustration in the woodblock-printed edition of *Yinshan Zhengyao* (1331) from the Yuan dynasty

period in the Northern Song dynasty (1238), *Meihua Xishen Pu* ("Picture Copybook of Plum Blossom") by Song Boren from the Song dynasty introduced various methods of painting plum blossoms, and effectively served as an illustrated album. Illustrations also appeared in literary works such as operas and novels. The existing illustrated books published in the Zhiji period (1321–1323) of the Yuan dynasty, such as *Sanguozhi Pinghua* ("Stories about the Three Kingdoms") and *Wuwang Fazhou Pinghua* ("The Plain Tale of King Wu's Conquest of King Zhou"), employed the same styles as similar books published in the Song dynasty did.

After more than 300 years of development since the Tang and Five Dynasties, remarkable achievements in religious mono-prints, represented by various Buddhist

Fig. 5.7 The New Year painting *Simei Tu* (Collected by Hermitage Museum, Saint Petersburg)

scriptures, had been made in the Song and Yuan dynasties. For instance, in 1069, *Miaofa Lianhua Jing* ("Lotus Sutra") was found in the Zi County of Shandong Province, which is now being held by Shandong Museum. The Sutra, published in Hangzhou during the second year of the Xining period of the Northern Song dynasty, contained various illustrations depicting scenes from the sutra, and was characterized by numerous figures, complex backgrounds, compact and rational layout, delicate lines, varied scenes, and ingenious design. *Kaibao Zang* ("Kaibao Edition of the Buddhist Canon"), republished in the second year of the Daguan period of the Song dynasty (1108) and currently housed in the Fogg Art Museum, Harvard University, is another example. In the publication, there were four exquisite illustrations designed to support the preface from Emperor Taizong of the Song dynasty to *Yuzhi Micang Quan* ("Imperially Commissioned Exegesis of the Secret Treasury").[6] In the painting, while religious narratives are depicted, the figures themselves occupy a minuscule portion of the composition, with the grandeur of the natural landscape taking center stage, thus pioneering a novel style within the realm of religious art. Emperor Huizong

[6] Loehr M, *Chinese Landscape Woodcuts: From an Imperial Commentary to a 10th Century Printed Edition of the Buddhist Canon* [M]. Cambridge: Belknap Press, 1968.

Fig. 5.8 One of the four illustrations to *Catalog of Buddhist Treasures* reissued in the second year of the Daguan period in the Northern Song dynasty (1108)

of the Northern Song dynasty, Zhao Ji, who was renowned for his proficiency in the arts, personally ordered the republication of the Buddhist canon, which included this exquisite set of woodblock prints. Drafted by court painters and executed with utmost finesse in both drafting and printing (Fig. 5.8), these prints stand as paragons of their craft. Not only were the Song's publications exceptional, but the Liao and Jin dynasties also produced Buddhist scriptures replete with illustrations of remarkable quality. The woodblock prints from the Song dynasty have even been discovered in the twentieth century, a testament to their enduring legacy.

5.2.3 Mono-printing in Ming and Qing Dynasties

Chinese mono-printing entered its period of great prosperity in the Ming and Qing dynasties. Non-religious and illustrated works became the mainstream in that period, and the popularity of operas, novels and other popular books among the readers was the driving force for the great prosperity of mono-printing. In the sixteenth century and after the middle of the Ming dynasty, an opera or a novel could have different variations, and in order to strengthen the competition, the engraving-service providers tried their best to increase the number of illustrations in the books in order to enhance the interest and artistry of the books. Sometimes, there were dozens of illustrations or even hundreds of illustrations in a book. The illustrations had different sizes ranging from small ones (illustrations above and text below) to large ones (accounting for a full page or half page). In addition, there were also picture-story books or picture-based albums. This period was also characterized by overriding of the pictures in Confucian classics, history, science and technology and literature and art works over the religious paintings in terms of quantity and quality.

The exceptional quality of Ming dynasty woodblock prints can be attributed to the collaboration between professional artists who provided the designs and a new

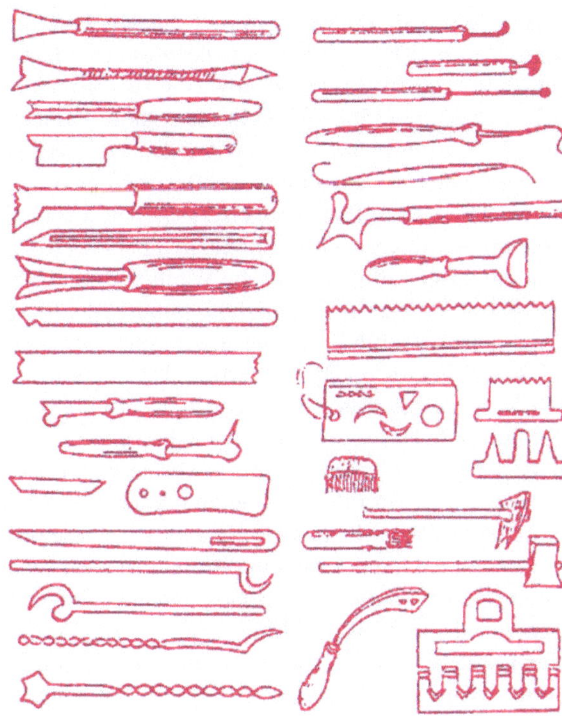

Fig. 5.9 Tools used by engravers of the famous Hui School in the Ming dynasty[7]

generation of highly skilled engravers who specialized in this art form. This synergy resulted in a monopoly over the creation of most of the significant woodblock pieces, a movement that became known as the prestigious Hui School of woodblock printing. Through hands-on experience, they perfected the craft by developing specialized tools (Fig. 5.9) for fine-line engraving and honing their carving techniques, which were the keys to their mastery.

By the late Ming dynasty, the number of woodblock print works had become too vast to enumerate; it is only possible to cite a few examples, which include *Zhongguo Banhua Shitu Lu* ("Record of Illustrations in History of Chinese Prints") by Zheng Zhenduo (1940–1947) and Chang Bipeter's *Mingdai Banhua Xuan* ("Selected Prints of the Ming Dynasty", 1969). In the twenty-seventh year of the Wanli period (1599), Huang Yingzu (1563–ca.1633) from Xiuning county completed the engraving of illustrations (supplied by local artist Wang Geng (1556–1621)) for *Renjing Yangqiu* ("The Mirror of Human Nature in Autumn"). In the illustrations, the vivid and lifelike figures and objects with a sense of three-dimensionality were depicted in meticulous line-drawing style. In the thirty-eighth year of the Wanli period (1610), during the production of *Zuoyin Tu* ("Secluded Meditation"), artists and engravers did the same

[7] Wang Bomin. *History of Chinese Prints* [M]. Shanghai: Shanghai People's Fine Arts Publishing House, 1961.

thing as before. Paintings for *Wang Fengzhou Li Zhuowu Ping Bei Xixiang Ji* ("Commentaries on 'The Romance of the Western Chamber' by Wang Shizhen (Wang Fengzhou) and Li Zhi (Li Zhuowu)") was provided by Wang Geng and engraved by Huang Yikai (1580–1622) in Hangzhou. Painter Chen Hongshou (1599–1652) created the albums *Bogu Yezhi* ("Ancient Objects Leaflets"), Shuihu Yezhi ("Water Margin Leaflets"), and Jiuge Tu ("Illustrations of the Nine Songs"), with paintings engraved by Huang Jianzhong (1611–?). These albums were published in Hangzhou in the thirteenth year of the Chongzhen period (1640).

In the thirty-fourth year of the Wanli period (1606), Cheng Dayue (1541–1616), an expert ink maker from Huizhou, published *Chengshi Moyuan* ("The Ink Garden of the Cheng Family"), a fine multi-color print with 14 volumes of illustrations and 4 volumes of poetry and prose. The paintings were drawn by Ding Yunpeng (1547–1628), and engraved by Huang Li (1564–?) and Huang Yingtai (1588–1642). *Wanhuxuan* (a Hui School print workshop), owned by Wang Yunpeng from Huizhou, published nine volumes of *Liexian Quanzhuan* ("Complete Biographies of Famous Transcendents") in the twenty-eighth year of the Wanli period (1600), which was authored by Wang Shizhen (1526–1590). The publication contained 581 figures and 222 accompanying portraits in different shapes, each occupying half a page. These drawings were engraved by Huang Yiben (1576–1641). Huang Yiben also collaborated with Huang Yifeng (1583–?) and Huang Yikai on *Yuanren Zaju* ("Dramas of Yuan Dynasty"), which was later published by Gu Quzhai in Hangzhou in the forty-seventh year of the Wanli period (1619), and contained 80 illustrations of exquisite craftsmanship (Fig. 5.10). The above-mentioned *wanhuxuan* also published another masterpiece of the Huizhou engraving tradition, *Pipa Ji* ("Cheng's Ink Garden"), in the twenty-fifth year of the Wanli period (1597), which contained 38 illustrations. *Jinpingmei Cihua* ("The Plum in the Golden Vase"), published in the Chongzhen period (1628–1644), also contained hundreds of illustrations engraved by Huang Jianzhong. There were over 30 engravers surnamed Huang, and 47 editions of illustrated books have survived to this day.

Engravers surnamed Wang and Liu also produced some exquisite works. For example, *Tang Shihua Pu* ("Tang Poetry Painting Manual") engraved by Wang Shiheng, and published in the twenty-eighth year of the Wanli period (1600), paired poems with paintings, with half the page dedicated to hand-written Tang poems and the other half to paintings. This type of book was favored by scholars. Another example is *Zhongyi Shuihuzhuan* ("All Men Are Brothers") engraved by Liu Junyu, which contained 120 illustrations depicting the deeds of heroes from Liangshan. Sometimes, engravers surnamed Wang, Liu and Huang worked together to produce books including the aforementioned *Jinpingmei Cihua.* The engravers of the Hui School were renowned for their exquisite craftsmanship, characterized by fine and smooth lines, and an unwavering dedication to the minutiae that captured the essence and spirit of the artists' original drafts. By establishing themselves in key printing hubs, they propelled the advancement of mono-printing techniques across various regions, elevating the field to new heights during the Ming dynasty. Their significant historical contributions deserve recognition.

Fig. 5.10 Illustrations in Wutong Yu, a poetic drama published by *guquzhai* (a printmaking workshop) in the Ming dynasty (1619)

Among Ming dynasty albums, *Gaosong Huapu* ("Gao Song's Painting Manual"), painted by Gao Song (1510–1575) and published in the 34th year of the Jiajing period (1555), depicted the methods of drawing chrysanthemums, birds, bamboo, etc. In *Lidai Minggong Huapu* ("Paintings by Famous Masters Through the Ages") produced by the court painter Gu Bing (1555–1615), 106 paintings from 106 famous painters in the past dynasties with themes ranging from landscapes to birds, flowers and figures were reproduced in ink, which made them accessible to people who had no opportunity to appreciate them. This masterpiece was published by *Shuangguitang* (a printmaking workshop) in Hulin (Hangzhou) in the thirty-first year of the Wanli period (1603), and each picture took up half of a page (Fig. 5.11). The aforementioned

Chengshi Moyuan ("The Ink Garden of the Cheng Family"), and *Fangshi Mopu* ("Fang's Ink Manual") published by another Huizhou ink maker Fang Yulu (1548–1613) in the seventeenth year of the Wanli period (1589), could be categorized as picture albums in a certain sense, in which all drawings and paintings were engraved by professionals of the Hui School.

Owing to the development of science and technology, the Ming dynasty also witnessed emergence of a series of scientific and technical books with highly specialized illustrations in the fields of botany, pharmacology, medicine, agriculture, crafts, mathematics, and military technology. For instance, in Zhu Su's *Jiuhuang Bencao* ("Edible Herbs and Other Plants Needed in Disaster Relief") published in the fourth year of the Yongle period (1406), there were drawings of 414 wild plants for famine

Fig. 5.11 Illustration of the study room in *Lidai Minggong Huapu*, published in Hangzhou in 1603

relief, making it the earliest of its genre in the world. Each plant is meticulously depicted with detailed renderings of its roots, stems, leaves, and fruits, all drawn to scale from the actual specimens, which reflects a high level of scholarship. Li Shizhen's monumental work *Bencao Gangmu* ("Compendium of Materia Medica"), published in Nanjing in the twenty-fourth year of the Wanli period (1596), contained 1160 illustrations of animals, plants, and minerals. Kuang Fan's *Bianmin Tuzuan* ("Illustrated Handbook for the Commoners"), published in the seventh year of the Hongzhi period (1494) in Wu County, contained 31 illustrations of agricultural and sericultural techniques, with two illustrations per plate. Each illustration was captioned with an explanatory text at the bottom. Maoyuan Yi's comprehensive military encyclopedia *Wubei Zhi* ("Treatise on Military Preparedness"), published in the first year of the Tianqi period (1621) in Nanjing, consisted of 240 volumes with hundreds of illustrations, each occupying half of the page, and engraved by Gao Liang. This comprehensive work is a significant resource for the study of military strategies, tactics, and equipment during the Ming dynasty. Song Yingxing's *Tiangong Kaiwu* ("Exploitation of the Works of Nature"), published in the tenth year of the Chongzhen period (1637) in Nanchang, was a comprehensive technical encyclopedia, with 123 illustrations of high scientific value.

Sancai Tuhui (an encyclopedic work of China's Ming dynasty) compiled by Wang Qi (1540–1615) and his son and published in Nanjing in the thirty-seventh year of the Wanli period (1609), consisted of 106 volumes, and served as an encyclopedic compendium covering both humanities and natural sciences. According to the author's statistics, the book contains over 4000 illustrations, an unprecedented number that was unrivaled by publications of previous eras and also unmatched by foreign publications of the same period. Among them, there are 596 portraits of historical figures, engraved by Nanjing engravers such as Tao Guochen and others. In the history of global mono-printing, this book has set a record for the abundance and breadth of its illustrations. Many illustrations in the Qing dynasty's massive encyclopedia *Gujin Tushu Jicheng* (Complete Library of Books Old and New) were sourced from *Sancai Tuhui*. This also illustrates that, aside from the Hui School, the Jinling (Nanjing) School of engravers also held a significant position in the realm of mono-printing during the Ming dynasty. In the late Chongzhen period (1642), Zhu Yishi (1600–1664) wrote in the postscript to *Shuguo Zhengqi* ("The Marvels of Fruits and Vegetables") authored by Deng Zhimo (1565–1630) that.

The art of engraving and printing today has reached a pinnacle, with exquisite editions as numerous as clouds, offering an endless feast for the eyes. Such printed works truly enhance the ambiance of a study and bring a touch of leisure to teahouses. A lady of grace, on an outing, holds an illustrated book in her carriage as if it were a precious gem. Medical practitioners, with their expertise, find it easier to consult these tomes, using diagrams to illustrate conditions to their patients. All these myriad functions are made possible by the diligent efforts of the engravers, whose contributions accumulate into a legacy of virtue that is no less worthy than the deeds of sages who have imparted knowledge and scriptures through the ages.

This lively description extolled the virtues of engravers in creating cultural wealth and reflected the widespread popularity and influence of illustrated books in society.

Among over 4000 types of illustrated books in China, those from the Ming dynasty accounted for almost a half, surpassing the total of such books published before the Ming dynasty.

During the early Qing dynasty, particularly in the reigns of Emperors Shunzhi and Kangxi, which spanned nearly 80 years, the art of woodblock printing evolved from the foundation laid in the late Ming dynasty. The painting style of Chen Hongshou, a painter with anti-Qing sentiments at the transition of the Ming and Qing dynasties, greatly influenced the woodblock prints of the early Qing period. His work *Shuihu Yezi* ("Outlaws of the Marsh Illustrations") depicted the defiant spirits of the heroes from Liangshan, and it was reprinted in the early Qing era. Under this influence, a new trend emerged in woodblock printing at the beginning of the Qing dynasty, focusing on the celebration of heroes, valiant women, and distinguished historical figures. Another trend was the depiction of landscapes, inspiring patriotic sentiments and nostalgia for the prosperous landscapes of the Ming dynasty. *Lisao Tu* ("The Illustrations of *Li Sao*"), painted by the painter Xiao Yuncong (1596–1673), was published in the second year of the Shunzhi period (1645), and consisted of 64 illustrations engraved by Tang Fu from the Hui School. Xiao's another masterpiece *Taipingfu Shanshui Tu* ("Taiping Prefecture Landscape Paintings") (Fig. 5.12), published in Dangtu in the fifth year of the Shunzhi period (1648), consisted of 43 illustrations engraved by Liu Rong, Tang Shang, and Tang Yi from the Hui School.

Fig. 5.12 Xiao Yuncong's masterpiece *Taipingfu Shanshui Tu* engraved in Anhui in the fifth year of the Shunzhi period (1648) in the Qing dynasty (Housed in National Library of China)

In 1668, the seventh year of Emperor Kangxi's reign, *Lingyange Gongchen Tu* ("The Illustrations of the Heroes of Lingyan Pavilion") was published by *guihutang* (a printmaking workshop) in Suzhou. This publication was crafted under the artistic influence of Chen Hongshou, with the illustrious painter Liu Yuan (1621–1689) contributing the artwork and the skilled engraver Zhu Gui (1643–1718), also from Suzhou, handling the woodblock carving. The piece is a tribute to the meritorious figures from the inception of the Tang dynasty. Furthermore, *Baiyue Ningyan Tu* ("White Mountain Enshrouded in Smoke"), a landscape print in the style of Xiao Yuncong, was published in Xiuning in 1714, during the 53rd year of Emperor Kangxi's reign. This exquisite collection of 40 paintings captures the essence of the White Mountain scenery, situated 15 km beyond the boundaries of Xiuning County. The collection was illustrated by the esteemed painter Wu Rong (1674–1734), with engravings expertly executed by Liu Gongchen, a native of Xiuning. By the mid-eighteenth century, during the reigns of Emperors Kangxi and Qianlong, the Qing dynasty's rule became increasingly consolidated, leading to a decline in anti-Qing publications and a surge in works that celebrated the golden age of the Qing Empire. The "Imperially Commissioned" prints, crafted by court painters and circulated by the Imperial Household, were a phenomenon rarely seen in the Ming dynasty. In 1696, during the 35th year of Emperor Kangxi's reign, *Yuzhi Gengzhi Tu* ("Imperially Commissioned Farming and Weaving Illustrations") were published by the Imperial Household. These were redrawn by the court painter Jiao Bingzhen (1650–1727), based on *Gengzhi Tu* ("Pictures of Farming and Weaving", 1145) by the Song dynasty artist Lou Xu (ca. 1090–1162). The collection comprises 23 illustrations, each accompanied by a poem inscribed by Emperor Kangxi himself, and the woodblocks were carved by Zhu Gui from Suzhou.

In the 52nd year of Emperor Kangxi's reign (1713), *Wanshou Shengdian* ("Grand Celebration for Emperor Kangxi's Birthday") was published by the Imperial Household with a total of 120 volumes, depicting the grand celebration for Emperor Kangxi's sixtieth birthday. The tome features 148 exquisite illustrations, masterfully crafted under the supervision of the court painter Wang Yuanqi (1642–1715), with the woodblocks skillfully carved by Zhu Gui, an artisan of the Imperial Household. Later, in the 31st year of Emperor Qianlong's reign (1766), *Nanxun Shengdian* ("Great Canon of the Southern Tours") was issued by the Imperial Palace, documenting the emperor's four southern inspections with scenes of grandeur. Furthermore, *Huangchao Liqi Tushi* ("Imperial Qing Ceremonial Paraphernalia") published in 1759, offered a glimpse into the imperial collections, while *Huangqing Zhigong Tu* ("Imperial Qing Tributary Maps") completed in 1751, showcased the diverse attire of 600 individuals from various ethnicities and foreign nations, painted by the skilled artists Men Qing'an and Dai Ji. These illustrations, each a testament to the meticulous documentation of the era, possess significant historical value, providing a visual chronicle of the Qing dynasty's cultural and diplomatic engagements (Fig. 5.13).

During the Qing dynasty, the folk portrait album *Wanxiaotang Huazhuan* ("Wanxiaotang Illustrated Biographies") by the Min School painter Shangguan Zhou (1665-ca.1749) was published in 1743. This masterful collection captures 120 historical figures in individual, half-page portraits. The realm of drama and literature, including

Fig. 5.13 *Yuzhi Gengzhi Tu*, published by Imperial Household Department in the thirty-fifth year of Emperor Kangxi's reign (1696) in the Qing dynasty

esteemed works like *Outlaws of the Marsh*, *Dream of the Red Chamber*, *Investiture of the Gods*, *Romance of the Sui and Tang Dynasties*, and *Romance of the Three Kingdoms*, all boast of illustrative accompaniments. This practice extends to scientific and technological tomes as well, such as *Qinding Shoushi Tongkao* ("Imperially Commissioned Comprehensive Examination of Timekeeping", 1742), a farming book ordered by Emperor Qianlong, and *Qinding Wuyingdian Juzhenban Chengshi* ("Imperial Printing Office Manual for Movable Type", 1776). Notably, *Gujin Tushu Jicheng* ("Complete Library of Books Old and New", 1728) transcends its contemporaries with its vast array of meticulously engraved illustrations, marking a zenith of historical documentation in the art of printmaking. At the end of the Qing dynasty, Wu Qiju (1789–1847), a preeminent botanist, authored the botanical work *Zhiwu Mingshi Tukao* ("An Illustrated Book on Plants", 1848), a work celebrated for its exquisite and precise illustrations, which are so detailed that they could serve as a taxonomic reference for botanists. Given the sheer volume of such publications, it is impracticable to provide an exhaustive list; however, a selection of exemplary pieces can be mentioned. Moreover, during the Ming and Qing dynasties, there was a considerable evolution in the creation of New Year paintings across different regions. These paintings, which were affixed to walls during festive seasons to augment the joyous ambiance, were initially printed with ink outlines and subsequently hand-colored in their respective sections. The southern hub for these paintings was Taohuawu in Suzhou, while the northern counterpart was Yangliuqing in Tianjin.

5.3 Origin and Development of Color Printing

5.3.1 Origin and Early Development of Multi-color Mono-block Printing

Early printings were mostly monochromatic with ink serving as the sole coloring agent, yet the demand for aesthetic and functional diversity has always driven the human pursuit of text and imagery in dual or multiple colors, thus precipitating the invention of color printing. During the Tang and Five dynasties, there emerged a desire to enhance the aesthetic beauty of woodblock prints in Buddhist scriptures. This was achieved by meticulously applying a spectrum of two to three, or even a greater array of colors, within the etched lines and along the silhouettes of the depicted subjects. Moreover, within this chromatic selection, there existed a subtle gradation from deep to light shades, allowing for an enriched and nuanced visual representation. The Musée Guimét in Paris houses a collection that includes a woodblock print of *Daci Dabei Jiuku Guanshiyin Pusa Xiang* ("Great Compassion and Mercy Avalokitesvara Bodhisattva") from the fourth year of the Kaiyun period during the Later Jin dynasty of the Five dynasties period (947AD), unearthed from the Dunhuang Mogao Grottoes. This intricate piece is adorned with meticulous artistry, with each section of the image being hand-tinted with a vivid array of colors including red, blue, yellow, green, and orange, exemplifying the rich chromatic diversity of the period. As elaborated in Chap. 4, by the Western Han dynasty, the technology for multicolor printing on silk had reached a significant level of sophistication, with unearthed printing plates serving as testament to this fact. Simply by replacing the printing medium from silk to paper and employing the same intricate printing and dyeing techniques, the achievement of multicolor printing on printed materials became feasible. Consequently, it is evident that the roots of this technology extend far back in time, suggesting that its inception was the result of a deliberate and calculated progression, rather than a mere coincidence. This strategic shift underscores the intentional evolution of the printing methods of the period, mirroring the profound depth of technological advancement at the time, rather than an unforeseen occurrence.

One of the methods to render monochrome printed matter in polychromatic hues involves applying different colorants to distinct areas of the same printing plate and then imprinting it onto paper in a single pass. This technique is referred to as "multi-color mono-block printing". Prints produced using this technique were firstly only in black and red, and later in three or four colors. It is evidently more convenient than hand-coloring, but the downside is that color bleeding and discoloration can easily occur at the junctions of different colors, and the complexity of rendering more than four colors presented a formidable challenge. On the other hand, the manual application of color allowed for unfettered adornment of any hue to any area. These two approaches, each with their unique merits and demerits, have coexisted for an extended period. By the early eleventh century, during the Northern Song dynasty, the practice of multi-color mono-block printing had reached a state of practical refinement. Guo Ruoxu (1039–1095), an art historian during the Song dynasty, wrote in his *Tuhua Jianwen Lu* ("Record of Painting Observations", ca. 1075) that:

In the first year of Jingyou (1034), Emperor Renzong of the Song dynasty commissioned Gao Keming and others to depict the glorious achievements of the three emperors (Emperor Taizu, Emperor Taizong, and Emperor Zhenzong of the Song dynasty). The figures were about an inch tall, and the palaces, mountains, imperial carriages, and ceremonial guards were all presented. Scholars such as Li Shu were then ordered to compile, annotate and praise them. The book consisted of a total of one hundred events, divided into ten volumes, and was later entitled *Sanchao Xunjian Tu* ("Illustrations of Three Dynasties' Admonitions and Examples"). Upon its consummation, the work was subjected to the meticulous process of replication and transference onto printing blocks, thenceforth to be disseminated in its printed form to the echelons of the bureaucracy and the scions of the imperial household.

Wang Mingqing (1127–ca. 1216) of the Southern Song dynasty mentioned in the section entitled *Empress Dowager Zhangxian Orders Compilation of Books and Carving of Printing Blocks within Forbidden Palaces* of his book *Huizhu Lu·Huizhu Houlu* ("Records of Whisking away Dust·Further Records", 1195) the following. When Renzong ascended to the throne in 1023 at the age of ten, the Empress Dowager Liu took the regency. Known for her wisdom and strategic acumen, she commanded the Confucian scholars Feng Yuan, Sun Shu, and Song Shu to compile a collection of historical accounts of emperors and their ministers, which was titled *Guanwen Luangu* ("Viewing Antiquity Through Literature"). Subsequently, she also oversaw the creation of *Sanchao Baoxun* ("Treasures of Three Reigns"), a ten-volume work that compiled stories from the early Song dynasty's first three reigns, with each volume dedicated to a single account. Furthermore, "she decreed that scholars of the Hanlin Academy (also known as Imperial Academy), such as Gao Ke Ming, should illustrate these volumes with exquisite artistry…The printing plates were then meticulously carved within the forbidden palaces. After Emperor Zhezong (1085) ascended the throne at the age of 9, those books were copied, then printed and distributed to officials and members of the imperial family". It is clear that after the manuscript of *Sanchao Baoxun* ("Treasures of Three Reigns") was completed in 1023, the painter Gao Keming was commissioned to add illustrations to the book, which was "copied, engraved onto blocks, and printed in 1034". The terms "copied" and "engraved onto blocks" refer to the process of carving the illustrations onto wooden blocks and then printing them using multiple colors. The color-printed edition of *Sanchao Baoxun* ("Treasures of Three Reigns") from 1034 was reprinted using the same method in 1085. It is difficult to determine the exact number of colors used in this Song dynasty imperial colored illustration book as the original is not present; however, it is estimated that it does not exceed four colors.

Emperor Huizong of the Song dynasty was a painter. In the first year of Daguan (1107) period during his reign, the paper currency called *qianyin* was issued. According to *Chubi Pu* ("Study on Collection of Banknotes", ca. 1360) by Fei Zhu, as banknotes should be decorated with patterns and pictures on both sides in red, blue, and black colors, double-sided printing was applied. Li You (1101–1171) of the Song dynasty mentioned in Volume 15 of his *Songchao Shishi* ("Facts of the Song

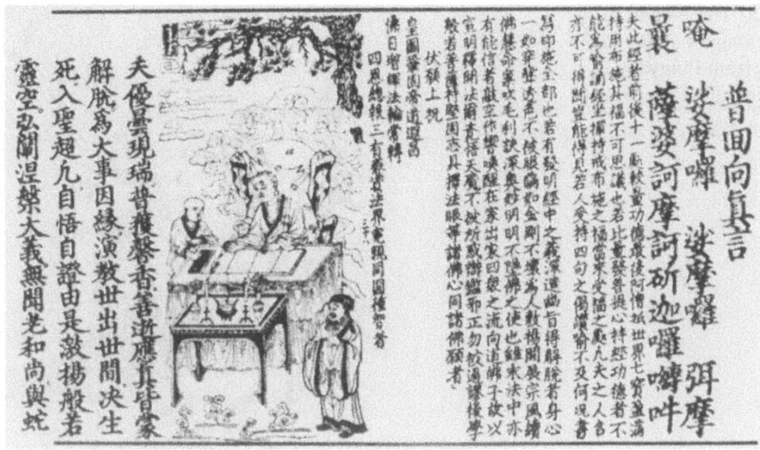

Fig. 5.14 Vajracchedik Prajna Ranita Sūtra

Dynasty", ca. 1130), that during the period from the thirtieth year of the Shaoxing period to the seventh year of the Jiading period (1160–1214), the paper currency called *huizi* was issued, which was "printed in one color, and decorated with houses, people, and shops; each bears a hidden serial number made in red and black." Since it was troublesome to print the three-color *qianyin* during Emperor Huizong's reign, a shift into a "mixture of red and black" was made, suggesting that two colors were applied to a single block which was then printed.

Based on historical records, it is certain that multi-color mono-block printing was used for woodcuts and paper currency during the early eleventh century of the Northern Song dynasty. However, no surviving double or triple color prints from that period have been found. What has been passed down is a Yuan dynasty print created based on this technique. In 1947, the Nanjing Library acquired a version of *Vajracchedik Prajna Ranita Sūtra*, which was annotated by the monk Wu Wen and engraved and printed by Zifu Temple in the sixth year of Emperor Shun of Yuan (1340). In the version, there were illustrations made in black and red (Fig. 5.14). After 1949, this version was transferred to Taipei, where Mr. Qian Cunxun found after careful check that the two colors blended at the junction,[8] suggesting that it was a product of multi-color mono-block printing. The conventional assumption has been that the color blending was a hallmark of double-color double-block printing, a hypothesis that may be misconstrued. If each of the two plates had been independently colored, the intermingling of hues could have been readily circumvented.

The earliest surviving examples of three-color prints were discovered more than 40 years ago. In 1973, archaeologists from the Shaanxi Provincial Museum found folk woodcuts, remnants of Jurchen manuscripts, rubbings of inscriptions, and copper coins in the empty niches of the Stele Forest in Xi'an. The latest of these copper coins

[8] Qian Cunxun. *Collection of Papers on the History of Books, Paper, Ink and Printing in China* [M]. Hong Kong: The Chinese University Press, 1992: 144.

Fig. 5.15 Three-color print of *Dongfang Shuo Stealing Peaches* from Pingyang prefecture of the Jin dynasty

were minted in the third year of the Zhenglong period (1158) of Jin, indicating that these artifacts could be dated back to the Jin dynasty.[9] One woodcut entitled *Dongfang Shuo Toutao Tu* ("Dongfang Shuo Stealing Peaches"), originally painted by the Tang dynasty painter Wu Daozi, was printed in black, red, and light green on light yellow hemp paper. The print, 108 cm high and 55.4 cm wide, was supposed to be a folk woodcut from Pingyang prefecture (modern-day Linfen, Shanxi Province) of the early twelfth century (Fig. 5.15). The woodcut was printed using the same process as the double-color version of *Annotations on Vajracchedika Prajna Paramita Sutra* from the Yuan dynasty did, which also demonstrated the presence of multi-color mono-block printing in the Northern Song dynasty.

During the process of multi-color mono-block printing, to reduce the overlap of coloring agents, a particular pigment was usually concentrated in a specific area. For text-based books, red ink was used for the main text, and black ink for annotations, making reading more convenient. However, this was not the case for woodcuts, as this would render a rigid and unnatural effect. To add more colors and avoid color bleeding at junctions, ancient artisans came up with the idea of separate-block coloring and batch printing, giving rise to multi-color multi-block printing,

[9] Liu Zuichang, Zhu Jieyuan. Discovery of Jurchen Documents and Prints in the Stele Forest of Xi'an [J]. *Cultural Relics*, 1979 (5): 1–5.

or today's so-called water-color block printing. Multi-color multi-block printing involves applying different pigments to several differently sized printing blocks, which are then printed several times on the same sheet of paper. Each block is carved out against certain parts of the image, and printed in a different color. After drying, another color is applied to another block, and this process continues until the full image was completed. To ensure that the pigment was placed in the correct position on the paper, the blocks in different colors had to be positioned correctly and aligned precisely.

Chinese woodblock color printing went through three stages of development: single-color mono-block printing, multi-color mono-block printing, and multi-color multi-block printing.[10] The transition from the second stage to the third stage occurred no later than the occasion of the Yuan and Ming dynasties, as multi-color multi-block printing had already seen significant development in the Ming dynasty. The latter two stages of printing also corresponded to the stages of technological advancement. All three forms coexisted during the Ming and Qing dynasties, when selection of a proper printing technique depended on the nature of the publication and economic considerations. Although multi-color multi-block printing was the most aesthetically pleasing, it could not totally replace the other two because not all publications should be made in multiple colors, and single-colored publications still dominated the market. The Chinese color printing has profound historical significance, as it opened the colorful world of splendor for printed materials.

Color printing in the Ming dynasty gained further development on the basis of the Song and Yuan dynasties, and reached its peak during the Wanli period (1573–1619). Hu Yinglin stated in Volume 4 *Jingji Huitong Si* ("A General Survey of Classical Books, Part Four") of *Shaoshi Shanfang Bicong* ("Notes from the Shaoshi Mountain Studio"), that "Prints are made in red or black, or indigo using double-color printing, or single-color printing. Prints made in red using double-color printing are more valuable." Prints made in blue indigo included *Ling Qi Jing* ("Ancient Miscellany on Divination") produced as early as in the fourteenth year of the Chenghua period (1478), and *Anlao Huaiyou Shu* ("Book on Nurturing the Aged and Cherishing the Young.") created in the eleventh year of the Hongzhi period (1498). Publishers such as Min Qiji (1580–ca. 1650) and Ling Mengchu (1580–1644) issued many double-color publications. For example, in the forty-fourth year of the Wanli period (1616) when Min Qiji printed *Chunqiu Zuozhuan* ("Zuo's Commentary on Spring and Autumn Annals"), the main text was rendered in black while contemporary scholars's comments were rendered in red and small characters and placed at the top of the page. *Mengzi* ("Mencius") printed in the forty-fifth year of the Wanli period (1617) by Min Qiji was printed in black, red, and indigo. The application of more than three colors was deemed superfluous due to the exorbitant costs and the potential for visual distraction in the reading experience. On occasion, the exegetical text was produced in a distinct red hue and juxtaposed with the primary text, a technique that, upon close examination, appeared to involve the use of dual plates rather than a

[10] Pan Jixing. *History of Chinese Science and Technology: Papermaking and Pring* [M]. Beijing: Science Press, 1998: 322–324.

singular one. Historical accounts reveal that the Min and Ling families printed nearly a hundred colorful books in Wuxing during the sixteenth and seventeenth centuries.[11]

5.3.2 Development of Multi-color Mono-block Printing

For readers, the practicality and affordability of double-color prints rendered them the preferred choice, consequently leading to a more substantial production run. In contrast, the opulent three-color and four-color tomes, burdened by their costly nature, found their appeal confined to a select few aficionados. The technique of overprinting, while innovative, was not without its flaws, particularly when it came to the reproduction of text. Despite the advancements in printing technology that have transpired to the present day, there remains an absence of necessity to produce text-laden tomes in a multitude of colors; the clarity and cost-effectiveness of single or double-color printing suffice for the majority of literary works. However, overprinting technique has great prospect in applying to block prints because colorful prints become more attractive for readers, especially instructional works on painting techniques which would be purchased by newcomers as well as connoisseurs and collectors regardless of the price. Existing examples include the album *Hua Shi* ("Flower History") printed in the twenty-eighth year of the Wanli period (1600). The album was divided into four volumes for spring, summer, autumn, and winter, respectively. Each depicted flowers and plants in a freehand style, with pictures made in different colors. There was color blending at the junctions, indicating utilization of multi-color mono-block printing. Nevertheless, this was not the case for *Chengshi Moyuan* ("The Ink Garden of the Cheng Family") meticulously created by Cheng Dayue in the thirty-third year of the Wanli period (1605). In the book, fifty illustrations were rendered in four or five colors, such as *Juzchuan Zhouji Tu* ("Picture of Huge Mountains and Boats") from the painter Ding Yunpeng, which was provided in black, green, blue, brown, or yellow. However, no color blending was found at the junctions, indicating application of multi-color mono-block printing or overprinting. *Fengliu Juechang Tu* ("Erotic Scene") created by Huang Yiming was another example of adopting the five-color mono-block printing.

It was natural that Huizhou was a center for both woodblock printing and overprinting. Multi-color multi-block printing also had a specific name in the Ming dynasty, namely "assembled blocks". This term was originated from a kind of pastry food, in which a variety of colored cakes of different shapes were piled on a plate for display. Assembled blocks, as a multi-color multi-block overprinting technique, involved color separation based on the tones, shades, intensity, and direction of the colors on the manuscript. Technicians outlined the color parts on paper, and then transferred them from the paper to wooden blocks. Carvers then carved out multiple blocks, each containing a part of the image to be printed in one color. Pre-prepared pigments were applied to different printing blocks, and each color was printed on the

[11] Tao Xiang. *Minban Shumu* ("Five-color Overprinting") [M]. Beijing: The Palace Museum, 1933.

paper one by one or in an overlaid manner. Finally, the colored parts were integrated to form a complete colored image. The number of color tones required the same number of printing blocks, and the printing process involved layering from light to dark and from thin to thick. The pigments, often soluble in water or in a suspension, are meticulously formulated to match the colors of the original painting, leading to the later nomenclature of "woodblock water printing". However, this term fails to accurately encapsulate the essence and characteristics of the technique, nor does it sufficiently differentiate it from mono-block multi-color printing. Technically, it seems more appropriate to refer to it as "multi-block multi-color overprinting".

The Ming dynasty also witnessed the integration of multi-color multi-block overprinting and the so-called "pattern embossing". Pattern embossing, originally used for papermaking, appeared as early as the Tang and Five Dynasties periods. Patterns or text in raised relief were carved on wooden blocks, which were then pressed onto the paper to obtain colorless patterns in intaglio. Introducing this technique into printing could create a three-dimensional effect by forming embossed patterns on paper. During the late Ming dynasty, Hu Zhengyan (ca. 1582–1674), a calligrapher and painter from Huizhou who resided in the suburbs of Nanjing and named his residence Ten Bamboo Studio due to the presence of more than ten bamboo trees in his courtyard, oversaw the publication of the renowned *Shizhuzhai Shuhua Pu* ("Ten Bamboo Studio Collection of Calligraphy And Painting") over a period of six years from the seventh year of the Tianqi period (1627) to the sixth year of the Chongzhen period (1633). The Collection, bound in four volumes in a butterfly-style form, allowed viewers to see complete pages. It included 180 paintings and 140 calligraphy works from Hu himself, ancient masters, and contemporary artists, covering various themes such as flowers and birds, landscapes, peculiar rocks, bamboo groves, antiques, and calligraphy. Most of the illustrations were printed in five colors (Fig. 5.16).

During the printing process of the above-mentioned Collection, Hu Zhengyan also took charge of the editing and publishing of *Shizhuzhai Jianpu* ("Ten Bamboo Studio Letterhead Collection"). In this book with a total of 4 volumes, various patterned letterheads collected were printed using the assembled blocks and embossing techniques. Each volume included a total of 289 paintings which were colorful or colorless and of which the theme ranged from antiques to rocks, flowers, trees, landscapes, and figures. The book began with a preface written in the seventeenth year of the Chongzhen period (1644), indicating that the book was published in the last year of the Ming dynasty. It can be considered a sister work to *Shizhuzhai Shuhua Pu*, both of which represented the pinnacle of overprinting. Hu Zhengyan's fellow townsman, Li Kegong (1595–1665), commented in the preface to *Shizhuzhai Jianpu* ("Ten Bamboo Studio Letterhead Collection") that "Before the Jiajing period (1522–1566) and Longqing period(1567–1572), letterheads were a simple and rustic. By the middle stage of the Wanli period (1573–1619), they became slightly complicated but did not yet reach their peak. Letterheads flourished until the later stage of the Wanli period, and reached their zenith during the Tianqi period (1621–1627) and Chongzhen period (1628–1644)". Besides, Li believed *Shizhuzhai Jianpu* as a remarkable job that gathered masterpieces from ancient and modern times, and made an unparalleled achievement in themes, carvings, and printing.

Fig. 5.16 *Shizhuzhai Shuhua Pu*, a multi-color multi-block print created at the end of the Ming dynasty in Nanjing (1627–1633)

Although *Shizhuzhai Jianpu* was mainly discussed above, observations of the calligrapher and engraver Li Kegong were equally applicable to *Shizhuzhai Huapu* ("Manual of the Ten Bamboo Studio"). Li deemed Hu Zhengyan's work as a scarce masterpiece in terms of themes, carvings, and printing, because excelling in all three aspects was incredibly challenging. Firstly, it was imperative to ascertain that the themes selected were replete with artistic creativity. Additionally, it was essential to preserve the essence and mood of the original artwork with unwavering fidelity during the processes of engraving and printing. Subsequently, a thorough analysis of the artwork from both artistic and technical standpoints was required to achieve accurate color separation, thereby laying a solid foundation for the subsequent engraving and printing stages. Finally, the engravers were expected to possess exceptional proficiency, and the printers were tasked with ensuring that the modulation of colors matched the original manuscript with precision prior to printing. The combination of assembled blocks and pattern embossing techniques was essential. These elements, indeed, constituted the very secrets of success in the intricate art of multi-color multi-block printing.

Meanwhile, Yan Jizhu (ca. 1590–1639), a Fujian native living in Nanjing, edited and published *Luoxuan Biangu Jianpu* ("Luoxuan Variation of Ancient Letterhead Collection") in the sixth year of the Tianqi period (1626). Composed of two volumes, the book described 182 letterhead patterns, each accompanied by poetry, calligraphy, and painting using assembled blocks and pattern embossing techniques, similar to those in Hu Zhengyan's letterhead collection. Those patterns were engraved by Wu Faxiang (1598–ca. 1652), a native of Jiangning in Nanjing, while Hu, on the other hand, enlisted the help of more than 10 engravers such as Wang Kai (1592–1666) from Huizhou. Owing to the limited content scope, Yan's letterhead collection was published 18 years earlier than *Shizhuzhai Jianpu* and one year earlier than the *Shizhuzhai Huapu*. A comparison of the two revealed that Yan's illustrations were less varied in color tone, with outlines drawn before coloring, while Hu adopted the Song dynasty's boneless painting technique when illustrating his book, and faced more challenges as the book had no outlines and featured more significant variations in color tone. Hu's work is now being held in major libraries worldwide, while Yan's work is being housed in the Shanghai Library. After *Luoxuan Biangu Jianpu* was reproduced using traditional methods in 1981, Shanghai Duoyunxuan (A Publishing House in China) reproduced *Shizhuzhai Huapu* in 1986.

Color printing continued to develop in the Qing dynasty. *YongZheng Zhupi Yuzhi* ("The Imperial Edicts of Emperor Yongzheng with His Red Revisions"), published in the eighteenth century, was a red and black print. In the fourth year of the Daoguang period, Lu Kun from Hebei Zhuozhou published 25 volumes of *Dugongbu Ji* ("Collected Poems of Du Fu"), in which the comments from the scholars of the Ming and Qing dynasties were printed in purple, blue, red or green and the text was printed in black according to the record in Volume 8 of Ye Dehui's *Shulin Qinghua* ("A Plain Talk on Books", 1920). This record is not certain because the original book can not be found, but the cost must be high. In the eighteenth year of the Kangxi period (1679), Shen Yinchu (1631–1701) published the book *Jieziyuan Huazhuan* ("Mustard Seed Garden Manual of Painting") in Nanjing (Fig. 5.17), which revived the technique of assembled blocks that emerged at the end of the Ming dynasty. This book had its name from the fact that it was engraved and printed in Nanjing's residence *Jieziyuan* of Shen Yinchu's father-in-law Li Yu, (1611–1679). Originally with 5 volumes, it contained 133 landscape paintings and was printed in a colorful style, serving as a reference for beginners. In the fortieth year of the Kangxi period (1701), another eight volumes were published, covering the four themes of orchids, bamboo, plums, chrysanthemums. Besides, another four volumes with the themes of flowers, trees, birds and insects were published at the same time. As for figure paintings, *Jieziyuan Huapu Siji* ("Four Series of Mustard Seed Garden Manual of Painting"), a supplement issued by a different group of professionals called *Xiaoyoushanfang* in the twenty-third year of the Jiaqing period (1818), was also a product of color overprinting, and used as a reference for generations of painters.

5.3.3 Techniques for Multi-color Multi-block Printing

Since the twentieth century, ancient printing techniques such as pattern embossing or multi-color multi-block printing have experienced a revival through the replication and reprinting of works like *Shizhuzhai Jianpu*. This resurgence has shed light on the

Fig. 5.17 *Jieziyuan Huazhuan*, a multi-color multi-block print created in the fortieth year of the Kangxi period in Nanjing (1701)

detailed craft process of this unique printing technology and allowed it to flourish. Generally, the technique includes the following basic steps:

1. Copying the Original Artwork: Skilled artists create a replica of the original artwork upon analysis of the brushstrokes, atmosphere, and colors of the original. This hands-on experience provides a basis for understanding the use of lines and colors in various parts of the artwork and facilitates color separation and plate making.

2. Color Separation and Plate Making: Artists analyze the original artwork from both artistic and technical perspectives and propose a color separation scheme based on their copying experience. Under the principle of one color per plate, the total number of color tones or plates necessary to reflect the original color scheme is determined according to the variety of colors, the shade of the same color, etc. The figure often ranged from a dozen to several dozens. Then, the parts to be colored are depicted on several sheets of paper, which is crucial for success of the project.
3. Separation and Tracing: Based on the color separation scheme proposed by artists, place transparent paper on the original artwork and depict individual small scenes on the paper against the original. Each scene should accurately correspond to a specific part or parts of the original artwork with the same color tone so that the original artwork is decomposed into various segments.
4. Verification: Artists carefully compare the tracing sketch with the original artwork to ensure that the lines are complete and the color segments decomposed corresponds to the original.
5. Mounting the Board: Transfer the corrected mirror images to the planed wooden boards, with one color per board. The number of boards required varies from a dozen to several dozens.
6. Plate Engraving: Engravers carve out images using different-shaped engraving tools, and rotate the plates periodically to create different lines, a process which is similar to the process of engraving bookplates but more challenging.
7. Pigment Preparation: In addition to ink, other pigments identical to those used in the original artwork must be prepared. For example, cinnabar needs to be finely ground and mixed with thin glue; ochre (including Fe_2O_3) is soaked in warm water to remove impurities, and then mixed with glue; gamboge is soaked in cold water to remove residue; safflower pigment is obtained from the Carthamus tinctorius, which is soaked in water and mixed with a little glue; indigo, the main content of which is indigotin and which is a wood dye, is soaked in cold water; verdigris, composed of basic copper carbonate [$CuCo_3 \cdot Cu(OH)_2$], needs to be finely ground and mixed with glue. Once prepared, all pigments are tested on paper to ensure consistency with the original artwork. If necessary, adjustments are made until no discrepancy is found.
8. Print Alignment: This involves fixing the paper and multiple plates on appropriate positions of the printing worktable so that the individual images on the plates are accurately printed onto predetermined positions on the paper. This process is called "print alignment". The worktable consists of two boards with a gap of about 10 cm between them. On the board on the right of the gap, there are two notches at both the upper and lower ends, and an internal bar clamp that tightly secures a stack of paper with a long bar. The stack of paper is placed on the board horizontally (see Fig. 5.18). Besides, there are also rectangular printing blocks. On the left side of the gap, there are printing plates, cylindrical ink brush, ink, color palettes, coloring pens, brush cleaning containers, and small printing blocks. A thin wooden board is placed vertically on the board on the left side of the gap, serving as a platform to flatten the paper during printing.

Fig. 5.18 Worktable of multi-color multi-block printing

The printing worktable is waist-high, allowing the printer to sit or stand. After fixing the paper in place, fix the plates of different sizes and shapes, each with its own number; use melted resin and wax to secure them in the appropriate positions of the boards and ensure that they coincide with the locations on the paper. Adjust such positions as needed, and once the wax has dried, the plates are firmly fixed in place.

9. Test Print: Before formal printing begins, several test prints are made to evaluate the results. When applying colors, apply black color first, followed by lighter colors. For Chinese paintings with a high need for ink, coat the plates with a cylindrical ink brush saturated with ink and then place the paper over the ink-stained plate. A rectangular impression brush is used to wipe the back of the paper. The paper with ink marks is then unfolded, and the process is repeated for the next sheet. If the ink color matches that of the original artwork, formal printing begins. Similarly, other pigments are applied one by one. Color adjustments are made as needed, and several test prints are typically required to reproduce the original artwork.
10. Brush Printing: Once the test prints are approved, formal printing begins. After one color is applied for one plate, another plate is applied with a different color. This process is repeated multiple times until the entire image is printed. After each sheet is printed, it is flipped to hang in the gap and left to dry naturally. For colors used in smaller quantities, a brush is used to dip into ink and apply them to the plate to avoid color mixing caused by the same brush. After printing, the plates are cleaned and stored.
11. Post-Printing Processing: After printing, remove the papers from the bar and check them one by one to remove defective products. Then, the paper is flattened, mounted, or bound into books. In recent years, there are specialized publications[12] on this topic, which can provide valuable references.

[12] Feng Pengsheng. *An Overview of Chinese Woodblock Watermark Printing* [M]. Beijing: Peking University Press, 1999: 113–137.

Chapter 6
Origin and Development of Non-metal Movable-Type Printing

6.1 Origin and Development of Wooden Movable-Type Printing

6.1.1 Wooden Movable-Type Printing Originating from Early Northern Song Dynasty

Woodblock printing had undergone steady development for four to five hundred years before the invention of movable-type printing in the eleventh century during the Northern Song dynasty. It has been a time-honored practice to apply movable-type notions to printing since ancient times. In fact, as early as in the Zhou and Qin dynasties, the casting of bronze vessels at that time followed the principle similar to that of movable-type printing. For instance, *Qingonggui* (also known as Duke Qin's *gui*), a bronze food container made in the Eastern Zhou dynasty in the seventh century BC, was unearthed in Gansu, the inscription of which consisted of 50 characters. As seen from the rubbings, it is evident that each character was cast from a separate mold, and the inscriptions of various molds were combined to form the text, with clear boundaries among the characters,[1] suggesting that each individual mother mold with a single character was imprinted on the pottery mold first, followed by casting. Such practices probably inspired the application of the movable-type principle in printing in later generations. However, it is not strictly right to consider the above-mentioned as movable-type printing, just as it is not accurate to regard ancient seals as woodblock printing due to the lack of social atmosphere and technology necessary for the development of printing.

[1] Luo Zhenyu. *Manuscripts from Songweng: Vol. 1* [M]. Stone Inscription by Luoshi from Shangyu, 1925: 32. Rong Geng. *A General Study of Yi Vessels of Shang Dynasty and Zhou Dynasty: Vol. 1* [M]. Beiping: Harvard-Yenching Institute of Yenching University, 1941: 88, 158, Fig. 35.

P. Jixing, *History of Printing Techniques in China*,
https://doi.org/10.1007/978-981-96-6119-0_6

The Song dynasty marked the golden age of woodblock printing, from which movable-type printing evolved as an inevitable outcome of technological development. Woodblock carvers, after days of hard work, would produce a set of printing plates that could only be used to print a single type of book. Even if reprinted, the content remained the same. What's worse, there would be enormous consumption of wood. If the printing plates could be used for multiple printings of different contents at the same time, it would save efforts of repeated carving and reduce material consumption as well. By detaching the movable-type components from the woodblock, a new set of printing plates could be created, enhancing the utilization rate of the plates and thus leading to the advent of movable-type printing. Once printing is completed, these blocks can be removed from the plate and rearranged to form a new one. As a single set of character blocks could achieve the same effect as multiple sets of printing plates do, static printing plates were then transformed into dynamic ones, thus giving rise to movable-type printing. From the perspective of technological development, wooden movable-type components are one of the earliest forms of such movable-type components. As long as a large number of component-bearing wood plates were produced, it would be easy to create individual wooden movable-type components for printing.

Since movable-type components were prefabricated, assembling them into a printing plate improved typesetting efficiency while conserving time, materials, and labor costs. After printing, movable types could be removed, recycled, and reused, owing to their typical feature of easy storage. Movable-type printing revolutionized the plate-making process in the printing industry, as well as some of the following craft procedures, including the addition of disassembling plates, recycling movable types, storage of movable types, type selection, and the use of new equipment, which already constituted an invention. This invention marked another milestone in the history of printing following the advent of woodblock printing, and had an epoch-making significance. Modern printing, evolving from movable-type printing, arose from China and later spread worldwide. Although the exact time when China first used wooden movable types for printing needs further exploration, it is certainly not as late as people thought in the past. Movable-type printing was invented in China because China was the birthplace of woodblock printing, which was the technological prerequisite for movable-type printing. Without the ideas and practices of woodblock printing, movable-type printing could not have emerged out of thin air. For the Chinese, who had long engaged in woodblock printing, it was natural to invent movable-type printing given enough experimentation.

By the end of the Five dynasties (907–960) and the beginning of the Northern Song dynasty (960–1127), when woodblock printing was greatly developed, experiments on wooden movable-type printing had already been conducted. By the early Northern Song dynasty in the eleventh century, wooden movable-type printing had been practically applied to print contracts and instruments issued by the government. For example, land deeds, due to their widespread issuance across various provinces, were standardized by the central government and printed using wooden movable-type characters. Ma Duanlin (1254–1323) in his *Wenxian Tongkao* ("Comprehensive Textual Research of Historical Documents", 1309), Volume 19, recorded that in the fifth year of Shaoxing period (1135), "Initially, it was ordered that *tongpans* (*tongpan*, a governmental official in ancient China serving as a deputy or assistant to a higher-ranking official, such as a prefect or governor of various prefectures or provinces) were responsible for printing and selling land deeds... The county clerks were in charge of the printing plates, and they often printed more deeds for unauthorized sale secretly. Now it is intended to entrust a *tongpan* to establish a *Qianziwen* (Thousand-character Classic) code to be printed on the official deeds, and to provide them to the counties every month. When people intended to buy houses or land, they should be given the official deed by the government." A *Qianziwen* code was printed on the land deed, and each deed changed to a different number, so the quantity of printed deeds was monitored to prevent county clerks from selling unauthorized deeds for their own benefit. Since the *Book of a Thousand Characters* originally written by Zhong Yao (151–230) of the Wei state during the Three Kingdoms period and rhymed by Zhou Xingsi (ca. 470–521) of the Liang dynasty is composed of 1000 characters without any repetition, it is suitable for consecutive coding or numbering.

According to Volume 30 *Jingzongzhi* ("General Administration System") of *Qingyuan Tiaofa Shilei* ("Classification of Laws and Regulations in Qingyuan Period", 1202) by Xie Shenfu (1145–1210) of the Southern Song dynasty, "Once someone intends to purchase land or house deeds, it is hoped that the old practice can be followed, allowing the *tongpan* of each prefecture to establish standards and issue land deeds coded or numbered *liao* (a character from *Qianziwen*). Deeds are distributed to the subordinate counties every quarter, managed by the county assistants, and made available for purchase by the people. This practice prevents unauthorized printing and sale of deeds, ensuring the formality and authority of official documents. The term *liaoli* (material code) or *liaohao* (material number) here refers to the sequential codes or numbers printed on official deeds according to the order of *Qianziwen* characters. According to the *Song Huiyao·Shihuo Nianwu Zhi Shisan* ("Memorials to the Throne of the Song Dynasty: Economics of Finance Records") updated on 1st February of 1171, "a *tongpan* was ordered to take charge of the affairs related to land and house deeds, which should be numbered according to *Qian Zi Wen* before being sold to each county. The local administrative offices are asked to determine *liao* (a character from *Qian Zi Wen*), and register the land deeds numbered by *liao* before selling them to the potential buyers." From above, we can see that the land deeds numbered by *Qianziwen* should be registered for late inspection, before being sold to intended purchasers. According to *Song Shi·Shihuo Zhi* ("History of Song Dynasty: The Chronicles of Food, Silks Clothing or Currency"), such a system could be traced back to the years of Yongxi (984–987) and Duan'gong (988–989) during the reign of Emperor Taizong of the Song dynasty (976–997), when trading license had to be obtained by salt and tea dealers. In the eighth year of the Qingli period (1048) and the seventh year of the Xining period (1074) both in the Northern Song dynasty, salt-trading and tea-trading licenses were also numbered by *Qianziwen* (characters from the book *Qianziwen*). The government charged merchants money for licenses, which were called *yin* (a type of trading licenses), numbered by *Qianziwen,* and then archived. Then, merchants could carry out sales of salt and tea by virtue of these licenses.

During the Northern Song dynasty, the land deeds issued by a *tongpan* were produced using woodblock printing, and each bore a different character from *Qianziwen*. As these characters could only be printed using wooden movable type characters and it was not feasible to reopen a mold due to slight variation in characters, some grooves were left in the corresponding parts of the printing plate designed for the official land deeds to accommodate the wooden movable-type components. After a land deed was printed successfully, the corresponding wooden movable-type components were removed from the plate which was then refilled with other components. Thus, most of the text on the printing plate remained unchanged except for the serial numbers. Documents such as land deeds, salt licenses (or coupons), and tea licenses, as products from movable-type printing, were widely applied under the control of the government as early as the first half of the eleventh century.

During the Song dynasty, wooden movable-type components were applied to not only official documents but also books such as various classics and Buddhist scriptures. Although wooden movable-type editions from the Northern Song dynasty failed to survive, those editions from the Southern Song dynasty have been passed down. For example, in May 1999, I found *Yili Yaoyi* ("The Essential Meaning of the Ceremonial Rites") printed in the 12th year of the Chunyou period in Emperor Lizong's reign (1252) when visiting the chamber of rare collections of Taipei Palace Museum. The skewed text and uneven ink distribution suggest that it was likely a wooden movable-type edition. After flourishing in the Northern Song dynasty, the wooden movable-type printing quickly spread to ethnic minority areas in the northwest. The Western Xia dynasty (1038–1227), which coexisted with the Northern Song dynasty, was a regime established by the Tangut people and had close political, economic, and cultural exchanges with the Northern Song dynasty. In 1036, Yeli Renrong (?–1042) imitated Chinese characters to create the Tangut language, which coexisted with Chinese characters in the region. Then, he also introduced woodblock printing and movable-type printing technologies from the Song dynasty for the application of book printing. In September 1991, a Tangut translation of the Tibetan edition of *Jixiang Bianzhi Kouhe Ben Xu* ("Mah-Laksmi-Dhrani-Sūtra") was discovered in a Western Xia pagoda in Baisigou, Helan County, Ningxia. This edition, made of locally-produced white hemp paper, consisted of nine volumes bound in butterfly-fold form, containing 220 pages and 100,000 characters (Fig. 6.1).

Each half-page frame of the aforementioned Buddhist text measured 23.6 cm in height and 15.5 cm in width, with double margins on all sides, a white edge and page numbers, but without fishtails. Upon investigation, Niu Dasheng asserted the text to be a wooden movable-type edition handed down from the second half of the twelfth century (1150–1180).[2] Such assertion was reaffirmed at an expert appraisal hosted by the Ministry of Culture in 1996. The text, identified as being published between the reigns of Emperor Gaozong (1127–1162) and Emperor Xiaozong (1163–1189) in the early Southern Song dynasty, features inverted Chinese numerals such as "four" and "twenty-seven", indicating that the carvers were Han people while

[2] Niu Dasheng. Mah-Laksmi-Dhrani-Sūtra, a Tangut Buddhist Script [J]. *China Print*, 1994, 12(2): 38–48.

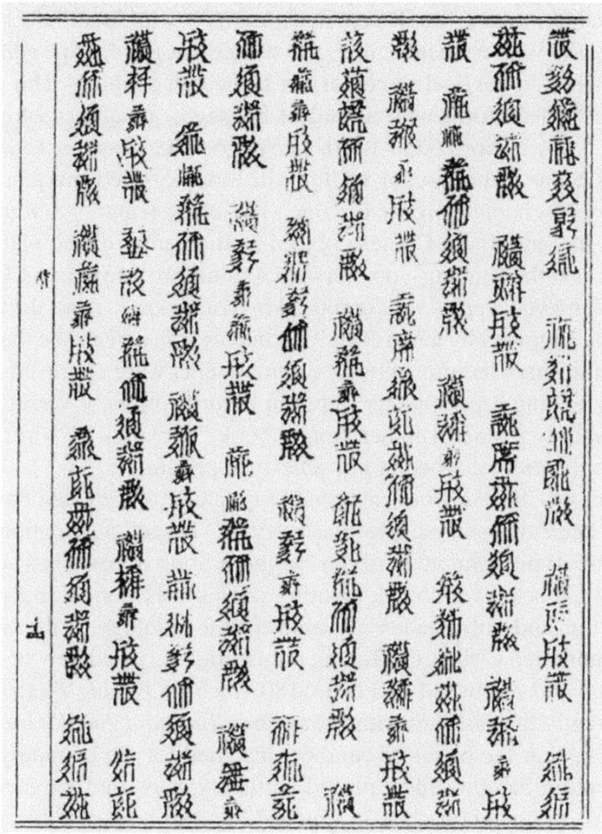

Fig. 6.1 The Tangut translation of *Jixiang Bianzhi Kouhe Ben Xu* discovered in Helan of Ningxia in 1991[3]

the typesetters were Tangut people, and that this text was a collaborative product of Han and Tangut. Besides, the text is characterized by disconnected box bar lines at the corners, uneven length of left and right lines on the type page, variable widths of characters (from 20 mm^2 to 6–7 mm^2), inconsistent stroke thickness, uneven ink distribution, occasionally reversed characters, missing heart lines, confusing or misplaced title abbreviations, typographical errors and omissions in the pagination, and presence of the remnants of the bamboo tablets that were used to make the boundary lines. These characteristics were typical of wooden movable-type editions, and the text was identified as the earliest surviving wooden movable-type work to date.

In 1909, the *Kozlov Expedition* (A Russian Expedition Team) discovered a large number of Tangut documents in Heishui City, among which *Sandai Xiangzhao Jiwen*

[3] Niu Dasheng. Mah-Laksmi-Dhrani-Sūtra, a Tangut Buddhist Script [J]. *China Print*, 1994, 12(2): 38–48.

("The Collected Works of the Three Generations") was identified by Jin Bo, an expert studying the Western Xia regime, as a wooden movable-type edition in Tangut language published in the twelfth century or thirteenth century.[4] This secular book, bound in butterfly-fold form with a total of 82 pages, measures 24 cm in straight height and 15.5 cm in horizontal width per page. The chase is 17 cm in straight height, and 11.5 cm in horizontal width, with double-sided margins, 17 lines on each page, and 16 characters on each line. The book features a white edge, with page numbers in Tangut and Chinese placed in the center of the plate. At the end of the volume are the making vow verses (Aspiration Prayer) in Chinese, which read, "The making vow verses are composed by Huizhao...... and the movable-type components are engraved by Chen Jijin." In this regard, we can see that the text was compiled by Hui Zhao, the movable-type components were engraved by Chen Jijin and typeset by a Tangut person. As stated in the making vow verses, the text was clearly a movable-type one. Another Tangut book, "Dexingji" ("Virtues"), is also a result of the application of wooden movable-type printing.

Although fewer Chinese wooden movable-type editions created during the Song dynasty have been discovered, the discovery of Tangut publications created by wooden movable-type technology of the Northern Song dynasty has added physical data regarding Chinese woodblock printing during this period and provided reliable clues for the study of wooden movable-type technology of the Northern Song dynasty. As mentioned earlier, the Tangut publications created by wooden movable-type technology and published in 1150–1180, such as *Jixiang Bianzhi Kouhe Ben Xu* ("Mah-Laksmi-Dhrani-Sūtra") and *Dajixiang Tuoluoni Jing* ("Great Auspicious Dharani Sutra"), bear the traces of bamboo designed for the boundary lines, which indicates that a thin bamboo slice, placed vertically, was used for each line of text. As the width of the bamboo slice is smaller than the height of the movable-type components, the movable-type components are squeezed together. Accordingly, as a bamboo slice is added once a line is filled with components, all lines of words are clamped by the bamboo slices. The empty space is filled by wood blocks of different sizes to restrict the movement of the movable-type components. After printing, the plate is dismantled and the movable-type components are recycled. The text is bound in butterfly-fold or wrapped-ridge form, and each plate has two pages which are then folded along the middle seam before being bound. This is in line with the characteristics of movable-type printing.

6.1.2 Wooden Movable-Type Printing After Song Dynasty

After the Song dynasty, wooden movable-type printing continued to advance during the Mongol-yuan period. In *Zao Huozi Yinshu Fa* ("The Methodology of Movable Type Printing") cited by the book *Nong Shu* ("Book on Agriculture", 1313 alive) from the early Yuan famous scientist Wang Zhen (1260–1330), a systematic summary

[4] Shi Jinbo. A Study of Tangut Movable-Type Editions [J]. *Beijing Library Journal*, 1997 (1): 67–80.

and detailed introduction were conducted for wooden movable-type printing. Wang Zhen, with the courtesy name of Boxiang, born in Dongping County of Shandong Province, served as the subprefect of Jingde (a county in Anhui) in the first year of the Yuanzhen period in the Yuan dynasty and was highly recognized by the locals because of his brilliant achievements during his tenure. In 1300, he was transferred to serve as the subprefect in Yongfeng of Jiangxi. While in office in Jingde (1295–1300), he carried out scientific research in his spare time, completed 39 volumes of *Nong Shu* ("Book on Agriculture", 1313), and produced 30,000 pieces of wooden movable-type components, which were used to produce *Jingde Xianzhi* ("Jingde County Records", 60,000 words in total) in 1298. Besides, he also wrote a paper on movable-type printing. Initially, he intended to make wooden movable-type printing components for his book *Nong Shu* (110,000 words in total) but failed because the movable-type printing characters were insufficient and the book had not finished yet. After he was transferred to Yongfeng, a person in Jiangxi was willing to publish this illustrated masterpiece using woodblock printing rather than movable-type printing. Later, Wang utilized movable-type printing technology to print other books but failed to pass them down. The movable-type printing technology discussed in *Zao Huozi Yinshu Fa* will be discussed below.

The wooden movable-type technology in the Song dynasty also spread to the Uyghur region in the northwest. In 1908, the Frenchman Paul Pelliot (1878–1945) found 968 pieces of wooden movable-type Uighur components (Fig. 6.2) in Mogao Grotto 181 (now 464) in Dunhuang, Gansu Province, which were used by the Uyghur people and could date back to the year 1300, namely the fourth year of Dade of the Yuan dynasty[5] or two years after Wang Zhen utilized the wooden movable-type printing to publish the county records. Carter published a photograph of this group of wooden movable-type printing characters (4 pieces) collected by the Metropolitan Museum of Art in New York. Each character was 2.2 cm high, 1.3 cm wide, and 1–2.6 cm long (Fig. 6.2).[6] In 1995, after studying a group of wooden movable-type components collected by the Musée Guimet in Paris, Dr. Gou Shouer, an expert in Uighur script, found that each character component consisted of letters, roots, affixes, etc., and was of different lengths due to the number of letters it contained. This batch of movable-type artifacts also contained column lines, movable-type components denoting the punctuation, holding strips and center wood used to fill in the blanks (Fig. 6.3).[7] It could be traced back to the period between the late twelfth century and the beginning of the thirteenth century. These characters, skillfully carved and elegantly lettered, were created by the Uyghur people through the integration of the

[5] Pelliot P. Une bibliothèque médiévale retrouvée au Kansou [J]. *Bulletin de l'Ecole Francaise d'Extreme-Orient*, 1908 (8): 525–527.

[6] Catter T F, Goodrich L C. *The Invention of Printing in China and Its spread Westward* [M]. 2nd ed. New York: Ronald, 1955: 146–147.

[7] Wu Shouer. *Excavations at Dunhuang of Migratory Writing and Its Significance in the Westward Spread of Movable Type Printing Techniques* [M]. // Ye Zaisheng. *Publishing History Research.* Beijing: China Book Publishing House, 1998: 1–2. Shi Jinbo, Wu Shouer. *Studies on Movable Type Printing in the Western Xia and the Huihu* [M]. Beijing: Social Sciences Academic Press, 2000: 89–105.

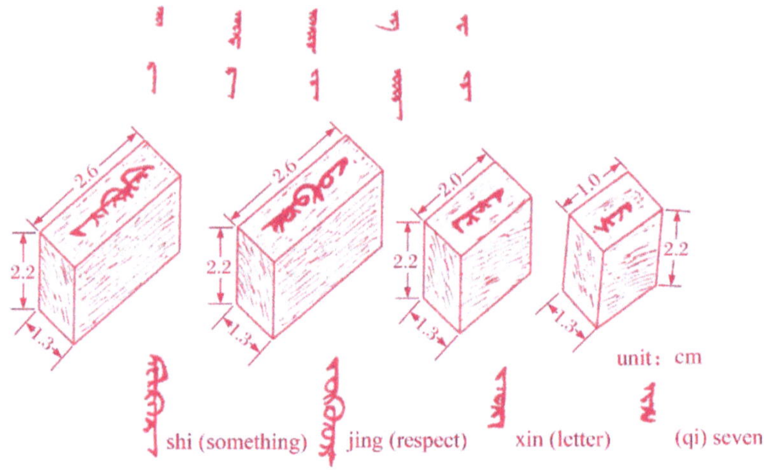

Fig. 6.2 Wooden movable-type Uighur components dated between the twelfth century and the thirteenth century and found in Mogao Grotto of Dunhuang in 1908 (These components were drawn by Pan Jixing)[8]

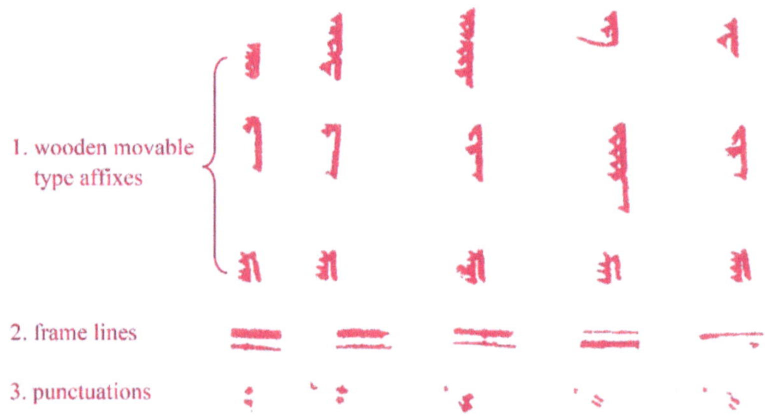

Fig. 6.3 Wooden movable-type components denoting affix, column lines, and punctuations found in Mogao Grotto of Dunhuang

mainland printing technology and the characteristics of their language. They served as an intermediary of movable-type components for the transition from Chinese characters to alphabetic letters.

Less than ten years after Wang Zhen employed the wooden movable-type printing to print books in the early Yuan dynasty, Ma Chengde (1279–1335), Mayor of

8 Catter T F, Goodrich L C. *The Invention of Printing in China and Its Spread Westward* [M]. 2nd ed. New York: Ronald, 1955: 146–147.

Fenghua in Zhejiang Province, also embarked on similar printing activities. Ma Chengde, with the courtesy name of Zhiyuan, was born in Guangping (today's Yongnian, Hebei Province). In 1319, he was appointed as Mayor of Fenghua in Zhejiang Province. During his three-year tenure (1319–1322), he, like Wang Zhen, was robustly devoted to the development of local agriculture and forestry and printed books using 100,000 wooden movable-type components to promote cultural development. Volume 11 of *Fenghuaxian Zhi* ("Fenghua County Records", 1686) compiled during the Emperor Kangxi period quoted the saying from *Zhizhou Machengde Quenbei* ("The Stele of Gratitude to Magistrate Ma Chengde") written by Li Yuansun, a scholar of the Yuan dynasty, that

Ma Chengde, with the courtesy name of Zhiyuan and born in Guangping, was appointed as Mayor of Fenghua. During his tenure, he was robustly dedicated to the development of local agriculture and forestry. In the third year of his tenure (1322), a successor reported that… under Ma's instruction, about thirteen hectares of the wasteland was reclaimed… the area of educational land was increased to one thousand and two hundred *dan* (a unit of capacity), and the number of movable type components reached one hundred thousand.

Furthermore, *Fenghuaxian Zhi* ("Fenghua County Records") also quoted the saying from *Jian Zunjingge Zengzhi Xuetian Ji* ("The Record of Building the Zunjing Pavilion and Increasing the Area of Educational Land", 1322) created by Deng Wenyuan, a scholar of the Yuan dynasty, that

Ma Chengde was appointed as Mayor of Fenghua County… During his tenure, he made great achievements such as building Zunjing pavilions…… He reprinted *Jiu Jing* ("Nine Classics"), as well as famous literary works from Hanyu and Liu Zongyuan, etc. In 1322, he applied the wooden movable-type printing technology to printing of books.

In the Ming and Qing dynasties, wooden movable-type printing continued to develop based on the foundation laid by the Yuan dynasty. Hu Yinglin, a bibliographer of the Ming dynasty, wrote in his *Shaoshi Shanfang Bicong* that "movable-type printing was invented by Bi Sheng of the Song dynasty, and the movable-type components were made of baked clay. Now (in the Ming dynasty), the movable-type components, made of wood rather than the originally baked clay, are called wooden movable-type components." Although there are wooden movable-type editions from the Ming dynasty, it is relatively difficult to authenticate them because the materials used for the movable-type components are rarely specified in these editions. In contrast, there are always clear statements about the materials used for the clay or metal movable-type components. Moreover, it's challenging to distinguish between wooden movable-type components and woodblock components in terms of carving or typesetting. Thus, it's important to carefully study the wooden, clay, or metal movable-type components and know their respective characteristics regarding font, typesetting, and ink. For example, the poetic collection named *Tangshi Leiyuan* ("Classified Collection of Tang Poems") (Fig. 6.4), compiled by Zhuo Mingqing (1552–1620) in 1586 and currently held by the National Library of China, is a wooden movable-type edition, with the characters *Songzhai Diaomu* (Wood Carving by Song's Studio) at

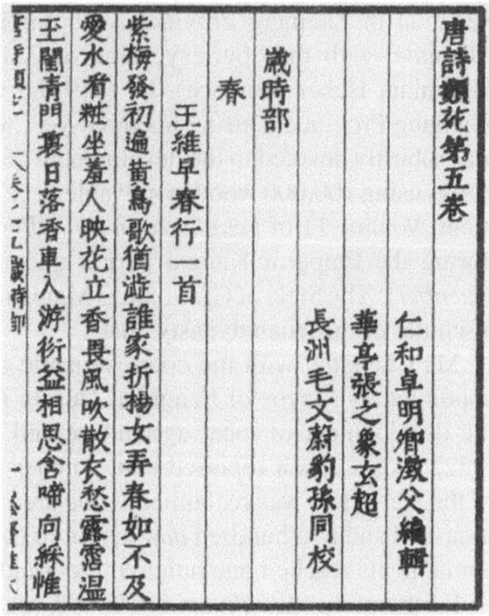

唐詩類苑第五卷

仁和卓明卿澂父編輯

華亭張之象玄超

長洲毛文蔚豹孫同校

歲時部

春

王維早春行一首

紫梅發初遍黃鳥歌猶澁誰家折楊女弄春如不及

愛水看粧坐羞人映花立香畏風吹散衣愁露霑溼

玉閨青門裏日落香車入游衍益相思含啼向綵帷

Fig. 6.4 The wooden movable-type edition of *Tangshi Leiyuan* compiled in the fourteenth year of the Wanli period of Emperor Shenzong (1586) and currently housed in the National Library of China

the bottom. *Songzhai* presumably refers to Zhuo Mingqing.[9] *Bishui Qunying Daiwen Huiyuan* ("The Assembly of Distinguished Scholars Awaiting Inquiry at the Bi Shui") with 90 volumes, published during the period from the tenth year of the Zhengde period to the ninth year of the Jiajing period (1515–1530) and currently held by Nanjing Library, is a kind of textbooks for the Song dynasty students to take the examination. At the end of the book lies these words like *Lizetang Huoban Yinhang// Gusu Husheng Shanxie//Zhangfeng Ke//Zhaosheng Yin* (Printed by Lizetang Printing House, Copied by Hu Sheng From Suzhou, Engraved by Zhang Feng, and Printed by Zhao Sheng.), which indicates that this is a wooden movable-type edition published in Suzhou.

Another contemporary work where the wooden movable-type printing technology was adopted is *Shimiao Shiyu Lu* ("The Records of Temple Periods") compiled by Xu Zhaoji, a native of Jiading, Zhejiang Province, in honor of his father Xu Xuemo (1522–1593). This 26-volume book contains various old practices in the reign of Emperor Shizong of the Ming dynasty (1522–1566) and says at the end that it was not until after more than ten years of efforts that the book was completed, and published in a movable-type form due to economic hardship. Some royal palaces and academies also issued wooden movable-type publications. For example, the prince Zhu Yiyin released in 1574 *Bianyi Lun* ("Discussion of Doubts") compiled by the atheist Xie Yingfang of the Yuan dynasty. At the end of the book bears the words "Yifan Movable-Type Printing House". *Xu Guwen Huibian* ("Sequel to the Collection of Ancient Prose"), completed in 1537, was published by East Lake Academy using

[9] Zhao Wanli, et al. *Catalogue of Chinese Engravings: Vol. 1* [M]. Beijing: Cultural Relics Press, 1961:101.

movable-type printing technology, with a line of words at the bottom of the page, indicating that the book was published by East Lake Academy.[10] According to the existing editions that survived over the generations, wooden movable-type printing prevailed at the end of the Ming dynasty (the sixteenth century). Gu Yanwu (1613–1682) mentioned in Volume 3 of his *Tinglin Wenji* ("Collection of Tinglin") that since 1638, the government applied wooden movable-type printing to release *Dibao* (a kind of newspaper in the Ming dynasty). According to incomplete statistics, there were over 100 varieties of wooden movable-type editions printed during the Ming dynasty.

During the reign of Emperor Qianlong (1736–1796) in the Qing dynasty (1644–1911), the wooden movable-type printing reached a climax. In the 38th year of Emperor Qianlong (1773), when the grand series *Siku Quanshu* ("Complete Library in the Four Branches of Literature") were being compiled, an order was issued to first publish the original books that failed to be handed down from past generations. After the list of books to be published was proposed, Jin Jian (1727–1791), vice deputy editor-in-chief of the grand series and director of the publication department, suggested to print this series with wooden movable-type methods. Such a suggestion was approved by the Emperor who also renamed this series as *juzhenban* (a precious edition) rather than *huoziban* (a movable-type edition). In the following year (1774), 253,000 wooden movable-type components were produced, and applied to print 134 series with a total volume number of 2389. For details, please check relevant book-lists.[11] The movable-type components made of Jujube wood were finely carved, and available in two sizes (1.28 cm × 0.9 cm and 1.28 cm × 0.64 cm; both with a height of 2.24 cm). The cost of manufacturing the movable-type components and tools was 2339 *yin* (a kind of currency unit in ancient China). Each book followed a unified format, with nine lines per half page, and 21 characters per line. At the bottom of the first page of each book were the words *wuyindian juzhen ban* ("Block-printed Juzhen edition of the Hall of Martial Valor"). This series was distributed throughout the country and was allowed to be reproduced. Additionally, other books published during the reign of Emperor Qianlong such as *Baxun Wanshou Shengdian* ("80th Birthday Grand Celebration") and *Xixun Shengdian* ("Western Tour Ceremony"), were also recreated by wooden movable-type printing. Jin Jian also provided standardized descriptions of the techniques employed during the publication of *Wuyingdian Juzhenban Chengshi* ("Regulations on Movable Type Printing Adopted at the Hall of Martial Valor", 1776) (Fig. 6.5). Afterwards, local governments, private printing houses, and individuals followed this technique to print books, leading to an increase in the number of publications. There were even professional printers who traveled with movable-type components to various places to provide services for family genealogies printing.

---

[10] Zhang Xiumin. *History of Printing in China* [M]. Shanghai: Shanghai Renmin Publishing House, 1989: 676–682. Xu Yingjian. *Series on Chinese Printing History: Vol. 1* [M]. Taibei: Taiwan Printing Association, 1997: 279.

[11] Tao Xiang. Movable Type Printing Series of the Hall of Martial Valor [J]. *Library Science* (quarterly), 1929, 1(2): 205–217.

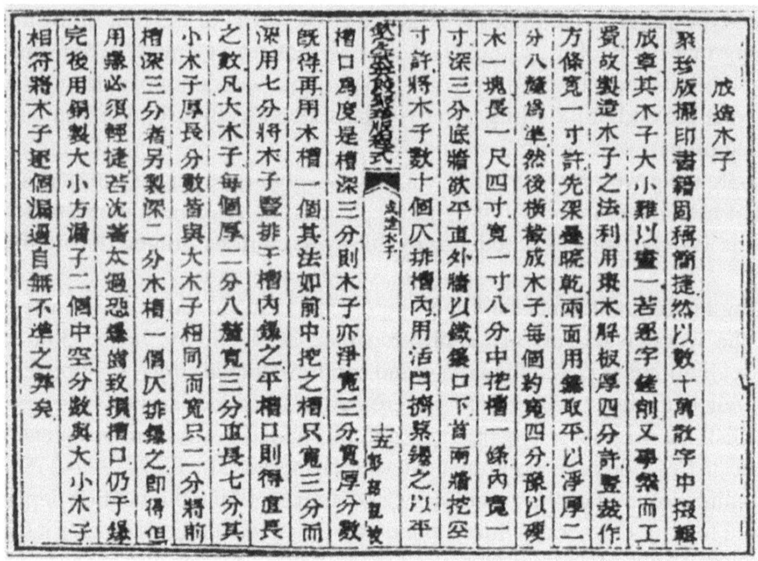

成造木子

聚珍版擺印書籍固稱簡捷然以數十萬數字中撿擺成章其木子大小雖以畫一若惡字維剞又事槃而工費故製造木子之法利用棗木解板厚四分許竪裁作方條寬一寸許先深疊曬乾兩面用鉋取平以淨厚二分八釐爲準然後横截成木子每個約寬四分餘以硬木一塊長一尺四寸寬一寸八分中挖槽一條內寬一寸深三分底牆欲平直外牆以鐵鑲口下首兩牆挖空寸許將木子數十個仄排槽內用活閂擠緊鑲之以平

武英殿聚珍版程式 成造木子 十五

槽口爲度是槽深三分則木子亦淨寬三分寬厚分數既得再用木槽一個其法如前中挖之槽只寬三分而深用七分將木子竪排于槽內鉋之平槽口則得直長七分其數凡大木子每個厚二分八釐寬三分直長七分其小木子厚長分數皆與大木子相同而寬只二分將前槽深三分者另製深二分木槽一個仄排鉋之即得但用鉋必須輕捷若沈著太過恐鉋齒致損槽口仍于鉋完後用銅製大小方漏子二個中空分數與大小木子相符將木子逐個漏過自無不準之弊矣

Fig. 6.5 The wooden movable-type edition of *Wuyingdian Juzhenban Chengshi* published in the 41st year of Emperor Qianlong (1776)

6.1.3 *Techniques of Manufacturing and Typesetting Wooden Movable-Type Components*

A systematic discussion of the wooden movable-type printing technology began with the classic paper *Zao Huozi Yinshu Fa* ("The Methodology of Movable-Type Printing") by Wang Zhen, a scientist of the Yuan dynasty. He summarized it as follows.

There is now a simple way to make movable-type components: using wood blocks as the printing form and bamboo strips as the boundary lines. Carve characters on the wood block, which are then separated with a fine saw, and trimmed with a small knife so that the components are identical in size or height. Then form them up into a line, and clamp them with bamboo strips. Once filled with movable-type components, the printing form is strengthened with wooden pins to keep the components in place. Finally, brush the printing form with ink before printing.[12]

The word "way" in the sentence "There is now a simple way to make movable-type components" refers to the method passed down from the Song dynasty and improved by Wang Zhen whose greatest achievement is an invention of a rotating turntable designed to store the movable-type components, which will be discussed below. The above-mentioned sentence mainly describes the techniques of manufacturing and typesetting these components. After skilled calligraphers write the characters

[12] Wang Zhen. *Book on Agriculture: Vol. 22, Method of Making Movable-Type Components* [M]. Shanghai: Shanghai Classics Publishing House, 1994: 759–762.

on paper, sprinkle the planed wood block with thin rice paste and then cover the wood block with the side of the paper with characters. After the paper becomes wet, the upright letters become inverted on the wood block. Rub the back of the paper with a brush to remove the paper antler, and carve the meshed wood block with a knife to produce the characters. Then, remove these characters from the block with a saw to obtain the movable-type components, whose dimensions are measured by standard tools and which are trimmed by a small knife or fine file. Before typesetting, use a rectangular frame made of wood as the printing plate, which is the same size as the layout. Implant row by row the movable-type components into the printing plate from the left side according to the manuscript, and add to each row a bamboo strip as a boundary line. Fill the empty space with wood wedges of various sizes. The movable-type components are sandwiched between bamboo strips and tightly squeezed by wooden wedges to avoid movement. Once the entire plate is filled, flatten the components with wood block, brush the ink onto the plate and place the paper on it. Then, rub the back of the paper vertically with a brush. At this point, the whole process is completed. Archaeological evidence confirms that this method of making and typesetting movable-type components was applied during the Song dynasty. As wooden movable-type components were made based on the character model, identical to wood carving components in terms of fonts, the finished products appeared more vivid with high ink absorbency.

Each Chinese character, especially for frequently used characters, should be prepared with several equivalents to meet the need for typesetting. In the Song dynasty, movable-type components were classified according to the phonetic system, and labeled and stored in drawers within large wooden cabinets. Furthermore, a location manual of movable-type components was prepared for easy pickup. The common practice for typesetting in the Song dynasty went as follows: one worker specified the location of a component in the cabinet, and another put components taken out into the tray and delivered them to the typesetters for typesetting. To ease the burden of workers shuttling among cabinets, Wang Zhen invented the rotating turntable to store the movable-type components. In *Wuyingdian Juzhenban Chengshi*, he described the way to manufacture the turntable (Fig. 6.6): there are two circular turntables, which are made of lightweight wood with a diameter of up to 7 *chi* (a kind of unit; 1 *chi* is equal to 0.333 m) and a turntable axle of about 3 *chi* (a kind of unit; 1 *chi* is equal to 0.333 m) in height. At the bottom is a thick plate on which 5 holes are drilled. The middle hole is designed for the turntable axle to support the turntable, and the rest four are used to place the horizontal brackets and fix the shaft seat. To rotate the turntable driven by the axle, the axle under the turntable should be gouged out and placed with round bits of wood. A slightly thinner shaft is placed in the groove and connected to the turntable. The turntable is divided into small partitions, each of which accommodates the movable-type components according to the phonetic system. These components are arranged from the inside out (Fig. 6.7). One turntable is used for common characters while the other is used for rare ones. The picker who sits between the two turntables rotates the turntable to pick out the components specified by his colleague. After printing and plate dismantling, the components can be put back in place by the phonetic system to reduce the workload. However, this labor-saving device is only suitable for the situation where there are few words in a book, because of the limited capacity. If there are many words in a book, the components must be placed in a large wooden cabinet.

Fig. 6.6 The rotating turntable invented by Wang Zhen in the first year of Emperor Yuanzhen

Fig. 6.7 Drawing of manufacturing movable-type component made according to *Nong Shu* from Wang Zhen

The method above[13] marks a major improvement in manufacturing movable-type components. During the Song and Yuan dynasties, components were removed with a saw one by one from the board engraved with words. To ensure the uniform dimension of the components and reduce the defective rate, the path of a saw must be straight. To solve this problem, Jin Jian proposed that the wood block with appropriate thickness (equivalent to the height of components) should be sawed into several long square wood bars which were then planed, and horizontally cut to obtain a number of movable-type components. A dozen of those components were placed in a hardwood fillister, squeezed with a live latch, and planed until they were level with the fillister mouth to ensure uniform dimension. Then, components were checked against the standard copper mold one by one to verify if they met the dimensional standard. Later, the components were tightly placed on the engraving bed. When the reverted characters were transferred to the surface of the components, it was time for carving. This method ensured consistent dimension and simplified the carving process.

In the past, storing, picking, and recycling of movable-type components were based on the phonetic system. However, Jin Jian proposed the use of radical, stroke, or character form rather than the phonetic system to complete those operations. This was more convenient for ordinary typesetters who were not familiar with phonetics. During the Song and Yuan dynasties, movable-type components, holding strips, and type pages were inserted one by one into the printing plate before inking and brushing, often resulting in gaps at the connecting lines. Furthermore, it was cumbersome and time-consuming to install hold strips. Jin proposed to first carve out the frame, type page, and grid lines on a wood block, and then print them onto paper to display all the grid lines. As only movable-type components that were uniform in dimensions were arranged on the printing plate, and the plate was covered with a piece of paper bearing grid lines, the boundaries would be clear and the connection between lines would be tight. Additionally, Jin Jian also proposed the "daily rotation system", an assembly-line method of carrying out typesetting, printing, disassembly, component recycling, and re-typesetting concurrently, which improved the utilization of movable-type components. By the eighteenth century, the traditional Chinese wooden movable-type printing technology had reached its historical peak (Fig. 6.8).

[13] Jin Jian. *Imperial Printing Office Manual for Movable Type* [M]. Wooden Movable-Type Edition. 1776 (The 41st Year of Emperor Qianlong in the Qing Dynasty): 15–32.

1. Make movable-type blocks

2. Carve out wooden movable types

3. Prepare the typesetting bed

4. Select and arrange characters for typesetting

Fig. 6.8 Drawing of applying the wooden movable-type printing mentioned in *Wuyingdian Juzhenban Chengshi*

6.2 The Origin and Development of Ceramic Movable-Type Printing

6.2.1 *The Origin of Ceramic Movable-Type Printing*

During the Northern Song dynasty, the invention of wooden movable type printing, although superior to woodblock printing, faced the challenge of dealing with nearly 20,000 Chinese characters. To print a book, one needed to carve 100,000–200,000 wooden movable type characters. This workload was not insignificant, and it demanded high-quality materials and substantial financial investment. As a result, the wooden movable type did not replace woodblock printing but rather developed in parallel, gradually improving over time. A significant breakthrough came when the Northern Song printer Bi Sheng (990–1051) developed a new method during the Qingli years (1041–1048), creating ceramic movable type. This innovation substantially reduced production costs. The contemporary scientist Shen Kuo (1031–1095), in his work *Mengxi Bitan* ("Dream Pool Essays", 1088), Volume 18, documented this new technology as follows:

In the middle of the Qingli era (1041–1048), a commoner named Bi Sheng created a movable type. His method involved carving characters in adhesive clay, as thin as the edge of a coin, with each character as a separate piece. They were then fired to make them firm. First, a metal plate was prepared, and on it, a mixture of pine resin, wax, and paper ash was applied. When printing was needed, an iron mold was placed on the metal plate, and the characters were densely arranged on the mold to form a complete plate. This plate was then placed over a fire to melt the mixture. Once the substance melted, a flat board was used to press down on the surface, making the characters flat as if polished. If only a few books were printed, this method was not particularly convenient. However, for printing tens, hundreds, or thousands of books, it was remarkably efficient.

Two iron plates were used, one for printing and the other already set with arranged characters. As the printing on one plate is completed, the second plate is ready, and they are interchangeably used, achieving instantaneous results. Each character has multiple imprints, such as function words "*zhi*" and "*ye*", each having more than twenty imprints, to account for potential repetitions within one plate.

Unused imprints are pasted onto paper, with each pasting forming a rhyme, stored in wooden compartments. For rare characters not readily available, they are promptly engraved and fired. Instead of using wood, which may have uneven grains and become uneven when wet, sticking to both water and ink, it is preferable to use fired clay. After use, the imprints are reheated to melt the substance, and a gentle brush causes the imprints to fall off without any residue.

Upon Bi Sheng's passing, the imprints were acquired by his nephew, and they have been preserved to this day (see Fig. 6.9).[14]

---

[14] Sheng Kuo, *Mengxi Bitan: Vol. 18, Techniques* [M]. Beijing: Cultural Relics Press, 1975: 15–16.

1. The 1305 reprinted edition from Dongshan Academy

2. Copy of Bi Sheng's clay movable type
(collection of the National Museum of China)

Fig. 6.9 Written account of Bi Sheng's invention of clay type in *Mengxi Bitan* (1088)

6.2.2 Raw Material Composition and Firing Technology of Ceramic Movable Type

The excerpt from Shen Kuo's account requires some clarification. The phrase "carving characters in adhesive clay, as thin as the edge of a coin" can be interpreted as follows: The edge of a coin is approximately 2 mm thick. Some incorrectly interpret this as referring to the height of the movable type, which would be impractical for clay type with a 2 mm height due to difficulties in manufacturing and typesetting, as well as low strength. In reality, "the edge of a coin" should be understood as the depth of the engraving, which makes more technical sense. The sentence "When printing was needed, an iron mold was placed on the metal plate, and the characters were densely arranged on the mold to form a complete plate" means that when printing is desired, the movable type is embedded into an iron plate, and then characters are densely arranged on the plate. Thin iron strips are used as spacers, adding one strip after each line is filled with characters. This process ensures the orderly arrangement of the movable type. Shen Kuo did not talk about the material of the adhesive clay. Some suggest it might be a material called *liuyini* (six-one clay)[15] used by ancient alchemists, and listed its 12 raw material formulas. The main ingredients are red ochre (**Fe_2O_5**), alum, talc, white lead carbonate [**$2PbCO_3 \cdot Pb(OH)_2$**], oyster shells (CaO), salt, vinegar, etc.

However, from a technical perspective, the likelihood of using *liuyini* is minimal, as alchemists primarily used it to seal alchemical furnaces without the need for firing. Even if fired, there are questions about whether it would have sufficient strength for printing. *Liuyini* is composed of 7 temporary ingredients, each alchemical text providing different recipes, making it impractical for mass-producing movable type. Additionally, the labor-intensive process of collecting and preparing these ingredients makes it unlikely that Bi Sheng, who came from a common background, would have used artificially prepared *liuyini* as a material for movable type. In reality, the material for adhesive clay movable type is likely regular pottery clay, a readily available and inexpensive natural material. Pottery clay comprises silicon dioxide (SiO_2), aluminum oxide (Al_2O_3), sodium oxide (Na_2O), calcium oxide (CaO), and potassium oxide (K_2O), among others. The process involves drying, grinding, sieving, mixing with water to form a slurry, filtration, precipitation, kneading when partially dry, and molding into consistent-sized clay blanks in a plastic state.

The phrase "carving characters in adhesive clay", when interpreted technically, involves a specific procedure. The actual operational sequence is as follows: (1) Craft a character mold with characters in the regular script; (2) Press the character mold into a clay blank, forming clay blocks with negative characters in reverse script; (3) Subsequently, the clay blocks with imprinted characters are fired in a kiln. During this process, a chemical transformation occurs, turning the clay into black ceramic movable type. While referring to it as "clay movable type" may not be scientifically precise, the colloquial acceptance of this term persists. The more accurate term is "ceramic movable type". In practice, clay materials, when fired at temperatures between 700–900 °C, exhibit a water absorption rate of 5%–10%,[16] making them suitable for use as printing movable type. Therefore, Bi Sheng's ceramic movable type would likely be fired at a temperature between 600–1000 °C, with 600–800 °C possibly being the most suitable range.[17]

After the movable type is fired, it is taken from the kiln and, after some refinement, becomes ready for use. The ceramic movable type appears as elongated cubes, gray-black in color, with a smooth surface, sufficient strength, and good ink absorption properties. Printing a book requires various sizes and quantities of movable type for each character, with commonly used characters requiring dozens of duplicates. The size of the characters should be comparable to those used in woodblock printing

---

[15] Feng Hanyong. Bi Sheng's Movable Type Clay Is Liuyini [J]. *Literature, History and Philosophy Monthly*, 1983 (3): 84–85.

[16] Feng Xianming, et al. *History of Chinese Ceramics* [M]. Beijing: Cultural Relics Press, 1982: 47–50.

[17] Pan Jixing. *History of Chinese Science and Technology: Papermaking and Printing Volume* [M]. Beijing: Science Press, 1998: 325–326.

during the Song dynasty, typically ranging from 0.5–1.0 cm in length and width and 1–1.5 cm in height. After firing, the ceramic movable type is categorized and numbered based on character rhymes, stored in wooden cabinets with labeled drawers for easy sorting and retrieval.

Considering the material and performance characteristics of ceramic movable type, the printing plate is made of an iron plate with borders on all sides, and the height of the borders is equivalent to the height of the movable type. To ensure that the movable type is firmly fixed to the plate, Bi Sheng first places pine resin and wax on the plate as an "adhesive substance". During typesetting, the iron plate is gently heated with fire, melting the adhesive substance, and then the ceramic movable type is planted into the printing plate. Starting from the left border of the plate, Bi Sheng plants the characters from top to bottom. After completing one line, an iron strip is placed as the boundary line. If the height of the iron strip is equal to that of the movable type, it can print boundary lines; if it is smaller, it cannot. Adding thin iron strips ensures that each character is aligned and serves as a clamping function. Of course, typesetting can also be done without iron strips. When the entire plate is filled, a flat board is used to press down on the flat movable type to ensure even ink absorption during printing. Two plates are used for printing books. When one plate has finished typesetting, ink is applied and printed. At the same time, the other plate is typesetting. After the first plate is printed, the bottom is heated with fire, and the adhesive substance melts. The plate is disassembled, and the used movable type is removed and stored in a cabinet for the next typesetting. The second plate, after completing typesetting, is then inked and printed. The two plates are alternately used. In the case of rare characters, temporary clay movable types or wooden movable types can be carved.

Bi Sheng used the above method to create movable type for printing books. The used movable type was passed down to Shen Kuo's nephew, and upon learning about it, Shen Kuo recorded this technique. Bi Sheng's technology includes the entire process of ceramic movable type manufacturing, picking characters, typesetting, printing, disassembling plates, and storing movable type. It is a complete invention with practical feasibility. He is recognized as the inventor of ceramic movable type printing technology in history. Ceramic movable type evolved from wooden movable type, but the two have different compositions and manufacturing methods. Ceramic movable type is formed through high-temperature firing, causing a chemical change in the raw materials, while wooden movable type is carved with a knife at room temperature without a chemical change in the raw materials. Using clay as the raw material for ceramic movable type expanded the source range of movable type raw materials and reduced production costs, thus promoting the development of movable type technology. It also served as inspiration for the casting of metal movable type. Bi Sheng's historical contributions deserve full recognition.

Bi Sheng produced a large quantity of ceramic movable type, estimated to be in the tens of thousands. After his death, his technique was inherited by his family and apprentices. Subsequently, the scientist Shen Kuo introduced the technique, leading to the rapid dissemination and continuation of ceramic movable type printing knowledge in various regions of the Song dynasty. Following Shen Kuo, the Zhejiang scholar Jiang Shaoyu (1036–1169) also made similar records of Bi Sheng's technique in *Huangchao Shishi Leiyuan* ("Dynasty Facts", 1145), Volume 52. During the Southern Song dynasty, Zhou Bida (1126–1204) used this method to publish books with ceramic movable type. In the fourth year of Shaoxi (1193), he wrote to his friend Cheng Yuancheng:

> I, with my humble pen name and growing age, combined with occasional mental vigor, have long been involved in these matters. Recently, following the method of Shen Kuo, I used adhesive clay and copper plates to transfer and copy imprints. Now, I complete twenty-eight entries of *Yutang Zaji* ("Miscellaneous Notes from the Jade Hall"), which I submitted for your review. There are still about ten entries awaiting further records to supplement and include in sequence. I humbly estimate that, upon reviewing the old, I cannot help but sigh at the passing years.[18]

The recipient Cheng Yuancheng (1120–1197), also known as Shuda, was a native of Yi (present-day Yixian County, Anhui). In the twentieth year of Shaoxing (1150), he passed the imperial examination along with Zhou Bida and held various official positions, eventually serving as a Cabinet Academician. He also authored the *Yutang Ji* ("Book from the Jade Hall"). In 1191, after resigning from his official position, Zhou Bida relocated to Tanzhou (modern-day Changsha, Hunan), where he worked as a local official. In his leisure time, Zhou published *Yutang Zaji* and recorded the historical events of the *Hanlin* Academy. The phrase "following the method of Shen Kuo" refers to the printing technique of Bi Sheng as described by Shen Kuo, specifically the method of using clay movable type for printing. The terms "transfer" and "copy imprints" refer to the processes of typesetting and printing. However, Zhou Bida replaced iron printing plates with copper ones, demonstrating a pursuit of refinement. Therefore, *Yutang Zaji* printed in 1193 in Changsha was likely produced using exquisite clay movable type, and the printed copy was sent to Cheng Shuda. Similar surviving examples from the Song dynasty reflect the use of clay movable type. In 1965, in Wenzhou, Zhejiang, remnants of the 12th-century *Aparimitāyur-sūtra* ("Immeasurable Longevity and Wisdom") were discovered at the White Elephant Pagoda. The dimensions were 13 cm × 10.5 cm, and the text was printed in Song typeface with a spiral arrangement, marked with "O" symbols at the spiral turns. It had 12 lines and 166 characters. The characters varied in size and stroke thickness, and exhibited a simple style. Many characters were missing, and the ink intensity varied. The archaeological significance of these findings, including the *se* character in *zase jingang* (Vajrapani) suggests that this sutra is a 12th-century clay movable type print[19] (see Fig. 6.10).

The printed edition discovered at the White Elephant Pagoda featured large characters with a width of 0.5 cm and a length of 0.45 cm, medium characters with a width of 0.4 cm and a length of 0.3 cm, and small characters with a width of 0.3 cm and a length of 0.15 cm. All characters were rectangular and considered small compared

[18] Zhou Bida. *The Complete Works of Zhou Yiguo Wenzhonggong: Vol. 49, Letter to Cheng Yuancheng* [M]. Ouyang Qi's reprint, 1851: 4.

[19] Jin Bodong. Introduction to the Northern Song Dynasty Buddhist Scripture Fragments Unearthed from the White Elephant Pagoda in Wenzhou [J]. *Cultural Relics*, 1987 (5): 15–18.

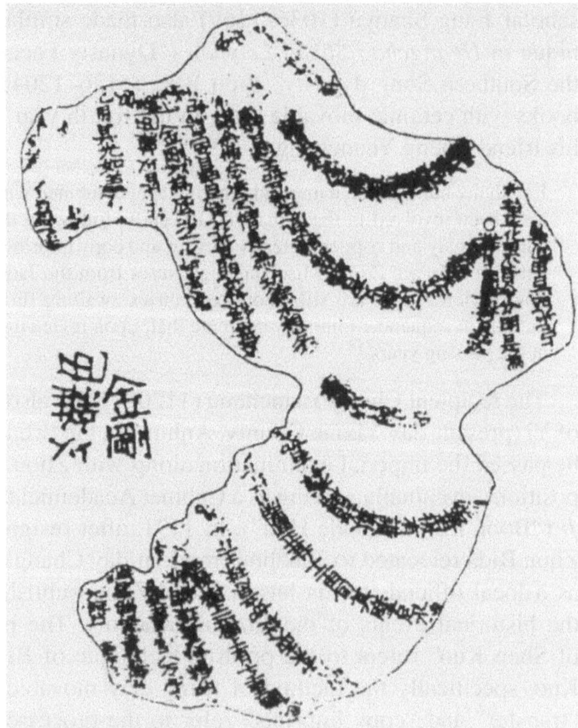

Fig. 6.10 Remnants of the clay movable type print of *Aparimitāyur-sūtra* (Preserved in Wenzhou Museum)

to the characters in the Song edition. Some observers noted that the narrow spacing between characters and the connected strokes suggested characteristics inconsistent with movable type, arguing that it might be a woodblock edition.[20] After a careful examination, printing historian Qian Cunxun concluded that the original conclusion was correct. Early movable type might have been carved according to the strokes of characters, resulting in varying sizes and no space left between strokes. When two characters were connected, the strokes would appear to merge, characteristic of movable type rather than woodblocks. The omission of characters in the text was also more likely with movable type. Some characters appearing inverted did not indicate the direction of the connected next sentence, as other turning points did not feature inverted text. The "O" symbols used at the turning points suggested that the inversion of the character *se* was a mistake in the movable type.[21] We agree with this assessment and the original conclusion. This particular movable type, distinct from standard horizontally printed books, was likely custom-made by workshops or temples. Its purpose was to use movable type to print various shapes of lines, which

[20] Liu Yun. Discussion on the Article "Physical Evidence of Early Movable Type Printing" [J]. *Cultural Relics*, 1988 (10): 95–96.

[21] Qian Cunxun. *Collection of Papers on the History of Books, Paper, Ink and Printing in China* [M]. Hong Kong: Chinese University of Hong Kong Press, 1992: 130–132.

could be combined to form different images or patterns as special reading materials. A set of movable type could be repeatedly used to print different patterns, and in this case, it was employed to print images alongside the text of the *Aparimitāyur-sūtra.*

Dr. Lu Guizhen (1904–1991) had in her possession a portrait of the longevity god, an image made from text. It appears that the use of text to create images has a long history. To accommodate curved lines, the movable type had to be made flat, with minimal space left above and below each character to ensure a tight arrangement. The movable type used for the *Aparimitāyur-sūtra* was specifically made into small, flat characters for this purpose. After applying adhesive to the iron plate, the movable type could be arranged into various shapes and then printed. If the layout was large, several plates were made, and the prints were later joined together. It is evident that composing with movable type is more accessible and convenient than carving woodblocks to create curved lines or images. Additionally, carving a woodblock requires significant effort and can only print a single image, while movable type allows for the printing of different images. The discovery in Wenzhou is an early historical testament to Bi Sheng's movable type technique. Regarding the phenomenon of characters closely juxtaposed, it is also present in horizontally set type books, such as the Korean copper movable type edition from 1403, which exhibits "overlapping of characters above and below",[22] indicating it is indeed a movable type edition.

The technology of ceramic movable type in the Song dynasty was also applied in the Western Xia (Tangut) region. In May 1987, a clay movable type printed edition of the *Vimalakīrti-nirde a-sūtra* ("The Vimalakirti Sutra") in Tangut script was unearthed at the Xuanmu site in Xinhuaxiang, Wuwei, Gansu Province (Fig. 6.11). It consists of 54 pages, each with 7 lines, and each line with 17 characters printed in Tangut script. Each page measures 28 cm in height and 12 cm in width. Each character is 1.4–1.6 cm square, and the printing was done in the first half of the thirteenth century (1223–1226).[23] Another Tangut script edition of the *Vimalakīrti-nirde a-sūtra* was discovered in 1907 at the Western Pagoda site in Heishuicheng. It is now housed at the St. Petersburg Branch of the Institute of Oriental Studies, Russian Academy of Sciences, with five volumes, each page measuring 27.5–28.7 cm in height and 11.5–11.8 cm in width. The height of the printing frame is 22.1 cm, with single borders at the top and bottom. Each page contains 7 lines with 17 characters, featuring varying character sizes and fonts, an irregular arrangement, varying ink permeability on the paper, and uneven strokes. This edition is identified as a clay movable type edition.[24] The colophon indicates the posthumous title of Renzong, the Tangut emperor (1140–1193), so the printing year falls in the mid-twelfth century, no later than the early thirteenth century.

---

[22] Cao Jiongzhen. *Comparison of Movable Type Printing Technology between China and South Korea* [M]. Taipei: Xuehai Press, 1986: 108.

[23] Sun Shouling. *Clay Movable-type Buddhist Scriptures in Xixia Dynasty* [N]. *China Cultural Relics News*, 1994–03–27.

[24] Shi Jinbo. A Study on Movable Type Printing in Xixia Dynasty [J]. *Journal of the National Library of China*, 1997 (1): 67–80.

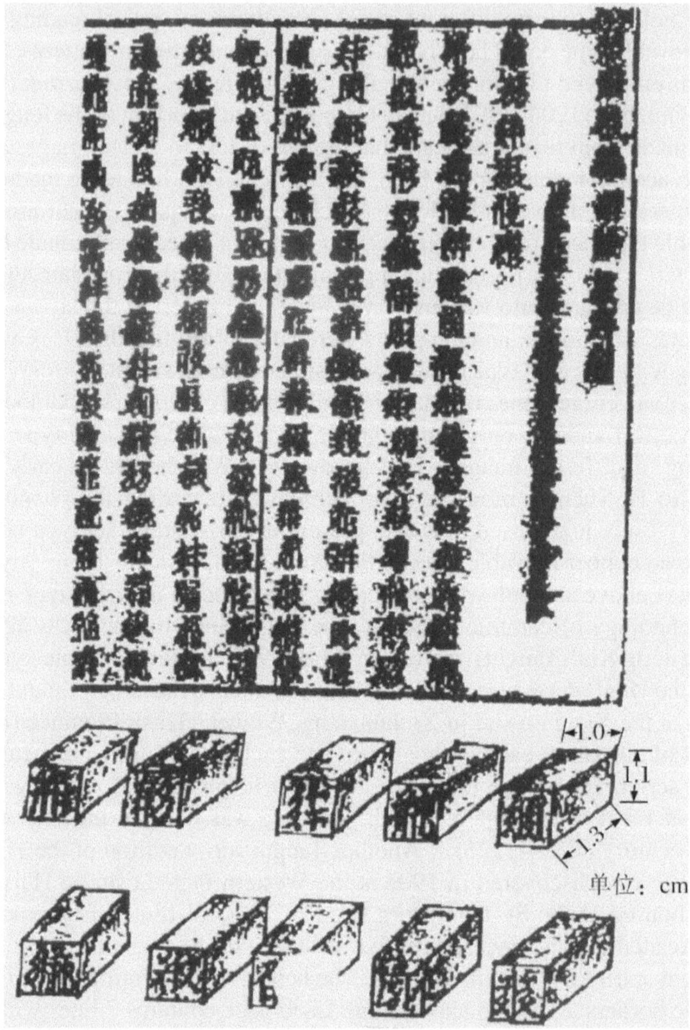

Fig. 6.11 A 13th-century Tangut clay movable type printed edition of the *Vimalakīrti-nirde a-sūtra* unearthed in Wuwei, Gansu in 1987 (The small block below is the clay movable type restored based on the text)

6.2.3 Ceramic Movable Type Printing in Yuan, Ming and Qing Dynasties

The development of ceramic movable type printing in the early thirteenth century is associated with the advocacy of the prominent Confucian scholar Yao Shu (1201–1278). He served as an advisor and minister to Emperor Shizu Kublai Khan (1215–1294). His nephew, Yao Sui (1239–1314), recorded the following information:

In the year Yiwei (1235), Emperor Taizong (Ogedei Khan) issued an edict for the Second Prince to undertake a southern expedition. Grand Master Yao (Yao Shu) and Chancellor Yang (Yang Weizhong) sought scholars proficient in Confucianism, divination, medicine, brewing, and music within the army... In the year Xinchou (1241), Yao was bestowed with garments and gold tokens... Subsequently, he brought his family to Hui... Diligently promoting the education of the people, Yao Shu authored elementary school books such as *Yu Meng Huo Wen* ("Dialogues on Mencius") and *Jia Li* ("Family Rites")... Furthermore, realizing that the dissemination of elementary school books was not extensive, he taught his disciple Yang Gu to use Shen Kuo's method to create movable type. This was employed to print books such as *Jin Si Lu* ("Reflections on Things at Hand") and *Donglai Jingshi Lunshuo* ("Discourses on Classics and History"), which were then distributed widely.[25]

This historical account indicates that in 1235, Yao Shu, along with Yang Zhongshu (1205–1259), accompanied Emperor Taizong Ogedei Khan on a southern expedition and sought scholars in various fields. In 1241, Yao Shu was rewarded, and he subsequently settled in Hui County, Henan Province, where he actively promoted education and published books. Considering that the dissemination of these "elementary school books" was not extensive, Yao Shu further instructed his disciple Yang Gu (1216–1281) to employ Shen Kuo's description of Bi Sheng's ceramic movable type printing method. This likely occurred between 1241 and 1250. Contemporary to Yang Gu, Wang Zhen wrote about movable type printing:

Some have ingeniously crafted techniques, using iron for the printing plate. They fill the spaces within the boundary lines with diluted bitumen, letting it cool to achieve a smooth surface. They then apply heat to melt it again. Burnt clay characters are placed within, creating a movable type printing plate. Finding this inconvenient, they use clay as a printing plate, filling the spaces within the boundary lines with thin clay. The burnt clay characters are arranged in it, and then it is fired in a kiln to form a single piece, which can also serve as a movable type printing plate.[26]

Wang Zhen's mention of "some" refers to individuals from the Song dynasty, particularly Bi Sheng and those who later developed ceramic movable type printing technique. Wang Zhen describes two methods of plate-making. The first involves planting burnt clay characters into an iron printing plate. Thin bitumen is poured into the space within the boundary lines, and clamps are placed on both sides of each line of characters. Bitumen, being an organic adhesive, is more viscous than the resin used by Bi Sheng. After heating, it easily separates from the movable type, representing an improvement over Bi Sheng's adhesive. The second method employs a thin clay as an adhesive. Clay characters are planted on a clay printing plate, and the entire plate is fired in a kiln to solidify. After printing, the clay plate is broken, and the movable type is retrieved. While this method may seem economical, its feasibility is questionable. From a technical perspective, it is deemed impractical to fire an entire

[25] Yao Sui. *Mu'an Collection: Vol. 15, The Monument to the Yao Wenxian, the Left Prime Minister of the Central Secretariat* [M]. Shanghai: The Commercial Press, 1929: 4.

[26] Wang Zhen. *Book on Agriculture: Vol. 22, Method of Creating Movable Type Printing* [M]. Shanghai: Shanghai Classics Publishing House, 1994: 759–762.

movable type plate in a kiln. There is no evidence to suggest that this method could be widely adopted.

During the Ming dynasty, there appears to be a lack of enthusiasm for ceramic movable type printing. However, the technique experienced a revival during the Qing dynasty. For example, Suzhou native Li Yao (1790–1855) published *Nanjiang Yishi Kanben* ("A Historical Account of Southern Territories") in Hangzhou in 1830. The colophon behind the title page bears two lines of seal script that read: "Seven treasures revolving wheel preserved edition. Imitating the Song dynasty method of ceramic movable type printing". The Preface also mentions, "This book follows the example of Bi Sheng's movable type and is arranged using plates", indicating the use of ceramic movable type. In 1832, Li Yao published *Jiaobu Jinshili Sizhong* ("Four Types of Corrected and Supplemented Inscriptions on Bronze and Stone"). He stated, "The characters are systematically arranged and displayed on the self-made clay plates". Following this, Zhai Jinsheng (1775–ca. 1860) also developed ceramic movable type and used it for printing. Zhai Jinsheng, also known as Xi Yuan and Wen Hu, was a native of Shuidong Village, Jing County, Anhui Province. Initially a scholar engaged in educational work, he later became passionate about researching movable type technology. The National Library of China holds a two-volume work titled *Niban Shiyin Chubian* ("The First Edition of Clay Plate Trial Printing",1844), consisting of half-pages with eight lines, 18 characters per line, and ruled lines, totaling 123 pages. The book includes inscriptions such as "Prepared and printed by Zhai Jinsheng of Shuidong in Jing", indicating his self-production of clay movable type for printing his own collection of poems[27] (Fig. 6.12).

Jing County scholar Bao Shichen (1775–1855) wrote in the preface to *Niban Shiyin Chubian*:

Mr. Zhai Xi Yuan from my hometown is an antiquarian. He dedicated thirty years of effort to creating clay movable characters, amounting to tens of thousands. He attempted to print his life's work of various poems, essays, and connected expressions in two volumes… Mr. Xi Yuan read Shen's *Mengxi Bitan*, saw the clay movable characters, and admired them, hence using clay to create them.

It is evident that Zhai Jinsheng, after reading Shen Kuo's *Mengxi Bitan* describing Bi Sheng's ceramic movable type technique, engaged in a similar practice. Between 1814 and 1844, he produced 100,000 movable characters using clay. In 1848, he used them to publish *Xianping Shuwu Chuji* ("Collected Works of Xianping Bookstore") written by his friend Huang Juezi (1793–1853). The cover bears a small inscription: "Jing Zhai Xi Yuan. Clay-type Typesetting", which is held in the National Library of China. Additionally, Peking University Library houses a two-volume set of *Niban Shiyin Xubian* ("The Second Edition of Clay Plate Trial Printing"), published in the same year. With half-pages, nine lines, and 21 characters per line, the font is smaller, representing a revised edition of the first edition, *Niban Shiyin Chubian.* In 1848, Zhai's younger cousin Zhai Tingzhen published *Xiuyetang Ji* ("Collected Works of Xiuye Hall"), a 20-volume set, indicating that other family members were also

[27] Ji Shuying. Newly Discovered Clay Movable Type Printed Books: The First Edition of Clay Plate Trial Printing [J]. *Library Work*, 1958 (1): 22–24.

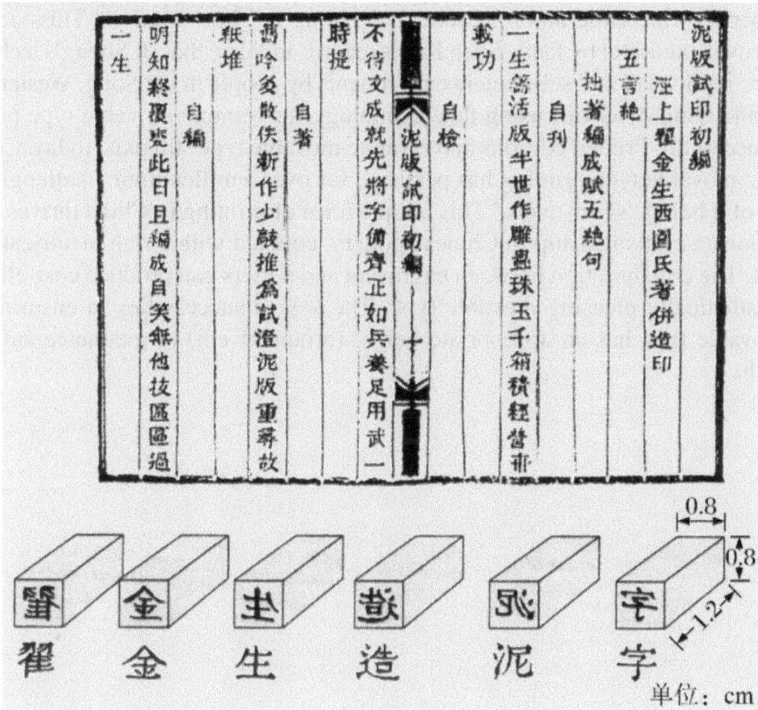

Fig. 6.12 The book made with clay movable type by Zhai Jinsheng in Jing County, Anhui, in 1844

engaged in movable type printing. In 1857, Zhai Jinsheng's grandson Zhai Jiaxiang used the family's collection of clay movable characters to print *Jingchuan Shuidong Zhaishi Zongpu* ("Genealogy of the Zhai Family in Shuidong, Jingchuan"), which is also housed in the National Library of China.

In the 1960s and 1970s, thousands of ceramic movable type characters produced by Zhai Jinsheng in his time were discovered in Jing County. These artifacts are currently housed in institutions such as the Institute for the History of Natural Sciences at the Chinese Academy of Sciences and Anhui Provincial Museum. The discovered movable type characters come in five different sizes, generally ranging from 0.5 to 0.9 cm in length, 0.35 to 0.85 cm in width, and 1.2 cm in height. They appear black, are relatively hard, and remain usable. In addition to the movable type characters for printing, there are negative characters for typesetting, with the same font, and also movable type blocks for filling blanks, providing tangible materials for the study of their craft and technology.[28] Zhai Jinsheng made significant contributions to the revival of the Northern Song ceramic movable type technique. Some people believed that the clay movable type created by Bi Sheng in the Northern

[28] Zhang Binglun. A Preliminary Study on the Clay Movable Type of Zhai Jinsheng [J]. *Cultural Relics*, 1979 (10): 90–92.

Song period was brittle and impractical, deeming it "short-lived".[29] This view has been proven incorrect by facts. Shen Kuo's records indicate that Bi Sheng's technique could be replicated, and subsequent experiments by people in the Song, Western Xia, Yuan, and Qing dynasties using this technology for ceramic movable type printing were successful. Printed editions and ceramic movable type still exist today. Chinese ceramic movable type printing has persisted for over a millennium, challenging the notion of it being "short-lived." This unique form of printing in China thrives due to the exquisite craftsmanship in Chinese pottery, coupled with a rich historical background. The collaboration between engravers and potters can produce cost-effective and aesthetically pleasing movable type. The key to success lies in ensuring that the movable type has an appropriate height (around 1 cm) to guarantee sufficient strength.

[29] Sohn P K. *Invention of Movable Metal-type Printing in Koryo: Its Role and Impact on Human Cultural Progress* [C]. Seoul: International Symposium on Printing History in East and West, 1997.

Chapter 7
Origin and Development of Metal Movable-Type Printing

7.1 Origin of Metal Movable-Type Printing

7.1.1 Copper Movable-Type Printing in Song and Jin Dynasties

The principle of movable type, used for casting inscriptions on bronze artifacts, has a long history in ancient China. Around 1925, artifacts from the Eastern Zhou period (seventh century BC) were discovered in Gansu Province. Among these artifacts was a bronze food container, the Qin Gong Gui, which features a 50-character inscription. Rubbings of this inscription reveal that each character was individually carved into templates, as the boundary lines where characters connect to each other are distinct.[1] These templates were then used to sequentially stamp the characters onto clay molds before casting the bronze. This technique of using movable components in the inscription process may have influenced later developments in movable type printing. However, strictly speaking, these ancient practices cannot be classified as movable type printing, just as ancient seals cannot be considered woodblock printing. The societal conditions and technological prerequisites for true movable type printing were not present in the ancient world. After 400 years of woodblock printing, movable type printing emerged in the eleventh century during the Northern Song dynasty. The earliest forms of movable type included wooden and ceramic types. Earlier, in the eighth century during the Tang dynasty, copperplate printing emerged. This technique involved pressing the templates of raised reversed script characters onto clay molds to create ceramic molds with recessed regular script characters, which were then cast in copper to produce printing plates. Copperplate printing saw further advancement during the Northern Song dynasty.

[1] Rong Geng. *A Comprehensive Study of Shang and Zhou Ritual Vessels: Vol. 1* [M]. Beijing: Harvard-Yenching Institute, Yenching University, 1941: 88, 158, Fig. 35.

P. Jixing, *History of Printing Techniques in China*,
https://doi.org/10.1007/978-981-96-6119-0_7

The development of movable type and copperplate printing during the Northern Song dynasty laid the technical groundwork for the emergence of metal movable type printing. The earliest form of metal movable type was copper movable type, more precisely, bronze movable type. We can consider metal movable type as a single-character copperplate. As the technology used for casting multi-character plates was similar, it is relatively easy to cast copper movable type. The process of casting copper movable type followed the same principles as the techniques used for casting copper coins, bronze mirrors, and copper seals. These casting practices have a long history in China, with a significant number of artifacts remaining as evidence. Although the technology for copper movable type may have been theoretically possible during the Tang dynasty and earlier, it was not until the Northern Song dynasty that movable type printing using copper was actually realized. For China, historical evidence indicates that the primary societal factor driving the development of metal movable type technology was the issuance of paper currency, rather than the demand for printed books. This is because China already had well-developed woodblock, copperplate, and non-metallic movable type printing techniques, which were sufficient to meet the societal demand for various high-quality printed materials. Of course, the ability to use metal movable type for printing currency also meant it could be used for book printing, but the emphasis on using metal movable type for currency in China was determined by the specific circumstances of the time.

The invention of metal movable type printing in China can be attributed to the country's advanced development in copperplate printing and non-metallic movable type printing, leading other nations by several centuries. China, the first to issue paper currency, had both the societal need and technological basis for developing metal movable type printing. By the time other regions were ready to adopt this technology, China had already cast copper movable type, which had spread to both the East and West. Historically, scholars both in China and abroad have lacked in-depth research on the origins and early history of metal movable type printing in China. Some have mistakenly dated its origin much later, such as Korean scholars proposing that copper movable type printing began in the late fifteenth century during the Ming dynasty.[2] Some Chinese scholars[3] also held similar views. However, based on our latest research, it is clear that China extensively utilized copper movable type printing technology as early as the 11th to twelfth centuries during the Northern Song dynasty, particularly for printing paper currency.[4] Therefore, to understand the early history of copper movable type technology in China, one should start with the history of currency printing.

[2] Chon H B. *Development Process of Movable Metal-type Printing in Korea* [C]. Seoul: International Symposium on Printing History in the East and West, 1997.

[3] Zhang Shudong, et al. *A Comprehensive History of Chinese Printing* [M]. Taipei: Foundation for the Promotion of Printing and Communication Materials, 1998: 343–344. Li Zhizhong. *On Ancient Printing Techniques* [M]. Beijing: Forbidden City Publishing House, 2000: 365.

[4] Pan Jixing. On the Origin of Metal Movable Type Technology [J]. *Scientific Bulletin*, 1998, 43(15): 1583–1594. Pan Jixing. *History of Metal Movable Type Printing Technology in China* [M]. Shenyang: Liaoning Science and Technology Publishing House, 2001: 38–46.

In the Northern Song dynasty, due to significant development in urban commerce and industry, merchants engaged in trade across various regions. Carrying a large amount of copper or iron coins was both inconvenient and unsafe. Therefore, drawing inspiration from the Tang dynasty's old practice of *feiqian* (flying money), during the reign of Emperor Zhenzong (1008–1016), 16 affluent individuals in Chengdu, Sichuan, jointly issued exchange vouchers for paper currency known as *jiaozi*. At that time, Zhang Yong (946–1015) governed the Shu region and offered support for this initiative by affixing the official seal of Yizhou, thus transforming privately issued *jiaozi* into precursors of paper currency. According to Li You (1101–1171), a Song scholar, in his work *Songchao Shishi* ("Historical Facts of the Song Dynasty", ca. 1130), under the section on finance, he mentioned:

The wealthy individuals would gather at regular intervals, using the same color of paper for printing. The printed patterns included images of houses and people, shop front seals, and confidential serial numbers. Each individual kept their own secret mark, mixing red and black ink as a private record. The denomination was not predetermined. When someone received the money, they would exchange it for *jiaozi*.

From the above, it is evident that the banknotes were printed on specially made paper, featuring not only text but also images of houses and people, shop front seals, and confidential serial numbers. Additionally, these banknotes were printed in red and black, and the denominations were filled in temporarily. During the Tianxi era of Emperor Zhenzong in the Northern Song dynasty (1071–1021), the shops managing *jiaozi* faced insufficient funds to honor the currency, resulting in customer runs and disputes. In response, Kou Jian, the regional governor of Yizhou, ordered a halt to private issuance and reported the matter to the imperial court. The *jiaozi* system was then taken over by the government, and an official institution was established to handle its issuance.[5] When Emperor Renzong (1023–1063) ascended to the throne, Kou Jian had left Yizhou. Zhang Ruogu, the Transport Commissioner, and Xue Tian, the new regional governor of Yizhou, were instructed to reconsider the matter. They proposed:

Now, if we abandon private *jiaozi* and have the government handle it, it would be much more stable and convenient. We propose to cast a copper seal for the Yizhou *jiaozi*, send it to Yizhou, and appoint local officials to manage it. The Yizhou Observer will also have a seal. Exchanging money requires a record in the account book. The amount of money issued will be noted on *jiaozi*, ranging from one to ten *guan*. After the *hetongyin* (denomination seal) is stamped and the records are made and sealed, officials will collect and manage the money. When individuals come to exchange the money, regardless of the amount, it will be duly collected, and stored in the

[5] *History of Song Dynasty: Vol. 182, The Chronicles of Food, Silks Clothing or Currency II // Twenty-Five Histories: Vol. 7.* Compact Edition. Shanghai: Shanghai Classics Publishing House, 1986: 5741.

treasury. According to the *hetongyin* and *zihao* (serial number), *jiaozi* will be given to individuals, facilitating convenient usage.[6]

Emperor Renzong approved the proposal, and official issuance of *jiaozi*, the earliest form of paper currency, commenced. *Jiaozi* had iron coins as the reserve and had a fixed circulation period of three years. After this period, old *jiaozi* could be exchanged for new ones. It circulated from 1023 to 1105, being limited to use in the regions of Sichuan and Shaanxi. From 1023 to 1038, *jiaozi* had various denominations ranging from one *guan* (equivalent to 1000 *wen*) to ten *guan*. The denomination needed to be filled in manually, which was quite inconvenient. Therefore, from 1039 to 1068, the denominations were limited to five *guan* and ten *guan*, and *jiaozi* with the two denominations were issued proportionally (20% for five *guan*, 80% for ten *guan*). The denominations and serial numbers were changed from being filled in temporarily to being printed.[7] Typically, 1.25 million *guan* of *jiaozi* was issued in each circulation period. From 1069 to 1105, the denominations changed to one *guan* and five hundred *wen*, and *jiaozi* with the two denominations were also issued proportionally (60% for one *guan*, 40% for five hundred *wen*). The term *hetongyin* refers to the rectangular copper seal, imprinted on the back of the banknote, confirming the denomination on the front and preventing alterations. The term *zihao* or *liaohao* refers to the use of *Qianziwen* numbering system, representing a consecutive serial number for each banknote. Each note was registered with a specific number, facilitating book-keeping and serving as a security measure to monitor the number of prints and prevent forgery.

The printing volume of paper currency was substantial, as seen in the case of the Sichuan *jiaozi* from 1039 to 1068, requiring the printing of 250,000 notes for five *guan* and 1 million notes for ten *guan*, totaling 1.25 million notes. Additionally, it was a matter crucial to national finance and people's livelihood, and woodblock printing could not meet the demand. Woodblocks were prone to distortion from changes in temperature and humidity, had poor durability, and with high print volumes, the lines on the blocks would wear down quickly, resulting in unclear impressions. Woodblocks were also susceptible to forgery. As a result, copperplate printing was adopted for banknotes. Surviving examples of early Song and Jin banknotes confirm the use of copper plates. Ma Duanlin, a Yuan dynasty scholar, mentioned in his work *Wenxian Tongkao* ("Comprehensive Textual Research of Historical Documents", 1309), Volume 9, *Qianbi Kao* ("Money"): "In the first year of Longxing (1163), an edict was issued to the Central Office to submit the copper plates used for printing to the Imperial Secretariat". This statement indicates that in 1163 Emperor Xiaozong of the Song dynasty ordered the copperplates used for printing Huguang *huizi* (a kind of currency) previously produced by the Central Office overseeing financial affairs in Hubei and Hunan to be handed over to the Imperial Secretariat. He mentioned later in the same volume: "In the third year of Chunxi (1176), an edict was issued extending

[6] Li You, *Historical Facts of the Song Dynasty: Volume 15, Finance* [M]. Beijing: Zhonghua Book Company, 1955.

[7] Liu Sen. *A History of Song and Jin Paper Currency* [M]. Beijing: China Financial Publishing House, 1993: 24–26.

the third and fourth terms of *huizi*, each for an additional three years. The Tea Trade and Tax-collecting Office was ordered to use the copper plates for the fourth term to continue printing two million *huizi*, which were sent to the Southern Warehouse for distribution." Copperplate printing for banknotes offered strong durability, resistance to distortion, and was difficult to counterfeit, making it the best material for printing plates.

Early banknotes typically included information such as the name of the currency, denomination, circulation area, issuing authority, printing date, and reward notices, along with decorative patterns and designs. Sometimes, certain details were omitted. All of the text was pre-cast onto the copperplate. However, to prevent counterfeiting and unauthorized printing, additional measures were taken. Besides affixing official seals, each banknote had a serial number, utilizing the *Qianziwen* numbering system, resembling the modern concept of serial numbers on currency notes. Additionally, the banknotes featured the official seal and sometimes the handwritten signature of the officials from the printing and issuing authorities. These details were not cast simultaneously with other text on the copperplate. Instead, recessed grooves were engraved on the copperplate. Prior to printing, the corresponding characters were typeset into these grooves using movable type, creating a complete plate. Since the printing plate was made of copper, the typeset characters naturally were made of copper as well. After printing one banknote, the original typeset characters were removed and replaced with new ones for subsequent prints. This process involved the entire sequence of copper typesetting printing, including the casting of copper movable type, typesetting, printing, disassembling the plate, retrieving the typeset characters, and re-typesetting. Thus, the process of banknote printing encompassed the full set of procedures involved in copper movable-type printing and was a specialized printing business operated by the government.

Following the issuance of *jiaozi*, the Northern Song dynasty introduced a new type of banknote called *qianyin*. According to the *Songshi Shihuozhi* ("History of Song Dynasty: The Chronicles of Food, Silks Clothing or Currency"), in the fourth year of the Chongning era of Emperor Huizong (1105), "a decree was issued to various provinces to adopt *qianyin*, following a new design, with Sichuan continuing its usage in the same manner as before... The use of *qianyin* was widespread across various provinces, except in Fujian, Zhejiang, and the regions of Huguang (Hubei and Hunan). In the first year of the Daguan era (1107), a decree was issued to change the Sichuan *jiaozi* service to *qianyin* service".[8] "Originally intended as a replacement for salt notes", *qianyin* was first issued in 1105, backed by iron coins as reserves, with a three-year circulation period. The denominations included one *guan* and five hundred *wen*, featuring a new design and circulated in areas such as Jingdong, Jingxi, Huainan, and the capital (Kaifeng). During this period, Sichuan continued to use *jiaozi* until the 43rd issuance in 1107 when it switched to *qianyin* service. *Qianyin* had the longest period of circulation among Song dynasty banknotes, lasting from 1105 to 1234, with

---

[8] *History of Song Dynasty: Vol. 181, The Chronicles of Food, Silks Clothing or Currency II // Twenty-Five Histories: Vol. 7.* Compact Edition. Shanghai: Shanghai Classics Publishing House, 1986: 569.

1.256 million *guan* issued per circulation period during the Northern Song dynasty. In *Chubi Pu* ("Guide to Paper Money", ca. 1360) by Fei Zhu, a Yuan dynasty author (ca. 1303–1363), the specifications of *qianyin* notes are discussed:

In the fifth month of lunar calendar in 1107, the service of *jiaozi* was changed to *qianyin*. Six seals were cast, namely the imperial edict seal, the serial number seal, the year limit seal, the back seal, all in black; the blue face seal, in indigo; the red round seal, in vermilion. All six seals were decorated with patterns, and the red round and back seals were adorned with illustrations.

The phrase "six seals" is commonly interpreted as "six copperplates", but integrating so many plates onto the banknote surface would be challenging. In reality, it should be understood as six main elements on the banknote design: (1) The imperial edict, indicating punishment for counterfeiting and reward for apprehension, as well as circulation decrees; (2) The serial number, signifying the serial number on the banknote; (3) The year limit, indicating the production date, circulation year, and expiration date; (4) The back seal, meaning the denomination seal on the back; (5) The blue face seal, depicting patterns and decorations on the banknote surface; and (6) The red round seal, showcasing decorative illustrations of narratives and stories. Seals (1) to (4) were stamped in black ink, (5) in indigo, and (6) in vermilion. However, Fei Zhu did not provide information on how these elements formed the overall pattern on the banknote. He only narrated the content of *qianyin* for the 70th to 79th issuances from 1161 to 1179 (Fig. 7.1), making it difficult to reconstruct the original appearance of the banknotes.

The serial number seal consists of five-character phrases expressing different terms related to politics and the economy, such as *zhi fu guo cai yong* (To enrich the country and use wealth for the people). Each *qianyin* issuance had different expressions, serving a similar purpose as the *Qianziwen* numbering system. Some believe it was used to distinguish between issuances,[9] but this might not be certain, as the "year limit" had already clearly indicated the division between issuances. In the Southern Song dynasty, another type of paper currency called *huizi* was issued. From 1160 to 1279, it circulated in locations such as Hangzhou, Huai, Zhe, Hubei, with denominations ranging from one to three *guan*, and later two hundred, three hundred, and five hundred *wen*. Each issuance had a three-year circulation period, with the first term involving the printing of 10 million *guan*.

Surviving physical artifacts of banknote printing plates from the Northern and Southern Song dynasties still exist. In the 1930s, a copperplate for Northern Song dynasty banknotes was unearthed in Henan, measuring 16 cm in height and 9.1 cm in width. Currently, only its rubbings remain. In 1938, Japanese numismatist Okudaira Masahiro first introduced this copperplate in his work *Toa Senshi* ("A Numismatic Record of Far Eastern Coins"), suggesting it might be related to *huizi* banknotes. There is no name of the banknote on this printing copperplate. The upper section has the design of ten copper coins. The middle section contains text indicating that, "apart from Sichuan, various provinces, counties, public and private entities were allowed

[9] Liu Sen. *A History of Song and Jin Paper Currency* [M]. Beijing: China Financial Publishing House, 1993: 59.

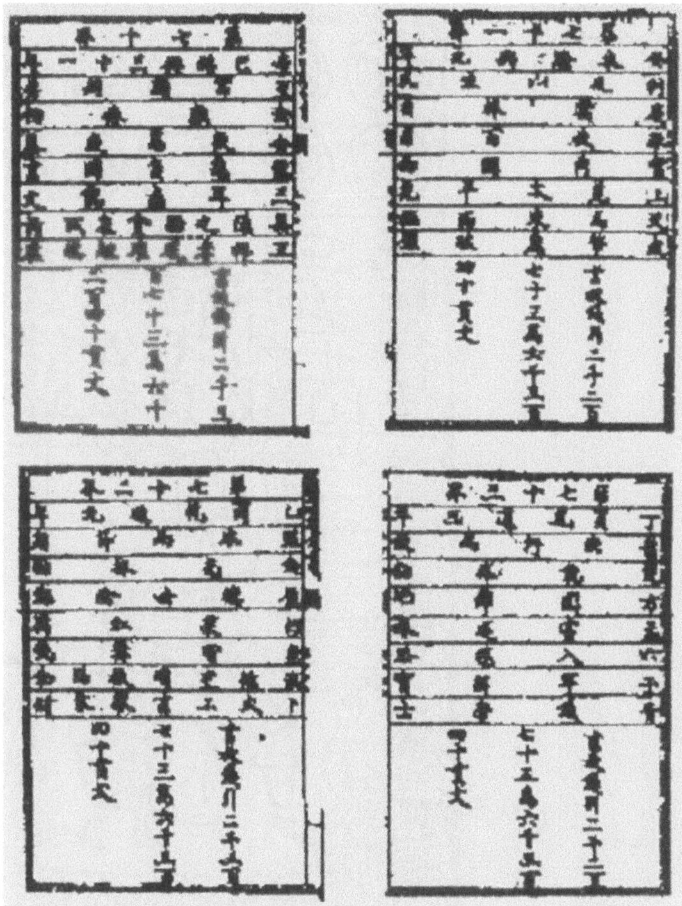

Fig. 7.1 Design of *qianyin* in the Southern Song dynasty described in *Chubi Pu*[10]

to manage and circulate this banknote, which equals to seven hundred and seventy *mo*". The lower section depicts houses and figures, with the words *Qian Si Cang* (Qiansi Warehouse) in the upper right corner. This copperplate was likely inscribed on both sides, but only one side is preserved in the extant rubbing (Fig. 7.2), and the other side is lost. Consequently, there are different opinions among scholars about its identification, with some suggesting it is related to *huizi*, while others propose it could be *jiaozi* or *qianyin.* In recent years, some scholars have suggested it might be associated with the Northern Song's *xiaochao* banknotes.[11] It seems unlikely to be *qianyin*, as its design does not align with the descriptions of *qianyin* in *Chubi Pu.*

[10] Cao Xuequan. *Local Gazetteer of Sichuan Province* [M]. Fuzhou, 1614.

[11] Liu Sen. *A History of Song and Jin Paper Currency* [M]. Beijing: China Financial Publishing House, 1993: 41–43.

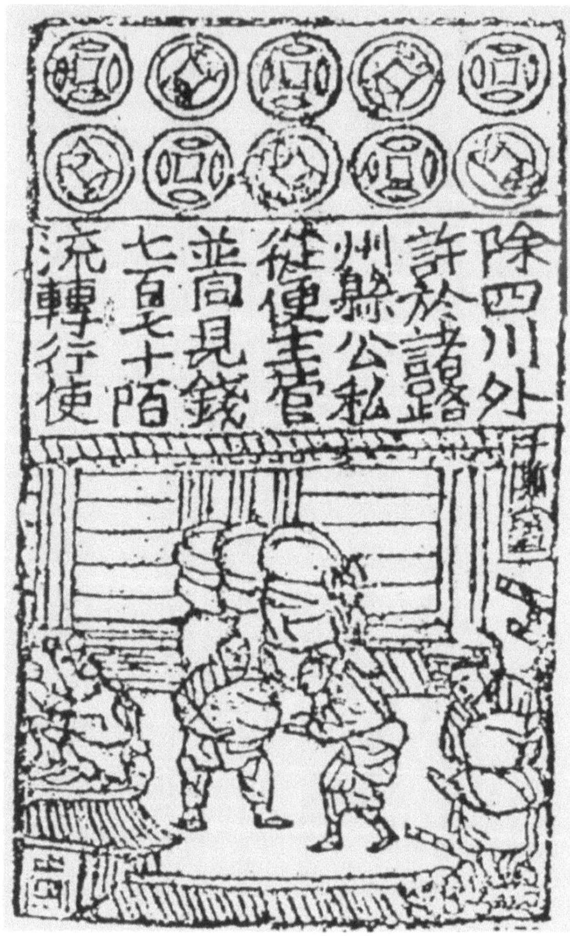

Fig. 7.2 Copperplate for Northern Song's xiaochao banknotes cast in 1106–1110

The presence of ten copper coins on the banknote implies circulation in regions that used *dangshi qian* (a monetary value equivalent to ten cash coins) during the reign of Emperor Huizong of the Northern Song (1102–1110), determining its production time and location. The phrase "which equals to seven hundred and seventy *mo*" corresponds to the *shengmo* system, established in the early Northern Song (977), where seventy-seven copper coins were equivalent to one hundred, indicating that the back side of this banknote should bear the inscription *Yi Guan Wen Sheng* or something similar.

Song dynasty official Xie Laibo (1176–1241) recorded in his work *Mizhai Biji* ("Secret Chamber Notes") Volume 1 that in the fifth year of the Chongning era (1106), an edict was issued to produce *xiaochao* banknotes, with a circulation period of three years. He noted that "the design was similar to the *huizi* banknotes of the Southern Song period". The banknotes featured prohibitions against counterfeiting and officials' seals. Due to the range of denominations from one hundred *wen* to one *guan*, they were referred to as *xiaochao* or "small banknotes". The first issuance occurred in the second year of the Daguan era (1108), so the aforementioned copperplate was likely related to the production of these small banknotes starting from 1108. In theory, the printing block should bear serial numbers, which would be engraved using copper movable type. The characters *qian si* appear, and these two characters are from *Qianziwen*. This suggests that they might be engraved using movable type, as otherwise, movable type would likely appear on the reverse side of the banknote. The Chinese History Museum houses a copperplate for *huizi* banknotes from the Southern Song, made in Hangzhou from 1161 to 1168. This plate clearly contains embedded copper movable type. It measures 17.8 cm in length, 12 cm in width, 1.7 cm in thickness, and weighs 2700 g. The upper section features the inscription "Those convicted of counterfeiting *huizi* banknotes will face execution, and informants will receive a reward of one thousand *guan*. Those who voluntarily forgo the reward may serve as trainee military officers. Accomplices or families that voluntarily confess will be pardoned, and a reward will be given for their information. Those who wish to supplement the aforementioned positions are welcome to do so." The left and right sides of the upper section feature the words "Denomination Equals to One *Guan*" and "Serial Number 110", respectively. The middle section contains "Operated in the Hangzhou Huizi Treasury", and the lower section depicts a mountain spring pattern with surrounding floral decorations (Fig. 7.3).

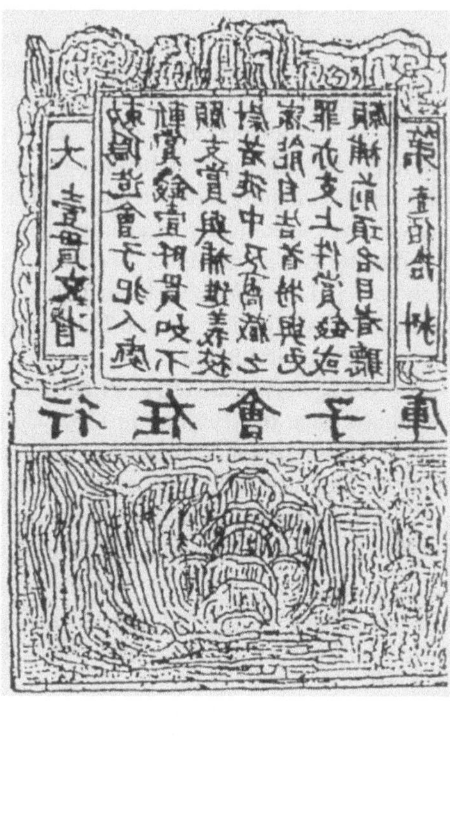

Fig. 7.3 Rubbing of the 1161–1168 Hangzhou-cast bronze printing plate of the *huizi* with one *guan* denomination (Collection of the National Museum of Chinese History)

The above-mentioned reward and punishment regulations are consistent with the records in *Songshi Shihuozhi* and Ma Duanlin's *Wenxian Tongkao: Qianbi Kao.* In the thirtieth year of Shaoxing era (1160) during the reign of Emperor Gaozong, Qian Duanli (1109–1177), a deputy minister of the Ministry of Revenue, was ordered to produce *huizi* banknotes. In the thirty-second year of Shaoxing era (1162), *Weizao Huizi Fa* ("Counterfeiting of *Huizi* Banknotes Law") was enacted, and the *huizi* banknotes were printed in Lin'an (present-day Hangzhou).[12] Hangzhou Huizi Treasury refers to the Huizi Treasury of the Tea Trade and Tax-collecting Office, which was responsible for the affairs related to coinage. Therefore, this printing plate was cast between 1161 and 1168. The three characters *yi bai shi* (Serial Number 110) in the upper section are arranged obliquely, and their shapes and sizes are noticeably different from other characters, indicating that they were embedded copper movable type. This plate serves as extant physical evidence of copper movable type from the early Southern Song period.

Contemporary with the Northern Song was the Jin dynasty (1115–1234), a regime established by the Jurchen people. Starting from the second year of the Zhenyuan era (1154), the Jin dynasty emulated the Northern Song system and issued paper money known as *jiaochao.* The circulation period was initially limited to seven years, after which the notes would be exchanged for new ones. After the twenty-ninth year of the Dading era (1189) during the reign of Emperor Shizong, the policy changed to allow unlimited circulation, lasting until the end of the Jin dynasty. However, the name of the currency underwent several changes. *Jinshi Shihuozhi* ("History of the Jin Dynasty: The Chronicles of Food, Silks Clothing or Currency") states:

In the second year of Zhenyuan (1154) after the capital was moved, Cai Songnian, the Minister of Revenue, resumed the issuance of *chaoyin,* then created *jiaochao...* Initially, during the period of Zhenyuan (1153–1155), when the issuance of *chaoyin* was implemented, they established the treasuries for *chaoyin* and *jiaochao.* Each had a director, deputy director, and an overseer. The deputy director of the *Jiaochao* Treasury was specifically in charge of matters related to signing and sealing. The denominations were one *guan,* two *guan,* three *guan,* five *guan,* and ten *guan,* collectively referred to as *dachao* (large notes)... With a limit of seven years, the old notes were exchanged for new ones, following the model of Zhang Yong's Sichuan *jiaozi,* thus easing the circulation.[13]

Wanyan Liang (1122–1161), Prince Hailing of Jin dynasty, moved the Jin capital to Yanjing (present-day Beijing) in 1153. Consequently, the issuance of *jiaochao* from 1154 to 1159 took place in Beijing and was managed by the Ministry of Revenue's Jiaochao Treasury. After relocating the capital to Nanjing (present-day Kaifeng) in 1160, this institution moved south accordingly. From 1189 to 1215, paper money circulated more widely, and local authorities were permitted to independently issue

[12] *History of Song Dynasty: Vol. 182, The Chronicles of Food, Silks Clothing or Currency II // Twenty-Five Histories: Vol. 7.* Compact Edition. Shanghai: Shanghai Classics Publishing House, 1986: 569.

[13] *History of Jin Dynasty: Vol. 48, The Chronicles of Food, Silks Clothing or Currency II // Twenty-Five Histories: Vol. 9.* Compact Edition. Shanghai: Shanghai Classics Publishing House, 1986: 7034.

money under the supervision of officials dispatched by the Ministry of Revenue. Starting in 1197, one, two, three, five, and seven hundred-*wen jiaochao*, also known as *xiaochao* or small notes, were issued. Regarding the design of the banknotes, *Jinshi* records:

The design of *jiaochao*: The outer part has railings with patterns. The denomination is written in the middle of the top. To its left is *ziliao* and to it right *zihao*, the serial number. The outermost column reads "Those who counterfeit *jiaochao* shall be executed", and "Those who report and capture will be rewarded with 300 *guan*". The bottom column reads "The Jiaochao Treasury in the Central Capital (modern-day Beijing), according to the instructions from the Ministry of Revenue, received the documents and credentials from the Central Hall. The Ministry of Revenue then reviewed and verified the names and seals of the clerks." Additionally, it says: "*Jiaochao* is issued by imperial decree and valid for use everywhere. In some treasuries, paper currency is exchanged for coins, while in others, coins are exchanged for paper currency. Both official and private parties can use this currency for transactions. The notes are valid without a fixed date of expiration... The controller of the treasury, the cashier, the vice-controller, and the assistant controller each bear a seal. Furthermore, regarding the printing and issuance of currency, on the specified date, the responsible officials from the currency printing repository—such as the chief of the repository, the deputy chief, and the assistant officials—are to sign off on the documents. Additionally, the officials from the Ministry of Revenue must also sign. Any use of other stamps must follow the standard procedures".[14]

The above records are from the entry below the twelfth month in the lunar calendar of the twenty-ninth year of Jin Shizong's Dading era (1189). Except for the item "valid without a fixed date of expiration", which belongs to the new policy in 1189, other contents reflect the *jiaochao* system before this time. Considering the situations where *yanyin*, *jiaochao*, *huizi* may have serial numbers or use *Qianziwen* as the serial number, and referring to the *jiaochao* styles and records after 1189, the *jiaochao* issued from 1154 to 1189 should also include serial numbers, denominations, anti-counterfeiting measures, reward notices, circulation periods, places of redemption, printing institutions, authorization texts, and the names and seals of officials. The surrounding patterns may include flowers, plants, dragons, and cranes.[15] Historian Jia Jingyan preserved rubbings of the fragment of a Jin dynasty *jiaochao* printing plate from the Taihe era (1201–1208). The upper part bears the words "serial number", "three hundred *guan*", "Taihe", "printed", and others (Fig. 7.4), making it the earliest surviving printing plate with an imperial era designation for Jin *jiaochao*. We have partially reconstructed it. It can be seen that the upper section for the serial number was designed to accommodate two copper movable types, while the section featuring the official's seal should have included an additional set of movable types.

---

[14] *History of Jin Dynasty: Vol. 48, The Chronicles of Food, Silks Clothing or Currency II // Twenty-Five Histories: Vol. 9.* Compact Edition. Shanghai: Shanghai Classics Publishing House, 1986: 7034.

[15] Liu Sen. *A History of Song and Jin Paper Currency* [M]. Beijing: China Financial Publishing House, 1993: 219.

Fig. 7.4 Rubbings of the fragment of a Jin dynasty *jiaochao* printing plate from the Taihe era (1201–1208)

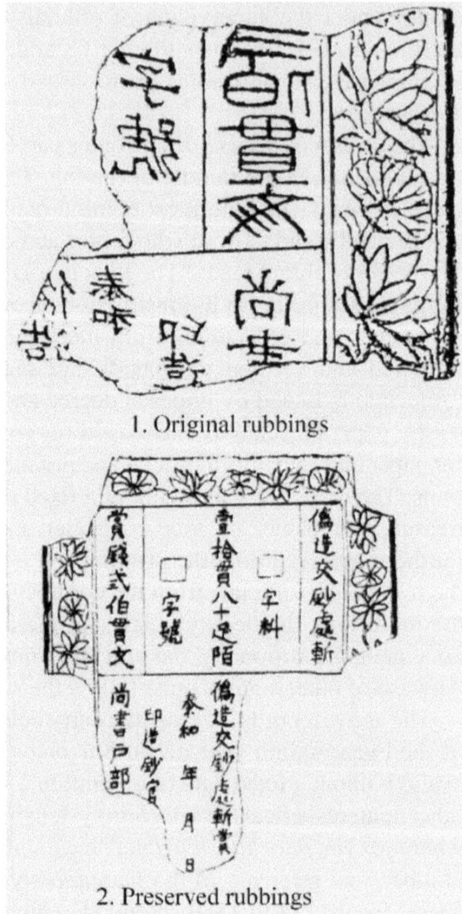

In the Qing dynasty, Xu Ziyin from Taicang, Jiangsu (1792–1857), possessed an imprinting plate for the Five *Guan* Zhenyou Treasure Voucher, issued during the Zhenyou era (1215–1216) of Jin Emperor Xuanzong (see Fig. 7.5, Image 1). A small character "輶" can be clearly seen on the right side for the serial number, which is a copper movable type inserted.[16] This character is taken from *Qianziwen.* There should have been another movable type for the serial number, but it is missing. The Shanghai Museum also possesses a similar imprinting plate, which is in the state

[16] Luo Zhenyu. *Illustrated Catalogue of Banknotes and Coins from Four Dynasties: Vol. 1* [M] // *Yongmu Yuan Series.* Shangyu Luo's Reprint Edition. Beijing: 1914. Wei Yuewang, et al. *Illustrated Collection of Ancient Chinese Banknotes* [M]. 2nd Edition. Beijing: China Financial Publishing House, 1992: 25–26, Fig. 2-10, 2-11.

before the insertion of movable types (see Fig. 7.5, Image 2).[17] In this version, only two officials' seals were cast with the original plate, while in Xu Ziyin's version, nine seals appear. This indicates that nine copper movable types were inserted, two of which were used for the *Qianziwen* numbering. According to *Jinshi Shihuozhi*, in the third year of Zhenyou (1215), the name of *jiaochao* was changed to Zhenyou Treasure Voucher. Although the name changed, the design and pattern remained the same as the previous *jiaochao*. The plate in the Shanghai Museum measures 35 cm in height, 21.5 cm in width, 3 cm in thickness, and weighs 6150 g. There are square grooves above the serial numbers,[18] each measuring 1.6 cm in height, 1.3 cm in width, and 1.5 cm in depth, corresponding to the size of a copper movable type. The grooves do not penetrate to the back of the plate, ruling out the possibility of piercing the movable types through the grooves to imprint on the surface of the voucher.

In 1956, the Inner Mongolia Autonomous Region Museum collected an unearthed copper imprinting plate for the ten *guan jiaochao* that circulated in the Eastern Route of Shaanxi during the Jin dynasty, cast in 1215. From the photos of the plate (Fig. 7.6), at least 8 grooves are clearly visible, designed for inserting copper movable types. Wang Lanxi's collection also includes a copper imprinting plate for the ten *guan jiaochao* that circulated in the Eastern Route of Shandong, cast in 1213–1214, which also has 8 grooves for placing movable types.[19] Since the issuance of *jiaochao* in 1154, the Jin dynasty consistently inserted copper movable types into imprinting plates to form complete patterns. Following the practice initiated by the Northern Song dynasty in 1039 when issuing Sichuan *jiaochao*, it is unequivocal that both Song and Jin paper currencies are the printed products of copper plates combined with copper movable types. Both historical records and unearthed artifacts confirm that the use of copper movable types for printing banknotes in China began in the eleventh century during the Northern Song dynasty and extended to the twelfth century in some minority-dominated areas under Jin rule.

Using copper movable types for banknote numbering, one could take two characters from *Qianziwen* as one set. Calculating based on the formula $m\ (m-n+1)/n$, where m is the total number of characters (1000 in this case) and n is the number of characters per set, we get approximately 499,500 different combinations or nearly 500,000 unique serial numbers, which would be sufficient to meet the demand. In addition to the serial numbers, each banknote varied with different denominations and circulation regions. The Song, Jin, and Yuan dynasties all used *Qianziwen* for numbering banknotes, while the Ming and Qing dynasties combined one character from *Qianziwen* with a numeral, such as "Heaven Character Number Fifty". Apart from characters for the numbers, various official seals were also created using

17 Luo Zhenyu. *Illustrated Catalogue of Banknotes and Coins from Four Dynasties: Vol. 1* [M] // *Yongmu Yuan Series*. Shangyu Luo's Reprint Edition. Beijing: 1914. Wei Yuewang, et al. *Illustrated Collection of Ancient Chinese Banknotes* [M]. 2nd Edition. Beijing: China Financial Publishing House, 1992: 25–26, Fig. 2-10, 2-11.

18 Pan Jixing. *History of Metal Movable Type Printing Technology in China* [M]. Shenyang: Liaoning Science and Technology Publishing House, 2001: 41–43.

19 Wei Yuewang, et al. *Illustrated Collection of Ancient Chinese Banknotes* [M]. 2nd Edition. Beijing: China Financial Publishing House, 1992: 12, Fig. 2-3.

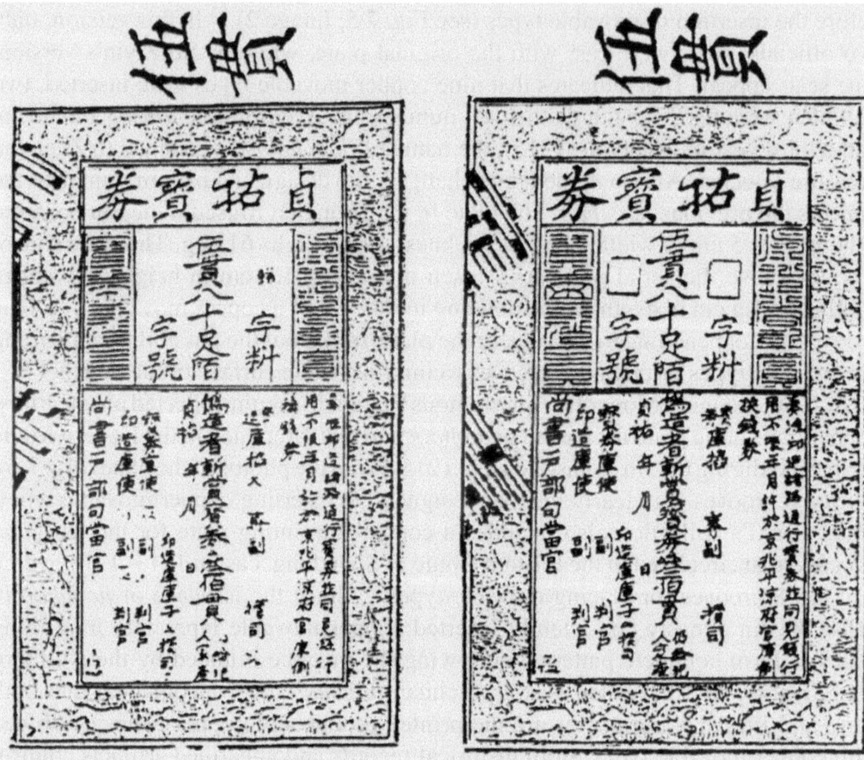

1. Xu Ziyin's collection of printing plate (with ten copper movable types including 輶)

2. Shanghai Museum collection of printing plate (without copper movable type)

Fig. 7.5 Printing plate for the Five *Guan* Zhenyou Treasure Voucher, issued during the Zhenyou era (1215–1216)

movable types. Hence, for every set of banknotes printed, a large quantity of copper movable types needed to be cast. Since the Song dynasty used copper movable types for banknotes, they could also be used for printing books. However, such printed books from the Song period have not been preserved in the same way as banknotes and printing plates, awaiting future discoveries.

Some argue that only using metal movable type for printing books qualifies as metal movable type printing, and using metal movable type for banknotes is not, as it lacks the typesetting process.[20] This claim lacks theoretical basis. As mentioned earlier, the act of implanting copper movable types into banknote plates itself is a typesetting process; without this step, usable banknotes cannot be printed. Prior to banknote printing, China had already invented the technique of using non-metallic movable type for typesetting books in the early eleventh century during the Northern

[20] Cao Jiongzhen. *A Letter to Pan Jixing* [Z]. 2000.

Fig. 7.6 Copper printing plate for the ten *guan jiaochao* that circulated in the Eastern Route of Shaanxi during the Jin dynasty[21]

Song dynasty. While the quantity of movable type used may differ between banknote plates and book plates, there is no fundamental difference in principle and operation. Besides, one should not equate the history of printing with the history of book printing alone. Printing encompasses not only book printing but also banknote printing, as well as the printing of bills, contracts, advertisements, and more, all of which are subjects of printing history. The development of metal movable type printing is based on an understanding of the principles of movable type technology and metal typecasting. Both of these appeared in the Northern Song dynasty in the eleventh century, with the initial application being in banknote printing. Using copper movable type for book printing is just a technological extension of using metal movable type for banknote printing.

[21] Wei Yuewang, et al. *Illustrated Collection of Ancient Chinese Banknotes* [M]. 2nd Edition. Beijing: China Financial Publishing House, 1992: 12, Fig. 2-3.

7.1.2 Tin Movable-Type Printing in Southern Song Dynasty

In the Southern Song dynasty (eleventh to twelfth centuries), there was a significant development in copperplate printing and copper movable type printing. However, during the twelfth and thirteenth centuries, the main copper-producing regions were gradually occupied by the Jin dynasty, leading to a scarcity of copper. As a result, the prevailing trend shifted to substituting tin for copper. Taking copper coins as an example, during the Northern Song, they generally contained about 64% copper, 9% tin, and 23% lead. In the Southern Song, the copper content in copper coins decreased, while tin and lead content increased, such as in the "tin-inlaid coins" with about 50% copper, and tin and lead each at about 25%. Correspondingly, tin printing plates replaced the copper plates. An example is the tin printing plate of *Dayuanman Tuoluoni Shenzhou Huiji Zhenyan* ("Great Perfection Dharani Mantra for Purifying Defilements") from the 13th year of Chunxi (1186) of Emperor Xiaozong, unearthed in 1957 at West Lake in Hangzhou. This tin printing plate, created by Hu Yan, measures 27.5 cm in length, 16.5 cm in width, and 0.65 cm in thickness. It has scripts on both sides and features another mantra on the back,[22] created by devotees Tangshiwuniang and others, who donated funds and conducted rituals to alleviate suffering for all sentient beings (Fig. 7.7).

In 1983, an engraved currency plate for the *jinyin jianqian guanzi* (*guanzi* banknote exchanged for gold and silver) from the 5th year of the Jingding era (1264) of Emperor Lizong of the Southern Song dynasty was discovered in Dongzhi County, Anhui. This plate was associated with the powerful chancellor Jia Sidao (1213–1275) during the period of his dominance. It is claimed to be made of a lead-iron alloy,[23] but from an alloy technology perspective, it is more likely composed of tin, lead, copper, and a small amount of iron. The plate measures 22.5 cm in height, 13.5 cm in width, 0.4 cm in thickness, and weighs 1000 g. The plate features decorative patterns at the top and bottom, with a header stating, "Banknote *guanzi* issued by the Ministry of Treasure". Below this, there are three vertical columns of text, with the center column containing large characters for the denomination. The left and right columns read: "All counties and prefectures should facilitate public and private transactions. The exchange rate between *guanzi* banknotes and copper cash coins was 1 *guan* for 770 wen. If there are officials, civilians, or those from government offices who dare to arbitrarily reduce the number of coins, in violation of the regulations, they shall be punished with imprisonment for two years. For serious offenders, a more severe punishment will be implemented. Exchange gold and silver for cash at the designated location is permitted." (Fig. 7.8).

Alongside the 1264 *guanzi* engraved currency plate, several other artifacts were discovered, including a *zhunchi* (preliminary edict) plate, a *banxing* (issuance) plate,

---

22 Jin Baidong. The Earliest Existing Tin Printing Plates [J]. *Oriental Antiquities*, 1996 (Inaugural Issue): 157–160.

23 Wang Benchu. Investigation and Study of the Southern Song Dynasty Guanzi Banknote Plates Discovered in Dongzhi County, Anhui [J]. *Anhui Financial History and Chronicle* (Supplement on Currency), 1987 (4): 58–64.

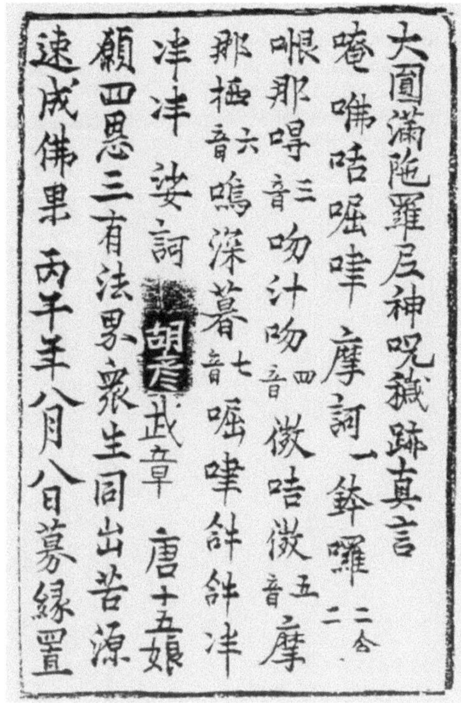

Fig. 7.7 Tin printing plate of *Dayuanman Tuoluoni Shenzhou Huiji Zhenyan* from 1186, unearthed in 1957 at West Lake in Hangzhou[24]

a *baoping* (treasure vase) plate, and four copper seals, making a total of eight items. In *Jiasidao Zhuan* ("Biography of Jia Sidao"), volume 474 of *Songshi* ("History of Song Dynasty"), and in Wang Qi's *Xu Wenxian Tongkao* ("Comprehensive Textual Research of Historical Documents", 1586), there are records of the issuance of the *jinyin jianqian guanzi*. Some believe that these eight items form a complete set of printing plates and can be assembled to create a *Jia* character-shaped printing plate, corresponding to the description in *Biography of Jia Sidao*.[25] However, this raises skepticism because a plate assembled in this manner would have a vertical height exceeding 50 cm, making it impractical for printing such large banknotes. It is also doubtful whether these eight items constitute a complete set, as there is no serial number. Nevertheless, they are authentic artifacts from the Southern Song dynasty and may have been prototype plates.[26]

In the Southern Song dynasty, due to the shortage of copper, tin alloy was used to make printing plates instead of copper plates. In this context, tin alloy would have also been used to cast movable type as a substitute for copper movable type.

---

[24] Jin Baidong. The Earliest Existing Tin Printing Plates [J]. *Oriental Antiquities*, 1996 (Inaugural Issue): 157–160.

[25] Zhang Jiqi. Song Dynasty Banknotes and Their Extant Printing Plates [J]. *China Printing*, 1994, 12(2): 34–37.

[26] Liu Sen. *A History of Song and Jin Paper Currency* [M]. Beijing: China Financial Publishing House, 1993: 149–151.

Fig. 7.8 The tin prototype plate of the Southern Song dynasty *guanzi* issued in 1264, discovered in Dongzhi County, Anhui in 1983[27]

In the Yuan dynasty, the scholar Wang Zhen discussed the technology of movable type printing in his work *Zao Huozi Yinshu Fa* ("Method of Creating Movable Type Printing"), which was completed in the second year of the Dade era (1298). In the text, he mentioned wooden movable type, clay movable type, and metal movable type, stating:

In recent times, tin has been cast to make movable type, with iron rods threaded through to form lines. The tin movable type blocks are inserted into the printing plate to print books. However, the movable type blocks in the above-mentioned method are difficult to ink, and many are prone to printing defects, making them unsuitable for long-term use.

Nowadays, there is a clever and convenient method. Wooden plates are crafted into printing frames, and bamboo strips are shaved to form lines. Wooden plates are

[27] Wang Benchu. Investigation and Study of the Southern Song Dynasty Guanzi Banknote Plates Discovered in Dongzhi County, Anhui [J]. *Anhui Financial History and Chronicle* (Supplement on Currency), 1987 (4): 58–64.

carved to create characters using small fine saws and chisels. Each character is then arranged in lines and carved into bamboo strips.[28]

The "clever and convenient method" refers to the ingenious method developed in the later half of the thirteenth century in the early Yuan dynasty, which was improvement upon the wooden movable type technology inherited from the Song dynasty. A significant improvement was the invention of a rotatable type storage mechanism. The term "in recent times" refers to an era not long before Wang Zhen's own time. When he mentioned "in recent times, tin has been cast to make movable type", he was likely referring to the casting of tin movable type for printing that occurred in the Southern Song dynasty (twelfth century to thirteenth century). In this practice, tin alloy movable type took the form of elongated cubes with a small hole in the body. During typesetting, they were threaded together with fine iron wire, forming lines of type that were then planted in the printing plate. Thin bamboo strips were used to clamp the lines of type, serving as borders to prevent the movable type from shifting. This method aimed to replace copper movable type in the Southern Song. The small hole in the bodies of the type allowed them to be threaded together, a tradition that continued into the eighteenth century when copper movable type used for the publication of *Gujin Tushu Jicheng* ("An Alphabetical Index to the Chinese Encyclopaedia") also featured small holes for threading with iron wire.[29]

However, Wang Zhen's concern about the ink adhesion issue with tin movable type, and the vulnerability of paper when pressed hard during printing, needs a more specific analysis. Judging from the successful practices of using copper plates, copper movable type, and tin plates for printing books and currency in the Song dynasty, the ink adhesion problem with copper-tin alloy materials had already been addressed. Additionally, corresponding artifacts have been unearthed, suggesting that this would not have posed a significant technological obstacle to the development of tin movable type printing. It should be considered feasible for the Southern Song dynasty in the twelfth to thirteenth centuries to use tin movable type as a replacement for copper movable type in printing books. The practical experience of Yangzhou's Guangling Ancient Book Engraving Society in 1998, where they used tin movable type to print *Tangshi Sanbaishou* ("Three Hundred Tang Poems"), demonstrated that ordinary ink could be used for printing, and there was no issue with difficult ink adhesion or paper tearing, as described by Wang Zhen. I witnessed this practice myself. Therefore, asserting that the use of metal movable type in China before the fourteenth century "ended in failure" and that metal movable type printing began only in the fifteenth century during the Ming dynasty[30] contradicts historical facts.

The use of tin plates and tin movable type by the Southern Song people as substitutes for copper plates and copper movable type was a necessity at that time. Many

[28] Wang Zhen. *Book on Agriculture: Vol. 22, Method of Creating Movable Type Printing* [M]. Shanghai: Shanghai Classics Publishing House, 1994: 759–762.

[29] Pan Jixing. *History of Metal Movable Type Printing Technology in China* [M]. Shenyang: Liaoning Science and Technology Publishing House, 2001: 52.

[30] Chon H B. *Development Process of Movable Metal-type Printing in Korea* [C]. Seoul: International Symposium on Printing History in the East and West, 1997.

copper-producing areas, which were under the control of the Southern Song, had already fallen, prompting an increase in the tin and lead content in alloys to reduce the consumption of copper. After the Mongols conquered the Jin dynasty in 1234 and the Southern Song dynasty in 1279, establishing the unified Yuan Empire, they gained control over mineral resources across the country. This allowed them to cast high-quality copper movable type again. Therefore, Wang Zhen's statement that the tin movable type is "unsuitable for long-term use" should be understood as the use of tin movable type no longer being a substitute for copper movable type during the Yuan dynasty. Despite this, tin movable type, due to its low cost, did not completely disappear from the historical stage and was later utilized by private printers during the Ming and Qing dynasties.

7.2 Metal Movable-Type Printing in Yuan, Ming, and Qing Dynasties

7.2.1 Metal Movable-Type Printing in Yuan Dynasty

In 1206, Genghis Khan established the Mongol Empire (1206–1270), which rose in the northern steppes and destroyed the Jin dynasty in 1234. In 1271, Kublai Khan founded the Yuan dynasty (1271–1368) and moved the capital to Dadu (present-day Beijing) the following year. In 1279, he defeated the Southern Song dynasty, establishing nationwide rule. According to *Yuanshi Shihuozhi* ("History of Yuan Dynasty: The Chronicles of Food, Silks Clothing or Currency"), in the first year of the Zhongtong era (1260), when Kublai Khan became Khan, he implemented a paper currency system similar to that of the Song and Jin dynasties, issuing *Zhongtong Yuanbao* notes in Kaiping Prefecture (now Zhenglan Banner in Inner Mongolia). This paper currency had denominations of ten, twenty, thirty, fifty, one hundred, two hundred, three hundred, and five hundred *wen*, as well as one and two *guan*, with silver as the standard; two *guan* equated to one tael of silver and had unlimited circulation. However, due to large-scale printing, the value of the notes depreciated. In the 24th year of the Zhiyuan era (1287), the Emperor changed the issuance to *Zhiyuan Tongxing Baochao* notes and stopped printing Zhongtong notes, although the old notes remained in circulation. In the second year of the Wuzong's reign (1309), the Zhongtong notes were discontinued. Two years later (1311), upon the ascension of the Renzong, he resumed printing Zhongtong notes alongside the Zhiyuan notes. In the tenth year of the Zhizheng era (1350), new Zhongtong notes were issued, with each *guan* equating to two *guan* of Zhiyuan notes and one thousand *wen* in copper coins.[31]

[31] *History of Yuan Dynasty: Vol. 93, The Chronicles of Food, Silks Clothing or Currency* [M] // *Twenty-Five Histories: Vol. 9.* Compact Edition. Shanghai: Shanghai Classics Publishing House, 1986: 7509.

The Yuan scholar Tao Zongyi (1316–1396) recorded in his *Chuo Geng Lu* ("Nancun's Notes in the Intervals of Ploughing", 1366) that Ye Li (1242–1292), a scholar from the Southern Song dynasty, submitted a memorial to Emperor Lizong in the fifth year of the Jingding era (1264), criticizing Jia Sidao for mismanaging public lands and issuing *guanzi* notes. As a result, Jia Sidao had him tattooed on the face and exiled to southern regions. In the fourteenth year of the Zhiyuan era (1277), Ye Li defected to the Yuan dynasty, and in the twenty-third year (1286), he was summoned to Dadu, where he presented a sample of the Zhiyuan notes. He noted that this sample had previously been submitted to the Song court, but was rejected, and he hoped it could replace the previous currency, *guanzi*. Kublai Khan accepted this proposal and decided to mint the notes, promoting Ye Li through the ranks for his contributions.[32] According to *Yuanshi Ye Li Zhuan* ("History of Yuan Dynasty: Biography of Ye Li"), Ye Li, whose courtesy name was Taibai and style name was Yiyu, was from Hangzhou. In 1264, he joined other scholars in condemning Jia Sidao, leading to his exile to Zhangzhou in Fujian. In 1286, upon hearing of Ye Li's reputation, Kublai Khan summoned him to Dadu to discuss governance. Ye Li advocated for establishing a national academy and formalizing the currency system. The emperor appreciated his suggestions, which were largely implemented, leading to Ye Li's rise to the position of Left Chancellor in the Central Secretariat, later promoted to Right Chancellor.[33] This demonstrates that the currency sample and regulations he had offered to the Southern Song court, which were rejected, were ultimately accepted by the Yuan court.

The *Zhiyuan Tongxing Baochao* was issued in eleven denominations: 500 *wen*, 300 *wen*, 200 *wen*, 100 *wen*, 50 *wen*, 30 *wen*, 20 *wen*, 10 *wen*, 5 *wen*, as well as 5 *guan* and 1 *guan*. Each *guan* was equivalent to 5 Zhongtong notes, with two *guan* equal to one tael of silver, and twenty *guan* equal to one tael of gold. Throughout the Yuan dynasty, paper currency circulated nationwide and became the primary form of currency. The design of these notes was similar to that of the currency in the Jin dynasty. From 1260 to 1276, notes were printed in Kaiping Prefecture using woodblocks and wooden movable type. Starting in 1276, all notes were printed using copper plates and copper movable type in Beijing. The numerous Yuan dynasty notes and printing plates discovered today are all made of copper, with copper movable type embedded in them. The printing plates measure approximately 33.2 cm × 25.5 cm and are printed on thick gray mulberry paper. The plates included the name of the currency, denomination, serial numbers, printing and issuing authorities, regulations for rewards and punishments, and the date of issue. The serial numbers were cast with grooves for embedding the copper movable type prior to printing, still using a numbering system based on *Qianziwen*. Each complete set of notes required at least 22,000 copper movable-type characters. From 1276 to 1350, the total number

[32] Tao Zongyi. *Nancun's Notes in the Intervals of Ploughing: Vol. 19, Sample of Zhiyuan Banknotes* [M]. Beijing: Zhonghua Book Company, 1959: 235.

[33] *History of Yuan Dynasty: Vol. 173, Biography of Ye Li* [M] // *Twenty-Five Histories: Vol. 9.* Compact Edition. Shanghai: Shanghai Classics Publishing House, 1986: 7702–7703.

Fig. 7.9 *Zhongtong Yuanbao Jiaochao* note with a denomination of one *guan*

of characters cast for printing notes was in the hundreds of thousands, demonstrating the extensive use of copper movable type in currency printing.

In 1907–1908, the Russian scholar Kozlov discovered a *Zhongtong Yuanbao Jiaochao* note with a denomination of one *guan* (1000 *wen*) in Heishui City (Fig. 7.9). The top section of the banknote features the inscription "Zhongtong Yuanbao Jiaochao", with the middle section displaying four large characters "Yi Guan Wen Sheng" (one *guan*). On both sides of this central section, there are two lines of text in Han seal script and Phags-pa script, reading "Zhongtong Baochao" and "Zhulu Tongxing" (valid across the country). Below that are the serial number inscriptions "Tao Ziliao" and "Tang Zihao", with the characters "Tao" and "Tang" taken from *Qianziwen*, made using copper movable type embedded in the copper plate. The original piece is housed in the Hermitage Museum in St. Petersburg. The back of the note features a rectangular seal reading "Zhizheng Yinzao Yuanbao Jiaochao."[34] The Zhizheng era (1341–1370) corresponds to the reign of the Yuan Shundi, marking this as the new Zhongtong notes issued from 1350 onwards. Additionally, there are surviving examples of Yuan dynasty *Zhiyuan Tongxing Baochao* printing plates

[34] Wei Yuewang, et al. *Illustrated Collection of Ancient Chinese Currency* [M]. 2nd ed. Beijing: China Financial Publishing House, 1992: 2, color illustration 2.

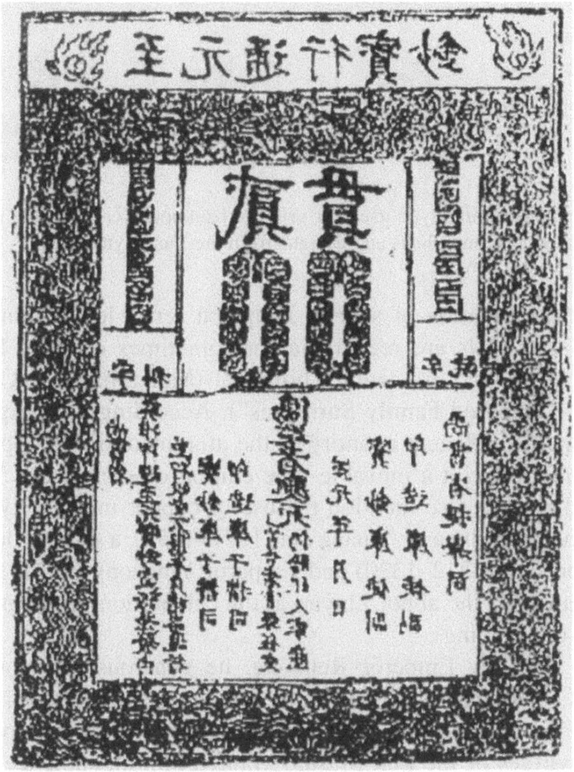

Fig. 7.10 Printing plate for *Zhiyuan Tongxing Baochao* note with a denomination of two *guan* issued in 1287 (Unearthed in Pingshan, Hebei, in 1993)

without embedded copper type in the serial number section (Fig. 7.10)[35] and notes with two copper movable type characters "Shao" and "Kou" embedded (Fig. 7.11),[36] which were issued starting in 1287. A significant number of similar notes and printing plates have been preserved.

The only difference between *baochao* banknotes in Yuan dynasty and *jiaochao* banknotes in Jin dynasty is the absence of official seals on the banknote surface. Therefore, the embedded copper movable type was limited to the characters required for the serial numbers. However, the Yuan dynasty still produced a large number of copper characters for seals. Examples of these artifacts have been unearthed in Xinjiang and are currently held in the Xinjiang Uygur Autonomous Region Museum. The seals include single Phags-pa or Chinese characters, essentially copper movable type used for documents. Sometimes, they include one Mongolian and one Chinese

[35] Wei Yuewang, et al. *Illustrated Collection of Ancient Chinese Currency* [M]. 2nd ed. Beijing: China Financial Publishing House, 1992: 58, illustration 3–26.

[36] Wei Yuewang, et al. *Illustrated Collection of Ancient Chinese Currency* [M]. 2nd ed. Beijing: China Financial Publishing House, 1992: 56, illustration 3–24.

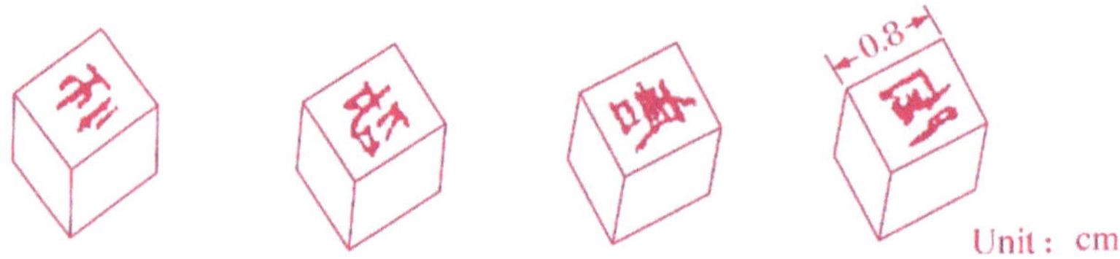

Fig. 7.11 The copper movable type used for printing banknotes, cast starting in 1276 (Restored based on the banknote printing plates and designs from the Yuan dynasty)

character, or a combination of several Phags-pa script letters, similar to Uighur movable type. The seals are rectangular or sometimes circular. Peking University Library possesses a 13th-century bilingual (Mongolian and Chinese) edition of *Baijia Xing* ("Hundred Family Surnames"). According to Mongolian historian Cai Meibiao, with a sufficient quantity of the aforementioned copper cast seals, it would be possible to print a movable type edition of such books.[37] On the other hand, historical records also mention the use of copper movable type for printing books during the Yuan dynasty. Huang Jin (1277–1357), a scholar during the reigns of Emperor Renzong (1312–1320) and Emperor Yingzong (1321–1323), wrote in a tower inscription for the abbot Zhiyan at the celebration of Qingshou Temple in Dadu (modern-day Beijing):

During the visits by Emperor Renzong, he generously bestowed dozens of precious items, including gold and jade Buddha statues, sutras, famous paintings related to Buddhist stories, and other exquisite artifacts. Emperor Yingzong, respecting the virtues of the Zen Master, granted him an audience on every visit, offering a seat and inquiring about the essentials. He ordered Zhiyan to collate and verify the Tripitaka with the senior abbots at Yongfu Temple and carve copper plates for transmission. Later, fearing that previously compiled books might be false or excessive, he ordered their revision.[38]

Huang Jin's record indicates that during the reign of Emperor Yingzong of the Yuan dynasty, he ordered Zen Master Zhiyan and other senior monks to gather at Yongfu Temple for the collation of the Buddhist scriptures—the Tripitaka, consisting of the sutras, vinaya, and shastras. The purpose was to use copper movable type to print a new edition of the Tripitaka. This historical account, discovered by the historian Gu Zuying (pseudonym for Deng Guangming, 1907–1998) in 1953, clarifies that the commonly misunderstood term *Jiaokan Sansui* (collation of the third year) should be corrected to *Jiaokan Sanzang* (collation of the Tripitaka) and should be punctuated accordingly.[39] Some argue that *Sansui* refers to the third year of the Zhizhi period (1323), suggesting that the planned publication was not the Tripitaka

[37] Cai Meibiao. On the Origin of Copper Movable Type Printing [N]. *Guangming Daily*, January 9, 1954.

[38] Huang Jin. *Collected Works of Mr. Huang from Jinhua: Vol. 41, Inscription of the Beixi Yangong Pagoda* [M]. Shanghai: Commercial Press, 1929.

[39] Gu Zuying. Issues with Copper Movable Type and Gourd Movable Type [N]. *Guangming Daily*, September 25, 1954.

but Buddhist scriptures and statues bestowed by Emperor Renzong.[40] However, this interpretation does not align with the original text, as dozens of scriptures and statues would not require the collaboration of monks from various temples for years of collation. Moreover, it contradicts the records in *History of Yuan Dynasty*. Therefore, it is evident that the senior monks were indeed collating the Tripitaka. Considering the magnitude of such a Buddhist scripture collection, it would not have been feasible to print it using copper plates. Therefore, the phrase "carve copper plates" must refer to the use of copper movable type for publishing the Tripitaka.

The *Yuanshi Renzong Ji* ("History of Yuan Dynasty – The Annals of Emperor Renzong") records that Emperor Renzong was devoted to Buddhism and Confucianism. He often visited Qingshou Temple for Buddhist activities and granted an additional 170 hectares of land to the temple. In the first year of the Yanyou era (1314), he decreed the establishment of the *Yinjing Tijusi*, an institution responsible for the printing and management of Buddhist scriptures. In 1316, he elevated it to *Guang Fu Jian*, an institution responsible for Buddhist affairs, often involved in the management of temples and the editing and printing of scriptures.[41] This suggests a clear initiative to publish the Tripitaka. His son, Emperor Yingzong, continued this effort, decreeing in the seventh year of the Yanyou era (1320) to increase the translation of Buddhist scriptures. In the following year, he bestowed 500 taels of gold, 2500 taels of silver, and 500,000 *guan* of *baochao* banknotes upon Yongfu Temple as financial support for printing the scriptures. Taking Yongfu Temple as a base, Zen Master Zhiyan and other monks were ordered to collate the Tripitaka with the intention of printing it using copper movable type. In the first year of the Zhizhi era (1321), Hong Yue was dispatched to Goryeo (Korea) to acquire paper for the Tripitaka,[42] indicating the intention to print using Goryeo paper. In the third year of the Zhizhi era (1323), the collation of the scriptures was completed. The Left Chancellor Bai Zhu (1298–1323) was ordered to supervise the transcription of two volumes in mud-gold.[43] During the transcription, it was discovered that some of the previously compiled books were possibly forged or excessively produced. Therefore, a new round of collation was ordered. However, before the final printing could take place, in August 1323, a political upheaval led by Tie Shi resulted in the death of Emperor Yingzong and Bai Zhu, causing the printing project to be abruptly halted. This historical record confirms that from 1314 to 1323, the Yuan dynasty court had indeed initiated plans to use copper movable type to publish a new and revised edition of the Tripitaka. A significant portion of the work had been completed, but due to the sudden political turmoil, the project was ultimately unsuccessful.

---

[40] Zhang Xiumin. The Invention of Printing in China and Its Influence [M]. Beijing: People's Publishing House, 1958: 87.

[41] *History of Yuan Dynasty: Vol. 26, The Annals of Emperor Renzong* [M] // *Twenty-Five Histories: Vol. 9*. Compact Edition. Shanghai: Shanghai Classics Publishing House, 1986: 75–76.

[42] *History of Goryeo: Vol. 35, The Family of King Chungsuk* [M]. Pyongyang: Academy of Sciences of North Korea, 1957: 539.

[43] *History of Yuan Dynasty: Vol. 27–28, The Annals of Emperor Yingzong* [M] // *Twenty-Five Histories: Vol. 9*. Compact Edition. Shanghai: Shanghai Classics Publishing House, 1986: 81, 83.

7.2.2 Metal Movable-Type Printing in Ming Dynasty

The Ming (1368–1644) and Qing (1644–1911) dynasties represent the culmination of traditional metal movable-type printing in China. During this period, not only did the quantity, variety, and distribution of printed materials surpass that of previous eras, but advancements in raw material selection, casting, and typesetting technology also reached unprecedented levels. At the establishment of the Ming dynasty in 1368, with its capital in Nanjing, metal movable-type printing began with the issuance of banknotes. According to *Mingshi Shihuozhi* ("History of Ming Dynasty: The Chronicles of Food, Silks Clothing or Currency"):

In the seventh year of Hongwu (1374), the Emperor (Ming Taizu) established the *Baochao Tiju Si* (Banknote Supervisory Office). The following year (1375), he ordered the Ministry of the Interior to produce the *daming baochao* banknotes for public circulation. The paper was made from mulberry bark, measuring one foot (31.1 cm) high and six inches (18.66 cm) wide, with a blueish tint. The exterior featured dragon patterns, and the top was inscribed with *Daming Tongxing Baochao.* On the inside, the top sides displayed the eight-character seal script reading *Daming Baochao* and *Tianxia Tongxing.* The center illustrated ten strings of coins as one *guan,* with the text below stating: "Approved by the Ministry of the Interior, the production of *daming baochao* banknotes shall circulate alongside copper coins. Those who forge shall be executed; informants will receive a reward of 25 taels of silver, and the offender's property shall be confiscated." For denomination of five hundred *wen,* the design depicted five strings of coins, and other denominations follow a similar pattern[44] (Fig. 7.12).

The *daming baochao* had denominations of one hundred, two hundred, three hundred, four hundred, five hundred *wen,* and one *guan.* One *guan* was equivalent to one thousand *wen* or one tael of silver, while four *guan* equaled one tael of gold. The banknotes followed the format of the Yuan dynasty's *baochao* but featured more concise text. A distinctive change was the relocation of the serial number from the front to the back of the banknote. The consecutive numbering of the banknotes was done using the characters from *Qianziwen* and numbers, with each character paired with a corresponding numeral. In the 1940s, copper printing plates for the one *guan* denomination of the *daming baochao* were unearthed in Nanjing. The copper plate measured 32 cm in height, 20.8 cm in width, and 1.0 cm in thickness, with a well-preserved surface. The plate was originally held by the Central Museum of Ancient Artifacts. After the outbreak of the counter-Japanese War in 1937, it was transferred along with other artifacts to Guizhou, and is now housed in the Guiyang Museum.[45]On the back of the plate were two rectangular grooves, each 0.5 cm deep, containing five copper movable type characters: "Quan Zi San Shi Hao"

---

44 *History of Ming Dynasty: Vol. 81, The Chronicles of Food, Silks Clothing or Currency V* [M] // *Twenty-Five Histories: Vol. 10.* Compact Edition. Shanghai: Shanghai Classics Publishing House, 1986: 216.

45 Wei Yuewang, et al. *Illustrated Collection of Ancient Chinese Currency* [M]. 2nd ed. Beijing: China Financial Publishing House, 1992: 82, 84, illustration 4-42A.

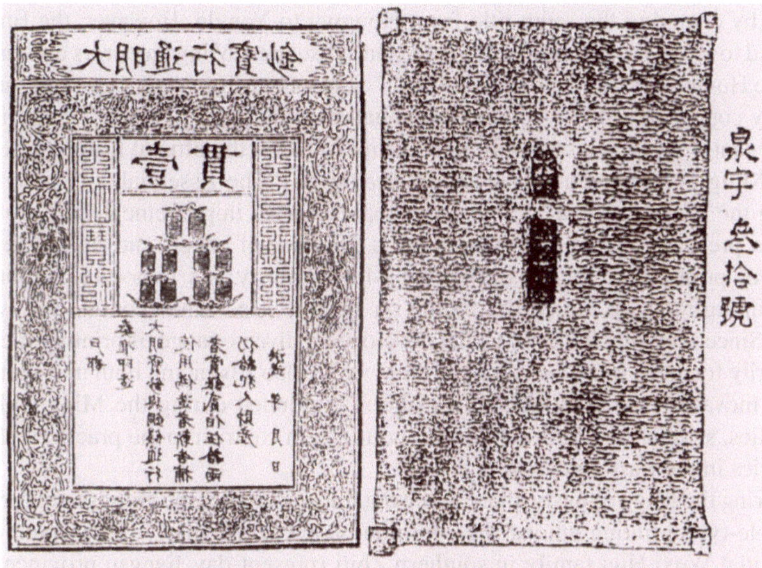

1. Front side (housed in the Guizhou Museum) 2. Back side (housed in the Guizhou Museum)

Fig. 7.12 The front and back of the copper printing plate for the Ming dynasty *daming baochao* note with a denomination of one *guan*, first cast in 1375

(Quan character, number thirty). These copper movable type characters were square, measuring 1.3 cm by 1.3 cm and standing 0.5 cm in height (Fig. 7.13). Additionally, a rubbing of a fifty-*wen* Ming banknote provided by Jia Jingyan showed a similar arrangement on the back, with two grooves containing "Yong Zi Shi Wu Hao" (Yong character, number fifteen), matching the font and size of the characters found on the one *guan* plate.

During the Ming dynasty, *baochao* circulated nationwide and had an indefinite circulation period. Large quantities, often in the millions, were printed with copper movable type, marking a significant era of widespread paper currency printing using metal type. When Ming Chengzu (the Yongle Emperor) ascended the throne in the first year of Yongle (1403), the Ministry of Revenue proposed updating the banknote

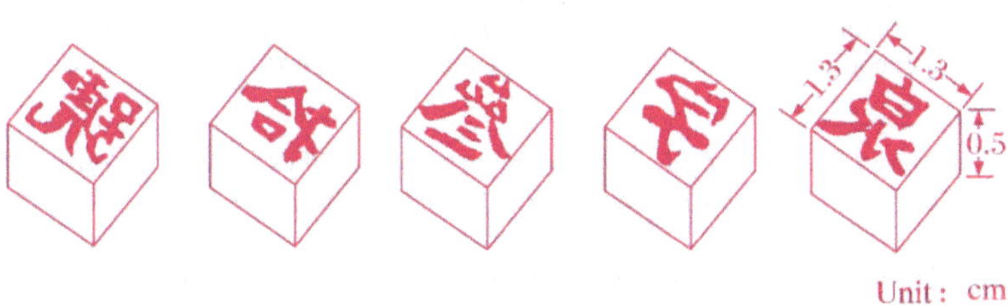

Fig. 7.13 The copper movable type font used for printing Ming banknotes (Cast in 1376 and illustrated by Pan Jixing)

plates by changing the reign title from Hongwu to Yongle. However, the Emperor decided to retain the Hongwu reign title, and subsequent Ming emperors continued to use the Hongwu reign title on banknotes.[46] Surviving Ming banknotes are characterized by copper movable type featuring a handwritten style of script. The characters are elegantly designed, resembling the script found in the official histories from the early Ming period, such as *Zhenguan Zhengyao* ("The Essentials of Government during the Zhenguan Era"). This script represents an improvement over the styles used in previous dynasties. While there is evidence of copper movable type being used for printing banknotes during the Ming dynasty, it's important to note that the printing methods varied depending on specific needs and economic considerations. Since the Song dynasty, China developed a diverse range of printing methods, primarily focusing on woodblock printing, while also advancing both non-metal and metal movable type printing. This tradition continued during the Ming and Qing dynasties, shaped by China's unique conditions, in contrast to the practices of other countries in the East and West.

During the Ming dynasty, there was a notable increase in the enthusiasm for metal movable-type printing among the common people, particularly represented by the influential Wuxi Hua family in southern Zhili (present-day Jiangsu province). One prominent figure from the Hua family associated with this development was Hua Sui (1439–1513). The family's achievements in the field of printing attracted the attention of both Chinese and foreign scholars after being introduced by Ye Dehui in his work *Shulin Qinghua* ("Conversations in the World of Books", 1911). The earliest biography of Hua Sui was written by Shao Bao (1460–1527). He wrote that:

Hua Sui, whose courtesy name was Wenhui, was from Wuxi. In his youth, he extensively studied classics and history. In his middle age, he developed a passion for comparing and analyzing texts, often hand-copying volumes after making corrections. When he encountered learned scholars, he would bring these writings to them for discussion, frequently engaging in deep conversations. Later on, he pioneered the use of copper plates and tin movable type, establishing a workshop named Huitong Guan (Society of Understanding and Communication) and earning the nickname Huitong, meaning "integration and thorough understanding". People subsequently referred to him as Huitong.[47]

Hua Sui's contemporary Qiao Yu (1457–1524) also wrote about him, stating:

> He meticulously compiled and integrated the essential ideas from a wide range of texts, earning him the title "Huitong". Concerned about the disorganization of his manuscripts, he created copper plates and tin movable type, allowing him to print and circulate rare and difficult-to-obtain works. His notable work, *Jiujing Yunlan* ("Verses and Reviews of the Nine Classics"), encompasses nearly all the classics and historical texts.[48]

---

[46] *History of Ming Dynasty: Vol. 81, The Chronicles of Food, Silks Clothing or Currency V* [M] // *Twenty-Five Histories: Vol. 10.* Compact Edition. Shanghai: Shanghai Classics Publishing House, 1986: 216.

[47] Shao Bao. *Posthumous Collection ofRongchuntang: Vol. 7, Biography ofHuitong* [M] // *Complete Library of the Four Treasuries: Collections, Separate Collections.* 1781.

[48] Qiao Yu. *Collected Works of Qiao Yu: Epitaph of Huitong* [M] // *Collected Works of Hua: Vol. 15.* 1572.

In summary, Hua Sui came from a scholarly family in Wuxi and had a rich collection of books. He engaged in scholarly pursuits from a young age, and in his middle years, he developed a keen interest in reviewing and correcting books. He hand-copied volumes, produced his own insights, and authored works such as *Jiujing Yunlan*. Later, he transitioned from textual criticism to the study of metal movable type printing, becoming adept at it and using it to publish high-quality books. Hua Sui was both a scholar and a printer, benefiting from his affluent background that allowed him to engage in this venture. Notably, he published 15 books using movable type,[49] with 11 of them having extant copies, many of which are preserved in the National Library of China. The earliest work is *Song Zhuchen Zouyi* ("Selected Memorials of Song Officials"), a collection of 150 volumes published in the third year of the Hongzhi era (1490). It features half pages with 9 lines per page, each line containing 17 characters, and bears the inscription "Huitong Guan Yin Zheng" (Printed by Huitong Guan). Following this is *Jinxiu Wanhua Gu* ("The Valley of Myriad Flowers"), a collection of 160 volumes, printed in the same format, with the title page indicating "Published in the seventh year of Hongzhi (1494)" and "Huitong Guan movable type copper plate print". The next year (1495), *Rongzhai Suibi* ("Random Thoughts from the Rong Studio", 74 volumes, see Fig. 7.14), *Wenyuan Yinghua Zuanyao* ("Essentials of the Literary Garden", 84 volumes), and *Gujin Hebi Shilei Qianji* ("The First Collection of Matters Uniting Ancient and Modern", 63 volumes) were published. In the fourteenth year of Hongzhi (1501), *Baichuan Xuehai* ("The Ocean of Learning from a Hundred Rivers", 177 volumes) was republished. This indicates that Hua Sui's publications primarily consisted of substantial scholarly works, reflecting his ambition for large-scale movable type printing. His endeavors in developing movable type began around the late Ming dynasty's Chenghua era (1465–1487) and achieved success in the early years of the Hongzhi era (1488–1505).

There are varying opinions about the materials used by Hua Sui for his movable type. Some claim they were made of copper,[50] while others believe he embedded tin type into copper casting plates.[51] We argue that the type was a copper-tin–lead alloy.[52] To determine the material Hua Sui used for casting type, his own statements should be considered. Hua Sui explicitly referred to his plates as "copper movable type plates", indicating that they were made of copper, not tin. From a technical standpoint, they must have been a copper alloy, similar to the composition of copper coins. The printing frames were often made of wood, but could also be made of metal; for instance, if the frame was made of copper, it could be forged. The phrase *fan tong wei ban* can be interpreted as "casting copper as plates", referring to using cast

[49] Qian Cunxun. *Collection of Papers on the History of Books, Paper, Ink, and Printing in China* [M]. Hong Kong: Chinese University Press, 1992: 181–183.

[50] Zhao Wanli, et al. *Illustrated Catalog of Chinese Woodblock Printing: Vol. 1* [M]. Beijing: Cultural Relics Press , 1961: 97–98. Zhao Wanli, et al. *Illustrated Catalog of Chinese Woodblock Printing: Vol. 7* [M]. Beijing: Cultural Relics Press, 1967: 599–602.

[51] Pan Tianzhen. The Printing of Huitong Guan in Wuxi during the Ming Dynasty Used Tin Movable Type [J]. *Library Science Communications*, 1980 (1): 51–64.

[52] Pan Jixing. *History of Metal Movable Type Printing Technology in China* [M]. Shenyang: Liaoning Science and Technology Publishing House, 2001: 87.

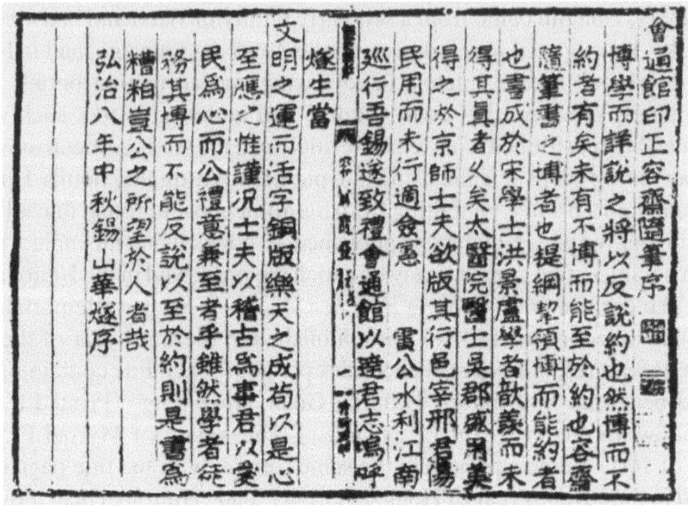
會通館印正容齋隨筆序
博學而詳說之將以反說約也然博而不
約者有矣未有不博而能至於約也容齋
隨筆書之博者也提綱挈領博而能約者
也書成於宋學士洪景盧學者歆羨而未
得其眞者久矣太醫院醫士吳郡盛用美
得之於京師士夫欲版其行邑宰邢君傷
民用而未行適僉憲　雷公水利江南
巡行吾錫遂致禮會通館以遂君志嗚呼
燧生當
文明之運而活字銅版樂天之成苟以是心
至憑之惟謹況士夫以稽古爲事君以愛
民爲心而公禮意兼至者乎雖然學者徒
務其博而不能反說以至於約則是書爲
糟粕豈公之所望於人者哉
弘治八年中秋錫山華燧序

Fig. 7.14 The copper movable type edition of *Rongzhai Suibi* printed by the Huitong Guan in Wuxi in 1495 (Housed in the National Library of China)

copper type, not that the frame itself was made from cast copper. The production of tens of thousands of metal movable type characters necessitates casting rather than individual engraving. It appears that the literati officials Shao Bao and Qiao Yu misused terms and misunderstood them literally, contributing to misconceptions. Hua Sui's copper movable types came in long, square, and flat shapes, and were produced in various batches, resulting in a significant total character count. The belief that metal movable type printing in China began in the fifteenth century with the Hua family in Wuxi is evidently incorrect. Claims that Hua Sui's research on copper movable type was influenced by Korea lack supporting evidence. In fact, during his era, China had a longstanding history of using copper movable type for over 300 years. Clearing up misunderstandings surrounding Hua Sui's casting of type is essential for a more accurate understanding of historical facts.

Hua Sui also passed his printing techniques on to his relatives. His uncle, Hua Cheng (courtesy name Rude, 1438–1514), printed *Weinan Wenji* ("Literary Collection from Weinan", 50 volumes) and *Jiannan Gao* ("Manuscripts from Jiannan", 8 volumes) using copper movable type at his Shang Gu Studio in the fifteenth year of the Hongzhi era (1502). Hua Sui's nephew, Hua Jian (courtesy name Yungang), at his Lan Xue Tang Studio, printed five works with copper movable type, including *Yuan Shi Changqing Ji* ("Collected Works of Yuan Zhen", 60 volumes) and *Bai Shi Changqing Ji* ("Collected Works of Bai Juyi", 71 volumes), published in the eighth year of the Zhengde era (1513). Two years later (1515), he published *Yiwen Leiju* ("A Collection of Arts and Literature", 100 volumes). The title page bore the studio name Lan Xue Tang, and the colophon stated "Hua Jian's movable type copper plate from Xishan Lan Xue Tang". Aside from the Hua family, the An family led by An Guo (1481–1534) represented another group involved in movable-type printing in

Wuxi. According to *Jiaoshan An Huang Family Genealogy* (1922) from Wuxi, the An family's ancestors were originally surnamed Huang. During the Hongwu period (1368–1398), a man named Huang Mao from Suzhou married into the An family and settled in Jiaoshan, Wuxi, passing down through four generations to An Guo. An Guo, whose courtesy name was Mintai and nickname was Guipo, was skilled in commerce and enjoyed collecting books and traveling, authoring works such as *Si Youji* ("Four Travels") and *Youyin Gao* ("Traveling Notes"). He also decided to invest in the printing industry, publishing over ten books using copper movable type. According to volume 14 of the genealogy, he published *Dongguang Xianzhi* ("Chronicle of Dongguang County", 6 volumes) in the sixteenth year of the Zhengde era (1521). An early edition of *Wuzhong Shuili Tongzhi* ("Comprehensive Gazetteer of Water Conservancy in Wuzhong", 17 volumes) is known to exist, featuring half pages with 8 lines and 16 characters per line. The book states "Xishan An Guo movable type copper plate published" and was printed in the third year of the Jiajing era (1524), now held in the National Library of China (Fig. 7.15). Additionally, the printing workers' names can be found in *Gujin Hebi Shilei Beiyao* ("Essentials of Matters Uniting Ancient and Modern", 69 volumes), which was published between 1524 and 1534.

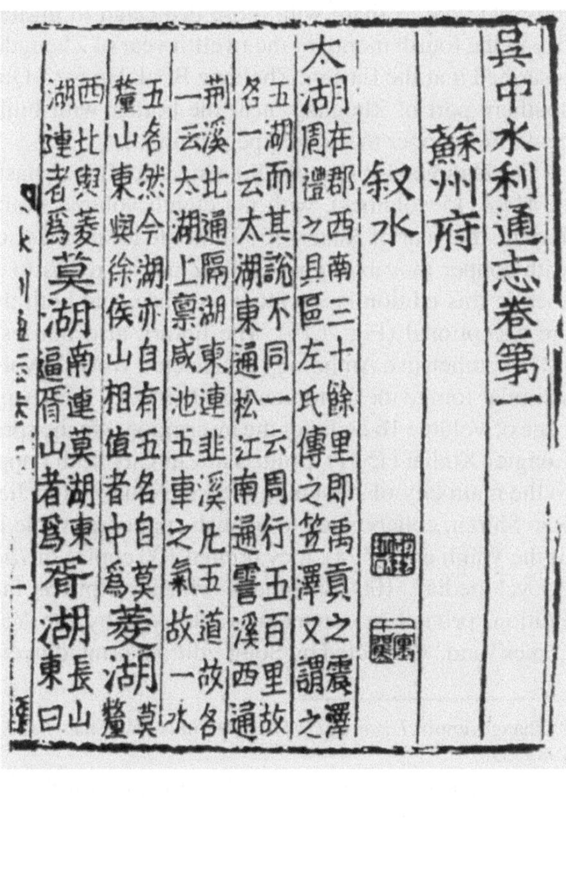
吳中水利通志卷第一
蘇州府
敘水
太湖在郡西南三十餘里即禹貢之震澤周禮之具區左氏傳之笠澤又謂之
五湖而其說不同一云周行五百里故名一云太湖東通松江南通霅溪西通
荊溪北通隔湖東連韭溪凡五道故名一云太湖上稟咸池五車之氣故一水
五名然今湖亦自有五名自莫釐山東與徐侯山相值者中為菱湖莫釐
西北與菱湖連者為莫湖南連莫湖東逼胥山者為胥湖長山東曰

Fig. 7.15 The copper movable type edition of *Wuzhong Shuili Tongzhi* printed by An Guo in Wuxi in 1524

According to historical records and existing editions, during the Ming dynasty, locations for the private publication of metal movable type works included not only Jiangsu but also major printing provinces like Zhejiang and Fujian, often funded by wealthy merchants. Ming scholar Lu Shen (1477–1544) noted in his *Jintai Jiwen* ("Chronicles of the Golden Terrace", 1505) that "recently, people from Kunling (Changzhou, Jiangsu) have been using copper and lead for movable type, which is much more convenient than woodblock printing, though it is easy to make errors in typesetting." This refers to the use of a copper-lead alloy for casting movable type. While proofreading errors can occur if not careful, the efficiency of printing surpasses that of woodblocks. The term "recently" pertains to the Hongzhi and Zhengde eras (1500–1508). The Peking University Library houses a copper movable type edition of *Shihu Jushi Ji* ("Collected Works of Fan Chengda", 34 volumes), published by Jin Lan Guan in the fifteenth year of the Hongzhi era (1502), originally authored by Fan Chengda from the Song dynasty. It has been verified that Jin Lan Guan refers to a hall name in Suzhou,[53] indicating that this edition was likely published there. The Shanghai Library holds a volume titled *Zhuge Kongming Xin Shu* ("Writings of Zhuge Kongming"), a copper movable type work published in Zhejiang in 1517. The book states, "Printed by Qiongtai Han Xifang, the teaching supervisor of Qinyuan County, Zhejiang." The preface includes a note from the publisher: "This is printed using movable type to share with those dedicated to military affairs... On an auspicious day in the fourth month of the twelfth year of Zhengde (1517), Qiongtai Han Xifang inscribed it at the Eastern Zhejiang Book House." Qinyuan County is located in the southern part of Zhejiang, near the border with Fujian, which is also known as a center for copper movable type printing.

The National Library of China holds a Ming dynasty copper movable type edition of *Mozi* (15 volumes), with the eighth volume stating, "In the thirty-first year of Jiajing, the year of Renzi (1552), in the auspicious summer, printed in the Zhi City with copper movable type." This Zhi City refers to Jianning in Fujian.[54] The font used in this edition is a printed typeface, and both the casting and printing quality are exceptional (Fig. 7.16). The library also houses *Tongshu Leiju Keze Daquan* ("Comprehensive Anthology of Selected Works", Volumes 16 to 19), which features a similar font to the aforementioned "Mozi", mostly printed in smaller type. The last page of Volume 16 includes the inscription, "In the spring month of the year of Jiajing Longfei Xinhai (1551), printed in Zhi City with copper movable type". In addition to the main city of Jianning, two individuals from Jianyang County, You Rong and Rao Shiren, collaborated to publish copper movable type works. In the second year of the Wanli era (1574), they printed 100 copies of *Taiping Yulan* ("Taiping Imperial Encyclopedia", 1000 volumes), with inscriptions indicating, "Corrected by Song Edition, printed in copper movable type by the You family of Fujian, over 100 copies" and "Corrected by Song Edition, printed in copper movable type by the Rao

[53] Zhang Xiumin. *History of Printing in China* [M]. Shanghai: Shanghai People's Publishing House, 1989: 695.

[54] Zhang Xiumin. *History of Printing in China* [M]. Shanghai: Shanghai People's Publishing House, 1989: 687–689.

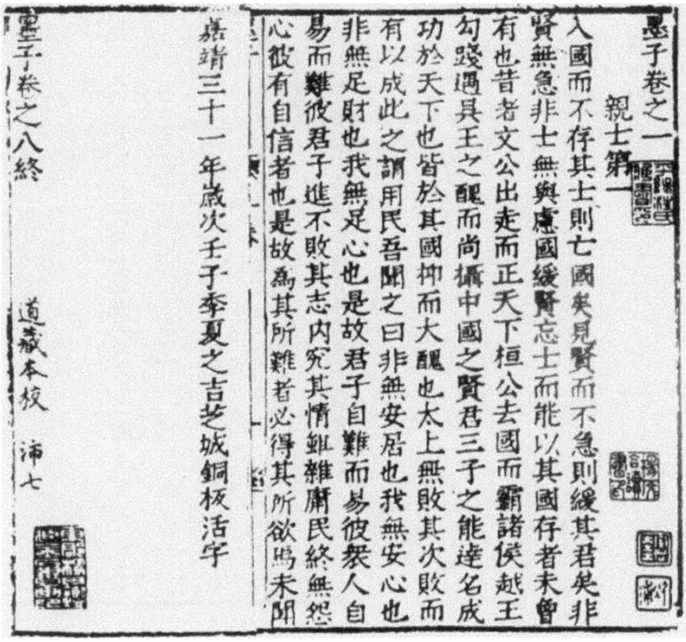
墨子卷之一
親士第一
入國而不存其士則亡國矣見賢而不急則緩其君矣非
賢無急非士無與慮國緩賢忘士而能以其國存者未曾
有也昔者文公出走而正天下桓公去國而霸諸侯越王
勾踐遇吳王之醜而尚攝中國之賢君三子之能達名成
功於天下也皆於其國抑而大醜也太上無敗其次敗而
有以成此之謂用民吾聞之曰非無安居也我無安心也
非無足財也我無足心也是故君子自難而易彼衆人自
易而難彼君子進不敗其志內究其情雖雜庸民終無怨
心彼有自信者也是故為其所難者必得其所欲焉未聞
嘉靖三十一年歲次壬子季夏之吉芝城銅板活字
墨子卷之八終
道藏本校

Fig. 7.16 The copper movable type edition of *Mozi* printed by Yao Kui in Zhi City, Fujian, in 1552

family, over 100 copies." The previous year (1573), they printed *Wenti Mingbian* ("Clarification of Literary Forms"), noting, "Printed using movable type by You Rong of Jianyang, Fujian," which was also a copper movable type edition. The publication of large-scale copper movable type works across various regions demonstrates the publishing capabilities of private printers during the Ming dynasty.

7.2.3 Metal Movable-Type Printing in Qing Dynasty

In the early Qing dynasty, the rulers showed little interest in issuing paper currency and continued to mint copper coins. However, the Qing government's large-scale copper movable type printing emerged as a way to publish massive books. After five years of effort, they compiled a significant work titled *Gujin Tushu Huibian* ("A Collection of Ancient and Modern Books"). Completed in the 45th year of Kangxi's reign (1706), it was presented for imperial review in the 55th year (1716) and was granted the title *Gujin Tushu Jicheng* ("An Alphabetical Index to the Chinese Encyclopaedia"), with an edict for the inner court to cast copper movable type for its publication. Bao Shichen (1735–1855) noted in *Anwu Lunshu* ("Discourse on

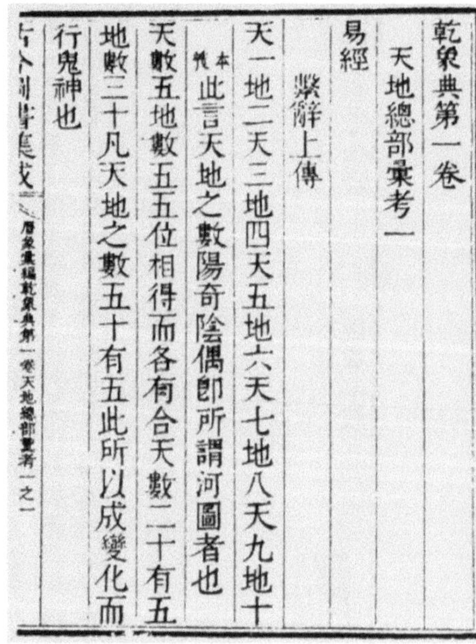
乾象典第一卷
天地總部彙考一
易經
繫辭上傳
天一地二天三地四天五地六天七地八天九地十
本義 此言天地之數陽奇陰偶卽所謂河圖者也
天數五地數五五位相得而各有合天數二十有五
地數三十凡天地之數五十有五此所以成變化而
行鬼神也
曆象彙編乾象典第一卷天地總部彙考一之一

Fig. 7.17 The copper movable type edition of *Qinding Gujin Tushu Jicheng* printed by the inner court in 1726

Calligraphy") that "during the Kangxi era, the inner court cast several hundred thousand fine copper movable types for printing books."[55] However, *Gujin Tushu Jicheng* had not yet been published when the emperor passed away. After a power struggle, Kangxi's fourth son, Prince Yong, also known as Yongzheng (1723–1735), ascended the throne. He ousted Prince Cheng's minister, Chen Menglei, and tasked Jiang Tingxi (1669–1732) with reworking the manuscript. By the fourth year of Yongzheng's reign (1726), the book was completed, and in the sixth year (1728), 66 volumes were printed using the inner court's copper movable type by the Wuying Palace, officially titled *Qingding Gujin Tushu Jicheng* ("The Imperial Edition of the Alphabetical Index to the Chinese Encyclopaedia", Fig. 7.17).

The *Qingding Gujin Tushu Jicheng* comprises over 10,000 volumes and 160 million characters, organized into 5020 books, making it the largest encyclopedia in the world at the time. Its length exceeds that of the famous 11th edition of the Encyclopaedia Britannica (1911) by more than four times. This work is of significant value for the study of traditional Chinese culture and science, and publishing such a monumental book using copper movable type is an unprecedented achievement in the history of printing. The book features a format of half a page with 9 lines per page and 20 characters per line, using Song typeface with large characters measuring 1 cm square and small characters 0.5 cm square. French sinologist Julien estimated that printing this work would require 250,000 movable types.[56] His estimation is

55 Bao Shichen. *Discourse on Calligraphy* [M]. //*Zhi Jin Zhai Series: Vol. 2*. 1883.

56 Julien S. Documents sur l'art d'imprimer, à l'aide des planches au bois, des planches au Pierre et des types mobiles [J]. *Journal Asiatique*, 1847, 4(9): 508.

based on *Wuyingdian Juzhenban Congshu* ("Movable Type Printing Series of the Hall of Martial Valor", 1776), which utilized 250,000 wooden types for fewer than 2400 volumes. However, given that *Qingding Gujin Tushu Jicheng* consists of over 10,000 volumes, it would require between 1 to 2 million copper movable types, aligning with Bao Shichen's assertion of "several hundred thousand". There are two theories regarding how the movable types were produced, and it is time to reach a final conclusion. Wu Changyuan (1743–1800) noted in *Chenyuan Shilue* ("Sketches of the Imperial Enclosure") that "the movable types used in Wuyingdian were cast in copper, designed specifically for printing *Gujin Tushu Jicheng*." Bao Shichen also supported this view.

Some have stated that "during the Kangxi era, *Gujin Tushu Jicheng* ("Compendium of Ancient and Modern Books") was compiled and copper characters were carved as movable type", suggesting that the type was hand-carved.[57] Others have pointed out that the same character appears differently within the same page of the book, arguing that if the movable types were cast, such discrepancies should not exist, and thus also asserting that the type was hand-carved.[58] We believe that the views of the Qing scholar Wu Changyuan are correct, as copper movable type has been cast since the eleventh century during the Northern Song dynasty. Neither the official nor the private sectors during the Ming and Qing dynasties would have abandoned established casting techniques, choosing instead to hire workers to hand-carve hundreds of thousands, if not millions, of copper characters. This fundamental principle applies universally. The technology of casting characters has over 400 years of history in China; how could there suddenly be such a technological regression, moving away from casting to hand-carving? The language used by ancient scholars may lack precision, but it should not lead to the misinterpretation that the originally cast movable type was hand-carved. When the number of movable types is large, the same character must be cast using multiple molds, and adjustments are often necessary. Therefore, variations in the shape of the same character are natural, and this phenomenon can also be observed in metal movable type books from other countries in both the East and West. Should we also conclude that those were hand-carved? Thus, when Emperor Qianlong stated "copper characters were carved as movable type", it does not alter the historical fact that cast copper characters were used as movable type. The Korean scholar Yi Gyu-gyeong (1788–ca. 1862) described the finely cast copper movable type of the Qing court in the eighteenth century as follows:

The movable types from the Central Plains (China), particularly those from *Collected Gems Editions of the Wuying Palace*, are the finest; their backs are flat, not recessed, and drilled through, which allows the characters to be aligned perfectly without slanting. In contrast, our (Korean) movable types vary in size and thickness,

---

[57] Zhang Xiumin. *History of Printing in China* [M]. Shanghai: Shanghai People's Publishing House, 1989: 718.

[58] Giles L. *An Alphabetical Index to the Chinese Encyclopaedia* [M]. London: British Museum, 1911: xvii.

with recessed backs and no drilled holes, leading to misalignment in the characters when printed, lacking elegance.[59]

Yi Gyu-gyeong's description closely resembles the one provided by Wang Zhen, a Chinese scholar from the Yuan dynasty, regarding the form of tin movable type used during the twelfth to thirteenth centuries in the Song dynasty. It can be seen as a continuation of the tradition.

In the 11th year of the Jiaqing reign (1806), a Manchu named Wu Long'a (ca. 1765–1831), serving as the Commander-in-Chief in Taiwan, imitated the casting of copper movable type from the Wuying Palace to publish *Shengyu Guangxun Zhu* ("Annotations of Sacred Edicts"). During the tenure of Yao Ying (1785–1853) as the governor of Taiwan, he personally witnessed these types and the printed books. In a letter to his friend, he mentioned:

Here in Taiwan, Wu Long'a also cast copper movable types and the characters are in Song font. However, each plate has only eight lines, which does not meet my preferences. Additionally, a Fujian man named Lin created wooden movable types, with ten lines per plate, each line having eleven characters. Both methods are acceptable, but the latter is slightly better than Wu's version.[60]

The *Wenyuan Yinghua Lvfu Xuan* ("Selected Works of Literary and Rhetorical Elegance") published during the Qing dynasty is a copper movable type edition (Fig. 7.18) currently housed in the National Library of China. This four-volume work features black borders, double fish tails, ten lines per half-page, with eighteen characters per line, showcasing a beautiful handwritten style. The title page includes the words "by Mr. Qian Xiangling of Yushan" and "Chuili Pavilion copper plate". The beginning of the book also contains a preface by the compiler, Qian Lucan, who states, "Thus I have slightly refined and presented it for movable type to circulate in the world." Qian Lucan (1612–ca. 1698), courtesy name Xiangling, was from Yushan (Changshu, Jiangsu). He was a candidate for the highest degree in the imperial examination in the Shunzhi era and had a passion for collecting books, teaching in the areas of Changzhou and Nanjing. The *Wenyuan Yinghua Lvfu Xuan* was compiled by him in Changzhou, indicating that the publisher, Chuili Pavilion, was likely located there.[61] In the first half of the nineteenth century, individuals with financial means in provinces such as Fujian, Zhejiang, and Guangdong also began to cast type and publish books. The Qing scholar Wei Song noted in his *Yishi Jishi* ("Validation of All Details from the Roots", 1834), Volume 9, that "movable type began in the Song… now using copper and lead for characters." Here, "now" refers to the Daoguang era (1821–1850), where "copper" refers to copper alloys, and "lead" denotes alloys including tin and lead. In the Ming and Qing periods, tin and lead were often used interchangeably, with "lead" referring to tin.

---

59 Yi Gyu-gyeong. *Oju yŏnmun changjŏn san'go (Scattered Manuscripts of Glosses and Comments by Oju): Vol. 24, Discourse on Casting Type for Printing* [M]. Seoul: Myeongmundang, 1982: 699.

60 Shen Wenzhuo. Letters of Scholars in the Qing Dynasty [J]. *Cultural Relics*, 1961 (10): 61–65.

61 Wei Yinru. *History of Ancient Chinese Classics Printing* [M]. Beijing: Printing Industry Press, 1998: 222.

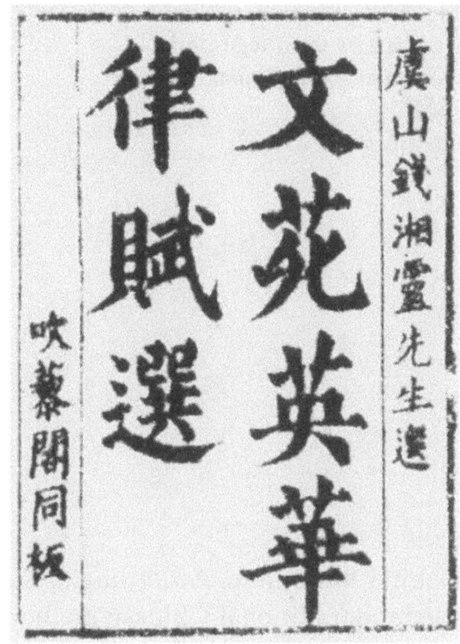

Fig. 7.18 The copper movable type edition of *Wenyuan Yinghua Lvfu Xuan* printed by Chuili Pavilion in Changzhou in 1686

In the 26th year of the Daoguang era (1846), Lin Chunqi (1808–?), a resident of Longtian, Fuqing County, Fujian, invested 200,000 taels of silver over 20 years to cast 400,000 pieces of copper movable type for printing. The surviving editions include *Yin Xue Wu Shu* ("Five Books on Music"), which contains *Yin Lun* ("On Music") and *Shi Ben Yin* ("Poetry and Music"), totaling 12 volumes, along with another work titled *Junzhong Yifang Beiyao* ("Essential Prescriptions for Military Use") in two volumes.[62] The first volume of *Yin Lun* features a preface written by Lin Chunqi titled "Copper Plate Account", explaining the details of his typecasting process. The back of the title page includes four lines stating "Futian Book Sea// Copper movable type plate//Fujian Houguan//Lin's precious collection", where he correctly employs the technical term "copper movable type plate". Towards the end of the Daoguang era (1850), a family surnamed Tang in Foshan, Guangdong, invested 10,000 taels of silver to cast three sets of movable type in tin alloy, totaling 300,000 characters (Fig. 7.19). The large characters were 1 cm × 1 cm in size, while the small characters were 0.6 cm × 0.8 cm, with a height of 1.32 cm. In the second year of the Xianfeng era (1852), they used these types to publish *Wenxian Tongkao* in 348 volumes, nearly 20,000 pages, bound in 120 books. American missionary Samuel Wells Williams (1818–1884) saw this set of movable type in Guangzhou and reported on it, disclosing samples of the characters.[63] Some people have suggested that the Tang family borrowed Western methods for casting types[64]; however, this

[62] Zhang Xiumin. *History of Printing in China* [M]. Shanghai: Shanghai People's Publishing House, 1989: 722.

[63] Williams S W. Movable Metallic-types in China [J]. *The Chinese Repository*, 1850 (19): 247–249.

[64] Hirth F. Western Appliances in the Chinese Printing Industry [J]. *Journal of the North China Branch of the Royal Asiatic Society*, 1886 (20): 166–167.

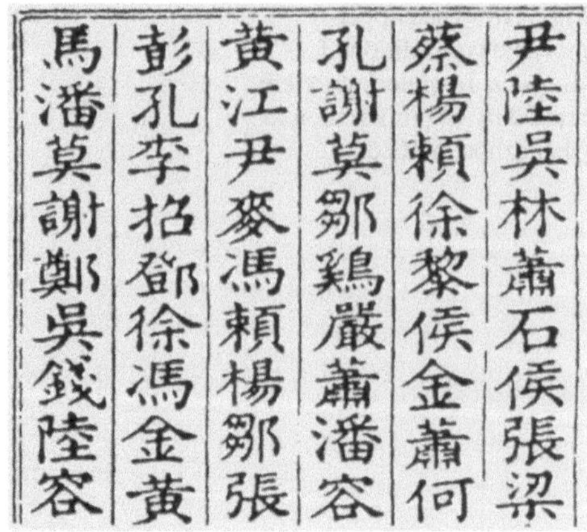

Fig. 7.19 The tin movable type cast by the Tang family in Foshan, Guangdong, in 1850[65]

claim is inaccurate. According to Samuel Wells Williams's account, the Tang family used the traditional Chinese method, employing wooden movable type as the mold and using the sand-casting method for typecasting.

7.2.4 Why Metal Movable-Type Printing Did Not Become Mainstream Before Qing Dynasty

Since the Song dynasty, China has developed a diverse printing tradition that primarily focused on woodblock printing, along with the development of non-metal and metal movable-type printing. This tradition continued through the Yuan, Ming, and Qing dynasties, reflecting the unique conditions of Chinese society. Although metal movable-type printing originated in China and showed great potential for advancement, it never became the dominant method, nor did it completely replace other forms of printing. Instead, it coexisted with woodblock printing, which remained prevalent until the late Qing dynasty, when metal movable type began to take over. This phenomenon warrants discussion. In a context where multiple printing methods are available, the choice between them by government or private printing institutions typically depends on the specific printing needs and economic considerations, rather than on technical limitations. For instance, when the government issues paper currency, it is necessary to embed copper movable type into cast copper printing plates. For large-scale publications with extensive text, metal movable type is more suitable, as it avoids the need to carve numerous woodblocks. Conversely, for shorter texts, woodblock printing remains more cost-effective.

Chinese characters are ideographic, with one character representing one syllable. Words and sentences are formed by combining different characters. Each Chinese character is composed of several strokes, and each movable type contains only one

[65] Williams S W. Movable Metallic-types in China [J]. *The Chinese Repository*, 1850 (19): 247–249.

character. Given the large number of Chinese characters, with over 20,000 commonly used ones and even more when considering rare and repeated characters, it typically requires 100,000–200,000 movable types to print a book. The process of casting metal movable types involves first carving hundreds of thousands of wooden movable types as molds, then casting an equal number of metal movable types. This process requires substantial capital, making it more costly than the production of woodblock or wooden movable type prints. The books produced are also priced several times higher. In order for metal movable type printing to be economically viable, the printed books need to be sold and the costs recovered promptly. However, once copper movable type editions were introduced to the market, they faced tough competition on pricing compared to other versions. Readers in bookstores had various options, and they generally preferred more affordable versions. For private printers, printing was a commercial activity, and the decision to cast movable types for printing depended on economic considerations, financial capacity, and manpower availability. They closely followed market trends. Woodblock printing had a long history in China, with mature engraving techniques, abundant wood resources, excellent print quality, and low prices. It satisfied the needs of a wide range of readers, and reprints were easy to make, making it a popular choice. Books with a high likelihood of reprints and those with illustrations were suitable for woodblock printing.

Some affluent scholars considered metal movable type printing as a scholarly pursuit and were willing to invest significant funds in publishing various rare and secret texts for personal collections. If private printers had sufficient financial resources and intended to publish multiple works with a small print run without planning for reprints, they could use metal movable types for printing. However, if they wished to avoid additional investments, they might choose wood or ceramic movable types. Movable types had the advantage of being detachable and reusable, reducing the time required for typesetting. However, due to the large number of characters, the costs and initial capital investment for metal movable types were higher, offsetting their advantages. On the other hand, characters carved on wood blocks were aesthetically pleasing, resembling handwritten manuscripts. Woodblock printing costs were relatively lower compared to metal movable types, compensating for their shortcomings. China's vast territory and large population resulted in a greater demand for a variety and quantity of reading materials than any other country. Both official and private printing institutions were more numerous than in foreign countries. They could publish various forms of editions according to needs, possibilities, and preferences, finding a market for all kinds of prints. Therefore, various printing forms could coexist for an extended period, and the dominance of woodblock with supplementary metal movable-type printing was determined by the market. Ultimately, the decisive factor was the preference of readers.

Goryeo belongs to the cultural sphere of Chinese characters, but it has emphasized the development of metal movable type printing while retaining woodblock printing. This choice was influenced by the country's specific circumstances. Since its establishment, the printing industry in Goryeo has been largely monopolized by the royal court and various government offices. Unlike China, private printing was not as widespread and developed, likely due to the smaller population and consequently lower demand for printed books. However, there was a diverse need for various types of books. The printing institutions operated by the royal court were the largest publishers in the country. The books they published were distributed to various government departments, and each edition typically consisted of 200–300 copies. Reprints were not a significant consideration, and new books from China were promptly reproduced. Government-operated printing offices, unlike profit-oriented private workshops, had sufficient financial resources and labor, allowing them to overlook the cost of book production. In such circumstances, the use of metal movable-type printing was more suitable. However, for publications with high demand, such as the calendars issued by the Chinese emperor, woodblock printing was still preferred.

After the metal movable type technology was introduced to Europe, it quickly replaced woodblock printing as the mainstream method. This was largely because European languages used phonetic scripts composed of twenty to thirty letters, allowing for the printing of all books with just a few thousand movable types. The investment required was less than 1/50 of that in China, and due to lower print runs, books could be sold quickly, enabling prompt recovery of production costs. In contrast, woodblock printing could only produce one type of book at a time, making it more labor-intensive and often resulting in higher costs for the final products. While Latin letters are simpler in structure, their numerous rounded strokes and smaller sizes make them more difficult to carve. Therefore, carving woodblocks was more laborious for Europeans than for the Chinese. Although wooden movable type has a low cost, it is difficult to carve smaller characters and lacks sufficient mechanical strength for printing. On the other hand, carving larger wooden movable types leads to wasted space and paper, which is why this method could not sustain its development. In a populous country like China, metal movable type printing only became mainstream in the nineteenth century with the introduction of machine production, which reduced costs. During the era of handicraft production, however, while woodblock printing was dominant, the total number of metal movable types produced in China still exceeded that of Goryeo and Europe, which is an important distinction to note.

7.3 Casting Techniques of Metal Movable-Types

7.3.1 Casting Methods and Materials for Patterns and Molds

Metal movable-type printing in China has evolved over 700 years, beginning in the eleventh and twelfth centuries during the Northern Song and Jin dynasties, and many artifacts from this period have been discovered. However, systematic records of this technology are scarce, representing a significant gap in the study of printing history. The process of metal movable-type printing is similar to that of wood or clay types and involves steps like type manufacturing, typesetting, printing, disassembling, and reassembling. Aside from the differences in type manufacturing, the remaining processes are fundamentally the same. In fact, there is abundant literature on non-metal movable type technology, particularly regarding wooden movable type, which shares even more similarities with metal movable type in terms of typesetting, disassembling, and collecting. Notable references include Wang Zhen's *Zao Huozi Yinshu Fa* ("Method of Creating Movable Type Printing", 1298) from the Yuan dynasty and Jin Jian's *Wuyingdian Juzhenban Chengshi* ("Regulations on Movable Type Printing Adopted at the Hall of Martial Valor", 1776) from the Qing dynasty. Due to the differences in composition and properties between metal and wood, the manufacturing methods for movable type made from these two materials are also clearly different. Therefore, the key to exploring metal movable-type printing technology is to clarify the type manufacturing techniques.

Theoretically, metal movable type can be produced by either engraving or casting. However, given the hardness of metals, individually engraving hundreds of thousands of metal types is impractical, necessitating casting methods similar to those used for coins and seals, a practice with a long history in China dating back to the eighth century BC. Texts such as Song Yingxing's *Tiangong Kaiwu* ("Exploitation of the Works of Nature", 1637) provide insights into these casting techniques. Historical bronze coins and their molds have been excavated, and analysis has provided valuable information on the materials and processes used in metal movable type manufacturing. A passage from Korean scholar Seong Hyeon's *Yongjaechonghwa* ("Assorted Writings of Yongjae", 1495) Volume 7 mentions that copper movable types were made using wood types as patterns through sand casting, aligning with traditional practices in both China and Korea. In 1980, Chinese scholar Feng Fugeng and others conducted experimental research on the casting technology of Shang dynasty bronze wares,[66] and in 1998, the Yangzhou Ancient Books Printing Society, led by Hou Guilin, attempted to print *Tangshi Sanbai Shou* ("Three Hundred Tang Poems") using traditional methods of metal movable type. These efforts revealed various details about the casting process. In recent years, Korean scholars have also made similar contributions in this area. Additionally, from American scholar Samuel Wells Williams's reports on the use of tin movable type for printing by the Tang family in Guangzhou in 1850,[67] we gain further insight into the technology of typecasting.

Traditional metal movable-type printing in China has largely fallen out of use since the late nineteenth century, having fulfilled its historical role by the twentieth century. It has been replaced by modern machine-produced lead type printing, which in turn has given way to computer typesetting. Today, ancient metal movable-type printing is regarded as a lost technology and a subject of research in the history of printing. The discussion of traditional Chinese metal movable-type printing essentially involves reconstructing a lost technology. The previous content offers essential technical information and clues for this research, but it is neither comprehensive nor systematic. Some sections are only briefly described, lacking in detailed operational insights. Certain techniques must be adapted to accommodate the characteristics of

[66] Feng Fugeng, Wang Zhenjiang, et al. Brief Report on the Trial Casting of Shang Dynasty Bronze Ware [J]. *Archaeology*, 1980 (1): 91–94.

[67] Williams S W. Movable Metallic-type in China [J]. *The Chinese Repository*, 1850 (19): 247–249.

metal type and validated through experimentation. Thus, there is still much work to be done in exploring traditional metal movable-type technology. First, a systematic examination of scattered historical records and scientific studies of unearthed artifacts is required. Second, this should be supplemented with simulated experiments and scientific evaluation. Through this research, we aim to clarify the processes involved in the casting of traditional Chinese metal movable type, the operational details of each step, and the necessary tools and equipment,[68] which will be outlined below.

There are two methods for movable type casting: sand casting and dewaxing casting (or investment casting). Both methods are discussed in *Tiangong Kaiwu* regarding the casting of copper coins and bells. Dewaxing casting is a precision technique that begins by mixing lime, clay, and fine sand to create an inner mold that matches the shape of the final piece. Once dried, a mixture of melted beef tallow (80%) and beeswax (20%) is applied to the inner mold, onto which inscriptions or designs are carved. A paste made from crushed, sifted refractory clay and charcoal is then layered over the wax, followed by a sprinkling of refractory sand. After hardening, the outer mold is baked slowly to melt and drain away the wax, creating a hollow casting cavity. Molten copper is then poured into this cavity, and once solidified, the piece is finished.[69] This method is complex and costly, primarily used for casting intricate items like statues, bells, tripods, or bronze seals, which are produced in small quantities. In contrast, copper coins feature a simple design and are manufactured in large volumes, making dewaxing casting economically impractical. As a result, sand casting is generally used for coin production.[70] Given that the design of copper movable type is even simpler than that of copper coins, it is reasonable to conclude that sand casting was also used for copper movable type after the Song dynasty.

Sand casting is a simple and cost-effective method, ideal for mass-producing coins and movable type. According to *Tiangong Kaiwu*, the process for casting copper coins is as follows. A coin pattern is made from tin, featuring raised characters. A mixture of clay, fine sand, and a small amount of charcoal powder is combined with water and packed into a wooden frame (37.32 cm long, 15.51 cm wide, and 3.73 cm high). The front and back of the coin pattern are alternately placed on the sand and clay surface inside the frame. Another frame is prepared in the same way and is placed on top of the first frame. The combined frames are flipped so that the pattern rests on the second frame. A rope secures the frames together, with a circular opening left at the seam for pouring in molten copper from a crucible. After cooling, the frames are opened, and the coins can be removed and finished by filing.[71] Thus, the sand-casting method for copper coins involves four main steps: pattern making,

---

[68] Pan Jixing. History of Metal Movable Type Printing Technology in China [M]. Shenyang: Liaoning Science and Technology Publishing House, 2001: 103–126.

[69] Song Yingxing, Pan Jixing. *Exploitation of the Works of Nature: Translation and Annotation* [M]. Shanghai: Shanghai Classics Publishing House. 1992: 107–110, 273.

[70] Zheng Jiaxiang. A Brief Discussion on the Casting Methods of Copper Currency Throughout History [J]. *Cultural Relics*, 1959 (4): 68–70.

[71] Song Yingxing, Pan Jixing. *Exploitation of the Works of Nature: Translation and Annotation* [M]. Shanghai: Shanghai Classics Publishing House. 1992: 112–115, 274–275.

Fig. 7.20 Illustration of coin casting[72]

mold making, pouring the molten metal, and sand removal and cleaning. The casting form is made within the rectangular sand frame (Fig. 7.20). The process for casting copper movable type is similar, but since the characters on the coin patterns are few and relatively simple, tin can be used. In contrast, the characters on copper movable type are numerous and varied, making it impractical to use a large amount of tin for the patterns; instead, wooden materials are used. In other words, the pattern for copper movable type is essentially a wooden version of the type.

In *Tiangong Kaiwu*, the description of coin casting technology lacks clarity on certain operational details, such as how the branched sprue is formed and how to remove the pattern from the sand frame after the casting cavity is created. These points need further explanation. Additionally, the formulation and composition of the casting materials are not mentioned and require further research. From the principles of casting, the casting material must be refractory, with fine and uniform granularity, plasticity, and permeability, while also being inexpensive and readily available. Studies of unearthed ancient molds indicate that they were primarily made from a mixture of clay, old mold soil, and fine sand, typically containing 21% to 33% clay and 67% to 74% sand, or 16% to 27% clay and 73% to 88% sand.[73] These materials can withstand temperatures up to 1580 °C and exhibit good permeability. The clay is composed of minerals with more than 30% of particles having a diameter smaller than 0.005 mm, with a plasticity index above 10. Among its components, 50% to 70% is silica (SiO_2), while 15% to 20% is alumina (Al_2O_3). The sand consists of mineral

[72] Song Yingxing, Pan Jixing. *Exploitation of the Works of Nature: Translation and Annotation* [M]. Shanghai: Shanghai Classics Publishing House. 1992: 112–115, 274–275.

[73] Hua Jue Ming, et al. *Collected Essays on the History of Metallurgy and Casting in China* [M]. Beijing: Cultural Relics Press, 1986: 262–263.

grains and rock fragments with diameters ranging from 0.05 mm to 2 mm, primarily including quartz, feldspar, and mica. Quartz sand contains about 95% quartz, mainly in the form of silica, while feldspar and mica contain silicates of potassium, sodium, calcium, magnesium, and aluminum. The sand must be crushed so that over 70% to 80% of its particles have diameters between 0.005 mm and 0.05 mm before it can be used.

7.3.2 Mold Manufacturing Techniques

Clay must be extracted from a depth of at least 1 m underground, then dried and finely ground into powder, which is sifted multiple times. The sifting process is illustrated in the upper part of Fig. 7.20. Sand is treated in a similar manner, but it needs to be processed more finely until it resembles a powder. The clay and sand are mixed in a specific ratio, combined with an appropriate amount of water, and then kneaded by foot. This mixture should be repeatedly worked until it reaches the right consistency. A small piece is taken out and tested by hand; it should feel soft and not gritty, allowing it to be easily shaped without cracking. If it does not meet these criteria, adjustments to the mixture should be made and the kneading repeated. Although the mold frame for copper movable type is also rectangular, it is typically made of iron. Two frames are placed side by side on a flat surface. The prepared clay and fine sand are placed into one of the frames, and once full, a wooden board is placed on top. A mallet is used to strike the board, compacting the materials. Next, wooden movable type is aligned according to a reference line, arranged in alternating orientations: the front of one pattern faces up, while the back of the next one faces up. Each row of patterns consists of an even number of characters, ensuring that the number of patterns in each row is equal and neatly arranged. Finally, the patterns are pressed into the sand to a depth equal to half their thickness.

To allow the molten metal to flow into the mold and cast the movable type, a pouring channel must be created within the mold. This channel consists of a gate, a main pouring channel, and branch channels, with the branch channels forming a "人" shape that connects to the cavity of the cast object. Excavated molds of Han dynasty *wuzhu* coins show that the pouring channels are tree-like (Fig. 7.21), and the pouring channels for metal movable type should follow a similar design. To create these branched pouring channels, an iron pattern must be carefully positioned between two rows of wooden type patterns. The height of the pouring channel pattern should match that of the type patterns, with half of it pressed into the sand. Alternatively, the pouring channel pattern can be placed in the sand first, followed by the insertion of the type patterns at the branch channel locations.

After the first sand mold is filled with sand and the pattern, it forms half of the casting mold. The second sand mold must then be placed over the first, filled with clay and fine sand, ensuring that the exposed type pattern and pouring channel pattern from the first mold are buried and compacted. The two combined sand molds are then flipped 180°, placing the first mold on top. Once removed, the type pattern and

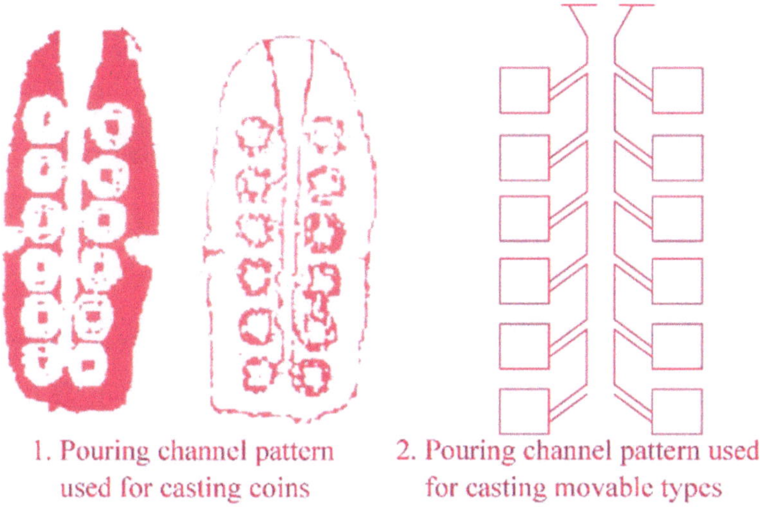

Fig. 7.21 Pouring channel pattern used for casting coins and movable types

pouring channel pattern will rest in the second mold (Fig. 7.22). They are extracted with tongs, cleaned, and air-dried for reuse. This process creates a hollow casting cavity in the two sand molds, featuring the negative relief characters. To facilitate the removal of the patterns from the sand, vegetable oil can be applied as a lubricant. If a small hole is desired in the type, a small iron rod can be placed in the type cavity before the two molds are closed to serve as a core. After casting the type, the iron rod can be removed. The two sand molds should be tightly bound with rope to form a complete mold set. Multiple mold sets should be prepared to allow for the simultaneous casting of many types. Each joint between the sand molds should include a pouring channel for casting purposes.

7.3.3 Casting Techniques and Storage of Movable Type

Before casting, the metal needs to be melted into liquid form in a crucible. For copper movable type, pure copper is avoided due to its high cost and melting point of 1083 °C; instead, copper alloys are typically used. Since the Spring and Autumn and Warring States periods (eighth to third centuries BC), a copper-tin–lead alloy has been the standard for casting copper coins in China, with variations in the alloy ratios over time. These alloys, containing tin and lead, help reduce both costs and melting points. For instance, an alloy of 80% copper and 20% tin has a melting point of 890 °C, while a mix of 70% copper and 30% tin melts at 755 °C. *Zhouli Kaogong Ji* ("Records on the Examination of Craftsmanship"), written during the Warring States period, documents six different copper alloy ratios and their applications; these alloys are collectively known as bronze. Chemical analyses of Northern Song (eleventh to

1. Place the clay and fine sand into the sand mold

2. Place the type pattern and pouring channel pattern in the mold, forming half of the casting mold

3. Place a wooden board on top and compact the material

4. Place the second sand mold over the first, fill with clay and fine sand, and compact to make a complete casting mold

Fig. 7.22 Diagram of metal movable type mold manufacturing

twelfth centuries) copper coins[74] show an average composition of 64% copper, 9% tin, and 23% lead, along with trace amounts of iron (2%) and zinc (1%). The alloy composition for copper movable type is generally similar. During the Southern Song period, due to a shortage of copper, the tin content in these alloys increased, ranging from 25% to 29% tin, 50% to 62% copper, and 8% to 25% lead. This resulted in what's known as tin movable type. While the alloy compositions varied over time, they were primarily ternary alloys.

As for crucible manufacturing, *Tiangong Kaiwu* provides a detailed description in the context of casting copper coins. The crucibles are made from a mixture of finely graded clay, crushed brick powder, and charcoal powder, typically in a 7:3 ratio of clay to charcoal. The clay acts as a refractory material, while the charcoal provides

[74] Xu Xiezhang, Huang Shaozhe. Quantitative Analysis of Coin Minting in China [J]. *Science*, 1921, 6(1): 1173.

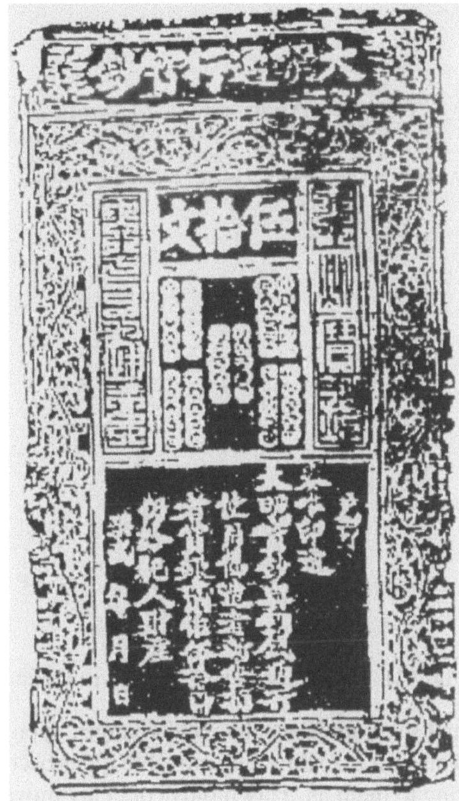

Fig. 7.23 Ming dynasty ceramic mold for the 50-*wen baochao* banknote featuring negative relief characters (Housed in the National Museum of China)

insulation. This mixture is combined with water and shaped into a flat-bottomed crucible. A typical crucible measures about 24.88 cm in height, 3.73 cm in diameter, and has a thickness of approximately 1.5 cm, allowing it to hold nearly 6 kg of material once fired. To prepare the alloy, copper, tin, and lead are weighed according to specific ratios. The process begins by placing copper blocks in the crucible and melting them in a furnace, using a piston bellows to provide airflow (Fig. 7.20). After the copper is melted, tin and lead are added in order of their melting points. At around 800 °C, the mixture becomes liquid, and the crucible is carefully lifted with iron tongs for pouring while still hot.

The reason why iron materials were used for the sand molds in casting is that, during large-scale production, the molds must be fired at 600 °C to withstand the rapid temperature changes from the high-temperature molten alloy. Additionally, they need to be preheated to about 200 °C before pouring. The Ming dynasty ceramic mold for the 50-*wen baochao* banknote, housed in the National Museum of China (Fig. 7.23), is made of clay and fine sand and features negative relief characters, designed specifically for casting. Similarly, all excavated ceramic molds for copper coins throughout history have also been fired. The materials used for movable type molds, such as clay

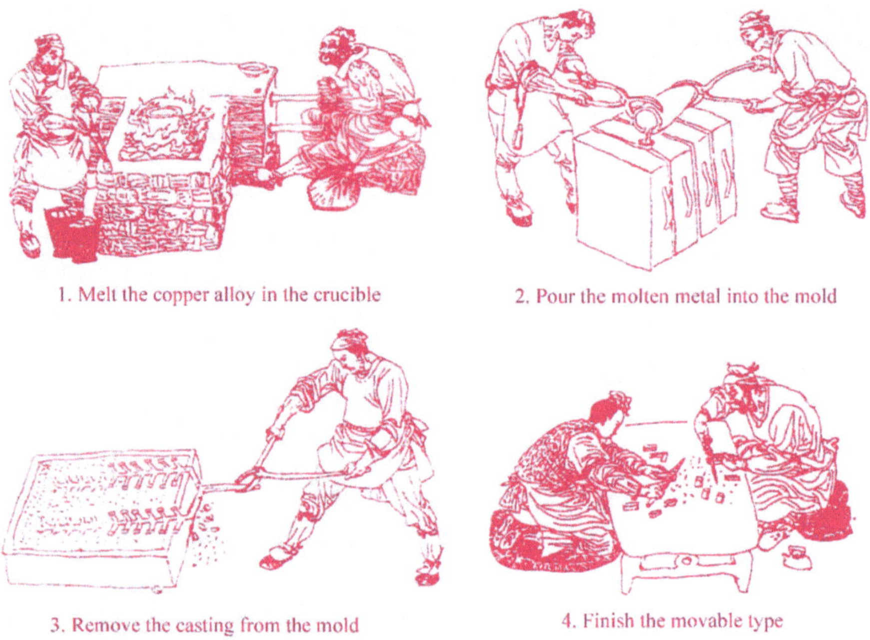

Fig. 7.24 Diagram of metal movable type manufacturing

and sand, undergo chemical changes upon firing, transforming into ceramic molds. This is akin to how clay movable types become ceramic types after firing; thus, referring to them as "clay molds" or "clay movable types" is technically inaccurate. The preheated molds are laid out on the ground with the pouring gates facing upward, tilted slightly for optimal flow. The pouring worker positions the crucible filled with molten alloy over the pouring gate, allowing the liquid to flow swiftly and smoothly into the main and branch channels, and then into the type cavity. Once one mold is filled, the next one is poured; as soon as one crucible is emptied, another follows. This process requires coordination among different pouring workers and must be executed in a tense yet orderly manner until an entire batch of molds is filled.

After pouring, the high-temperature molten metal cools in the mold, solidifying into copper movable type with raised characters. The casting is attached to the branch channels, which also fill with molten metal, forming a tree-like structure. Once the molds have cooled, they are broken apart to retrieve the branch-like castings. The movable types are then removed one by one, and any sand, burrs, or other debris clinging to their surfaces is cleaned off (Fig. 7.24). Uneven surfaces are smoothed using fine files, and a copper gauge is employed to check whether the dimensions of the movable type—length, width, and height—meet standard specifications. The copper gauge is a hollow rectangular tool designed to match the standard dimensions of the type. If any pieces are found to be oversized, excess material can be filed away; if they are undersized, they are set aside for remelting. Generally, movable type comes in two sizes: the larger type measures 1–1.3 cm in length and 0.7–1.3 cm in width,

Fig. 7.25 Diagram of sorting and typesetting metal movable type

while the smaller type ranges from 0.5–1 cm in length and 0.5–1.2 cm in width. The height of the type typically falls between 0.5–2 cm, with the letterforms taking on either a cubic or rectangular shape. The characters are either in a printed style or handwritten form, with the raised letters protruding 1.5–2 mm from the base. The long rectangular type features a small hole for threading with iron wire during typesetting, allowing it to be aligned within the printing plate. The dimensions, shape, and style of the metal movable type are determined by the original type pattern; whatever is crafted as a wooden type can be cast as a metal type.

The metal movable type is stored and organized by character rhyme or radicals, with the processes of typesetting, inking, and disassembly being largely the same as those for wooden movable type (Fig. 6.8). This has been discussed previously, so we will not elaborate further (Fig. 7.25). Next, we will provide a brief overview of the paper and ink utilized for printing. Because metal movable type is water-resistant, traditional water-based inks used with wooden type do not work well. To address this, it's necessary to reduce the water content in the ink and increase the amount of glue, which helps with ink adhesion. For example, adjusting the weight ratio of carbon black to animal glue to 100:30 or 100:40. Additionally, replacing water with plant oils—such as rapeseed oil, soybean oil, or sesame oil—yields even better results. For paper, it is best to use high-quality papers made from mulberry or paper mulberry. While bamboo paper is cheaper, it does not match the quality of bark paper and is mainly used for woodblock printing. Due to the high paper consumption during printing, high-quality bark paper is generally avoided. It's essential to have enough paper on hand to avoid using different paper for the same book. Experience suggests that allowing the paper and ink to sit for a while before use improves their performance.

Chapter 8
Spread of Chinese Printing Techniques Around the World

8.1 Spread of Chinese Printing Techniques in the Korean Peninsula and Japan

8.1.1 Beginning of Woodblock Printing on the Korean Peninsula

It has been proved that the plant fiber paper invented in China was the best writing material in ancient times, and printing with paper as the writing medium was the most advanced replication technology that replaced hand-copying. Therefore, when Chinese paper and printed materials spread to other countries, they were imitated, resulting in the external dissemination of papermaking and printing technology. Generally speaking, it is logical that the outbound spread of papermaking preceded that of printing, because paper had to be available before printing could be developed. So, when discussing the spread of printing, it is necessary to mention the spread of papermaking as well. China is located in the eastern part of Asia, and so, it's quite natural that the art of papermaking and printing first spread to the neighboring Korean Peninsula and Japan.

The Korean Peninsula and China are separated merely by a river. Since ancient times, there have been close economic and cultural exchanges between the ruling powers of the two regions, and they also shared the Chinese characters. In 206 BC, Liu Bang, Emperor Gaozu of the Han dynasty unified China and appointed Lu Wan as the King of Yan, whose territory bordered Korea. In 195 BC, Lu Wan rebelled against Liu Bang, causing turmoil in Yan. General Wiman led thousands of people to seek refuge in Korea, and the Korean King Jun allowed him and his followers to settle in the eastern part of the peninsula. The following year (194 BC), Wiman claimed the throne in place of King Jun, establishing the Wiman Joseon (194 BC–108 BC) in the former territory of the Gija Joseon in the north of the peninsula, with Wanggeom-seong (present-day Pyongyang) as its capital. The southern part of the peninsula was

P. Jixing, *History of Printing Techniques in China*,
https://doi.org/10.1007/978-981-96-6119-0_8

inhabited by three Korean tribal states: Mahan, Jinhan, and Byeonhan, collectively known as "Samhan".[1] Wiman Joseon prevented nearby tribes from contacting Han. In response, during the third year of Yuanfeng's reign (108 BC), Emperor Wu of the Han dynasty launched a joint land and naval expedition to annihilate Wiman Joseon and established four commanderies in the area: Lelang, Lintun, Xuantu, and Zhenfan, placing them under direct rule. In the fifth year of the reign of Han Shiyuan (82 BC), the four commanderies were merged into Lelang and Xuantu. During the establishment of commanderies on the peninsula (108–24 BC), a large number of Han officials, scholars, craftsmen, and farmers settled there, bringing with them Han culture, science and technology. In *Book of Han: Treatise on Geography*, it is recorded that Lelang commandery had 62,000 households with a population of 406,000 and it governed 25 counties, where most were Han people and Chinese was commonly spoken. Its administrative and cultural systems were similar to those in the interior China. Since the twentieth century, many cultural relics such as silk, copper, iron and lacquer ware have been unearthed from ancient tombs in the Lelang site, all of which came from the interior of China. At the same time, ancient paper from the Western Han dynasty has been unearthed in various northwestern regions, indicating that Lelang and other commanderies might have also used paper at that time.

During the transitional period between the Western Han and Eastern Han, China was in a turbulent state, too preoccupied to attend to other matters. During this period, three emerging states, Goguryeo (37 BC–668 AD), Silla (57 BC–935 AD), and Baekje (18 BC–660 AD), successively rose on the Korean Peninsula, ushering in the Three Kingdoms period (57 BC–668 AD). The former Xuantu commandery fell under Goguryeo's control, while Han's rule was limited to Lelang. At that time, the Governor of Liaodong, Gongsun Du, and his son Gongsun Kang seized the opportunity to establish their own power, taking control of Lelang commandery and its counties. To counter the rising power of Goguryeo, in the 10th year of the Jian'an era (AD 205), Gongsun Kang established Daifang commandery to the south of Lelang, which governed seven counties (in the territory of present-day Jinhai Province, South Korea) and accepted refugees from the Central Plains. After unifying northern China, the Wei dynasty destroyed the Gongsun family's separatist regime in Liaodong and controlled Lelang and Daifang. Following Wei came the new Jin dynasty, by which time, paper had been commonly used as a primary writing material in China. Additionally, many Han Chinese migrated to the Korean Peninsula, bringing with them the paper-making technique. At the end of the Western Jin dynasty (fourth century AD), Goguryeo took advantage of the situation and once again conquered Lelang and Daifang, unifying the northern Korean Peninsula. Therefore, in the fourth and fifth centuries hemp paper was already being produced in Goguryeo, and engaged in the production were craftsmen who had moved from Liaodong to Lelang. Baekje and Silla had frequent maritime interactions with eastern China, and there were tens

[1] Ban Gu. *Book of Han: Vol. 95, Biography of Korea: Vol. 278, Treatise on Geography* [M]. / / *Twenty-five Histories: Vol. 1*. Compact edition. Shanghai: Shanghai Classics Publishing House, 1986: 156, 358.

of thousands of Han Chinese living there. Although the time they started papermaking was later than in Goguryeo, it was not much later. The discovery of paper in a Silla ancient tomb in 1920 also indicates that by the mid-Three Kingdoms period, paper production had already begun on the Korean Peninsula.[2]

From the sixth to the ninth century in the Sui and Tang dynasties, the interaction between the three kingdoms on the Korean Peninsula and China became more frequent. During this time, the bark paper technology emerging in China facilitated the further development of mulberry bark paper production on the peninsula. However, in the sixth to seventh centuries, rivalry among the three kingdoms intensified, and Silla, located in the southeast, was squeezed by both Goguryeo and Baekje, so it sought help from the Tang Empire. Tang's mediation among the three kingdoms was unsuccessful, and then, Emperor Gaozong sent army and naval forces to aid Silla. In 660, Tang and Silla destroyed Baekje, followed by the defeat of Goguryeo in 668. Tang established the Andong Protectorate within Goguryeo's former territory and stationed in Pyongyang, while the former territory of Goguryeo was administered jointly by Tang-appointed officials and the local people. This marked the end of the Three Kingdoms period. The pro-Tang Silla unified the peninsula to enter into a new era, the Unified Silla dynasty (668–935), with its capital in Gyeongju (present-day Gyeongsangbuk-do, South Korea). Silla absorbed the Tang culture comprehensively, and sent a large number of students and Buddhist monks to study in Tang. The rulers of Silla advocated Confucianism and Chinese literature. In 682, they established a national academy, inviting scholars proficient in Chinese Five Classics to teach students. In 788, they implemented the imperial examination system and produced many literati. They also introduced different Buddhist sects from Tang, including Vinaya, Huayan, Faxiang, Pure Land, Tiantai, Chan, and Esoteric Buddhism. This period saw the construction of temples and pagodas, and a large number of Buddhist scriptures were brought in from the Tang dynasty, both manuscripts and printed editions, which were widely copied and disseminated.

The earliest surviving dated Buddhist scripture from Silla is the manuscript of *Avatamsaks Sutra* ("Great Flower Garland Sutra") dated to the 14th year of the Tang Tianbao era (755), now housed in the Hoam Museum of Art in Seoul, Korea. The inscription in clerical script at the end of the volume states that the sutra was transcribed with reverence on white paper made from the bark of mulberry trees which were planted with fragrant water. This is the earliest surviving record of making mulberry paper in Sill. The practices described in the inscription, like planting mulberry trees with fragrant water, making paper, and transcribing *Avatamsaks Sutra*, are consistent with those recorded in the third scroll of *Biography of the Avatamsaka Sutra* (ca. 702) by Fa-tsang (643–712), founder of Huayan Buddhism theory in early Tang dynasty. There are not many surviving Silla manuscripts available today, and almost all of them are Buddhist scriptures, with few non-religious works. Due to the impact of wars over the centuries, many texts were regrettably lost. It is worth noting that the entire Silla dynasty did not leave a single documentary record of printing,

[2] 関義城 (コレクション). *A Study on the History of Hand-made Paper* [M]. Tokyo: Mokujisha, 1976: 2.

but there were many records of manuscripts. To date, no printed materials with clear indication of being engraved in Silla or relevant physical materials have been found. Some printed materials once thought to be produced in Silla have been proved by research to be either products of Goryeo Kingdom (935–1392) or brought in from the Tang dynasty.

The ancient Korean chronicles (1392–1910) once recorded that the printing blocks for the 80,000-volume *Tripitaka Koreana* stored at Haeinsa Temple in Sacheon, Gyeongsangnam-do province, were carved in the year of Ding-Chou during the reign of Silla King Aejang (800–808 CE). However, further research revealed that they were all engraved during the Goryeo dynasty, as many of the printing blocks have inscriptions stating "Engraved by imperial decree by Director of Goryeo *Tripitaka* in so-and-so year"; also, there was no Ding-Chou Year during the reign of King Aejang. Therefore, over 100 years ago, Korean scholars already pointed out that the claim about the creation of the printing blocks was incorrect.[3] In 1966, a printed Buddhist sutra, *Mugujeonggwang Daedaranigyeong* ("Great Dharani Sutra of Immaculate and Pure Light"), was discovered inside the Seokga Pagoda at Bukguk Temple in Gyeongju, South Korea. Then, some claimed that it was the world's earliest existing printed material produced in Silla between 706 and 751,[4] and they further proposed that woodblock printing was invented in Korea.[5] However, after further research, scholars from the United States, Japan, and China have proved that the sutra was actually a Tang dynasty woodblock print brought into Silla.[6] Thus, the claim that the sutra was a woodblock print produced in Silla lacks solid evidence. It seems that since paper manuscripts and printed books brought in from the Tang dynasty could basically meet the country's needs, Silla found no need to develop printing technology. If any more claims appear about printing activities in Silla, we would like to see new evidence.

Physical evidence as recorded in literature shows that woodblock printing on the Korean Peninsula was introduced from the Northern Song dynasty during the early Goryeo period, starting in the early eleventh century and developing into officially organized large scale printing by the mid-eleventh century. The direct driving force for developing printing technology in Goryeo was the project of printing the massive 5049-volume Buddhist canon *Kaibao Tripitaka*, completed in the 8th year of the Taiping Xingguo era of the Northern Song dynasty (983). Professor Luther Carrington Goodrich from the United States said that the transmission of Chinese printing technology to Japan and the Korean Peninsula was facilitated through the

[3] Yi Gyu-gyeong. *Oju yŏnmun changjŏn san'go* (Scattered Manuscripts of Glosses and Comments by Oju): *Vol. 24*, Seoul: Myeongmundang, 1982: 685.

[4] Jin Mengshu. *The Discovery of the World's Earliest Woodblock Prints* [N]. Chosun Ilbo, 1966–10–16 (07).

[5] Sun Baoji. *History of Printing in Korea* [M]. Seoul: Institute of Ethnic Culture, Korea University, 1981: 974–975.

[6] Pan Jixing. Pure Light Dharani Sutra: A Focus of Sino-Korean Academic Debate [J]. *Publishing Journal*, 2000 (4): 33–41.

medium of Buddhism,[7] which is justified. Rulers of Goryeo were devout followers of Buddhism, and the completion of *Kaibao Tripitaka* coincided with the beginning of Goryeo King Seongjong's reign (982–997). Taking note of the incomplete collection of Buddhist scriptures in his country after the fall of the Silla dynasty, King Seongjoing dispatched Han Rin-kyung and other envoys to Song, and he also "sent monk *Ruke* with a formal request for the *Tripitaka*, which was granted along with a purple robe, and then he was allowed to take them back to his own country."[8] During the second year of the Chunhua era (991) of Song Emperor Taizong, King Seongjong once again dispatched Han Yangong (940–1004), the Minister of War, to Song, "expressing [king's] desire to print Buddhist scriptures. The Song court granted the *Tripitaka* along with the imperial edition of *Esoteric Xiaoyao Ode*, *Lotus Sutra* and *Wheel of Dharma*." Goryeo historical literature also records, "Han Yangong, the Minister of War and Imperial Censor, asked for the *Tripitaka*, and the emperor bestowed 481 cases of *Tripitaka*,[9] totaling 2500 volumes, and additionally granted the imperial edition of *Esoteric Xiaoyao Ode*, *Lotus Sutra* and *Wheel of Dharma*." Evidently, the Northern Song court presented two complete sets of *Tripitaka* to Goryeo in 989 and 991 respectively, and Goryeo's intention was to make its own copies. When the *Tripitaka* arrived in 991, King Seongjong welcomed it into his inner palace and "sent a Hanlin scholar named Baek Siori to Song to express his gratitude for the scriptures and imperial prints."

In the fourth year of Chunhua era (993), Emperor Taizong of the Song dynasty sent Chen Jing, the *Mishucheng* in charge of book publishing and Chief Archivist, and another *Mishucheng* Liu Shi to Goryeo, possibly bringing craftsmen to teach printing technology. They were rewarded by King Seongjong and returned with letters of gratitude. In order to cultivate local talents, King Seongjong also sent Wang Bin, Cui Han, and others to study at *Guozijian* (Imperial Academy) and provincial schools in the Song dynasty. In the third year of Chunhua era (992), Emperor Taizong of Song personally examined the candidates of all subjects and granted the overseas students from Goryeo titles of *Jinshi* (Metropolitan Graduate) and *Mishulang* (Secretariat Clerk) of the Imperial Secretariat before allowing them to return to their homeland.[10] Being awarded titles such as *Mishulang* and *Jiaoshulang* (Editing Clerk) indicated that they had received professional training and could serve as technical officials to work in book publishing institutions. They became the first backbone for Goryeo's printing industry. There were up to 40 Goryeo overseas students who studied not only printing but also astronomy, calendar calculation, Confucian classics, and history. After they returned, King Seongjong sent envoys to thank Emperor Taizong and also submitted a request for the Song *Guozijian* edition of "*Nine Classics*" to promote

[7] Goodrich L C. Printing: Preliminary Report of a New Discovery [J]. *Technology and Culture*, 1967, 8 (3): 376–378.

[8] *History of Song: Vol. 487, Biography of Korea* [M]. // *Twenty-five Histories: Vol. 8*, Compact edition. Shanghai: Shanghai Classics Publishing House, 1986: 6761–6762.

[9] *History of Korea: Vol. 93, Biography of Han Yangong* [M]. Pyongyang: Korean Academy of Sciences, 1958: 71–72.

[10] *History of Song: Vol. 487, Biography of Korea* [M]. // *Twenty-five Histories: Vol. 8*. Compact edition. Shanghai: Shanghai Classics Publishing House, 1986: 6762.

Confucian studies, and Emperor Taizong granted the request. This is a concrete demonstration of the Song Government's goodwill towards Goryeo. From 990 to 993, Goryeo imported from Song woodblock printing technology and the best printed editions of Buddhist and Confucian classics as blueprints for reproduction. By the time of King Seongjong, Goryeo had already had the technical capability of carving and printing books, but they lacked stable social conditions. At that time, the Liao dynasty (916–1125), founded by the Khitans in northern China, frequently ventured south to harass Song and even repeatedly invaded Goryeo, causing social unrest so that King Seongjong's desire to print Buddhist scriptures fell short. In 994, under the pressure from the Liao dynasty, Goryeo suspended its communication with the Song dynasty, and King Seongjong died of depression in 997.

Mokjong (998–1009) succeeded the throne, but indulged in alcohol and hunting, he paid little attention to state affairs. His reign was short-lived as he was assassinated by the treacherous court official Gang Jo, so the late king's wish to print the scriptures was not fulfilled. By this time, however, civilian printing activities had already begun, thus it is appropriate to set the origin of printing on the Korean Peninsula in the late tenth century. The earliest surviving printed text on the Korean Peninsula is the one-volume *Baoxie Yin Tuoluni Satra* (Dharani Sutra) (Fig. 8.1), its full title being *The Secret Whole-body Relic of All Tathagatas Stored in the Treasure Chest Dharani Sutra,* printed by the head temple in 1007. It was translated from Sanskrit by the Tang dynasty monk Bukong (Amoghavajra), and later printed in Hangzhou by Qian Chu (929–988), the King of Wu-Yue during the Five Dynasties in the second year of Xiande era (955) of the Later Zhou dynasty. This should be the prototype of the head temple's edition. The publication inscription states, "The abbot of the head temple in Goryeo, Master Shi Hongche, respectfully engraved the printing blocks for *Baoxie Ying Satra,* printed and consecrated in the Puan Pagoda for offerings. Recorded in the 25th year of Tonghe, the year of Ding-Wei." The scripture consists of five sheets of paper sewn together, with each sheet measuring 7.8 cm in height and 240 cm in width. The frame size is 5.4 cm high and 10 cm wide, with 9–10 characters per line. At the beginning of the scroll is a Buddha transformation image followed by the text. The twenty-fifth year of Tonghe followed the era name of Liao Shengzong, equivalent to the fourth year of Jingde in the Northern Song dynasty and the tenth year of King Mokjong in Goryeo. However, from the following year (1008), Goryeo resumed using the Northern Song's era name. The characters in this edition vary in size, the lines of the transformation image are blurry, and the carving technique is not as mature as in later editions, showing an antique and simple feature. It was initially collected by Kim Wan-seok and currently it is housed in Tokyo National Museum in Japan.

After the assassination of King Mokjong of Goryeo, King Hyeonjong (1010–1031) ascended the throne. Around this time, the treacherous court official Kang Zhao killed 95 people from the Liao's Jurchen tribe and Liao emissaries, bringing great disaster upon Goryeo. Taking this as a pretext, Liao Emperor Shengzong led the Khitan forces to invade on a large scale in 1010, killing Kang Zhao and capturing the capital city (present-day Kaesong). King Hyeonjong fled to the south for refuge and made a vow with his courtiers that if the Khitan forces retreated, they would

Fig. 8.1 *Baoxie Ying Sutra* (Dharani Sutra) printed on the Korean Peninsula in 1007[11]

swear to carve the *Tripitaka.* Yi Kyubo (1168–1241), a Goryeo *hanlin,* later wrote of this event:

> Tracing back to its origin, in the second year of King Hyeonjong's reign (1011), the Khitan forces launched a large-scale attack on Goryeo. Hyeonjong fled southward to take refuge and the Khitan forces settled in Songgyeong (Kaesong) without retreating. Thus, by divination his courtiers made a big vow. They swore to engrave the *Tripitaka,* and therewith the Khitan forces withdrew on their own.[12]

Coincidentally, after devastating Goryeo, Liao Emperor Shengzong returned to his court in January 1011. In February, King Hyeonjong of Goryeo returned to the capital city and ordered to start carving the *Tripitaka* immediately, indicating that printing of the official version began in 1011. The earliest official version of Goryeo's *Tripitaka* was largely printed during the reign of King Hyeonjong and completed in the fourth year (1087) of King Seonjong's reign (1083–1094),[13] totaling about 6000 volumes. During this process, the Chinese version of the Liao *Khitan Tripitaka* was completed around 1068 and it was also gifted to Goryeo. Therefore, the Goryeo *Tripitaka,* based on the Northern Song dynasty's *Kaibao Tripitaka* and supplemented by the *Khitan Tripitaka,* was completed after 76 years of effort. The boards are stored in the Furen Temple in Daegu, Gyeongsangbuk-do, and copies are available today (Fig. 8.2).

Wang Xu (1057–1101), the younger brother of King Seonjong of Goryeo, initiated the publication of another set of scriptures, historically known as *Supplemented Tripitaka.* Wang Xu, the fourth son of Emperor Munjong, was known by the courtesy name Xitian. He became a monk at the age of 11, and he was intelligent, studious, well-versed in various Buddhist sects and also knowledgeable about Confucianism. In the second year of King Seonjong's reign and the eighth year of Emperor Shenzong of Song's Yuanfeng era (1085), he traveled to Song China with his disciple, Shou Jie, on a ship owned by the Song merchant Lin Ning. When they arrived in Bianjing (now Kaifeng, Henan Province), they were introduced to the Chuigong Place to meet the newly enthroned Emperor Zhezong of Song. They were received

---

[11] Chon Hey-bong (千惠鳳). *Han'guk sojihak* [M]. Seoul: Min'eongsa, 1997.

[12] Yi Kyubo. *Complete Works of Minister Yi of the Eastern Country: Vol. 25, Great Treasury Engraved Prayers of Emperors and Subjects* [M]. Seoul: Choson Ancient Book Publishing Association, 1913.

[13] *History of Goryeo: Vol. 10, The Xuanzong Family* [M]. Pyongyang: Korean Academy of Sciences, 1957: 145.

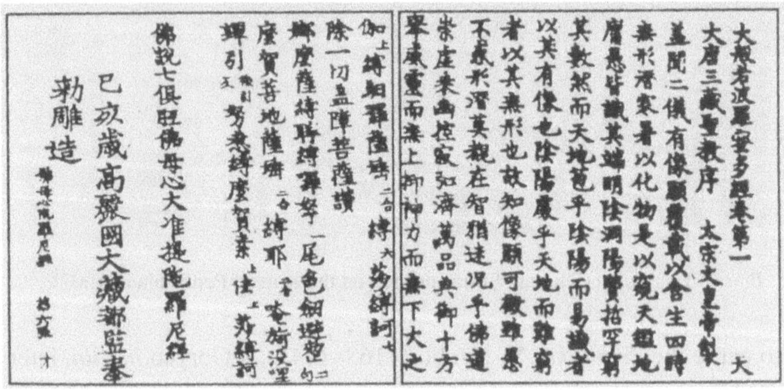

大般若波羅蜜多經卷第一 大
大唐三藏聖教序 太宗文皇帝製
蓋聞二儀有像顯覆載以含生四時
無形潛寒暑以化物是以窺天鑑地
庸愚皆識其端明陰洞陽賢哲罕窮
其數然而天地苞乎陰陽而易識者
以其有像也陰陽處乎天地而難窮
者以其無形也故知像顯可徵雖愚
不惑形潛莫覩在智猶迷況乎佛道
崇虛乘幽控寂弘濟萬品典御十方
舉威靈而無上抑神力而無下大之

除一切蓋障菩薩讃
佛說七俱胝佛母心大准提陀羅尼經
己亥歲高麗國大藏都監奉
勅雕造

Fig. 8.2 Goryeo version of *Tripitaka*

with preferential treatment and were granted permission to travel around seeking Buddhist teachings in different places. In the first year of Emperor Zhezong of Song's Yuanyou era (1086), Wang Xu returned to Goryeo and presented a thousand volumes of Buddhist scriptures from Song to King Seonjong. He also "requested the establishment of *Jiaozangdoujian* (ancient official title) at Xingwang Temple, and purchased books from Liao and Song, amounting to four thousand volumes, all of which were published".[14] He passed away in the sixth year of King Sukjong's reign (1101), and was posthumously awarded the title of *Guksa* (National Teacher). The *Supplemented Tripitaka*, which was reprinted under his supervision, was likely published between 1086 and 1100, with inscriptions stating "Engraved by the Da Xingwang Temple of Goryeo" and "Edited and corrected by Haidong missionary samana Yi Tian."

While printing Buddhist scriptures, Goryeo also printed Confucian classics (Fig. 8.3), historical and literary works, and works on science and technology. During the same period, many books were also shipped from China to Goryeo, and even book plates were trafficked from Hangzhou to Goryeo. Therefore, from King Jeongjong's reign (1035–1046) at the latest, the carving of non-religious works began. In the eighth year of Jeongjong's reign and the second year of Song Qingli era (1042), "Cui Hao, the deputy governor of Dongjing (Gyeongju), and Judge Luo Zhi said … newly revised editions of *Book of the Two Han Dynasties* and *Book of Tang* were produced by imperial order, and noble titles granted.[15] Three years later (1045), the Imperial Library published new editions of *Li Ji Zheng Yi* ("Orthodox Commentary on the Classic of Rites") and *Mao Shi Zheng Yi* ("Rectified Interpretation of the Book of Songs")." In the tenth year of King Munjong's reign (1056), the Governor of Xijing (now Pyongyang) requested that the newly printed *Nine Classics*, *Book of Han*, *Book of Jin*, *Book of Tang*, *Analects*, *Classic of Filial Piety*, works of history, collected

---

[14] *History of Goryeo: Vol. 97, Biography of National Teachers* [M]. Pyongyang: Korean Academy of Sciences, 1958: 34–35.

[15] *History of Goryeo: Vol. 6, The Jeongjong Family* [M]. Pyongyang: Korean Academy of Sciences, 1959: 89.

writings of various schools, medical and divination texts, works of geography, legal and mathematical books should be placed in various academies for scholars to study. In the twelfth year of King Munjong's reign (1058), Chungju Governor brought in a total of 95 volumes of newly carved books, including *Yellow Emperor's Classic on Eighty-one Difficult Issues*, *Chuanyu Collection*, *Shanghan Lun* ("Treatise on Cold Pathogenic Disease"), *Bencao Kuoyao* ("The Essentials of Materia Medica"), *Cao Shi Jing's Source on Children's Diseases*, *Eighteen Discussions on the Prescription and Diseases of Children's Medicine*, and *Zhang Zhongjing's Five Organs Theory.* These books were in the imperial palace depository by imperial edict. Thereafter, the work of printing books in various regions developed further and continued until the Joseon dynasty.

Goryeo editions have the same format as Song editions, using the Song editions as the base texts, and they were carefully proofread by Confucian officials, exquisitely engraved, and mostly printed on mulberry paper with a generally large font size. They were predominantly official editions. The printing locations were concentrated in Gaegyeong (Gaeseong), the eastern and western capitals, as well as Chungju, Haeju, and Nanyuan Prefecture. In addition to Chinese books, some works by Goryeo authors were also published, such as *History of the Three Kingdoms* and *Emperor's Rhyme Chronicle.* Although the number of each Goryeo edition was small, only dozens of copies generally, there were numerous varieties, and among them there were many good books, some of which were already out of print in China so that in the

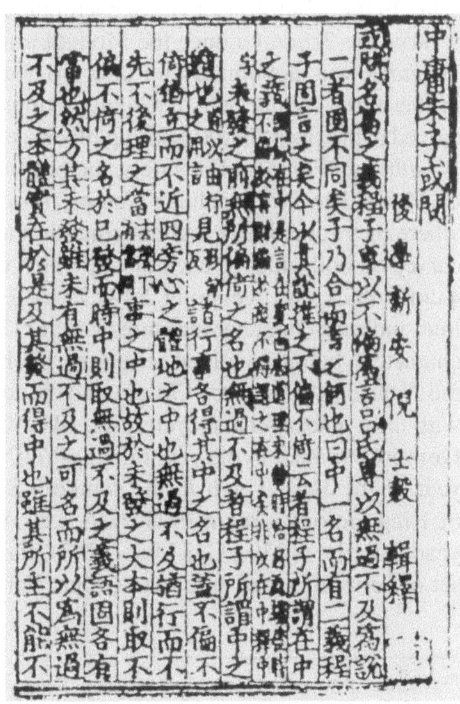
中庸朱子或問
後學新安倪士毅輯釋

Fig. 8.3 *Zhongyong Zhuzi Huo Wen* ("Answers to Questions on Zhongyong by Zhu Xi") published by Jin provincial governor in the 20th year of King Gongmin of Goryeo (1371) (Housed in Central Library of Korea University)

sixth year of Yuan You era (1091), Emperor Zhezong of Song ordered the acquisition of Goryeo edition books such as *Er-Ya Tu Zan*, *Goryeo Records*, *Zhou Chu's Local Records*, *Wei Lue* ("A Brief History of Wei"), *Shui Jing* ("Water Classic"), and so on.[16] The agency responsible for publishing books in Goryeo was the Imperial Library, which had officials such as supervisors, assistant officers, proofreaders, and scribes, similar to the practice in China, with an affiliated Book Division for storing book plates. The Imperial Library printed files of tabooed words or phrases and distributed them to various places for implementation. Important books such as *Cefu Yuangui* ("Prime Tortoise of the Record Bureau"), *Zizhi Tongjian* ("Comprehensive Mirror to Aid in Government"), and *Taiping Yulan* ("Taiping Imperial Encyclopedia"), were proofread by Confucian officials appointed by the king. Because Goryeo's monarchs put a premium on publishing, versions of books such as history and classics were abundant, sufficient to meet domestic demand, making Goryeo another publishing power in the East besides China.

In the 18th year of King Gojong (1231) of Goryeo, the Mongol ruler Ogedei Khan invaded Goryeo with a large army on the grounds that his envoy had been killed there. They captured over 40 cities, and the Goryeo *Tripitaka* and printing blocks stored at Bongnyeon Temple and Hyeongwang Temple were destroyed in the war. In 1232, King Gojong took refuge on Ganghwa Island, where he established a disaster-relief ritual space and ordered government officials to pray to Buddha for protection against the invading Mongol army. It was not until 1235 that the Mongol troops finally withdrew after continuing their attacks and making Goryeo a vassal state. In the 22nd year of Gojong's reign (1235), the king and his officials vowed to recarve the *Tripitaka* once the Mongol army retreated. Thus, the Director of Great *Tripitaka* was appointed and after 16 years of work, the project was completed in the thirty-eighth year of Gojong's reign (1251). A total of 6796 volumes were carved using over 80,000 printing blocks, earning it the name *The Eighty-Thousand-Woodblock Tripitaka*. For safety, the printing blocks were stored in Haeinsa Temple on Mount Kaya in Sanchuan County (now in South Gyeongsang Province, Korea), where they remain in excellent condition to this day. To conserve wood, texts were carved on both sides of each block, which was 24 cm in height, 65 cm in length, and 4 cm in thickness, weighing 2.4–3.75 kg. Each block has 23 lines, 14 characters per line, and each character measures 1.5 cm^2. The entire *Tripitaka* consists of 26 million characters, bound as folded scriptures. Each volume is numbered by a character from the *Thousand-Character Classic*, and the colophon does not include the year number, with only a sexagenary cycle year, like "Carved on Emperor order by Director of **Goryeo** *Tripitaka* in the year of Ding-You," where Ding-You corresponds to the first year of the Jiashi reign in the southern Song dynasty and the twenty-fourth year of King Gojong's reign in Goryeo, i.e., AD 1237. Similar colophons include the year of Ji-Hai (1239), Gui-Mao (1243), Jia-Chen (1244), and Wu-Shen (1248). This set of scriptures also spread to China. For instance, a volume of *Mahayana Sanju*

[16] *History of Goryeo: Vol. 7, The Xuanzong Family* [M]. Pyongyang: Korean Academy of Sciences, 1959: 150.

Confession Sutra stored in Beijing Library is inscribed with "Carved on Emperor order by Director of Goryeo *Tripitaka* in the year of Ren-Yin (1242)".

8.1.2 Early Development of Moveable Type Printing on the Korean Peninsula

After more than 200 years of developing woodblock printing during the Goryeo dynasty, moveable type printing technique was introduced to Korea from China at the end of the Goryeo dynasty (the end of the fourteenth century) when China was transitioning from the Yuan dynasty to the Ming dynasty. Korean scholar Yi Gyu-gyeong wrote about this:

> In the middle of the Qingli period (1041-1048) (Song dynasty), Bi Sheng, a commoner, invented moveable type...[in the East] our movable type also started from the Goryeo dynasty, and spread into the Joseon dynasty. In the third year of King Taejong's reign, the year of Gui-Wei (1403), he ordered the establishment of a type foundry to cast moveable type, using copper from the royal treasury. Note: Jin Zhi's *Annotations on the Great Ming Code* was printed with the wooden moveable type made by Xu Zan, the governor of Baizhou in the year Ji-Hai (1396) of Hongwu era, four years after the founding of our Joseon dynasty in 1392, indicating that moveable type had already existed in the Goryeo dynasty and spread into our country.[17]

Yi Gyu-gyeong was born into an academic family, well-versed in both Chinese and Korean academic knowledge and institutional systems. His belief that moveable type printing on the Korean Peninsula began in the late Goryeo dynasty and originated technically from the Song dynasty is historically accurate. To substantiate his view, some additional arguments are necessary. As discussed in Chaps. 6 and 7, woodblock printing in China was already in its golden age during the Northern Song dynasty, and on this foundation, non-metallic moveable type (wooden and clay) printing and metal moveable type printing were developed. The Northern Song dynasty had friendly relations with Goryeo, and during the reign of Emperor Taizong of Song, interactions between the two sides were frequent, allowing the woodblock printing technology to be first introduced to Goryeo.

The reason why movable type technique did not reach Goryeo in the early period of Goryeo dynasty is that Goryeo was threatened by its northern neighbor Liao and suspended its interactions with the Northern Song dynasty, and the Northern Song dynasty was also harassed by the Jin dynasty and eventually perished. The rulers of the Song dynasty then established a rump state known as the Southern Song dynasty, but as they continued to suffer successive defeats in the war against Jin, they were too preoccupied to maintain contact with Goryeo. The rising Mongols first destroyed Jin with the Southern Song and then destroyed the Southern Song, and established their rule over the entire country. In the Yuan dynasty, woodblock and moveable type

---

[17] Yi Gyu-gyeong. *Oju yŏnmun changjŏn san'go* (Scattered Manuscripts of Glosses and Comments by Oju): *Vol. 24*. Seoul: Myeongmundang, 1982: 686.

printing, based on the achievements in the Song dynasty, continued to develop and dominated Goryeo. Thus moveable type printing technology was introduced into the Korean Peninsula from China at the end of the Goryeo dynasty. *Mengxi Bitan* ("Dream Pool Essays"), first published in 1166 during the Southern Song dynasty, was unlikely to have been introduced to Goryeo at that time, but it was republished in the ninth year of the Dade era (1305) in the Yuan dynasty and would certainly have flowed into Goryeo. Korean scholars generally acknowledge that the Goryeo people grasped the concept of moveable type printing from this book. However, they interpreted Shen Kuo's description in *Mengxi Bitan* of Bi Sheng "carving characters on adhesive clay, as thin as the edge of a coin" as meaning that the height of the clay type was the same as the edge of a coin (2 mm), and they fixed the types onto an iron printing plate with wax-based adhesive. Such thin clay types not only lacked durability but also were difficult to detach from the plate, so the clay type printing failed to develop in Goryeo. Actually, the phrase "thin as the edge of a coin" mentioned in *Mengxi Bitan* refers to the depth of carving or the height of the character protruding from the concave surface, rather than the height of the type itself. The height of the moveable type should be 1.0 cm or higher to be practical, as used by Bi Sheng. Although Goryeo did not succeed in developing clay movable type printing, they did grasp the idea of moveable type printing and the technique of type setting.

Wooden movable type and metal movable type emerged almost simultaneously in Korea at the end of the Goryeo dynasty. Early records of metal movable type on the Korean Peninsula were found in a memorial submitted by the Korean *jinshi* scholar and official Jeong Dojeon (1335–1395) to King Gongyang (1389–1392) in 1391. The memorial proposed casting characters at the bookhouse for printing books on various subjects, including classics, historical works, religious scriptures, medical texts, and military codes, so that those who aspire to study have books to read. King Gongyang approved the proposal and "in the following year (1392), a printing bureau was established with Director responsible for typefounding and printing books."[18] However, the Goryeo dynasty collapsed in that year and the last King Gongyang also died, so the printing bureau did not operate after its establishment. Jeong Dojeon's proposal only confirms the existence of movable type printing activities in Korea during the latter half of the fourteenth century. The oldest surviving book printed from metal movable type on the peninsula is *Essential Passages Pointing Directly to the Mind*, printed at Xingde Temple in Cheongju in the seventh year of Xuanguang (1377). The book contains two volumes, but only the second volume has survived and is currently housed at the National Library of France[19] (Fig. 8.4). It was authored by **Paegun** (1298–1374), also known as Monk Baegun, who explained the teachings of the Buddha using stories from various Buddhist scriptures. The "seventh year of Xuanguang" is the era name of the Northern Yuan dynasty, by which time Yuan had been overthrown by Ming. It corresponds to the tenth year under the reign of Emperor

[18] *History of Goryeo*: *Vol. 79, Records of Hundred Officials* [M]. Pyongyang: Korean Academy of Sciences, 1958: 573.

[19] Courant M. Supplément a la Bibliographie Coréence: No. 3738 [J]. *Paris: Imprimerie Nationale*, 1901 (1): 70–72.

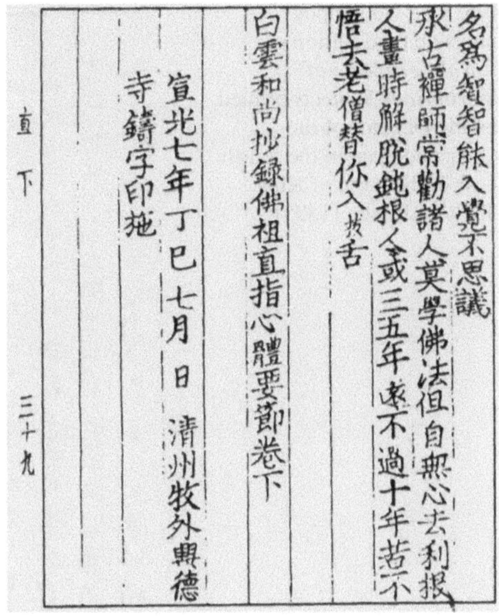
名為智智能入覺不思議
求古禪師常勸諸人莫學佛法但自無心去利根
人盡時解脫鈍根人或三五年遠不過十年若不
悟去老僧替你入拔舌

白雲和尚抄錄佛祖直指心體要節卷下

宣光七年丁巳七月 日 清州牧外興德
寺鑄字印施

直 下 三十九

Fig. 8.4 Copper movable type edition of *Essential Passages Pointing Directly to the Mind*, published in 1377 during Goryeo dynasty

Hongwu of the Ming dynasty and the third year of Goryeo U'wang. However, from the following year, Goryeo started using the Ming's era names.

As with the woodblock printing, the movable type printing in Goryeo began among the common people. The new dynasty established by General Yi Seonggye who overthrew Goryeo is the Li dynasty, with its title changed to Joseon. At the beginning of the Joseon dynasty, the system of the former imperial printing bureau was restored, but casting and printing did not start immediately. The initial official publications were printed from woodblock movable type. The Joseon dynasty and the Ming dynasty were both emerging dynasties, and they had close relationship with each other, sharing the same era name. In the twenty-eighth year of the reign of Hongwu Emperor of Ming dynasty and the fourth year of the reign of King Taejo of Joseon (1395), the Director of the Gongsindogam (government office of records for meritorious subjects) printed *Royal Certificate of Meritorious Subjects*, a woodblock movable type edition (see Fig. 8.5). There is a copy stored at the Hansung City Ancestral Hall Rare Books Museum. The typeface in this edition is uneven and the characters skewed in arrangement with uneven ink color, indicating that the early wooden movable type printing technique on the Korean Peninsula was still immature. The postscript by Jin Zhi in the 1456 edition of *Da Ming Lu Zhi Jie* ("Literal Explanation of the Great Ming Code") states, "printed by the printing bureau from the carved types made by Xu Zan, the magistrate of Baizhou. The first hundred copies were printed and distributed for trial, in hopes of not betraying the Emperor's benevolence. Carefully inscribed by Jin Zhi of Shang Youzhai at the beginning of Hongwu, the year of Ji-Hai (1395)." It is evident that the 1395 edition of *Da Ming Lu Zhi Jie* published by the printing bureau was also printed from wooden movable type.

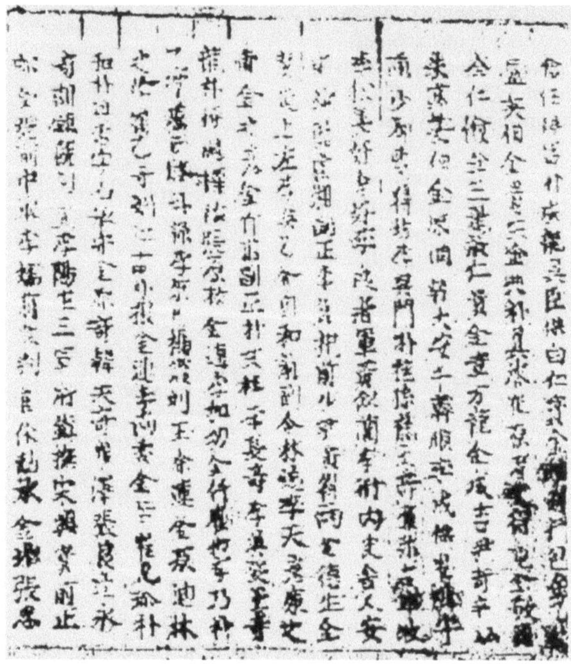

Fig. 8.5 Woodblock movable type edition of *Royal Certificate of Meritorious Subjects* printed by the Director of the Gongsindogam in the fourth year of the reign of King Taejo of Joseon (1395)

During the late Goryeo period, metal movable type printing in Goryeo was only sporadically used. Large-scale typecasting and printing began during the reign of the second ruler of the Joseon dynasty, King Taejong (1401–1417), in the early fifteenth century. In the third year of King Taejong's reign and the first year of the Yongle reign in the Ming dynasty (1403), a "Typecasting Foundry" was established in the capital (present-day Seoul) with funding from the royal treasury and government officials to cast copper movable type and conduct printing activities. In *Annals of the Joseon Dynasty: Annals of King Taejong* (1431), the entry in Volume 5 for the third year of Taizong's reign and the second month of Gengshen records that on March 4, 1403, the king ordered "the establishment of a new typecasting foundry. The king was concerned that the scarcity of books and scholarly materials in the country would limit the scholars' ability to broaden their knowledge. Therefore, he commanded the creation of this foundry. Yi Saek, the Director of the Royal Hall of Art and Literature, Min Tianji in charge of general affairs and Lee Yeong, an expert in ancient scripts, were appointed to oversee the project. They used copper and iron from the royal treasury and also requested donations from high and low-ranking officials to support the operation." At that time, the Minister of Rites and senior official of Baowen Hall, Kwon Geun (1352–1409), wrote a postscript for the typecasting work, which was later published at the end of the movable type edition of *Shi Yi Jia Zhu Sunzi* ("Eleven Commentaries to the Art of War by Sunzi") in 1409 and was also included in Volume 22 of Kwon Geun's *Yangcun Anthology*. The postscript also notes that the typecasting foundry was built in the spring of 1403 and used the Chinese editions

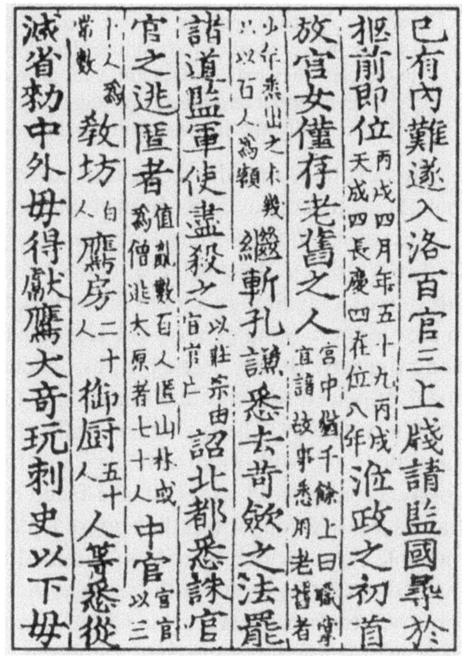
巳有内難遂入洛百官三上牋請監國事於
柩前即位丙戌四月年五十九丙戌天成四長慶四在位八年涖政之初首
放宮女僅存老舊之人宮中舊千餘上曰嘗掌宜諳故事悉用老舊者
少府亦出之木錢以百人爲額繼斬孔謙悉去苛斂之法罷
諸道監軍使盡殺之以莊宗由宦官亡詔北都悉誅宦
官之逃匿者值亂數百人匿山林或爲僧逃太原者七十人中官以三
十人爲常數教坊白人鷹坊二十人御廚五十人等悉從
減省勑中外毋得獻鷹犬奇玩刺史以下毋

Fig. 8.6 A copper movable type edition of *Essentials of Seventeen Dynastic Histories* issued in Joseon in 1403 (Housed in the National Library of Korea)[20]

of *The Book of Songs* and *Zuo Zhuan* from the royal inner court as templates for the characters. Within a few months, the foundry cast 100,000 copper movable types. That was the Year of Gui-Wei, and these first movable type characters cast by the government were referred to as "Gui-Wei characters", which were available in two sizes: large (1.4 cm × 1.7 cm) and small (1.1 cm × 0.8 cm). Some of the surviving books printed with this type include *Eleven Schools' Annotations on Sunzi*, *Essentials of Seventeen Dynastic Histories* (see Fig. 8.6), and *Collection of Appeals and Letters to the Emperors of the Song Dynasty*.

Since Taejong's reign, the Joseon royal court had a special fondness for metal movable type and intended to publish more kinds of books in a short period of time. With metal movable type, books could be printed on demand, and repeated typesetting and disassembly were feasible. However, the printing runs were small, usually only three to five hundred copies. From 1403 to 1883, there were 37 instances of character casting, including 30 for copper movable type, 2 for lead movable type, and 5 for iron movable type.[21] It should be noted that the use of iron for typecasting was an invention unique to Korea, and the hybrid typesetting of copper and wood movable types was also innovative. Typecasting was monopolized by the government, with little participation from civilians, which was different from the situation in other countries. As for the method of typecasting, Joseon scholar Song Hyon (1439–1504) wrote in his book *Yongjae Chonghwa* ("Yongjae Narratives"), Volume 3 (1495):

[20] XXX.

[21] Chon Hey-bong. *Korean Historiography* [M]. Seoul: Min'eongsa, 1997: 577–579.

> Generally like casting coins, first carve the movable type out of boxwood, and then lay soft clay from the sea coast on the printing plate. When the carved wooden characters are pressed into the clay, they make indentations for characters. Then put two printing plates together, pour molten copper through an opening and let the liquid flow into the indentations to form the types. Finally, the characters are scraped and smoothed out.[22]

This explains that the method of casting copper movable type was the same as that of casting copper coins, both employing the sand-casting technique as described in the session on metallurgy of *Tiangong Kaiwu* ("Exploitation of the Works of Nature") (1637). The casting of copper coins on the Korean Peninsula began in the 7th year of King Sejong's reign in Goryeo, the first year of Zhezong's Chongning era in the Northern Song dynasty (1102), with technology introduced from the Northern Song dynasty. Before casting types, wooden types were first carved as typeheads, and then placed soft clay and fine sand in the sand frame, and inserted the typeheads into the soft clay. Then, another sand frame was placed over the sand frame with the typeheads, with more soft clay and fine sand filled and compacted to form a hollow casting mold with concave-faced regular script types between the two sand frames. Melted copper was then poured into the cavity between the two frames to cast movable types, which were then trimmed and used for typesetting. So, the Korean method of casting type was the same as that used in the Song dynasty of China two to three hundred years earlier, but Chinese movable types were long cuboids, while the Korean movable types were short cuboids with concave characters, designed to save on copper usage. The *Yongjae Chonghwa* also mentions the Korean metal movable typesetting technique: "Initially, they did not know how to arrange characters, so they melted wax on the plate to fix the characters… Later, the technique of using bamboo or wood to fill the gaps was used, avoiding the cost of melting wax, which demonstrates the endless ingenuity of people."

It is evident that during the early period of Joseon dynasty (1403–1420), the way of using wax to fix copper movable types onto an iron printing plate was still Bi Sheng's movable typesetting method introduced during the Goryeo dynasty. There are writings about this in Volume 65 of *Veritable Records of the Joseon Dynasty: Veritable Records of King Sejoing* (1454): "However, the wax was soft, and the planted characters were not firmly fixed. After printing a few sheets of paper, the characters would move and often tilt, which required constant adjustments." "Only about 20 sheets could be printed a day." Sejong ordered the Vice Minister of Works, Li Pyeong (1375–1451), to improve the typesetting method. Then the "method of filling gaps with bamboo or wood" instead of wax was used to securely hold the characters, resulting in evenly spaced and secure characters and could "print more sheets without unevenness, and more than 40 sheets could be printed in one day, doubling the previous amount." In 1434, Sejong summoned Li Pyeong to the inner palace and rewarded him. This method became standard from then on. Improvements of typesetting technique took place between 1420 and 1434, and the method used was the wooden movable typesetting method described in the "Method of Printing

---

[22] Seong Hyeon. *Yongjae Chonghwa*: *Vol. 3* [M]. // *Ohashinomari*: Vol 1. Seoul: Korean Ancient Book Publishing conference, 1909: 158.

Books with Movable Type" in *Nong Shu* ("Book of Agriculture") (1313) by Wang Zhen, a scientist of the Yuan dynasty.

From the above description, it can be seen that the metal movable type printing technique on the Korean Peninsula was influenced by China, and the spread of Chinese metal movable type on the peninsula might have taken a long time with several steps. At the end of the Goryeo dynasty, the introduction of *Mengxi Bitan* brought in the technique of manusfacturing clay movable type and typesetting method, and with the issuance of paper money came the metal movable type technique. During the reign of Emperor Shizu (1280–1293) of the Yuan dynasty, the process of Mongolization in Goryeo was accelerated; the king often spoke Mongolian, took Mongolian names, and ordered the whole nation to wear Mongolian clothing. From at least 1276, Goryeo began to use Yuan paper money. In April 1287, "the Yuan court dispatched envoys to decree the circulation of Zhiyuan Baochao along with Zhongtong banknotes." From 1276 to 1290, Yuan banknotes circulated throughout the country. At this time, Yuan established the Eastern Expeditionary Province in Goryeo, and Baochao was also used by the hundreds of thousands of Mongol and Han troops stationed there. Between the thirteenth and fourteenth centuries, the Yuan government established a Supervisorate of Paper Money for the printing and issuance of Baochao in Goryeo, and the technique of printing banknotes was subsequently introduced. In the fourteenth century, during the Yuan-Ming transition period, the Ming issued Daming banknotes in the 8th year of Emperor Hongwu (1375), and from 1378, Goryeo began to use Ming era names again, and Ming Baochao was also in circulation in Goryeo. Since paper money was more convenient than other currencies previously issued in Goryeo, Goryeo decided to emulate the Yuan and Ming systems to print and issue paper money. In the third year of King Gongnyeong's reign and the 24th year of the Ming Hongwu era (1391), Jungnangjang (military officer in charge of the security of the royal palace) Fang Shiliang (방사량) proposed in March that mulberry paper money be used as currency, which was approved by the king. In July, the Office of Imperial Censors proposed abolishing the Hongfu Director for a *Jeohwa* treasury (mulberry currency depot) and requested the making of mulberry paper money, stating: From Han to the present, all dynasties had their own currency, such as Huizi of the Song dynasty and Baochao of the Yuan dynasty. Changing the currency system was actually the intention of ancestors. It is advisable for the officials to draw upon both ancient and modern practices, and, following the examples of Huizi and Baochao, to introduce mulberry paper currency for printing and circulation in Goryeo.[23]

Chubi or *Chuhuo* (*Jeohwa*) is paper money. The *Jeohwa* treasury in Korea was equivalent to the Supervisorate of Paper Money in the Yuan and Ming dynasties. When Goryeo imitated the Chinese system and issued *Jeohwa* in 1391, characters were printed from copper movable type on the face of the currency, and movable type could only be cast using the Chinese method, as mentioned earlier. Yi Yuwon, a scholar of Joseon dynasty, wrote in his book *Linxia Biji* ("Notes Under the

[23] *History of Goryeo*: *Vol. 79, Food and Currency* II: *Currency* [M]. Pyongyang: Korean Academy of Sciences, 1958: 608–609.

Trees"), Volume 22: "In the 8th year of Hongwu era (1375), Daming Baochao was first printed … and our *Jeohwa* modelled it". Just because copper movable type had been successfully cast by this time, Jeong Dojeon, a *jinshi* scholar, proposed to King Gongyang that a printing house should be established to cast type and print books. However, that was on the eve of the fall of Goryeo, and the work was only left to be completed during the reign of King Taejo of the Joseon dynasty. The early copper movable type was set with the clay movable typesetting method. Later, during the reign of King Sejong, the technique described by Wang Zhen of filling gaps with bamboo and wood was adopted for typesetting, thus completing the process of introducing Chinese metal movable type technology to the Korean Peninsula. The population of the Korean Peninsula was small and the demand for printed books was limited to two or three hundred copies. However, they required a wide range of books, and reprinting was rarely considered. With one set of movable type, they could print many different books, most of which were official publications. Therefore, while continuing to use woodblock printing, they focused on the development of metal movable type printing, and when copper was scarce, they used wooden movable type instead. In contrast, China had a large population with great demand for printed books, so reprinting was common. In addition to official publications, there were also private printing houses that printed books mainly for sale. Therefore, while retaining movable type printing, China focused on the development of woodblock printing. This difference resulted from different circumstances in the two countries. Studies on the spread of Chinese metal movable type technique to the Korean Peninsula have only started in recent years,[24] and some misunderstandings in the past need to be clarified. Some people claimed that metal movable type technique first appeared during the middle Goryeo period in the twelfth-thirteenth centuries, and one piece of physical evidence is a copper cast type block with the character "𥻳" purchased from a Japanese antique dealer named Akahoshi Sashichi by the Deoksugung Palace Museum in Keijo (modern-day Seoul) in 1913 (Fig. 8.7). It is said to have been excavated from a Goryeo royal tomb, but there is no concrete evidence to support this claim. The block is currently housed in the National Museum of Korea. Its chemical composition is similar to that of Goryeo's Haedong Tongbao copper coins, so it is believed to have been cast between 1102 and 1232, being the "oldest surviving metal movable type in the world."[25] Since the object lacks distinctive characteristics of its time, dating it based solely on its chemical composition is problematic, because copper objects with similar chemical compositions cannot prove they were produced in the same period. For example, copper mirrors from the Western Han dynasty are similar to those from the Tang dynasty in chemical composition, yet they differ by 900 years. Similarly, the chemical composition of copper coins from the Tang Kaiyuan period is close to that of copper coins from the Song Xining period, yet there is a difference of 300 years. Such examples are abundant. Besides, with the front four sides unequal in length, the block is not at 90-degree angles, a characteristic

[24] Pan Jixing. *History of Metal Type Printing Technology in China* [M]. Shenyang: Liaoning Science and Technology Publishing House, 2001: 156–172.

[25] Sohn P K. *Early Korean Typography* [M]. Seoul: Po-chin-chai Co. Ltd., 1982: 62–66.

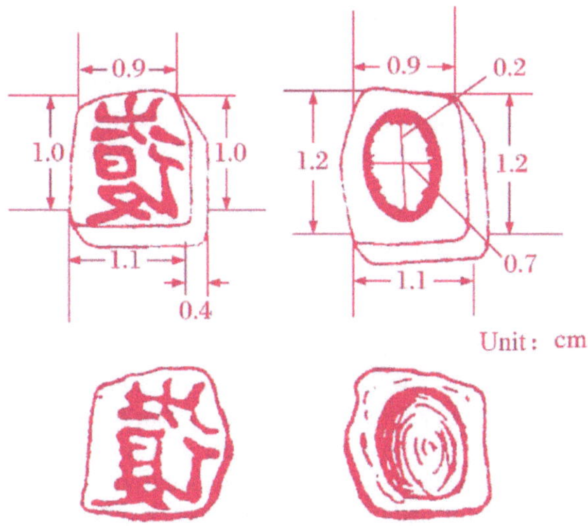

Fig. 8.7 Irregular copper character block with "複" and its dimensions of each part[26] (Collection of the National Museum of Korea)

of movable type, and so it is questionable whether it was movable type used for printing. Korean scholars unanimously believe that the development of movable type in Korea started after the introduction of Bi Sheng's movable type concept described in *Mengxi Bitan*, which was first published in China in 1166. So, before that time it was impossible for Goryeo to cast copper movable type.

Korean scholars believe that the earliest record of casting movable type and printing books on the Korean Peninsula is in the *Postscript to the New Printing Detailed Ritual Texts* drafted by the Goryeo *jinshi* Yi Gyubo on behalf of Prime Minister Choe U (1175–1249). The postscript mentioned that Choe U's father, Choe Chung-Heon (1149–1219), compiled two copies of *Detailed Ritual Texts of the Past and Present*, one kept at home and the other given to the official of Rites. In 1232, when the Mongol army attacked Goryeo, King Gojong (1214–1259) moved the capital to Ganghwa Island, but the official of Rites failed to bring the *Ritual Texts* to the island. Therefore, Choe U contributed the copy from home and "printed 28 copies using cast type and distributed them to various government offices." The postsctipt was titled "Postscript Written on Behalf of the Duke of Jinyang to the *New Printing Detailed Ritual Texts*" and was included in Volume 11 of *Posthumous Collection of Minister Yi of the Eastern Country* with no date at the end. Literally, the movable type was cast after the capital was relocated, and some speculate it to be in 1234 or 1241, which is difficult to ascertain. For instance, Yi Gyu-bo referred to Choe U as Duke of Jinyang, but Choe U was not conferred the title of Marquis of Jinyang until 1234 and later he became Duke of Jinyang in 1242. Yi Gyu-bo died in 1241, making it impossible for a decreased person to draft a postscript on behalf of another living individual. Thus, the credibility of this historical record is greatly reduced, and it is

[26] *History of Goryeo*: *Vol. 79, Food and Currency II*: *Currency* [M]. Pyongyang: Korean Academy of Sciences, 1958: 608–609.

challenging to determine the exact time of casting the metal movable type. Given the chaos of war, when the royal court sought refuge on Ganghwa Island, would it be worthwhile or even possible to cast movable type and print books there solely for distributing twenty-eight copies of the *Ritual Texts*? Moreover, even if the metal movable type was cast at this time, it would still be more than 100 years later than in China.

Another widely cited historical data is the postscript written by Choe U for the *Song of Enlightenment of Nanming*, its full title being *Ode to the Song of Englightenment by Nanming Monk Faquan*. The *Song of Enlightenment* was originally composed in rhymed verse by the Tang dynasty Zen Buddhist monk Xuanjue (ca. 643–713) to elaborate on the Zen teachings. Later, Monk Faquan of Nanming Mountain (now within the territory of Xinchang, Zhejiang Province) annotated it with rhymes, and it was published by Zhu Kuang, a resident of Yicang (now Lishui, Zhejiang Province) in the ninth year of Xining of the Northern Song dynasty (1076). After the Northern Song edition was introduced to Goryeo, it became an introductory book for understanding the essence of Zen. In Choe U's postscript, he stated that he "recruited craftsmen to recarve and cast movable type for the book to ensure its dissemination. Sincerely inscribed by Choe U, head of the Imperial Secretariat and Duke of Jinyang, in early September of the year of Ji-Hai" The postscript is dated to the year Ji-Hai, which corresponds to the 26th year of King Gojong's reign in Goryeo and the third year of Jiaxi Emperor Lizong of the Southern Song dynasty (1239), and it bears the words "Choe U, Duke of Jinyang". Some people interpret the phrase "recruited craftsmen to recarve and cast metal movable type" to mean that Choe U recruited workers to engrave wooden blocks to reprint the type-cast version of *Song of Enlightenment of Nanming* on Ganghwa Island in 1239 and assume that Goryeo had already possessed the technique of metal movable type before the royal court moved to the island in 1232.[27] However, neither the so-called republished book by Choe U nor the original "metal movable type printing version" has been handed down, and we do not know what they looked like or whether this postscript actually existed. The current surviving edition is later woodblock print, and it still lacks evidence as to whether it reflects the original appearance of Choe U's work.

If the existing edition with Choe U's postscript is believed to reflect the original appearance of the Goryeo version, there is a problem. In the year of Ji-Hai (1239), Choe U was still Marquis of Jinyang, how could he have called himself the Duke of Jinyang? Also, it is unimaginable that the postscript, which is only 72 words long, can contain some basic mistakes. The *Postscript to the Ritual Texts* and the *Postscript to the Song of Englightenment of Nanming*, both signed by Choe U, advanced the title for him before he became Duke of Jinyang, which contradicts historical facts. People cannot help but question the authenticity of both postscripts, and it is not convincing to believe that there were printing activities using cast type on the Korean Peninsula in the first half of the thirteenth century. Even if the two postscripts are reluctantly treated as records of cast type, the suggested timing of the Goryeo cast type is still one or two hundred years later than in China, and metal movable type technique is

[27] Sohn P K. *Early Korean Typography* [M]. Seoul: Po-chin-chai Co. Ltd., 1982: 129.

still an earlier Chinese invention. We acknowledge the contribution of the Korean people to the development of metal movable type, but the fact is that they developed this technique after China, though more than 70 years ahead of Europe. Facts show that printing activities using cast type only started in the late Goryeo period (late fourteenth century). Those who still insist that the Korean Peninsula invented the metal movable type must provide new evidence of cast type printing on the peninsula before the eleventh-twelfth century, which is actually impossible.

8.1.3 Beginning of Woodblock Printing in Japan

China and Japan have a history of two thousand years of cultural exchange and interaction, as they are separated only by a strip of water. The term "Japan" first appeared in Chinese historical records in the *Old Book of Tang: History of Japan* (945); before that, Japan was called Wa. *The Book of the Former Han: Geography* (83) states: "There are Wa people in the region of Lelang and Haining, divided into over a hundred states, and they come to offer tribute every year." So, it can be seen that during the Western Han dynasty, some tribal states on the Japanese archipelago had constant interactions with the Han Empire throughout the year. Through the intermediaries established by the Han Empire in Lelang and other counties on the Korean Peninsula, they imported advanced agricultural and handicraft technologies from China so their social productive forces developed rapidly and the primitive clan commune system gradually disintegrated. In the first century BC, to seek development, elevate status, and annex other small states, some tribal states in Japan established contact with the Western Han Empire. With the acceleration of annexation process, the number of Japanese tribal states in contact with Wei decreased to 30, and eventually only a few states such as Ya Tai Guo in Kyushu and Yamato in Honshu survived. By the time of the Wei and Jin dynasties, paper had already become prevalent in China, and with personal interactions, paper and paper documents could be transmitted to Japan. Records about papermaking in Japan can be found in Nihon Shoki ("Chronicles of Japan") (720). Volume 22, *Chronicle of Empress Suiko*, states:

> In the spring, march of the 18th year (610), the Goguryeo king offered up Buddhist priests Damjing (Donchō) and Hojo (Hōjō) as tribute [to Japan]. Damjing had a good knowledge of *Five Classics*, and was able to produce colors, paper, ink, and he also made watermill. Has making watermill presumably started ever since?[28]

Since the seventeenth century, scholars have generally cited the aforementioned records to suggest that papermaking in Japan began in 610 AD, and they regard Monk Damjing, who introduced the papermaking technique, as from Goguryeo. This view was once widely accepted, but upon closer examination, it raises many doubts. The *Nihon Shoki* only mentions that Damjing, who arrived in Japan in 610 AD, was able to make paper and ink; it does not say that papermaking in Japan did not start until he

[28] Prince Toneri. *Nihon Shoki* ("Chronicles of Japan")*: Vol. 22, Annals of Suiko* [M]. Tokyo: Iwanami Shoten, Publishers, 2000: 118, 464.

arrived. It does suggest that making the watermill, i.e. water-powered trip-hammer, might have started around that time. Earlier than that time, Japan had already used wooden pestles. Considering Damjing's knowledge of both Confucian classics and paper and ink technology, he was most likely a Chinese monk living in Goguryeo. The nineteenth-century scholar Igarashi Atsuyoshi mentioned that Damjing came to Japan more than a hundred years after the introduction of printed books, and that papermaking had already existed in Japan before he arrived. In recent decades, some Japanese paper history scholars have also expressed similar opinions.[29] The *Nihon Shoki: Chronicle of Emperor Richū* records that in the fourth year of Emperor *Richū*'s reign (the first half of the fifth century), "local recorders were appointed for the first time in various states, who noted down statements and communicated common aspirations."; the *Chronicle of Emperor Kinmei* in the same book states that in the first year of Emperor Kinmei's reign (540), orders were given to compile household registers for various naturalized tribes throughout the country. The Yamato court required a large amount of paper for recording histories and compiling registries in various states, which might not have been sufficient through imports. So, it is possible that papermaking in Japan existed between the fifth and sixth centuries.

After the fourth century, the Yamato regime had completed the unification of Japan, and by the reign of Emperor Homudawake (between the fourth and fifth centuries), he became the hereditary ruler, later known as Emperor Ōjin. He extended Japan's influence to the Korean Peninsula, where the Three Kingdoms of Goguryeo, Silla, and Baekje were competing for dominance. Internally, he strengthened political, economic, and cultural development. To achieve this goal, he urgently needed various talents and recruited many skilled Chinese people living on the Korean Peninsula to Japan. According to the original ancient records from the Nara period (710–794), in the 14th year of Emperor Ōjin's reign (403), a large number of people led by Yuzuki no kimi, a descendant of Qin Shi Huang, came to Japan from the Korean Peninsula. They were called "Qin people", settled in Yamato (now Nara) and engaged in sericulture, civil engineering, brewing, and agriculture. In the 16th year of Emperor Ōjin's reign (405), Wang Ren, a descendant of Han Gaozu and a scholar of *Five Classics* in Baekje, came to Japan with his people. Four years later (409), the great-grandson of Emperor Ling of Han, Azhi, and others arrived one after another. They were called the "Han people" and engaged in spinning, garment making, metallurgy, writing, and diplomacy. In the fifth century, the arrival of a large number of Qin and Han people marked the first upsurge of massive import of Chinese culture, science, and technology into Japan. Along with these people, papermaking technology was introduced to Japan. After discussions with Japanese paper historians, we agree that papermaking in Japan started in the fifth to sixth centuries, but further investigation is needed to determine whether it was introduced by the Qin people or the Han people.

The early work that explicitly discusses the origin of papermaking in Japan is the anonymous *Ku Hang Ji* (*Collection of Withered Hang*), published in the eighth year

---

[29] Bunsho Jugaku. *Journey of Japanese Paper* [M]. Tokyo: Unsodo, 1973: 28. プロフィール久米康生 *Cultural History of Japanese Paper* [M]. Tokyo: Mokujisha, 1977: 7–10.

of the Kanbun era (1668) during the Edo Period. The second volume of this book states: "In ancient times in our country (Japan), someone called Kishi-shi first began to make paper. Before this, wooden tablets were used for writing, ..." The washi paper historian 関義城 (コレクション) was the first to cite this historical record in his work "Japan's earliest papermakers",[30] and he pointed out that there were similar accounts in other artifacts from the seventeenth and eighteenth centuries. However, the identity of "Kishi-shi" needs to be clarified. Kishi-shi, along with Yoshishi and Yoshijin, are transliterations of ancient Japanese "きし" (kishi), which in Japanese refers to an honorary title given to learned individuals or diplomats who came from abroad. The "kisi" in Korean refers to royal or noble family members or official titles, and though it sounds the same as "kishi", its meaning is different from that in Japanese. So, the meaning of "kishi-shi" should be based on the Japanese definition. This person was mentioned in the *Kojiki* ("Records of Ancient Matters") (712): "And, [the Emperor Ōjin] ordered [Geunchogo, the King of] Baekje, if there is any wise man, offer him up as tribute. Therefore receiving the [Imperial] commands, [the King] sent as tribute a man named Wani-Kishi, and along with him, ten volumes of *Analects* and one volume of *Book of a Thousand Characters*, altogether eleven volumes."[31] The Wani Kishi mentioned here is the scholar Wang Ren (369–440 AD) recorded in *Nihon Shoki*, Kishi-shi or Yoshishi being his title, and Wani is a transliteration of *Wang Ren*'s name. Wang Ren brought to Japan *Analects* annotated by Zheng Xuan and *Thousand-Character Classic* (ca. 210) by Zhong Yao (151–230), marking the start of Confucianism's transmission to Japan. He also worked on government documents and served as a tutor of Confucianism for the imperial princes. These tasks required paper, so he organized Han artisans to make paper locally in Japan.

Wang Ren, as the most knowledgeable scholar of *Five Classics* in Baekje, was recruited by Emperor Ōjin to Japan, where he was highly valued and honored with the title "Kishi." His role was related to paper. In ancient Japanese books, he was not the only person referred to as "Kishi"; there were others such as Naniwa Kishi, Kunji Kishi, but they were royal nobles of Goguryeo or Baekje rather than scholars engaged in paper-related work in Japan. Moreover, they arrived in Japan several decades later than Wang Ren, after papermaking had already started in Japan. Among the Qin people there were no individuals titled "Kishi", and their engagement in paper-related library work was also after the advent of papermaking. The paper produced during Wang Ren's era was from hemp, and in China after the Northern and Southern dynasties and Sui and Tang dynasties, there was great development in the techniques of making bark paper, which were thereafter introduced to the Korean Peninsula and Japan. During the reign of Prince Shotoku (592–622) in the Asuka period (592–710), there were orders that mulberry trees be planted and mulberry paper produced around the country, which were all conducted under the guidance of the Chinese monk

[30] 関義城 (コレクション). *A Study on the History of Handmade Paper* [M]. Tokyo: Mokujisha, 1976: 2.

[31] Ono Yasumaro. *Kojiki*: *Middle Volume, Records of Emperor Ōjin* [M]. Tokyo: Iwanami Shoten, Publishers, 1999: 145, 276.

Damjing (Donchō), who came from Goguryeo in 610.[32] The rise of the papermaking industry provided a material basis for the development of printing and also promoted cultural and educational development.

After the Taika Reforms during the Asuka period (646), Japanese economy and culture developed rapidly and reached a peak during the Nara period. At this time, communication between China and Japan was widespread as a large number of emissaries, students and learned monks visited Tang China and brought back everything they had learned to their home country. Tang emissaries also visited Japan, resulting in a close cultural exchange between the two countries. Buddhist monks who visited Tang China brought back numerous printed Buddhist scriptures, so the mechanical reproduction technology that could replace manual copying was introduced to Japan during the Nara period. Empress Kōken (746–758), like Tang Empress Wu Zetian, was a devout Buddhist. After abdicating in favor of Emperor Junnin in 758, she proclaimed herself as "Cloistered Empress"; she shaved her head to become a nun, and appointed the monk Dōkyō (710–772) as the Imperial Preceptor to practice Buddhism. At that time, the Chief Minister Fujiwara no Nakamaro, an imperial in-law, was domineering under the protection of Emperor Junnin. In September 764, he launched a rebellion but was successfully suppressed by the Cloistered Empress and Fujiwara no Nakamaro was killed. In the same year, the Cloistered Empress deposed Emperor Junnin and reassumed the throne, known as Empress Shōtoku (764–770) in history. At the beginning of the rebellion, the Cloistered Empress made a great vow to build a million pagodas, each with a Buddhist incantation to be enshrined and worshipped in various locations. This was similar to Empress Wu Zetain's practice of having the esoteric Buddhism scripture *Great Dharani Sutra of Immaculate and Pure Light* printed and enshrined in pagodas in 702. Even the sutras used were the same. In Volume 30 of *Shoku Nihongi* ("Chronicle of Japan Continued") (797), the entry dated April 26, the first year of the Hōki era (May 25, 770), Sugano no Mamichi (741–815) recorded:

> Initially, in the eighth year of Empress Shōtoku's reign (764), after suppression of the rebellion, the Empress made a grand vow to construct one million small pagodas with three tiers each, measuring four and a half inches in height with a base diameter of three and a half inches. Beneath the "Dew Plate", *The Root*, *Merciful Heart* (also known as *Self-Mind Seal*), *Image Wheel*, *Six Perfections*, and other dharanis were placed until the project was completed, and then distributed to various temples. Titles of nobility with varying ranks were granted to 157 individuals, ranging from officials in charge of the religious services to official servants.[33]

The Nara document *Todaiji Yoroku* ("The Digest Record of Todaiji Temple") states in Volume 4:

> In the first year of the Jingo-keiun era (767), the east and west pagoda halls were built by Monk Jicchu (じっちゅう 実忠). On September 1, 764, during the reign of Empress Kōken,

[32] 関義城コレクション. *A Study on the History of Handmade Paper* (手漉紙史の研究) [M]. Tokyo: 木耳社, 1976:2.

[33] Sugano no Mamichi. *Shoku Nihongi: Vol. 30, The First Year of the Hōki Era* [M]. Tokyo: Kodansha, 2000: 33

one million pagodas were built and distributed to ten major temples, each containing a printed *zheben* (ancient folded book made of long scrolls) of *Great Dharani Sutra of Immaculate and Pure Light*.[34]

Fujiwara no Nakamaro (706–764), who was given the name Emi no Oshikatsu by Emperor Junnin, rebelled in September 764 and was suppressed within ten days. In the 9th year of the Tenpyō Hōji era (765), Empress Shōtoku, in celebration of the victory over the rebellion and her restoration to the throne, changed the era name to Tenpyō Jingo and appointed Dōkyō as Daijō-daijin (Chancellor of the Realm). Under the guidance of Dōkyō, the construction of pagodas with Darani sutras was promptly carried out. It took six years to complete, starting from 764 to April 770. Dōkyō (710–772), whose secular family name was Yuge, was a descendant of Han Chinese immigrants who had been naturalized in Japan. He was proficient in both Sanskrit and Chinese, especially in Esoteric Buddhism, and taught esoteric scriptures to the Empress. The construction of the Hyakumanto darani might have been his idea. The Buddhist pagodas were miniature wooden structures, mostly triple-tiered, measuring an average height of 21.5 cm. The tower body averages 13 cm in height and the phase wheel 8.5 cm in height. In the center of the upper part, there is a cavity with an inner diameter of 2.2 cm and a depth of 8.5 cm, used for placing the Darani sutras. The base has a diameter of 9–10.4 cm, with each pagoda weighing an average of around 200g, and the tower body is coated with white powder (as shown in Fig. 8.8). Since the pagoda was small, the "dew plate" (phase wheel) could not accommodate the entire sutra scroll, and only daranis could be placed inside. The Hyakumanto darani included four out of the six daranis in the *Muke Joko-kyo* ('Pure Light' sutra), namely *Konpon darani*, *Jishin'in darani*, *Sorin darani*, and *Rokudo darani*, all transliterated from Sanskrit into Chinese characters. Copying out one million daranis would take a lot of time, so they were printed to stay in sync with the construction of the pagodas.

The construction was completed with one million pagodas along with one million printed darani prayers. Each pagoda had one darani placed inside, and they were distributed to ten major temples including Daianji, Kohfukuji, Yakushiji, Todaiji, Saidaiji, Horyuji, Shitennoji, Sufukuji, and Hongfukuji. However, due to successive wars and other historical factors, most of them have been lost, with only over 40,000 remaining in Horyuji in Nara, which are now housed in different places in Japan and abroad. The "*zheben*" ("すりほん" (sirihon) in Japanese) mentioned in *Todaiji Yoroku* means printing, equivalent to "printed version" in Chinese. Based on the existing physical evidence, the four mantras were printed using two types of printing plates: one long edition and one short edition. The plate size varied because of the difference in the number of characters used. Relevant data of each printing paper are shown in Table 8.1.

Table 8.1 is based on the measurement of 40 material objects stored in the Seikado Bunko Library in Tokyo. These data reflect the variation range of the size of each darani because they were placed and operated in different locations across the country, resulting in slight differences in size. The four daranis were printed in regular script,

[34] Yusho Tokushi. *Research on the History of Oriental Printing* [M]. Tokyo: 青裳堂書店 (せいしょうどうしょてん).

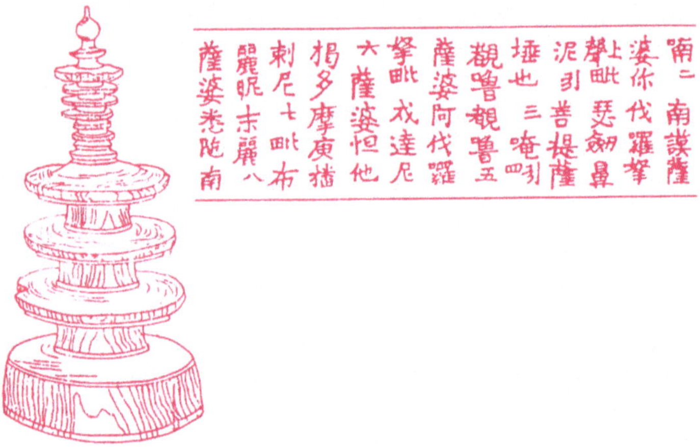

Fig. 8.8 The One Million Pagodas and One Million Printed Dharani Installed Inside, constructed in Japan from 764–770 AD[35]

Table 8.1 Relevant data of Hyakumanto Darani printing paper

		Number of lines	Height (vertical) (cm)	Total length (cm)	Character height (cm)	Character length (cm)
Long version	Konpon darani	40	5.4–5.9	54–58	4.7–4.8	49.2–49.6
	Sorin darani	23	5.4–5.5	39.7–41.1	4.5–4.6	28.1–29.9
	Jishin'in darani	31	5.4–5.6	45.8–46.4	4.7	35.6–35.8
	Rokudo darani	13	5.4	45.6	4.5	17.85
Short version	Konpon darani	40	5.4–5.7	51.3–57.4	4.5–4.6	42.5–43.3
	Sorin darani	23	5.4–5.8	38.8–41.9	4.5–4.6	28.1–28.2
	Jishin'in darani	31	5.4–5.6	45.8–46.4	4.7	31.8–32.3
	Rokudo darani	13	5.7	23	4.5	16.1

with a general height of 1.2 cm and a width of 0.7 cm for each character, five characters per line. Small characters in double lines were used for annotations. One page is printed from one plate, and each page has one darani (Fig. 8.9). The number of

[35] 増尾信之(マスオ・ノブユキ). *History of the Printing Ink Industry* (印刷インキ工業史) [M]. Tokyo: Japan Printing Ink Industry Association (日本印刷インキ工业联合会), 1955.

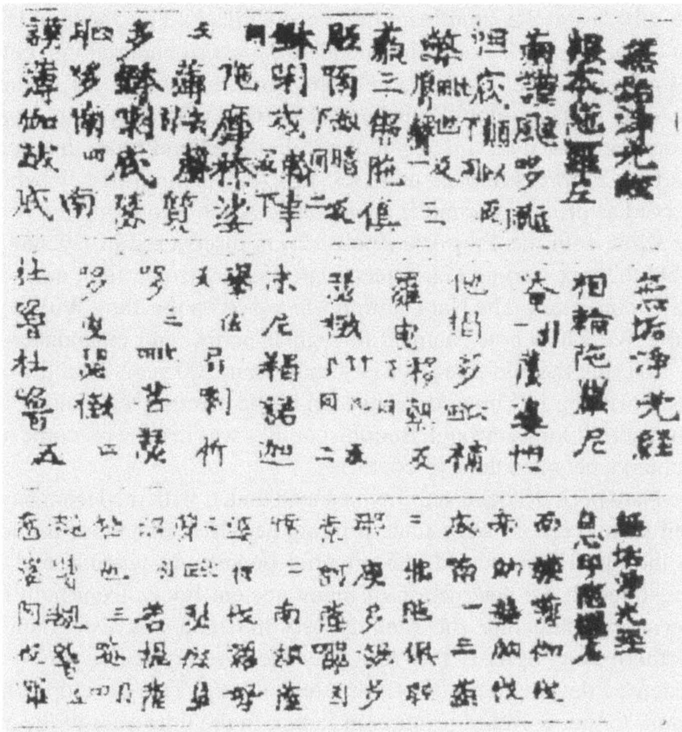

Fig. 8.9 Hyakumanto Darani printed in Japan in 764–770

characters varies from 79 characters (*Six Degrees Rokudo darani*) to 200 characters (*Konpon darani*). Tests show that the paper used for printing was mostly made of hemp or mulberry paper dyed yellow with yellow paulownia. Assuming the paper produced by the government paper mill in the Nara period was 1.2 feet × 2.2 feet (about 36 cm × 66 cm), 114,000 sheets of paper would be required. The printing plates used were mainly made of prunus pseudocerasus (sakura wood). Based on existing relics, two sets of plates were used, each set consisting of eight plates. The Nara period had a large-scale woodblock printing industry from the beginning, which can be attributed to the courage of Empress Shoken, leaving behind a precious legacy for future generations. After fulfilling her wish of building the pagodas, she died in the same year (770) when the project was completed.

Where did the technology used for the sudden large-scale printing during the Nara period come from? Dr. Kimiya Yasuhiko, a historian of printing, wrote: "Judging from the transportation and cultural exchange between Japan and Tang China at that time, I believe it was imported from the Tang dynasty."[36] Another Buddhist printing historian, Dr. Yusho Tokushi (1879–1960), also stated: "In view of the prevalence of communication with the Chinese mainland from the Nara period to the Heian

[36] Kimiya Yasuhiko. *Cultural History of Old Japanese Printing* [M]. Tokyo: Fuzambo, 1932: 17–29.

period (794–1192) and the significant influence China had on Japan, it is clear that the printing of the Darani is not an independent Japanese undertaking, but rather an imitation of a practice that China had already implemented."[37] He further pointed out that the Tang dynasty monk Jianzhen (687–763), who arrived in Japan in 754, and his entourage passed on this technique.[38] Jianzhen had more than ten years of interactions with Empress Shoken and Dōkyō, so it is natural that Jianzhen and his disciples served as printing technical consultants. After comparing the Nara edition of *Dharani Sutra* with the Tang Wu Zhou edition discovered in Gyeongju, Korea, we found that the text, variant characters, plate frame format, font, and paper color were basically consistent. The Nara edition was based on the Tang Wu Zhou edition, with only the Wu Zhou font changed to regular script, and emendation of a few characters.[39] During the mid-Tang dynasty, more than 100 years after the emergence of woodblock printing in China, transmission of the printing technology from Tang China to Nara period Japan through Buddhist monks was just the outcome of frequent cultural exchange between the two countries.

After the Nara period, social stability in Japan ended with the Heian period (794–1192), when there were no large-scale printing activities like those in the previous periods. In the late Heian period, the imperial power was waning, and long-term civil wars resulted in the destruction of many ancient books from both Heian and previous periods. Meanwhile, the Tang dynasty in China was also going downhill, and the Northern Song dynasty that rose after the Five Dynasties period witnessed an unprecedented development of printing technology. This development brought foreign stimuli to the revival of printing in Japan. In the third year of the Yongxi era of the Northern Song dynasty (986), Chonen (ca. 951–1016), a Japanese nobleman and monk who disliked wealth and fame, led his disciple Josan and others to sail to Song China on board a ship owned by the Chinese merchants Ch'en Jen-shuang and Hsu Jen-man to seek Buddhist teachings. They got gifts from Emperor Taizong of Song dynasty including a Song block-printed edition of *Kaibao Tripitaka* and sixteen Arhats statues,[40] which they brought back to Japan. These gifts provided exquisite rare editions for publishing Buddhist scriptures in Japan. Kimiya Yasuhiko, citing the diaries and literary collections of the nobility during the late Heian period, listed a table of Buddhist texts published from 1009 to 1169, totaling 8601 individual works with a total of 2058 volumes. For example, according to *Mido Kanpakuki* ("Diary of a Hojoji Regent") by Fujiwara no Michinaga (966–1027), one thousand copies of *Lotus Sutra* were printed in 1009. The *Hyohanki* ("Diary of Taira no Nobunori") records that in 1169, the Retired Emperor Goshirakawa had one thousand copies of *Lotus Sutra*, three hundred and fifty copies of each of *Guanyin Xianjing*, *Amitabha Sutra*,

[37] Yusho Tokushi. *Research on the History of Oriental Printing* [M]. Tokyo: 青裳堂書店 (せいしょうどうしょてん), 1981: 166.

[38] Yusho Tokushi. *Research on the History of Oriental Printing* [M]. Tokyo: 青裳堂書店 (せいしょうどうしょてん), 1981: 182.

[39] Pan Jixing. *A comparative research of early printing technique in China, Korea and Europe* [M]. Beijing: China Science Publishing & Media Ltd, 1997: 240–242.

[40] *History of Song*: *Vol. 491, Biography of Japan* [M]. // *Twenty Five Histories: Vol. 8*. Compact edition. Shanghai: Shanghai Classics Publishing House, 1986: 1600.

and *Prajna Paramita Heart Sutra* engraved and printed to pray for the happiness of his deceased son in the afterlife. Within one year nearly 2400 Buddhist texts were printed, and all of these works were printed in Heian-kyo (present-day Kyoto).[41]

During the Heian period, Buddhist scriptures were also printed in Nanto (present-day Nara) in Japan. For example, in the second year of the Kanji era (1088), the Kohfukuji printed the ten-volume classic *Cheng Weishi Lun* ("Discourse on the Perfection of Consciousness-only"), which is now housed in the Nara National Museum. In the second year of the Genei era (1119), *Cheng Weishi Lun Shu Ji* ("Explanatory Notes on the Cheng Weishi Lun") by the Tang dynasty monk Kuiji was also printed in Nanto. This is an annotated version in scroll format. Each scroll measures 26.7 cm × 51 cm, with a frame height of 24.3 cm. There are 40 lines per page, each line having 21 characters engraved in the regular script. The inscription reads "molded by the monk Guan Zeng", indicating that the monk Guan Zeng engraved it. Later, during the Kamakura period (1190–1335), Zen Buddhism and Neo-Confucianism of the Song dynasty were introduced to Japan. Neo-Confucianists in the Song dynasty also worked on Buddhism scriptures while many Buddhist monks were familiar with the "external classics" (Confucianism). Japanese monks who returned from Song China brought this academic tradition back to Japan, and some temples printed Confucian classics based on Song editions. For example, in the first year of the Hoji era (1247), a man named *Louxiangzi* printed the ten-volume *Lunyu Jizhu* ("Collected Commentaries on the Analects of Confucius") based on a Song edition, and in the second year of the Genkō era (1322), the Song edition of *Guwen Shangshu Kongshi Zhuan* ("The Commentary on the Old Text of Book of Documents by Kong") was reprinted, which typifies the practice of Buddhist monks printing Confucian classics. This tradition continued until the fourteenth century, as proved by the printing of *Lunyu Jizhu* by the lay Buddhist Doyu in the 19th year of the Shohei era (1364), which is now in the Nanshuji Temple (Figs. 8.10 and 8.11).

In the latter half of the fourteenth century, during the turmoil at the end of the Yuan dynasty in China, printing workers from Fujian and Zhejiang provinces along the southeast coast fled to Japan with others to seek refuge, bringing with them the highly developed printing techniques of the Song and Yuan dynasties. For example, the Fujian engravers Yu Liangfu (1340–1400), Chen Mengrong, Chen Boshou, and others worked on book engraving in Kyoto. In addition to assisting in engraving books for Kaiganji Temple, Yu Liangfu also set up his own printing business (Fig. 8.12). Chinese engravers revitalized Kyoto's printing culture, firstly by diversifying the types of printed books. In addition to Buddhist texts, non-religious works such as literature and history which were rarely published previously in Japan were also printed. Secondly, Japanese printed books generally used engraved handwritten characters with varied fonts; after the arrival of the Chinese printers, books printed in Kyoto commonly adopted the imitated Song-style font, the printed font characterized by uniform and neatly aligned strokes. The book layout also adopted the format of the Yuan editions, i.e., bound with threads, called 袋綴じ in Japanese, which made books easier to use and read. These new forms of engraving and printing brought by the

[41] Kimiya Yasuhiko. *Cultural History of Old Japanese Printing* [M]. Tokyo: Fuzambo, 1932: 34–37.

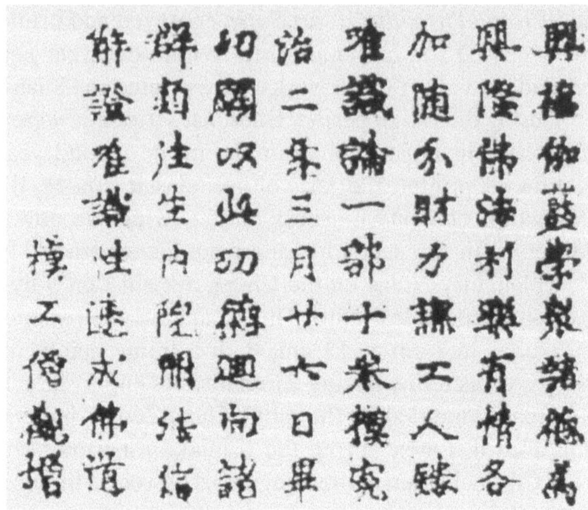

Fig. 8.10 The 1088 edition of *Cheng Weishi Lun* printed by the Kohfukuji in Nara, Japan[42]

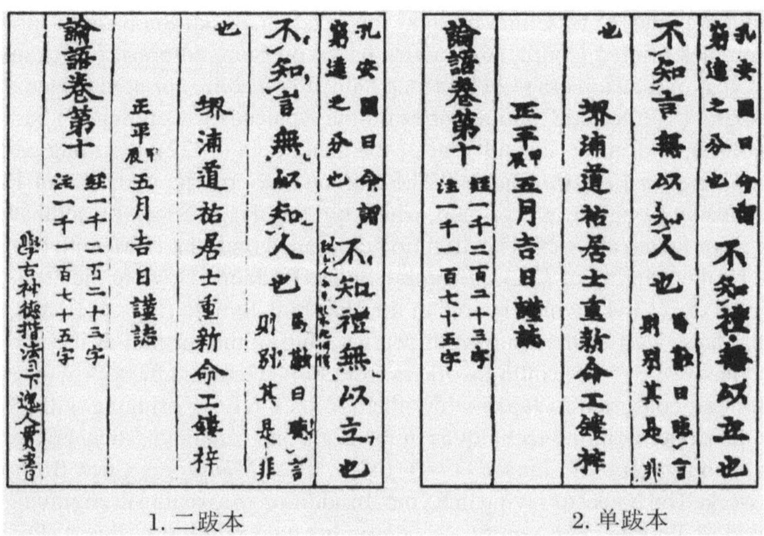

Fig. 8.11 *Lunyu Jizhu* printed by the lay Buddhist Doyu in the 19th year of the Shohei era (1364)

Yuan people to Kyoto had a far-reaching impact on later Japanese printing. Before the Muromachi period (1336–1408), most Han Chinese manuscripts and printed books in Japan had no punctuation marks. To facilitate reading for young people, the kunten with punctuation marks and katakana annotations indicating syntactic relationships appeared in *Lotus Sutra* published in the fifth year of the Ōei era (1398) by the

42 The engraver was the monk Guan Zeng, and the word '蓝 (blue)' was blurry, thus supplemented today.

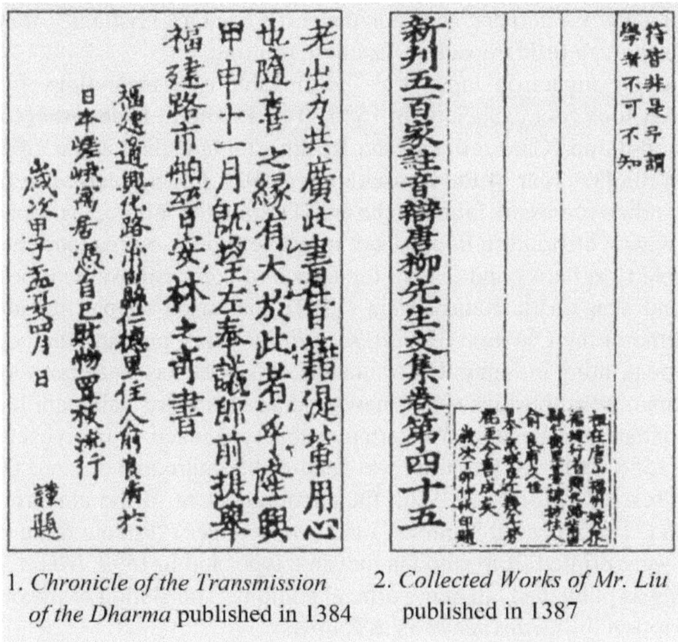

Fig. 8.12 Books published by the Fujian native Yu Liangfu in Kyoto[43]

lay Buddhist Daojian. After the seventeenth century, the kunten edition became the major popular prints.

8.1.4 Beginning of Moveable Type Printing in Japan

Woodblock printing was developed in Japan in the eighth century, but after the Kamakura period, internal conflicts occurred frequently and trade with China was reduced for a time, which hindered the timely development of moveable type printing technology. It was not until the late Azuchi-Momoyama period (1573–1600) that China's moveable type technology was introduced through the Korean Peninsula. Prior to this, the Italian Jesuit Alexandre Valignani (1538–1606) came to Japan as a missionary from Macao in 1590–1592, bringing with him Western printing workers, Western movable type, and printing presses, and they operated in Kyushu and Nagasaki. In Japan, Valignani was called "bateren" as a pun on "Valignani" and in Japanese, Christianity is known as "Kirishitan (キリシタン)", phonetically derived from "Christian". Valignani used movable type to print some Western and Japanese books, known as Kirishitan editions. However, printing technology was in

43 中山久四郎 (Yoshiro Nakayama). *A General History of World Printing* [M]. Tokyo: 三秀舎について, 1930.

the hands of a few Westerners, and soon under the ban on Christianity, they departed quickly, thus having little impact on Japanese printing.

What had an impact on Japan was the movable type technology from China, which quickly took root in society. In 1586, after Toyotomi Hideyoshi (1537–1598) became Daijō-daijin (Chancellor of the Realm), he launched a war of aggression on Korea in the first year of the Bunroku era (1592). Facing resistance from Ming and Korean allied forces, he failed in the end. Despite the failure, the Japanese army found a new way of printing books from movable type in Korea, and they brought back movable type books and tens of thousands of copper movable types to Japan. In the second year of the Bunroku era (1593), they used copper movable type to print a volume of the Old Text of *Classic of Filial Piety*, marking the beginning of movable type printing in Japan. Unfortunately, this book has since been lost. At that time, the Japanese authorities were unaware that Alexandre Valignani had secretly printed Christian books using a Western movable type machine in Kyushu. Emperor Go-Yōzei (1586–1610) of that time was fond of literature and ordered the printing of the Old Text of *Filial Piety*. With the encouragement of the emperor, in 1597, *Jinxiu Duan* ("The Splendid Sections") and *Quanxuewen* ("Encouraging Learning") (Fig. 8.13) were printed from wooden movable type, and in 1599, *Nihon Shoki: Age of the Gods* was published, all being official editions. Translation of the original text of the inscription for *Quanxuewen* is as follows:

The craftsmen were commanded to carve each character on a wooden movable type block and then arrange the movable types on a printing plate for printing. This method, which came from Korea, is extremely convenient and thus used to print this book. Late August, the 2nd year of the Keichō era (1597).

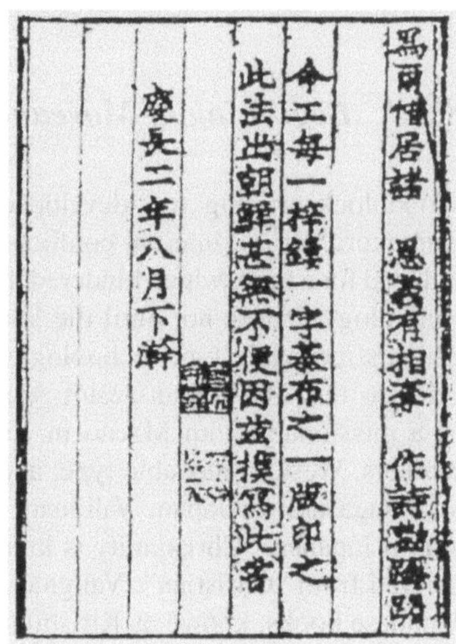
命工每一梓鏤一字棊布之一版印之
此法出朝鮮甚無不便因茲摸寫此書
慶長二年八月下澣

Fig. 8.13 The wooden movable type edition of *Quanxuewen* published in Japan in 1597

Fig. 8.14 *Qunshu Zhiyao* published in Japan in 1616 using copper movable type from Korea[44]

The term "*moxie* 模写" in the inscription is a technical term used in ancient Japan, meaning "printing" in Chinese; so, it should not be taken literally as "copying". During the Edo period (1603–1868), both woodblock printing and movable type printing flourished in Japan. Under the rule of the Tokugawa shogunate led by Tokugawa Ieyasu (1542–1616), who was committed to economic and cultural development, society was relatively stable, and traditional frequent exchanges with China were restored. Tokugawa Ieyasu emphasized Confucianism, especially Neo-Confucianism, making it the official doctrine. He established a national library and built a school in Fushimi Castle. In the fourth year of the Keichō era (1599), 100,000 wooden movable types were used to print *Kongzi Jiayu* ("Family Sayings of Confucius"), *Liu Tao* ("The Six Secret Teachings") and other books, totaling 80 copies of 8 books within 8 years, which are called Fushimi editions. During Tokugawa Ieyasu's reign, copper movable type was also used to print books during 1603–1616. In 1607, for example, Naoe Kanetsugu (1560–1619) used copper movable type to print *Liu Chen Zhu Wen Xuan* (Monzen annotated by six retainers) in 61 volumes at the Yobo-ji Temple in Kyoto. In 1615, while overseeing the government in Sunpu or Fuchu (present-day Shizuoka), Tokugawa Ieyasu ordered the renowned Confucian scholar Hayashi Razan (1583–1657) to supervise the printing of *Dazang Yilan* ("The Tripitaka at a Glance") in 125 volumes, using large and small copper movable type. The following year (1616), *Qunshu Zhiyao* ("The Governing Principles of Ancient China") was published in 60 volumes (Fig. 8.14). The copper movable type used mostly came from Korea, with shortfall supplemented with newly casted ones, 13 thousand copper movable types in different sizes, by Lin Wu-guan, a Chinese man who came to Japan from Fujian in 1574. Some people claim he was Korean, which is a misunderstanding.

In the 12th year of Keicho era (1605), Suminokura Soan and Honami Koetsu published *Ise Monogatari* ("The Tale of Ise"), a classical literature masterpiece of the Heian period, using hiragana wooden movable type. The printing frame was 19.2 cm × 26.4 cm in size, with occasional upper and lower characters joined-up, and also accompanied by woodblock illustrations (Fig. 8.15). The paper used was quite exquisite, mica paper with embossed floral patterns. Other books were also published in the years that followed. The Edo period was different from the previous

[44] Printing Museum. *Printing in the Edo Period* [M]. Tokyo: Toppan Printing Co., 2000: 88–89.

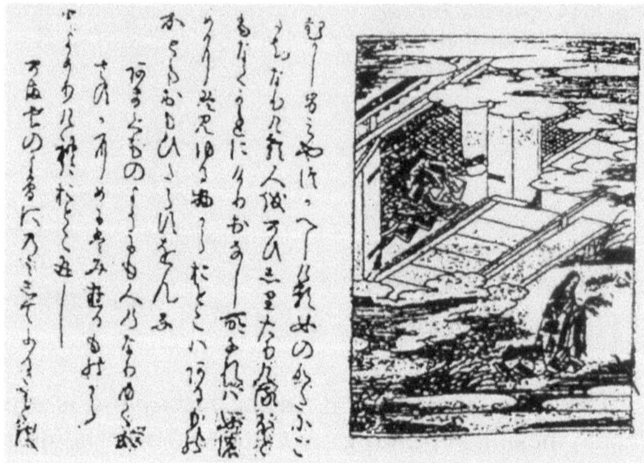

Fig. 8.15 *Ise Monogatari* from Tsugawabun wooden movable type published in the Edo period

times in that publishing Buddhist scriptures was no longer the mainstream of printing, but the enormous projects of engraving and printing *Issai kyo* ("Complete Buddhist Scriptures") or *Daizo-kyo* ("Tripitaka") cannot go unmentioned, as they were the long-cherished wishes of generations of Buddhist monks that had not been realized until then. In the 14th year of Kan'ei era (1637), Daisojo Tenkai (1536–1643) was commissioned by the third shogun of the Shogunate, Tokugawa Iemitsu, to supervise the publication of *Issai kyo* at Kan'ei-ji Temple in Tokyo. Funded by the shogunate, the project took 12 years to complete, and in March of the first year of Keian era (1648), a total of 1453 books, 6323 scrolls, and 665 cases were produced with exquisite craftsmanship and paper quality. This edition is commonly referred to as the Tenkai or Kan'ei-ji edition. This is the first official edition of *Tripitaka* in Japan, marking the beginning of printing Buddhist sutras in Japan. The Tenkai edition was also printed from wooden movable type,[45] bearing even greater significance. Prior to this, although China and Goryeo had repeatedly printed Buddhist sutras, they were all carved editions, and the moveable type printing of Buddhist sutras began in Japan.

8.2 Spread of Chinese Printing Techniques in Other Asian Countries and Africa

8.2.1 Early Development of Printing in Vietnam

Vietnam is connected by land to China's Guangxi and Yunnan provinces and has had close ties with China since ancient times. The Vietnamese ethnic group was one of the Baiyue tribes originally spread out in the middle and lower reaches of the Yangtze River, who later migrated to the northern part of what is now Vietnam, where they

45 川濑一马 (*かわせ かずま*). *Study of the Early Typographic Edition of Japan* (Kokatsuji-ban no kenkyu): *Vol. 1* [M]. Tokyo: Antiquarian Booksellers Association of Japan, 1967: 32–38.

formed various tribes and established the states of Au Lac and Nanyue. In the sixth year of the Yuanding era of Emperor Wu of Han (111 BC), the Han dynasty conquered Nanyue and established prefectures of Jiaozhi, Jiuzhen, and Rinan on its territory. From then on to the early Northern Song dynasty, for over one thousand years Vietnam was under the rule of the same feudal dynasty as China, using the same era names and Han Chinese characters. During the Eastern Han dynasty, Shi Xie (137–226), a learned man from Guangxi, was appointed as Jiaozhi's governor, serving for 40 years (187–226), during which he further developed culture and education. He also provided refuge to a large number of Han artisans, farmers, and scholars. As a result, culture and education in the region flourished, and the technology of producing hemp paper was then introduced from the mainland of China to Jiaozhi. During the Wu dynasty (222–280) of the Three Kingdoms period, Sun Quan conferred the titles of General of the Guards and Marquis of Long Bian on Shi Xie, and the regions of Jiaozhi, Jiuzhen, and Rinan were merged into Jiaozhou, with Shi Xie and his brother Shi Yi in charge. At this time, the technology of producing bark paper was also introduced to Jiaozhou from Guangzhou.

After the Tang dynasty, printed materials including Buddhist scriptures and almanacs were gradually introduced to Vietnam, and after the Song dynasty, printed books continued to flow in. According to History of Song: Chronicles of Emperor Zhenzong, in the third year of the Jingde era (1006), "Jiaozhou offered tributes and [Emperor Zhenzong of Song] granted Lê Long Đình Nine Classics and Buddhist scriptures." Lê Long Đình was the ruler of the Early Ly dynasty of Vietnam (980–1009), and the bestowed *Nine Classics* and Buddhist scriptures were all official Song engraved editions, the latter possibly being *Kaibao Tripitaka.* Starting in 1075, Vietnam implemented the imperial examination system, which increased the demand for reading materials. Relying solely on imported books could no longer meet the demand, so printing technology was introduced from China. The starting time of printing in Vietnam is still under research, but at the latest, woodblocks were used to print household registries in the early thirteenth century in the Tran dynasty (1225–1400). Vietnamese historian Ngô Sĩ Liên (1439–1499) wrote in *Chronicles of the Trans II* of his *Complete Annals of Great Viet* (Đại Việt sử ký toàn thư) (1479):

In the third year of the Daqing (Đại Khánh) era (1316) of Emperor Minh Tông of Tran dynasty, civil and military officials noted discrepancies in the household registry. When the officials saw the woodblock-printed papers, they deemed them as fake and therefore rejected them. Upon hearing this, the retired emperor said: "This is an old practice from the Yuanfeng era (Nguyen Phong) (1251–1258), and they are official records." Thus, there was a decree to government officials: Those who hold positions in the government but are not familiar with historical precedents often hold things up.

The "woodblock-printed papers" mentioned here refer to the household registers created using woodblock printing during the Yuanfeng era (1251–1258) in the Tran dynasty. Later, in 1295 Emperor Yingzong (Anh Tông) (1293–1313) sent the Senior Official Cheng Keyong as an envoy to the Yuan court and acquired a copy of *Tripitaka.* In 1299, some of the Buddhist scriptures were printed in Tianchang Prefecture (present-day Nanding). According to *Yueshi Tongjian Gangmu* ("Việt

Sử Thông Giám Cương Mục",1884) revised by Phan Thanh Gian, Director of the National History Museum of the Nguyen dynasty (1802–1945), it is stated in Volume 8 *Chronicles of the Tran Dynasty*:

"In the seventh year of Emperor Yingzong's reign (1299), a decree was promulgated for the spread of Buddhism around the country. Initially in 1295, Cheng Keyong was sent as an envoy to the Yuan court to request *Tripitaka*, which was brought back and kept in Tianchang Prefecture, and copies were printed for publication. Then in 1299, an edict was also issued that Buddhist scriptures for religious ceremonies and formal documents for public use be printed and distributed throughout the country".

According to Volume 37, *Records of the Le Dynasty* in *Yueshi Tongjian Gangmu*, during the reign of Emperor Le Thai Tong of the Le dynasty (1428–1527), in the second year of Thieu Binh era (1435), *Complete Works of the Four Books* was officially printed based on an edition from the Ming dynasty. In the eighth year of Quang Thuan era (1467) under the reign of Emperor Lê Thánh Tông (1460–1497), *Complete Works of Five Classics* and other printed materials such as historical works, poetry collections, and dictionaries were also issued. The printing locations were mainly in the traditional capital city of Hanoi. In addition to official prints, there were also privately published books that catered to the needs of the general public. The majority of Vietnamese editions were books in classical Chinese and those annotated in Chu Nom. Chu Nom was square-shaped, phonetic Vietnamese characters created in early 13th Century at the end of the Ly dynasty, using Chinese character strokes and character-making methods. Chu Nom annotated editions often appeared in later publications (as shown in Fig. 8.16). In the nineteenth century, during the Nguyen dynasty (1802–1945), when the capital was moved to Hue, the city also became another printing center where cultural, religious, and scientific works were published, such as *Tân truyên hải Thượng y tông tâm lĩnh toàn trật* ("Medical Understandings of the Medical Master Hải Thượng") by Lê Hữu Trác (1720–1791), now in the collection of the National Library of China. Vietnamese printed books have the same format as Chinese prints, but relatively few have survived to the present day.

Volume 8, *Annals of Tran* in *Đại Việt Sử Ký Toàn Thư* ("Complete Annals of Great Viet") also records the issuance of paper currency in late Tran dynasty:

> In April, summer of the ninth year of Emperor Thuan Tong's reign (1396), *tong bao hui chao* initially circulated. The denominations were designed with specific images: ten-wen notes with seaweed, thirty-wen with mangrove trees, one-hundred-wen with clouds, two-hundred-wen with turtles, three-hundred-wen with qilins, five-hundred-wen with phoenixes, and one-thousand-wen with dragons. Counterfeiters were punished with execution and property confiscation. After printing, people were required to exchange their old money for this new currency.

The ninth year of Tran Thuan Tong's reign (1396) corresponds to the 29th year of Emperor Hongwu's reign in the Ming dynasty. The paper currency, named *tong bao hui chao* (Thong bao hoi sao), was issued following the official Wang Ruzhou's proposal and modelling after the Ming dynasty's currency system. After printing, it was decreed that copper coins should be collected from the capital and various places and stored in the national treasury; private hoarding and use of copper coins were

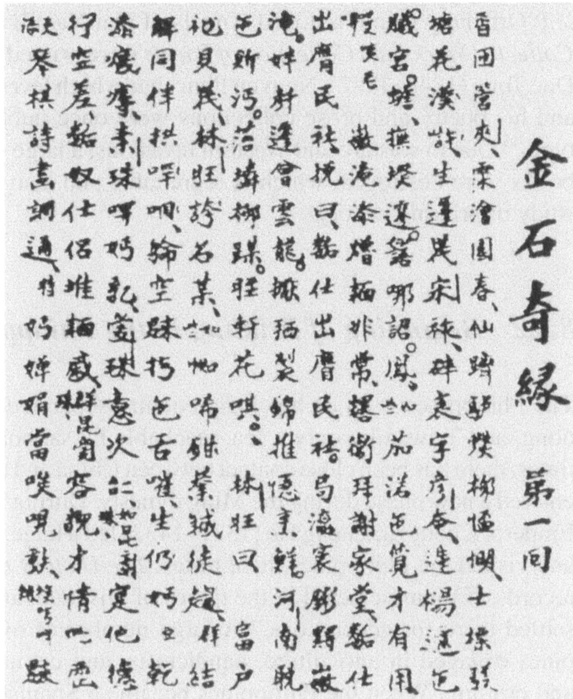
金石奇緣
第一回

Fig. 8.16 A carved edition of *Jin Shi Qi Yuan* ("The Marvels of Stones and Ink"), with chu nom and Chinese characters interspersed, and published in Vietnam during the nineteenth century

prohibited and counterfeiters of the new currency were executed. There were seven denominations, including 10-wen, 30-wen, 100-wen, 200-wen, 300-wen, 500-wen, and 1000-wen, each with different designs, with the highest denomination of 1000-wen featuring a dragon. The notes were stamped with official seals and inscribed with serial numbers based on combination of a character from *Qianziwen* and a number, such as "Character Heaven Number Fifty 天字伍拾号". The character and number were created by embedding the movable type into the engraved grooves of the printing plate. From the Song to Ming dynasties, Chinese banknotes were typically printed with copper plates and copper movable type, and Vietnamese paper money was likely the same. Thus, from 1396, copper movable type was undoubtedly used in Vietnam to print currency. However, excessive issuance later caused economic chaos, and copper coins were reintroduced during the later Le dynasty.

During the reign of Emperor Thanh Thai (1889–1907) of the Nguyen dynasty, manuscripts of both ***Record of Sacred Miracles*** and *Records of the Dharma Rain* had the phrase "copied from copper plates by decree, without a single mistake, stored at the Jiāfú Chéngdào temple." Here, "copper plates" should be understood as copper movable type plates. This indicates that books such as ***Record of Sacred Miracles*** were once printed from copper movable type but were not preserved. The Nguyen dynasty was the last feudal dynasty in Vietnam. During the reign of Emperor Thieu Tri (1841–1847), a set of wooden movable type was bought from the Qing dynasty, which was used in 1855 to print the 96-volume *Khâm định Đại Nam hội điển sự* *lệ* ("Officially Commissioned Annals of Dai Nam") and in 1877, *Tu Duc Imperial Collected Works* and *Collection of Poems* were printed, totaling 68 volumes. Emperor Duc Tong (1848–1882), Nguyen Phuc, had a high level of Chinese literary cultivation and his poetry and prose collections were once published using wooden movable type.[46] Due to warfare and Western invasions, a large number of Vietnamese ancient books were destroyed, which is regrettable and also has brought difficulties to the study of printing history.

8.2.2 Beginning of Printing in the Philippines and Thailand

The Philippines is an archipelagic country separated from China's Fujian, Guangdong, and Taiwan by only a sea, reachable by sailboat in three days. Since ancient times, there has been close contact between China and the Philippines, and the contact entered a new phase during the Ming dynasty. During the Yongle era (1403–1424) of Emperor Chengzu, Zheng He (1371–1433) led a huge fleet to sail to the Western Seas and visited the Philippines three times. *The History of Ming: Biography of Lusong* records, "Fujianese heard of the riches of this land, and tens of thousands of traders settled there for generations." A large number of overseas Chinese in the Philippines engaged in agriculture, handicrafts, and commerce, including papermaking and printing. When the Philippines became a Spanish colony in 1565, the colonial rulers persecuted the overseas Chinese, but they had to acknowledge that without them, the economy would hardly function. During the Spanish rule of the Philippines, Catholicism was brought over, but printing was not developed right away. Instead, it was the local overseas Chinese who initially laid the foundations of the printing industry. The most famous overseas Chinese involved in printing and publishing in the capital city of Manila was a Fujianese named Kong Yong (1538–1603), whose Spanish name was Juan de Vera. The earliest Chinese-language book printed in the Philippines is *Doctria Ckristiana* ("Christian Doctrine",1593), engraved and printed by Kong Yong in the twenty-first year of Wanli of the Ming dynasty, which is now housed in the National Library of Madrid (Fig. 8.17).[47]

The original Chinese title of this book is *Xin Ke Seng Shi Gao Mu Xian Zhuan Wuji Tianzhu Zhengjiao Zhenchuan Shilu* ("A Printed Edition of the Veritable Record of the Authentic Tradition of the True Faith in the Infinite God"), and its original Spanish title is *Rectificacian y Mejora de Principios Naturales* ("Rectification and Improvement of Natural Principles"). The first three chapters are related to religion, while the last six chapters introduce geographical and biological knowledge, including the spherical Earth theory. The author, whose Chinese name was Gao Muxian (Juan Cobo), was a Spanish Dominican friar who preached in the Philippines from 1588

---

[46] Zhang Xiumin. *The Invention of Printing in China and its Influence* [M]. Beijing: People's Publishing House, 1958: 157–158.

[47] Henri B M. Les Origines Chinoises de L' imprimerie aux Philippine [J]. *Monumenta Serica*, 1942 (7): 312.

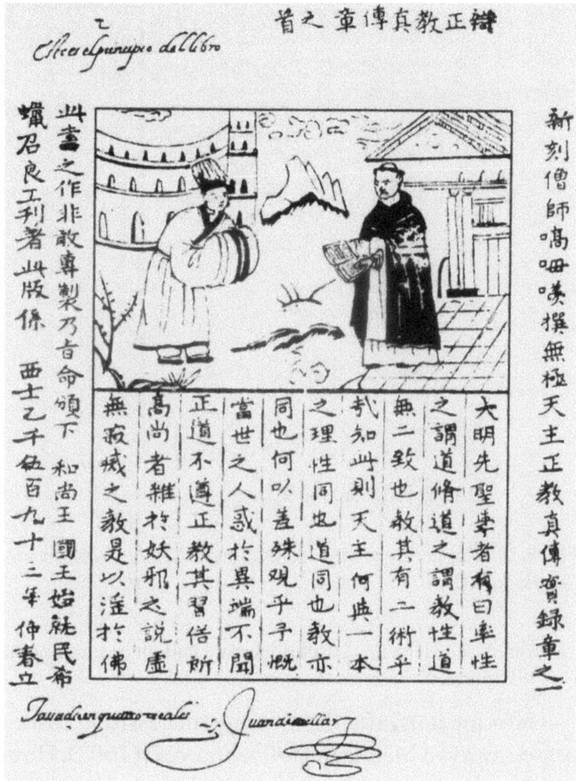

Fig. 8.17 Chinese woodblock edition of *Doctria Ckristiana* (*Christian Doctrine*) printed by Kong Yong in Manila in 1593

to 1592. He learned Chinese from the overseas Chinese[48] and wrote this book. In the same year (1593), Kong Yong published the book again in the local Tagalog language also using woodblock printing.[49] The Tagalog language was later made the national language of the Philippines in 1962. Therefore, it can be said that printing in the Philippines was pioneered by Kong Yong. After woodblock printing, he successfully cast copper movable type during his later years in 1602, which was used to print books in Chinese and foreign languages. In 1640, the Spanish missionary Aduarte wrote about Kong Yong:

> He devoted himself to inventing a printing machine in the Philippines, where there were no printing machines for him to model after nor any European printing techniques significantly different from those of the Chinese Empire for him to learn from ....Kong Yong (Juan de Vera) worked tirelessly and spared no effort to achieve his goal...So, this Chinese Catholic

---

48 Pelliot P. Notes sur Quelques Livres ou Documents Conserés en Espagne [J]. *T'oung Pao*, 1929 (26): 48.

49 Van Der Loon P. The Manila Incanabula and Early Hokkeins Studies [J]. *Asia Major*, 1966, 12(1): 2–8.

1. Title page in Spanish 2. Text in Chinese

Fig. 8.18 Copper movable type edition of *Newly Printed Orthodox Doctrine of God* published in Manila in 1606

> Kong Yong is the first practitioner and half an inventor of metal movable type printing in the Philippines.[50]

Unfortunately, after Kong Yong successfully cast copper movable type in 1602, he passed away in Manila the following year (1603). However, his brother Pedro de Vera and his apprentices continued his work by using the copper movable type to print books. The Spanish colonizers forced the Filipinos and local Chinese to convert to Catholicism and adopted Spanish names, so we currently cannot find out the Chinese name of Kong Yong's brother. The Imperial Library of Vienna (Biblioteca Imperial de Vienna) has a Chinese edition of *Xinkan Liao Shi Zhengjiao Bianlan* ("A Printed Edition of the Guide to the True Faith in God") (Fig. 8.18)[51] written by 罗明敖·黎尼妈 (Domingo de Nieva) in Chinese (which should now be standardly translated as Domingo Neva). "*Liao Shi*" is transliteration of the Spanish word "Dios" (God), so the title of the book might better be translated as *Newly Printed Orthodox Doctrine of God*. The title page is printed in Spanish, with the meaning in Chinese that "The *Orthodox Doctrine of God* was compiled in Chinese by Dominican friar Domingo Neva and printed by Pedro de Vera at Sagley Bookshop in Binondoc in 1606." The following pages consist of the Chinese preface and main text, formatted as half-pages with 9 lines each, and 15 characters per line.

Some people believe that this book was printed from woodblock, but after careful research, we argue that it is a copper movable type edition.[52] Comparing the 1593

---

50 Fernandez P. *History of the Church in the Philippines*: 1521–1878 [M]. Manila, 1979: 358–359.

51 Retana W E. *Origines de la Imprenta Filipina* [M]. Madrid, 1911: 71.

52 Pan Jixing. *History of Science and Technology in China*: *Papermaking and Printing* [M]. Beijing: Science Press, 1998: 556–557.

edition with this edition, it is found that the former has smooth printing, mainly with traditional characters, minimal distortion, and orderly arrangement of each line of characters; the font used for the text and illustrations is the same, indicating that it is a woodblock print edition. In contrast, the 1606 edition features dull and rough strokes, mostly simplified Chinese characters similar to those used today, with some distorted characters and irregular arrangement of each line of characters; some characters are slanted with unwanted blank spaces beneath the characters, and the book has no illustrations. All these indicate that this edition was printed from the metal movable type developed by Kong Yong, as described by Aduarte in 1640, and it should not be a woodblock print edition. The use of simplified characters was likely for casting the type more easily, and the stiff font was a characteristic typical of the early experimental stage in typecasting. The Spanish letters on the title page are relatively well-formed, as they have simpler shapes and strokes, easier to cast.

From 1593 to 1608, printing in the Philippines was monopolized by the Chinese, who ran bookstores as well as printing shops. Before 1603, publications were printed using woodblocks, but afterward, both woodblock and copper movable type editions were printed, each requiring individual authentication. In 1911, the Spanish writer W. E. Retana listed the names of eight Chinese printers in the Philippines between 1593 and 1640 in his book *The Origin of Printing in the Philippines*,[53] but he only provided their Spanish names instead of their Chinese names. They published books in Chinese, Spanish, and Tagalog, forming a powerful publishing group. Undoubtedly, the Ming dynasty Fujianese Kong Yong was a leading figure among them, and his brother owned the largest bookstore in the commercial district of Parian in Manila. As we know, copper movable type printing was quite developed in Fujian during the Ming dynasty, which deeply impressed the Spanish Martin de Rada (1533–1577) who visited in 1575 (the third year of the Wanli reign).[54] Therefore, Kong Yong's typecasting techniques obviously came from his hometown. Only after receiving technical instructions from the Chinese, did the Filipinos begin participating in printing work after 1608–1610.[55]

Thailand is another country in Southeast Asia that developed printing based on traditional Chinese technology. During the Ayuthaya dynasty (1350–1767), Thailand had frequent and friendly exchanges with the Ming dynasty, with 131 envoys sent between the two countries, including 112 from Thailand to Ming, an average of once every two years. According to *Xu Wenxian Tongkao* ("Supplement to Comprehensive Textual Research of Historical Documents", 1586) by Wang Qi (1540–1615), the king of Siam sent young students to study at the *Guozijian* in Nanjing in 1371 (the fourth year of the Hongwu reign of Ming), and thereafter the two countries exchanged students to learn each other's languages. Craftsmen from Fujian and Guangdong also went to Thailand with merchant ships to make a living, and after the first year of

[53] Retana W E. *Origines de la Imprenta Filipine* [M]. Madrid, 1911: 71.

[54] De Mendoza J G. *The History of the Great Empire and Mighty Kingdom of China* [M]. Parke R, tr. London: Hakluyt Society, 1853: 131–134.

[55] Boxer C R. Chinese Abroad in the Late Ming and Early Manchu Periods, compiled from contemporary sources 1500–1750 [J]. *T'ien-Hsia Monthly*, 1939, 9(5): 459.

the Longqing era (1567) they operated businesses in Thailand such as agricultural tools, copper and ironware manufacturing, sugar, tea, papermaking, and printing. In his book *Haiyu* (1536), Huang Zhong (1474–1553) of the Ming dynasty mentioned that Ayuthaya, capital of Siam, had Milk Street as an overseas Chinese district where Chinese run paper stores and bookstores.

During the Thonburi dynasty (1767–1781), more than 50 Chinese merchant ships from Songjiang (present-day Shanghai), Ningbo, Xiamen, and Chaozhou arrived annually, with thousands of people on board. The travelers read *Sanguo Yanyi* ("Romance of the Three Kingdoms") as a pastime during their journey, resulting in the introduction of the book to Thailand. Rama I (1782–1809), the founder of the Chakri or Bangkok dynasty (eighteenth-nineteenth century), was very interested in *Sanguo Yanyi* and ordered the officials to translate it into Thai, and so both the Thai and Chinese versions were published simultaneously. During the reign of Rama II (1809–1825), other Chinese novels such as *Shuihu Zhuan* ("Water Margin"), *Xiyou Ji* ("Journey to the West"), *Dongzhou Lieguo Zhi* ("Records of the Eastern Zhou Dynasty"), *Fengshen Yanyi* ("Creation of the Gods"), *Liaozhai Zhiyi* ("Strange Stories from a Chinese Studio"), and *Honglou Meng* ("Dream of the Red Chamber") were also translated into Thai and became popular in Thailand.[56] After the era of Rama V (1868–1910), Bangkok had three royal printing houses and a specialized printing house for publishing ancient Chinese classics. At this time, books were already being printed using mechanical printing methods. Prior to this, publishing and printing in Thailand were still undertaken by local Chinese bookstores.

8.2.3 Beginning of Printing in Central and West Asia

The countries in Central and West Asia, collectively known as the Western Regions in ancient Chinese books, introduced papermaking technology from China in 751 during the early period of the Abbasid Empire (751–1258), and they mainly produced hemp paper. The first paper mill was established in Samarkand located in present-day Uzbekistan, and subsequently, new paper mills were built in present-day Iraq, Syria, Iran, and other areas. However, the Muslim world did not timely introduce printing technology from China, presumably due to differences in religious and cultural backgrounds. Some people say that the brush used by the Chinese for inking was made of pig bristles, which the Arab people regarded as blasphemous if they used it to print *Quran*.[57] But this is not a sufficient explanation, as the Chinese traditionally used soft palm brushes made from the husk of the palm tree (Trachycarpus fortunei) and

---

[56] Ge Zhilun. *Cultural Exchanges between China and Thailand before 1949* [M]. // Zhou Yiliang. *History of Sino-foreign Cultural Exchanges*. Zhengzhou: Henan People's Publishing House, 1987: 516–517.

[57] Carter T F. *The Invention of Printing in China and its Influence* [M]. Translated by Wu Zeyan. Beijing: Commercial Press, 1957: 129.

never used brushes made of pig bristles. It is probable that unlike the Buddhist scriptures, *Quran* does not require believers to repeatedly transcribe verses or mantras, so printing did not have the same stimulating effect as it did on Buddhism. We note the fact that when the Mongols who believed in Buddhism took control over the previously Islamic regions of Central and West Asia, the technology of printing began to develop quickly there.

After establishing the Mongol Empire in northern China in 1206, Genghis Khan (1162–1227) led his army westward, defeating the Western Liao dynasty (1124–1211) in 1211, and from 1219 to 1223, he advanced into Central Asia, conquering Khwarizm, Bukhara, and Samarkand, and pushing all the way to the Caspian Sea. The Western Regions, previously taken by the Abbasid Empire from the Tang Empire, all fell under the control of the Mongol Empire. After Ogedei Khan (1189–1241) succeeded the throne, he sent Genghis Khan's four grandsons including Batu (1209–1256) on a second western expedition, during which they occupied large areas of Russia and invaded Poland and Hungary, establishing the Kiptchak Khanate (1240–1480) with its capital in Sarai (present-day Astrakhan, Russia) downstream from the Volga River, and approaching Bohemia in the Holy Roman Empire. During 1253–1259, Mongke Khan (1208–1259) sent his brother Hulagu (1219–1265) on a third western expedition. They captured Baghdad with artillery and destroyed the Abbasid Caliphate, thus ending the rule of the once-glorious Arab Empire in the Middle Ages.

In 1260, Kublai Khan (1215–1294) ascended to the throne and appointed Hulagu as the ruler of the Mongol Il-Khanate (1260–1353) in the conquered territories, which included present-day Iran, Iraq, Syria, and Anatolia (the Asian part of Turkey), with its capital in Tabriz, Iran. The Il-Khanate stretched from the Amu Darya in the east to the Mediterranean Sea in the west, across the sea from Europe, bordered by the Caucasus or ancient Scythia in the Kipchak Khanate in the north, and the Indian Ocean in the south. The Mongol army used force to open up the Silk Road, which had been blocked between the East and West, making Il-Khanate a hub for trade and cultural exchange between the East and West. During the reign of Kublai's grandson, Arghun Khan (1284–1291), Il-Khanate's economy and culture further developed. At the time of his brother, Gaykhatu Khan (1291–1295), Il-Khanate took a significant economic measure, which was to adopt the Chinese paper money system to issue paper money. In 1311 under the commission of Ghazan Khan (1295–1304), the Persian scholar Rashid al-Din compiled *Jami' al-Tawarikh* ("Compendium of Chronicles"), written in Persian. The story of the issuance of paper money under Gaykhatu Khan was narrated in this book:

> On the 6th of December in the year 691 AH (November 10, 1292), at the winter camp in Arran, Sadr-ad-Din Jackhan was appointed as the Grand Vizier and Minister of Finance. At the beginning of June, 695 AH (early May, 1294), a meeting was convened to discuss paper money. Sadr al-Din Jackhan and some of the governors, inspired by the widespread use of *baochao* in China, discussed how to introduce *baochao* in this country (Il-Khanate). They brought this matter to the attention of the monarch, who ordered Pulad-insang to give details of the matter. Pulad-insang reported that, '*Baochao* is paper stamped with the imperial seal, in place of minted coins throughout China. The Chinese coins and silver ingots are sent to the state treasury…' Then a decree was issued to immediately produce paper money. On August the 27th (July 23), Akbuka, Togacar, Sadr al-Din, and Tamaci Inak went to the

> city of Tabriz to print paper money. On September the 19th (August 13), they arrived to announce the decree and printed a large number of banknotes. At the same time, it was decreed that anyone refusing the paper money would be immediately executed. Within a week or so, people accepted the paper money, afraid of being executed, but it could hardly buy anything. Most residents of Tabriz had to leave. Eventually, the implementation of paper money failed.[58]

The paper money printed by Il-Khanate in Tabriz in 1294 imitated the format of the Yuan dynasty's *zhiyuan tongxing baochao* used in China, with denominations ranging from half a dirham to 10 dinars. The notes featured Mongolian and Arabic script, indicating the year of issuance and content of beheading counterfeiters as well as the Chinese character "钞" (chao) and its transliteration "caw". Besides, a unique serial number was printed on each note. Considering that the Yuan dynasty banknotes were printed using copper plates and copper movable type from the 14th year of Zhiyuan reign (1277), it is likely that the paper money issued in Tabriz followed the same method. If not, they would have used wooden blocks and wooden movable type. In any case, paper money had to be printed with movable type. The printing was carried out by both local Chinese and Persian artisans. Prior to the issuance of paper money, in the 21st year of Zhiyuan reign (1284), Kublai Khan sent his Prime Minister Pulad-insang from Beijing to Il-Khanate, possibly for the enthronement of Arghum Khan to his father's throne, and he was retained there. Pulad-insang was a witness to or participant in the issuance of the *zhiyuan baochao* in the 17th year of Zhiyuan era (1280) of Yuan Emperor Shizu and now he became an advisor on the issuance of paper money in Il-Khanate. However, Il-Khanate lacked sufficient gold reserves to back the currency, had little experience in managing the currency, issued excessive amounts of paper money, and faced resistance from the Arabs who were not used to this new form of currency, ultimately leading to the failure of the initiative.

Although the implementation of paper currency in Il-Khanate failed, its success in printing banknotes had significant influence in the history of printing in the West Asian region. Rashid al-Din also covered Chinese printing technology in the world history section of his *Jamiʿ al-Tawarikh.* In 1317, the contemporary Persian poet Abu Sulayman Dau'd referred to Rashid al-Din's account in *Rawdatu 'uli- 'l-Albab fi Tawarkhi 'l-Akabir wa 'l-Ansab* ("On the History of Great Men and Genealogy"). This book, abbreviated as *Tarikh-i-Banakati* ("Garden of the Intellegents"), is on par with Rashid al-Din's *Jamiʿ al-Tawarikh* and attracted the Europeans' attention during the Renaissance. Rashid al-Din's original account was translated into French in 1834, while Dau'd's citation was rendered into English in 1920. After comparing the French and English translations, Carter deemed the contents to be essentially the same. When translating into Chinese, we found some inaccuracies with certain terms, which have been corrected. Below is the excerpt of our translated text:

The Chinese, according to their custom, have employed an ingenious way to reproduce a book exactly as it was written, which is still in use today. When people

---

[58] Rashit. *Collection of Chronicles: Vol. 3, Biography of Kiheedu Khan* [M]. Translated by Yu Dajun. Beijing: Commercial Press, 1986: 225–229. This is a translation of the Russian edition, with the names of people quoted from Russian to Latin transliteration, and correction of a few words made.

want to accurately copy a well-written and valuable book without changing anything, they hire skilled scribes to hand copy the text, and then transfer the words onto wooden blocks page by page.[59] They also invite learned people to carefully proofread the text and sign the back of each wooden block. Skilled engravers then carve the words onto the wooden blocks, mark each block with the book's page number, and then number the entire set of blocks. Much like the coin molds used in a mint, the blocks are then placed in bags with a special seal and entrusted to a trustworthy person and kept in an office designated for this purpose. If someone wants to obtain a printed copy of the book, they must apply to the preservation office and pay a fee to obtain the wooden blocks. Then paper is pressed against the wooden blocks to "print" the book, just like the way that coins are cast with the molds. The resulting printed books are free from any errors or omissions and are completely reliable. This is how Chinese history books have been handed down.[60]

The above description of the traditional Chinese woodblock printing technique provided by Rashid al-Din in 1311 is the earliest account of printing technology in regions west of China. This author once served as Grand Vizier when Ghazan Khan ruled the Ilkhanate. In 1313, Rashid al-Din compiled in Persian *Tanksuq-nāmah-i-Ilkhan dar Funūn-i 'Ulūm-i Khitāi* ("Treasure Book of the *Ilkhan* on Chinese Science and Techniques"*)*, which introduced Chinese medical knowledge. This book was translated into Turkish and published in 1934.[61] Since Ghazan Khan's reign, Mongol rulers accelerated their Islamization process, and Ghazan Khan himself converted from Buddhism to Islam. With his scientific literacy, extensive knowledge, and linguistic talent, he recognized the role of printing technology in promoting cultural development. Therefore, he ordered the printing of *Quran* and other works in Arabic, further developing the printing industry in the Khanate. Meanwhile, Persian language books also appeared, though early printed copies are rare. A 1491 Persian print can be found in the British Museum in London (Fig. 8.19).

According to an article dated October 8th, 1982 from *The Jewish Weekly*, a 14th-century Hebrew print in the Taylor-Schechter Genizah Collection at the Cambridge University Library was also identified as woodblock print.[62] This finding suggests that the Jewish people who believed in Judaism and resided between the Ilkhanate (present-day Iran) and Egypt also possessed the printing technology. The Jews were known for their business acumen and trade operations between the Mediterranean, Europe, and Western Asia. During their travels, these Jewish merchants needed to

---

[59] The word "writing" in the original text is mistaken, because the Chinese do not write the characters on the block, but transfer the characters on the paper to the wooden block in reverse, and then carve the characters.

[60] Klaproth H J. *Lettre a M. le Baron Alexandre de Humboldt sur l'invention de la boussole* [M]. Paris, 1834: 131–132. Browne E G. *Persian Literature under the Tartar Dominion* [M]. Cambridge: Cambridge University Press, 1920: 100–102.

[61] Sarton G. *Introduction to the History of Science*: *Vol. 3* [M]. Baltimore: Williams & Wilkins Co., 1947: 969.

[62] Needham J. *Science and Civilization in China*: *Vol. 5*. Cambridge: Cambridge University Press, 1985: 307.

Fig. 8.19 Persian edition of a religious book published in 1491

recite biblical scriptures, which ultimately led to the printing of the Bible, a technology that apparently originated from the Ilkhanate of the Mongol Empire. By that time, residents across vast areas of Western Asia could access printed works in Arabic, Persian, and Hebrew, particularly religious texts. Although printed materials at that time could not fully replace handwritten manuscripts, there was a transitional period of using both printing and handwriting, a situation common in both Eastern and Western nations. The Ilkhanate was ruled by the Mongols, and many of its local high officials appointed by the Mongol administration were proficient in the Mongolian language, thus making it one of the official languages in those regions. Official documents, such as household registers, were printed in both Mongolian and Arabic, as with the paper money. In addition to his native Mongolian, Ghazan Khan was proficient in Chinese, Tibetan, Arabic, Persian, Sanskrit, and Latin, and he also had a deep study of Buddhism and Islam. The development of printing technology during Ghazan Khan's reign was likely related to the ruler's personal qualities and interests.

The Central Asian region, geographically closest to China, has a long history of papermaking. So, it's likely that this region was exposed to Chinese printed materials and acquired printing knowledge before West Asia. During the Yuan dynasty, the Chagatai Khanate's territory covered a large area of what is today's Xinjiang (China), Kazakhstan, Uzbekistan, Tajikistan, and other regions, with its capital located in Almaliq (present-day Huocheng, Xinjiang). During the thirteenth and fourteenth centuries, printing industry was well developed in the Turpan region of eastern Xinjiang, where woodblock editions and wood moveable type editions were published. Since the twentieth century, Buddhist prints in Chinese, Mongolian, Tibetan, Tangut, and the Lantsa style of Sansckrit common in Central Asia, dating from the thirteenth century to the fourteenth century, have been unearthed there.[63] In 1908, 960 pieces of twelfth-thirteenth century Uyghur woodblock movable type used by the Uyghur people in Xinjiang to print books were found in Dunhuang. Out of Xinjiang were countries of Central Asia, where paper resources were abundant and many Chinese merchants lived and did business. Using local paper for publishing was unsurprisingly more profitable than transporting materials from the mainland. As previously mentioned, Chinese people were engaged in printing in the Philippines and Thailand. Central Asia was also a vital part of the East–West overland Silk Road, serving as a transit point for the westward spread of Chinese prints and printing technology.

8.2.4 Beginning of Printing in Africa

Africa is far from China, and although there were some interactions between China and Africa during the Tang and Song dynasties, they were mostly sporadic short-term visits, thus impossible to transfer papermaking and printing technologies there. It was only after the development of papermaking and printing in the Arab region that it became possible to introduce these two technologies to Africa, because Northeast Africa borders the Arab region and is also Islamized.

In 641, Umar ibn al-Khattab (581–644), the second caliph of the Arab Empire, sent troops to conquer Egypt, an ancient civilization in North Africa, bringing with them Arab legal systems and Islamic culture. Since ancient times, papyrus flakes abounding along the Nile were used as a writing material, but during the Abbasid era, paper produced in West Asia was brought to Egypt since 800, and the first paper mill in Africa was established in Cairo before 900. In 909, the Muslim Shiite Moez ad-Din Allah founded the Fatimid dynasty (909–1170) in North Africa, breaking away from Abbasid rule. In 986, the Fatimid caliph Al-'Aziz (975–996) conquered another ancient civilization in northwest Africa, Morocco, and in 1100, a new paper mill was established in the city of Fez in Morocco where most papermakers were Arabs. At its peak in 1202, this paper mill had 472 water-powered pestles for pulping[64] and exported paper to Europe in addition to supplying local demand. With the expanding

[63] Carter T F. *The Invention of Printing in China and its Spread Westward* [M]. Translated by Wu Zeyan. Beijing: Commercial Press, 1957: 122–125.

[64] Hunter D. *Papermaking: The History and Technique of an Ancient Craft* [M]. New York: Dover, 1978: 470.

use of paper, the market share of papyrus gradually diminished and eventually disappeared from history, paving the way for the advent of printing technology.

Printing in Africa first appeared in Egypt, and the technology was introduced through the Ilkhanate. In 1878, a large number of paper manuscripts and over 50 pieces of printed fragments were unearthed from ancient tombs in el-Faiyum in northern Egypt, along with many papyrus manuscripts. These artifacts were later owned by Archduke Rainer of Austria-Hungary, and subsequently after his death, they were included in the Austrian National Library's Rainer Collection. Heidelberg University Library in Germany holds six such printed items, which were identified as woodblock prints by Adolf Grohmann in 1922. One of them is printed on sheepskin and the remaining five are on paper. Most of the prints in Vienna vary in size, the larger being about 30 cm × 10 cm and the rest being smaller fragments. Some of the prints are finely engraved and printed with lines, and others are roughly made without lines. In addition to black ink, red ink was also used in printing. After research, Carter wrote, "There is various evidence to suggest that they were not printed by pressing but rather in the Chinese way, by placing the paper on the woodblock and brushing it lightly with a brush."[65]

This batch of unearthed prints bears Arabic script in different fonts but there are no parts bearing dates. Experts of Arabic studies Joseph Karabacek and Grohmann believed that these printed materials were Islamic prayers, amulets, or texts from *Quran.* They also dated these prints between 900 and 1350 based on the Arabic script style. However, this time range should only be understood as upper and lower time limits. The lower limit of 1350 is correct because the manuscripts of chronicles excavated at the same time were written up to this point of time. However, setting the upper limit at 900 is likely too early. For example, what is believed to be the earliest printed item (No. 946, Rainer Collection) is a fragment of verses 1–6 from Chapter 34 of *Quran* (Fig. 8.20), measuring 10.5 cm × 11 cm, and its font style is the oldest. Carter pointed out that dating the prints by font style has limitations, because prints with earlier font style might have been engraved using fonts from earlier manuscripts. He suggested that it is improbable for Egypt and other Arab regions to have had printing technology as early as the tenth century. In 1954, Grohmann revisited Egypt and began to have doubts about the original dating in view of the fact that more woodblock prints from the tombs of el-Ushmunein and Ikhmin in Upper Egypt since 1925 all date from after the tenth century.[66]

Therefore, the age of the unearthed fragments of the Quranic texts as mentioned above needs to be re-examined. From the perspective of the westward transmission of Chinese printing technology, the prints unearthed in Egypt, regardless of the font, are all products of the period after the Mongol Westward Expedition and cannot be earlier than 1294, as printing activities in the Arabic cultural area only started from that year. On that account, we believe that the discovered prints in Egypt should be

---

[65] Carter T F, Goodrich L C. *The Invention of Printing in China and Its Spread Westward* [M]. 2nd ed. New York: Ronald, 1955: 176–178.

[66] Carter T F, Goodrich L C. *The Invention of Printing in China and Its Spread Westward* [M]. 2nd ed. New York: Ronald, 1955: 181.

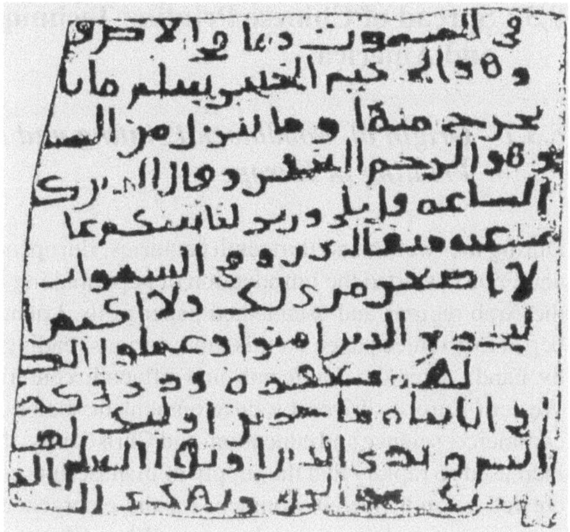

Fig. 8.20 Fragment of Arabic Quranic text carved and printed in Egypt during 1300–1350

dated between 1300 and 1350. It was the time when Egypt was under the rule of the Mameluke dynasty (1250–1519) established by the Turkic military general Baybars (al-Marik al-Zehr Rakn-al-Din Baybars, 1233–1277). The term "Mameluke" means "slave" in Arabic, so the dynasty is also known as the Slave dynasty. Before the thirteenth century, Arab rulers enlisted Central Asian Turkic slaves as soldiers, who were brave, good at fighting, and became the backbone of the regular army and palace guards holding military power. Baybars, a slave of Turkic origin, made great contributions in resisting the Mongolian invasions of Egypt and the European Crusades, and he later overthrew the Arab rules and established his own dynasty. The Turkic people were one of the ancient Chinese ethnic groups who once roamed around the Altai Mountains in what is now Xinjiang, and after the Tang dynasty, they migrated to Central Asia, so they have Chinese lineage. Although the Turkic rulers converted to Islam, they did not ban the printing of *Quran*, so religious books were printed in Egypt under Turkic rule, following the Ilkhanate under Mongol rule. From the above discussion, it is evident that there were already printing activities along the eastern and southern coasts of the Mediterranean and other surrounding areas of Europe during the thirteenth and fourteenth centuries, and Europe could not have been unaware of this.

8.3 Spread of Chinese Printing Techniques in Europe and America

8.3.1 Origin of Woodblock Printing and Movable Type Printing in Europe

During the twelfth and thirteenth centuries, European countries such as Spain, Italy, and France, started the introduction of papermaking technology from China through the Arab regions and established paper mills. Later, other European countries also began to produce paper. At this time, various reading materials were still transcribed by hand. After the fourteenth and fifteenth centuries, during the Renaissance in Western Europe, driven by development in society, economy, urban industry and commerce, science and education, and Christianity, the demand for reading materials increased so rapidly that the supply of manuscripts could not meet social needs. This created a conducive environment for the emergence of printing technology, which in turn promoted social development. When Europe needed printing technology, the Yuan and Ming dynasties in China were entering a new stage of comprehensive development in woodblock, copperplate, and movable type printing techniques, coinciding with an unprecedentedly active time of direct contacts between China and Europe. In addition to acquiring printing knowledge through West Asia and North Africa, Europeans also had the possibility of directly importing printing technology from China via the overland Silk Road. The Western expeditions of the Mongol army in the thirteenth century reopened the once-blocked Silk Road, creating conditions for economic and technological exchanges as well as the movement of people between the East and the West. Transportation between the Yuan's capital (now Beijing) and European cities such as Rome and Paris was unimpeded. Regarding the pathways for the spread of Chinese printing technology to Europe, it can be said that "all roads lead to Rome."

In the Yuan dynasty, emissaries, tourists, merchants, clergymen, craftsmen, and scholars from both China and Europe visited each other, and Europeans might have seen printed materials and related technical knowledge in China. For example, in 1245, the Pope dispatched Italian clergyman Jean Plano de Carpini (1182–1252) to the Mongol Empire, accompanied by Polish clergymen and Austrian merchants. They arrived in Karakorum (present-day Inner Mongolia) in 1246 and were received by the Mongol Emperor, Güyük Khan. After returning to France in 1247, Plano wrote in Latin *Libellus Historicus*, in which he introduced China, stating that the Chinese were masters of craftsmanship, and their skills were unparalleled in the world, and they had written characters and historical records of their ancestors.[67] He also mentioned that China had scriptures similar to the Bible, which likely refers to the printed Buddhist scriptures. At the same time, French King Louis IX sent the Franciscan friar Guillaume de Rubrouck (1215–1270) to visit China along with the Italian Bartolome da Cremona and others. They were received by Mongke Khan

[67] Dawson C. *The Mongol Mission* [M]. London: Sheed & Ward, 1955: 22.

in Karakorum in 1253 and returned to Paris in 1255. In Rubrouck's *Itinerarium ad Orientales*, he wrote about the paper currency issued at the time:

> The commonly used currency in China was made of bark paper, each piece measuring about one palm in length and width (7.5 cm x 10 cm), stamped with several lines of scripts resembling those on the seal of Mongke Khan. They used fine paintbrushes to write characters, each composed of several strokes.[68]

During the Yuan dynasty, paper currency was printed with mulberry paper, and Guillaume de Rubrouck was the first European to introduce the use of printing technology for issuing paper money in China. He also saw skilled people from Germany, Russia, France, Hungary, and England serving the court of the Great Khan in Karakorum. During the reign of Emperor Shizu of Yuan, Venetian merchant Nicolo Polo and his brother came to China and were summoned for an audience. When they returned home, emissaries were sent to accompany them, carrying a letter to the Pope. After completing their mission, they returned to China in 1271, bringing the young Marco Polo (1257–1324), who was left in the palace by Emperor Shizu and later entrusted with important duties. He stayed in China for 17 years. In 1292 Marco Polo returned to Europe from Quanzhou and arrived in Venice in 1296. His *Travels of Marco Polo* (1299) opened Europeans' eyes to China's wealth and highly developed material civilization. He mentioned in his book that there was a money printing factory in Beijing and paper money made of mulberry paper was circulating throughout the country, and long-used old notes could be exchanged for new ones.[69]

Italian merchants from Venice and Genoa were keen on trading with China during the Yuan dynasty, as witnessed by the Venetian silver coins unearthed in Guangzhou in 1952 and the Latin tombstones of two merchants from Genoa dated 1342 and 1399 respectively unearthed in Yangzhou.[70] The Venice municipal archives also record a lawsuit in 1341 against a local defendant who had taken a large sum of money to China for business.[71] So many merchants came to China that the book *Practica della Mercantura* ("The Practice of Commerce"), written in Italian in 1340 by the Florentine merchant Francesco Balducci Pegolotti (1305–1365), had a special chapter on how to trade with China.[72] At that time, there were two transportation routes between China and Europe: the northern route started from Xinjiang via the Kipchak Khanate through Russia, Poland, Bohemia to the Holy Roman Empire, basically along the overland Silk Road; the southern route was of both land and sea, starting from Quanzhou or Guangzhou via the Ilkhanate through Armenia, Persia, Turkey to Italy. When traveling between China and Europe, one had to use either the northern or southern route for entry and exit. Sometimes the northern route was

---

[68] Dawson C. *The Mongol Mission* [M]. London: Sheed & Ward, 1955: 171–172.

[69] Polo M. *Travels of Marco Polo* [M]. Translated by Li Ji. Shanghai: The Oriental Book Company, 1936: 159–160.

[70] Xia Nai. Latin Tombstones in Yangzhou and Venetian Silver Coins in Guangzhou [J]. *Archaeology*, 1979 (6): 552.

[71] Larner J. *Culture and society in Italy*: 1290–1420 [M]. London, 1971: 30.

[72] Yule H, Cordier H. *Cathay and the Way Thither* [M]. London, Hakhuyt Society Publication, 1914: 137, 171.

taken for the outbound journey and the southern route for the return journey, or vice versa. Generally, a one-way journey took 1–2 years, including the time needed for stops in cities along the way.

While Europeans travelled east to China, there were also Chinese traveling west to Europe during the thirteenth century. The *History of Yuan: Annals of Emperor Xianzong* records that in 1253, a group led by Bieluobie'erge was sent to Russia to conduct a census. Russian history also records that in 1257, Mongol officials went to Ryazan, Suzdal, and Murom in Russia to register households and collect taxes. In 1259, Bieluobie'erge and Gosachieke, along with their families and subordinates, went to Valkhov to conduct a census. The household registers during the Yuan dynasty were uniformly printed by the Ministry of Household, filled in at any time, and then bound into books before being archived. Russians often saw printed materials from Beijing. The *History of Yuan: Annals of Emperor Yingzong* records that in 1320, after Russia and other regions submitted to the imperial court, a total of 14,000 guan in paper money was granted for them to take back, indicating that paper money issued by the Great Khan was also in circulation in Russia, and when they learned how it was printed, they would pass it on to the Western Europeans who traveled through the region.

Among the Chinese who visited Western Europe during the thirteenth and fourteenth centuries were Rabban Bar Sauma (1225–1293), a Uyghur Nestorian Christian born in Beijing, and his Mongolian disciple Marcos (1294–1314). During 1275–1276, they left Beijing and traveled westward through Xinjiang to Jerusalem for a pilgrimage. In 1280, they reached Baghdad, where Rabban Bar Sauma was appointed as the chief inspector of the Nestorian Church and Marcos as the bishop of the Baghdad diocese. In 1285, the Mongol Khan Arghun sent Marcos as an envoy to Rome. In 1287, dispatched by Arghun, Rabban Bar Sauma went to Rome and also visited Genoa and Paris where he was received by the King of France and paid a visit to the University of Paris. In 1288 when he was in Rome he presented a letter of credence to the Pope. After returning to Baghdad, Marcos wrote a travelogue in Persian,[73] which was later translated into Syrian and French in the nineteenth century and into English in the twentieth century. Rabban Bar Sauma was the earliest known Chinese traveler to Western Europe. He studied in Beijing from a young age and was knowledgeable about printing technology. On his journey, he passed through his ancestral homeland Xinjiang, the center of printing, and when asked by Europeans about how books were printed in China, he would be happy to share his knowledge.

In 1288, just one year after Rabban Bar Sauma's visit, Pope Nicholas IV (1288–1292) sent the Italian priest Giovanni da Monte Corvino (1247–1328) from Rome to China, accompanied by the priest Nicholas da Pistoia and an Italian merchant named Pietro da Lucalonga. When they arrived in India via Ilkhanate Nicholas died of illness. In 1293, Monte Corvino and Lucalonga arrived in Quanzhou and in the following year they reached Beijing at a time when Kublai Khan, the Yuan Emperor died. They were received by Emperor Chengzong (1295–1307), and after presenting a letter

[73] Chabot J B. *Relations du Roi Argoun avec l'Occident* [M]. Revue l'Orient Latin, 1894: 57. Moule A C. *Christians in China before the Year 1550* [M]. New York, 1930: 106.

from the Pope, they were granted permission to preach Christianity. The National Library of Paris holds two letters written by Monte Corvino in Latin to the Roman Curia. The first letter, dated May 18, 1305, describes how he faced discrimination from the Nestorian Christians upon his arrival and how he was prohibited from printing any doctrine different from Nestorian beliefs. Only due to the protection from Emperor Chengzong was he able to preach independently. In 1298 and 1305, two churches were built in Beijing, and in 1303, the German friar Arnold de Cologne came to Beijing to assist him in the work, baptizing 6000 people and recruiting 40 children to learn Latin and religious rituals to form a choir to sing hymns. By this time, Monte Corvino had already learned Mongolian and translated the New Testament and Psalms into Mongolian,[74] indicating that many of the followers were Mongolians.

In Monte Corvino's second letter, written on February 13, 1306, he mentioned that "I created six paintings of sacred images based on the Old and New Testaments to educate people with low literacy. The paintings had texts in Latin, Tursic, and Persian so that anyone who could read any of these languages understand them."[75] Carter suggests that the religious paintings with words provided by Monte Corvino were actually prints because "in China at that time, the printing of all important works was quite natural." He also noted that 50 years after Monte Corvino printed his religious paintings in Beijing, similar prints appeared in Europe, which "may not be a complete coincidence",[76] suggesting that Europeans might have adopted the printing techniques used by this Italian in China and adapted them to their own use in Europe.

Carter's view is correct because when Monte Corvino first arrived in Beijing, he discovered Nestorian Christian prints, and when he wanted to adopt the method to print Christian doctrines, he faced opposition from Nestorian Christians. It was not until he received support from Emperor Chengzong of Yuan and the Mongolian prince Korguz (1234–1298) that he was able to print religious images independently. The purpose of printing in three languages was to distribute them to a wide range of believers within and outside the region to compete with Nestorian Christianity. Given the large number of copies, it was impossible to hand-paint each one in the Yuan dynasty of China, so they had to be engraved and printed. The printing should be done during the Dade period (1297–1307) and was fulfilled by Chinese craftsmen. For Monte Corvino and Arnold de Cologne, they accomplished a feat of participating in printing work in China that Europeans had never experienced before. These printed materials could easily be taken to Europe and replicated following the same way. So European visitors to China from 1300 to 1368 became the medium of transmission.

---

[74] Mould A C. *Christians in China before the Year 1550* [M]. Translated by Hao Zhenhua. Beijing: Zhonghua Book Company, 1984: 193, 199.

[75] Mould A C. *Christians in China before the Year 1550* [M]. Translated by Hao Zhenhua. Beijing: Zhonghua Book Company, 1984: 203–205.

[76] Carter T F. *The Invention of Printing in China and its Spread Westward* [M]. Translated by Wu Zeyan. Beijing: Commercial Press, 1957: 139.

8.3.2 Beginning of Woodblock Printing in Europe

In addition to paper currency, religious paintings, and printed books, the Europeans in the thirteenth and fourteenth centuries also encountered popular entertainment products like playing cards. In the twentieth century, playing cards printed in the Yuan dynasty in the fourteenth century (Fig. 8.21), measuring 9.5 cm × 3.5 cm, were unearthed in Turpan, Xinjiang. As the Mongol armies marched westward, these playing cards were brought into Europe, where they quickly became popular in several Western European countries. Therefore, these printed products became the forerunners for the introduction of printing to Europe. The period between 1350 and 1400 was the initial stage of the development of woodblock printing in Europe, and the early printed products were exactly popular playing cards and religious paintings for the general public. Germany and Italy were the first European countries to have such printed products. Early municipal records of Augsburg and Nuremberg, both in southern Germany, frequently mention "card makers" (Kartenmacher) in the years 1418, 1420, 1433, 1435, and 1438. Playing cards could be made by hand-drawing, stamping, or printing, and obviously printing was the most cost-effective and thus more widely used method.

Some of the surviving early Italian playing cards were printed, but their exact time periods are difficult to determine. A historical record with a clear date is a proclamation issued by the Venetian authorities in 1441, which states:

> Given that the production of large quantities of printed cards and painted images in places outside Venice has resulted in the decay of the techniques and secret methods of making paper and printing icons originally intended for use in Venice, this deplorable situation must be remedied ... It is stipulated henceforth that all the above-mentioned products printed or

Fig. 8.21 Playing cards printed in the Yuan dynasty (1300), discovered in Turpan in 1907 (collection of the Berlin Museum of Ethnology)

> painted on cloth or paper, i.e., paintings behind the altar, icons, cards ... are not allowed to be imported into this city.[77]

This historical record indicates that Venice was a center for the printing of playing cards and religious images, and the municipal authorities took such measures to protect local interests because of the threat of competition for local product markets from outside. This move may have been a response to the dumping of German products, as records from the German city of Ulm during the same period show that the city shipped printed playing cards in wooden barrels to Sicily and Italy. The seventeenth century Italian scholar Valere Zani (ca. 1621–1696) claimed that playing cards were introduced to Venice from China, and he wrote:

> When I was in Paris, a French priest from Palestine, Abbé Tressan (1618-1684), showed me a deck of Chinese playing cards, telling me that a Venetian first brought playing cards from China to Venice, and that Venice was the first place in Europe to know about playing cards.[78]

Considering frequent interactions between Chinese merchants and those from Venice and Genoa of Italy during the Yuan dynasty, Zani's claim is well founded. For Venetian merchants on their long journeys, playing Chinese cards aboard ships was a best way to pass the time. As for the transfer of playing cards from Yuan dynasty China to Europe, it seems to have various channels. The increase in the urban population in European cities provided new business opportunities for card manufacturers. Manufacturers able to produce playing cards were also capable of printing religious images, which should be similar in style to those printed by Monte Corvino in Beijing. It is no accident that these two types of early printed materials in Europe appeared in Italy and Germany before spreading to other countries. Being the seat of the Roman Pope and the birthplace of the Renaissance, Italy had developed overseas trade, and as the hometown of Marco Polo, it had frequent interactions with people from Yuan dynasty China, thus making it natural to first introduce the printing technology. Germany, located in central Europe with well-established trade routes, was closer to Italy and the Mongol Empire, serving as a hub for goods and information from various regions. The earliest Europeans who used Chinese techniques to print religious images in Beijing came from Italy and Germany, which is a fact sufficient to speak for itself.

The earliest dated European woodcut religious image is *St. Christopher Carrying the Christ Child across the River* (Fig. 8.22)[79] printed in 1423. This woodcut measures 28.5 cm in height and 20.5 cm in width. It was discovered in 1769 in the library of a monastery in Augsburg, Germany, affixed to the cover of a manuscript, and now is housed in the Rylands Library in Manchester, England.[80] After being printed, the image was hand-colored. It depicts St. Christopher carrying the young

---

[77] Carter T F. *The Invention of Printing in China and its Influence* [M]. Translated by Wu Zeyan. Beijing: Commercial Press, 1957: 161, 165–166.

[78] Carter T F. *The Invention of Printing in China and its Influence* [M]. Translated by Wu Zeyan. Beijing: Commercial Press, 1957: 166.

[79] De Vinne T L. *The History of Printing* [M]. New York: F. Hart, 1876.

[80] Oswald J C. *A History of Printing: Its Development Through 500 Years* [M]. New York, 1928.

Fig. 8.22 Woodcut single-page religious painting *St. Christopher and the Child Jesus Crossing the River* printed in Germany in 1423

Jesus across a river, and also includes a waterwheel imported from China in the lower left corner. At the bottom of the image are two lines of Latin verse, which translate as "Whenever you see this image, you will be spared from the plague of death", somewhat like the talismans found in Chinese printed materials.

From 1400 to 1450, woodblock printing was popular in Germany, Italy, the Netherlands, Flanders (which is now part of present-day Belgium) and other countries. During this period, the property inventory of the Bethany Monastery in Liege, Flanders, where the German priest Jean de Hinsberg (1419–1455) and his sister resided, listed "one instrument for printing scriptures and images" (Unum instrumentum ad imprimendas scripturus et ymagines) and "nine wooden blocks for printing sacred images and fourteen stone blocks for printing" (Novem printe lignee ad imprimendas ymagines cum quatuordecim aliis lapideis printis),[81] explicitly stating the use of woodblocks for printing sacred images. Early printed materials were mostly roughly

[81] Oswald J C. *History of Western Printing Culture* (セイヨウインサツブンカシ) [M]. Translated by Tamaki Hajime (タマキ, ハジメ). Tokyo: Aoi shobo, 1943: 365.

Fig. 8.23 A European religious woodcut print *Apocalypse* published in 1425 (Housed in the British Library)

carved religious images based on stories from the Old and New Testaments, sometimes colored after printing, and short handwritten texts, often in Latin, were included in the images. When multiple related images were printed, they were grouped together and bound into volumes.

Many such collections can be found in places like the British Library in London, but most do not include information on the publication date, location, and the engraver's name. Based on the format, cutting and painting techniques and style, it is possible to determine which ones were published earlier. Apart from the aforementioned 1423 sacred image, the most notable work is *Ars Moriendi* ("The Art of Dying") published in Germany in 1450, consisting of 24 pages, each printed on a separate wood block and bound together into a volume, explaining how to die peacefully. A slightly earlier publication is *Apocalypse* from 1425, though the place of publication is unknown (Fig. 8.23). Holland's *Biblia Pauperum* ("Paupers' Bible") also belongs to early works. In the late fifteenth century, woodcut editions with both text and images began to appear, along with printed books with only text. The first popular reading book was *Ars Grammatica*, a Latin grammar book written by the Roman Aelius Donatus (320–370) in the fourth century, used as a compulsory school textbook. Later, the number of printed books primarily with text gradually increased.

Early European woodcut prints are very similar to those of Yuan dynasty China in terms of format and manufacturing techniques. According to the research of American printing historian Theodore Law de Vinne (1828–1914), Europeans first wrote the manuscript on paper with a pen, then fixed the paper with ink marks onto a wooden block with rice paste to form a mirror image. The carver used a knife to cut characters into the wood along the grain facing towards themselves, and each block had two pages separated by a center seam. After the carving, paper was placed on the inked surface of the woodblock and rubbed with a brush to print on one side only. Finally, the paper was folded along the center seam with the printed side facing outwards, the folded edges were aligned, and holes were punched on the other side before binding the pages together with thread.[82] It can be seen that early European

[82] De Vinne T L. *The History of Printing* [M]. New York: F. Hart, 1876: 119–120, 203.

woodcut prints in the fifteenth century followed traditional Chinese techniques in format, carving, inking, printing, and binding, thus making it unsurprising that European printed books from this period have a similar appearance to the Yuan dynasty thread-bound printed books. The only difference is that the text in European books runs horizontally rather than vertically.

European woodcut prints adopted a format of two pages per plate, with text printed on one side and folded along the center seam, which is contrary to the traditional format of European books and entirely in line with the Chinese practice. Prior to de Vinne, other scholars had also noted the similarities in carving and printing techniques between Chinese and European prints, believing that this technology was introduced from China. For example, after comparing Eastern and European woodblock prints, British orientalist Robert Curzon (1810–1873) wrote in 1860, "we must acknowledge that the printing process of European woodblock printed books was surely modeled after the ancient Chinese book samples brought back from China by early travelers, but the names of these travelers have not been passed down."[83]

These travelers who transmitted printing techniques were likely Europeans who came to China during the Yuan dynasty. Whoever they were, what's important is that Europeans adopted China's existing printing techniques for the development of their own printing industry. Although Europe already had seals and textile printing in ancient Roman times, it did not transform them into printing for reproducing documents until the spread of Chinese printing technology between 1350 and 1400. Therefore, Dr. Carter asserts that "Chinese influence was actually the most important determining factor at the inception of woodblock printing in Europe."[84] This is an objective view in line with historical facts.

8.3.3 Early Wooden Moveable Type Printing in Europe

1. **Early Wooden Moveable Type Printing in Italy**

The view that European woodblock printing came from China has been widely accepted in the academic community. This can be seen when we browse relevant Western publications or visit museum exhibitions. However, whether European moveable type technology was influenced by China is not clear to many people and there are few relevant studies on it. Therefore, this issue requires careful study. Historically, after the invention of woodblock printing, moveable type printing was bound to appear sooner or later, and this was true in both China and foreign countries. China took four to five hundred years to develop from woodblock to moveable type printing, while Europe achieved this transition in just a few decades, clearly taking a shortcut by drawing on China's ready-made experience. European scripts, which

---

[83] Curzon R. The History of Printing in China and Europe [J]. *Philobiblon Society Miscellanies*, 1860, 6(1): 23.

[84] Carter T F. *The Invention of Printing in China and Its Spread Westward* [M]. Translated by Wu Zeyan. Beijing: Commercial Press, 1957: 180.

feature many rounded characters such as a, e, d, g, o, p, r, s, etc., require tight arrangement of letters, making it difficult to engrave them into the woodblock. European engravers must have had more difficulties and worked longer time than their Chinese counterparts. Unlike Chinese, which has tens of thousands of ideographic characters, the commonly used Latin and various national languages in Europe generally use 26 letters (28 letters in the Slavic alphabet and 24 in the Greek alphabet) to form all words and sentences. This characteristic of European languages makes moveable type printing more suitable than woodblock printing. When Europeans learned that China also had moveable type, they quickly introduced this printing method.

During the thirteenth and fourteenth centuries, clay moveable type, wooden movable type, and metal movable type printing in Yuan dynasty China continued to develop based on the foundations established during the Song dynasty, and they were used in parallel with woodblock printing. In 1313, Wang Zhen, a scientist in the early Yuan dynasty, published an article on wooden moveable type technology along with his book *Nong Shu* ("Book of Agriculture"). Shen Kuo, a scholar of the Song dynasty, included an article on clay moveable type printing in his book *Mengxi Bitan* ("Dream Pool Essays"), which was republished in 1305. Both books were readily available in bookstores in Beijing and other major cities. When Europeans entered China from Xinjiang, they would immediately encounter printing workshops as well as editions printed from wooden moveable type in Uighur, also a phonetic script. During the Yuan dynasty when transportation between China and Europe was smooth, Europeans traveling to China would not be completely unaware of moveable type printing in China. When they learned about this technology, they would pass it on, thereby facilitating the spread of the technology. Among various types of moveable type, wooden moveable type was the easiest to produce, and Europe's earliest moveable type happened to be wooden, which undeniably is a product of Chinese printing culture and Europeans' preferred option. In 1548, Theodor Buchmann (1500–1564), a theology professor and orientalist at the University of Zurich in Switzerland, wrote about wooden moveable type printing in Europe: "Recently, people have been engraving texts onto full-page wooden blocks, but this method is quite laborious and costly. So, people have made wooden moveable type, which can be assembled piece by piece to create printing plates."[85]

This is an important record of the use of wooden moveable type for printing books in Europe. Judging from his surname, Buchmann was a German-speaking Swiss. In German, Buchmann means "book man", so his surname is often translated by the Greeks as Bibliander. Though both carry the same meaning, Buchmann and Bibliander should not be regarded as two different people. Buchmann's academic activities date no more than a hundred years from Europe's initial use of wooden moveable type, making his records credible. When Europeans made wooden moveable type, they likely first transferred the text written on paper onto a wooden board or strip in reverse, engraved the raised reverse characters, and then used a fine saw to cut out each character into individual moveable type. All the moveable types were then trimmed and used for typesetting. After a book was printed, the plates were

[85] Oswald J C. *A History of Printing: Its Development Through 500 years* [M]. New York, 1928.

disassembled and the moveable types were collected for printing other books. This is exactly the technique used by the Chinese 400 years earlier. Wooden moveable type was a bridge from woodblock printing to metal moveable type, and its use enabled the Europeans to master the concept of moveable type printing for the first time.

European countries like Italy, Nederland (now the Netherlands), and Germany that developed woodblock printing earlier were most likely to be the first to engage in wooden moveable type printing in Europe. The aforementioned British orientalist Robert Curzon reported that the Italian physician and printer Pamfilo Castaldi (1398–1490) published some folded books using large wooden moveable type in Venice in 1426. These books were reportedly preserved in the archive of Feltre, his hometown, in Belluno Province.[86] Feltre was formerly known as Feltria, located northwest of Venice. Castaldi has thus been regarded as the founder of European moveable type technology, and a bronze statue was erected in his memory in Lombardy in 1868. According to Curzon's records, Castaldi printed wooden moveable type books in 1426, just a few years after the earliest European woodblock printing appeared. Therefore, Dr. Joseph Needham said that Europe introduced moveable type printing technology from China shortly after developing its woodblock printing.[87]

2. **Early Wooden Moveable Type Printing in the Netherlands**

In addition to Italy, the Dutch also engaged in wooden moveable type printing, as Hadrian Junius (1511–1575), a contemporary of Buchmann, recorded similar activities. Hadrian Junius was a physician in Haarlem near Amsterdam. He noted that a local named Laurens Janszoon (1395–1465) printed books such as *Latin Grammar* and *Horn Book* using large wooden moveable type in 1440.[88] Janszoon was a sacristan for the Catholic parish in Haarlem, a role referred to as "koster" in Dutch, equivalent to "marguillier" in French. Hence, he was also called "Koster" or "Coster" even though it was not his actual name. Early German literature, such as *Cologne Chronicle* published in 1499, also mentioned Janszoon. Therefore, the Dutch regarded him as the founder of European moveable type technology and erected a bronze statue in his honor in Haarlem. In the early nineteenth century, a large conference was held in commemoration of him.

Apparently, it is an indisputable historical fact that during the first half of the fifteenth century, various countries in Europe experimented with wood moveable type printing as a replacement for woodblock printing and achieved initial success. We have noticed that there were debates in the European academic community on when and where moveable type technology originated in Europe. Some argued that it started from the wooden moveable type in Italy or the Netherlands, while others claimed it began in Germany with metal moveable type. This involves the technical definition of "moveable type". Since the sixteenth and seventeenth centuries, metal moveable

[86] Curzon R. A Short History of Libraries in Italy [J]. *Philobiblon Society Miscellanies*, 1854 (1): 6. Yule H. *The Book of Ser Marco Polo* [M]. 3rd ed. London: Murry, 1903: 138–140.

[87] Needham J, Pan Jixin. *The Collected Works of Joseph Needham* [M]. Shenyang: Liaoning Science and Technology Press, 1986: 262–263.

[88] Oswald J C. *History of Western Printing Culture* (セイヨウインサツブンカシ) [M]. Translated by Tamaki Hajime (タマキ, ハジメ). Tokyo: Aoi shobo, 1943: 365.

type has been prevalent in Europe and people have come to equate moveable type with metal moveable type. Today, in the study of printing history, we should adopt a developmental perspective to understand the meaning of "movable type". Wooden moveable type is certainly a form of moveable type, and discussions about the origin of moveable type should start with non-metal moveable type. The contributions made respectively by the Italians, Dutch, and Germans to the development of European moveable type technology are complementary rather than mutually exclusive, and all should be acknowledged. However, the view that Europe went through a phrase of wooden moveable type and was influenced by China has been criticized, because accepting this view would challenge the claim that moveable type technology was an independent European invention. On the other hand, some of the evidence presented by proponents of China's influence on Europe is weak, thus providing ammunition to those opponents.

Given the historical context of great opening of transportation between China and Europe and the transmission of numerous Chinese inventions to the West at that time, combined with early documentary records by European writers, it cannot be denied that Europe went through a phrase of wooden moveable type and was influenced by China. Before 1420, Europeans had no idea of moveable type printing and moveable type technology. For them, these were foreign concepts unknown to their ancestors. Due to their lack of knowledge of Chinese history, they did not know where this new technology came from. Some people are reluctant to acknowledge that European printing evolved through three stages: woodblock, wooden movable type, and metal movable type, assuming that it jumped directly from woodblock to lead moveable type. They deny that figures like Pamfilo Castaldi and Laurens Janszoon once engaged in wooden moveable type printing on the grounds that modern simulated experiments using precision equipment and tools to produce small-sized Western wooden type all failed.[89] While this may be the true state of the experiments, modern researchers cannot deny the fact that creating large-sized Western wooden type does not require precision equipment and tools, and the types can be used for typesetting and printing books. In fact, their ancestors did so long ago.

In 1908, the French sinologist Paul Pelliot discovered in Dunhuang of Gansu, on the Silk Road, a large number of Uyghur wooden movable types used by the Uighurs during the Yuan dynasty. This discovery provides tangible evidence for the feasibility of making large-sized phonetic wooden type, and it also reveals the route of the spread of movable type technology from the mainland of China to Xinjiang and then to the West. French printing historian Pierre Gusman therefore believed that Chinese movable type technology entered Europe during the thirteenth and fourteenth centuries through two routes in the Yuan dynasty: first, through contact with the Uighurs of the Mongol Chagatai Khanate, with the Armenians living in the Netherlands later transmitting it to Europe during Pamfilo Castaldi's activities; second, through Johannes Gutenberg (1400–1468) who, in Prague, the capital of Bohemian, learned the movable type technology that had been introduced into Europe through

[89] Reed T B. *A History of the Old English Letter Foundries* [M]. London, 1887.

the Central Asian and Russian overland route.[90] In hindsight, both possibilities exist and they are consistent with the earlier accounts from European authors.

8.3.4 *Early Metal Movable Type Printing in Germany*

1. **Early Attempts at Casting Metal Movable Type**

As previously mentioned, Europeans successfully printed books using large-sized wooden moveable type in the first half of the fifteenth century, taking a decisive step from woodblock printing to movable type printing. However, using large type required more space and paper, resulting in thicker and heavier books, and paper at that time was also very expensive. In order to fit more text onto a page, smaller type had to be used. Europeans encountered technical difficulties when making small wooden type: it was difficult to cut and strength was not sufficient. This was different from making the larger Chinese wooden type, so movable type was limited in Europe. In this situation, Chinese metal type gained favor among Europeans. Thereupon, some people began to explore in this field. It is recorded that when the Dutchman Laurens Janszoon from Haarlem was making wooden movable type, he experimented with movable type using lead and tin. Prokop Waldfoghel (1367–1444), a German-born silversmith, also conducted similar experiments. He settled in Prague, the capital of Bohemia under the rule of Charles IV von Luxemberg (1347–1378), the emperor of the Luxemburg dynasty (the Holy Roman Empire). Prague was well-known for making tableware between 1367 and 1418. During the Hussite Wars (1419–1434), a Bohemian rebellion against the Roman Catholic Church, he worked at a metallurgical plant in Nuremberg from 1433 to 1441. In 1439, he became a citizen of Lucerne, not far from Basel, Switzerland, and in 1441 he moved to Avignon in present-day southeastern France. Avignon was the papal residence and center of bookselling during1309-1417 and 1439–1449.[91]

According to the research of Wolfgang von Strömer, a historian of technology from Nuremberg, Waldfoghel had already obtained information about typecasting for printing books from the East during his stay in Prague, as Prague was one of the major destinations for Chinese silk transported to Europe, and information about metal movable type was further spread from Prague to German industrial and commercial cities such as Nuremberg, Strassburg, and Mainz. After moving to Avignon, Waldfoghel used the technical knowledge he learned in Prague, combined with his own expertise to explore a new profession for living. Between 1441 and 1444, he invented an "art for writing artificially" (ars scribendi artificialiter) for book production and shared this technique with his partners, the Jew David Caderousse, French churchman

[90] Gusman P. *Le Gravure sur Bois et L'épagune sur Métal* [M]. Paris, 1916: 37–38.

[91] Martin H J. *The History and Power of writing* [M]. Cochrane L G, tr. Chicago: University of Chicago Press, 1994: 215.

Manuel Vitalis, his friend Arnaud de Coselhac, and the Avignon merchant George de la Jardin.[92]

According to the 19th-century French priest Pierre Henri Requin, in 1446, David Caderousse ordered 27 iron Hebrew letters and tools of wood, tin, and iron. Within two years, he acquired 408 Latin letter blocks, which he used as collateral for loans. Records from July 4, 1444, indicate that when Vitalis partnered with Waldfoghel, he kept two steel letters, two iron casting frames, 48 tin frames, and "other items related to the art for writing artificially", which he later sold.[93] This suggests that those items were valuable at the time, and that the so-called "art for writing artificially" refers to the use of type blocks instead of hand writing to create texts that resemble handwritten words. In other words, Waldfoghel and his partners in Avignon were involved in the production of metal movable type between 1441 and 1446, obviously for printing purposes. However, they did not continue their efforts, but soon disbanded and sold the type and tools.

2. **Earliest European Metal Movable Type Printing Shop Established in Germany**

Ten years later, another German, Johannes Gensfleisch Zum Gutenberg, did similar work and achieved great success. Born in the industrial city of Mainz in 1400, Johannes Gutenberg studied at the University of Erfurt from 1418 to 1420 but dropped out to learn metalworking after the death of his father who had worked in a mint. From 1434 to 1444, Gutenberg went to Strasbourg, where he signed a contract with the locals Andreas Drizehn, Hans Riffe, and Andreas Heilmann to jointly process gemstones and make mirrors with new techniques. With the contract, the project was funded by the others while Gutenberg contributed his technical expertise, and they shared profits. When Andreas died in 1436, his brother, as his heir, demanded that Gutenberg hand over the technology, but Gutenberg refused, thus leading to a lawsuit. The court records state that in 1436, Gutenberg was involved in "matters related to printing" and he paid 100 Gulden in gold coins to Hans Dünne, a goldsmith in Frankfurt. The testimony includes words like "type".

It is evident that Gutenberg suddenly changed his profession from mirror-making and gem-processing to secret experiments with printing, but he failed. From 1444 to 1448, he travelled to the Netherlands, Switzerland, Basel, or Venice in Italy, probably to conduct technical investigation for solutions to his problems.[94] Some sources even say that he may have visited Prague. These travels opened his eyes to new possibilities and ways to solve his problems. In 1448, he returned to Mainz and borrowed money from a wealthy merchant named Johan Fust (ca. 1400–1466), using his developed technology and equipment as collateral. The contract stipulated a five-year term, with profits split equally and the principal and interest to be repaid at the end of the

---

[92] Von Strömer W. *Hans Friedel von Seckingten, der Banker der Strassbourger Gutenberg-gesellschaften* [M]. Mainz: Gentenberg-Jahrbuch, 1983: 45–48.

[93] Requin P H. *Documents Inédits sur les Origines de la Typographie* [M]. Paris: Bulletin de Philologie et d'Histoire du Ministère de l'Instruction Publique, 1890: 328–350.

[94] Martin H J. *The History and Power of Writing* [M]. Cochrane L G, tr. Chicago: University of Chicago Press, 1994: 219.

Fig. 8.24 The Latin edition of *42-line Bible* printed by Gutenberg using lead movable type in Mainz in 1455[95]

term. Gutenberg made a breakthrough in his experiment, successfully casting large moveable type in 1450 and printed a Latin Bible, the *36-line Bible* (36 Linne Bible), in bold Gothic script font based on a handwritten manuscript. In 1454, he printed indulgences issued by Pope Nicholas V (1447–1455). The greatest achievement in his technological career, however, was the publication of the small type hardcover edition of the *42-line Bible* (42 Linne Bible) in 1455. The book has a layout size of 30.5 cm × 40.6 cm, with two pages per sheet, printed on both sides, totaling 1286 pages bound in two volumes (Fig. 8.24). The page borders feature decorative woodcut floral patterns, making it a rare edition that combines movable type with woodblock printing.

[95] Oswald J C. *A History of Printing: Its Development Through 500 Years* [M]. New York, 1928.

8.3.5 *Chinese Background of Metal Movable Type Printing in Europe*

The metal movable type printing technology developed by Gutenberg was undoubtedly the best at the time but it was not the earliest. Today we know that the entire process of metal movable type casting, typesetting, proofreading, printing, dismantling, recycling, and storage was put to practical use in China during the Song and Jin dynasties from the eleventh to the thirteenth centuries. Therefore, from a global historical perspective, Gutenberg was not the inventor of metal movable type printing technology. Comparing the European technology represented by Gutenberg with the traditional technology in ancient China, we can see that he was an important technological innovator and also made original contributions in the process.

Firstly, both China and Europe used ternary alloys including lead and tin as materials for moveable type, which is a similarity. However, there is a slight variation that China added copper while Europe used antimony. European type had a higher lead content (83%), thus called "lead type", while Chinese had more copper (50%–62%), called "copper type". European type had a lower melting point and was cheaper, whereas Chinese type had a higher melting point and was more expensive, so it was once improved by adding zinc. Both Chinese and European types were rectangular in shape with a small hole in the body for connecting them into lines with iron thread to prevent movement when placed in the printing plate, which is another similarity. Chinese metal type was larger due to the complex strokes of Chinese characters. Chinese metal type was made by the sand-casting method, using wooden type as the mold, fine sand and clay as casting materials. Carter and others believed that the early European movable type was cast in a similar manner,[96] but some people argue that Europe used steel type as the mold and by a hard-casting method, with copper as a casting material.[97] As China cast tens of times more types than Europe, sand-casting was practical and cost-effective. Europe's hard-casting, though more laborious, was not a problem because the amount of typecasting was small. While casting principles and procedures were the same for both, there were differences in the casting materials due to distinctive features of the characters used. In ancient China, metal molds and frames were used for casting coins and weapons, but since the Song dynasty, the number of characters cast often reached thousands or tens of thousands, making this method impractical though not impossible.

Secondly, the ink used for Chinese metal movable type was thick liquid made by mixing pine soot or oil soot charcoal with animal glue in a weight ratio of 100:30 to 100:50 and then fermenting it, sometimes with a small amount of vegetable oil added.

---

[96] Carter T F. *The Invention of Printing in China and its Spread Westward* [M]. Translated by Wu Zeyan. Beijing: Commercial Press, 1957: 200. Oswald J C. *History of Western Printing Culture* (セイヨウインサツ ブンカシ) [M]. Translated by Tamaki Hajime (タマキ, ハジメ). Tokyo: Aoi shobo, 1943: 235.

[97] Hanebutt-Benz E. *Features of Gutenberg Printing Process* [C]. Ch'ongju: International Forum on the Printing Culture, 1997. Fussel S. *Gutenberg and Printing in the Western Culture* [C]. Seoul: International Symposium on Printing History in the East and West, 1997.

Fig. 8.25 The screw-type press device for European movable type printing[98]

Gutenberg created oil-based ink by boiling flaxseed oil, adding distilled pine resin to obtain products such as turpentine, and then fermenting it. This was an innovation not used in China. However, Chinese ink was glossy, while European ink was not.

Thirdly, though China and Europe had the same principles of typesetting and printing, the tools used were different. In China, paper was placed over the inked printing plate and then a palm brush was used to gently rub the back of the paper. As Chinese paper was thin, little force was required. Printing was always done on one side of the paper, and the printed sheets were then folded in half along the center and bound into a book. Gutenberg modified the screw press used for crushing grapes or moistening paper in Europe to create a printing press (Fig. 8.25). The press had a wooden frame and the printing plate was fixed on the base; the printing plate was controlled by an iron screw rod that could move up and down, and there was also hard felt underneath the plate. Ink was applied to the printing plate with a sheepskin-covered, wool-wrapped pad, and then paper was placed over it. Finally, turn the screw rod to make the printing plate press down on the paper to print the text. Although it was single sheet printing, each sheet of paper was printed on both sides, eliminating the need to fold the paper. The printing was designed so because European hemp paper was hard and thick, making it difficult to apply ink with a brush, and in addition to paper, books in Europe were also printed from sheepskin boards, requiring more pressure. Double-sided printing was necessary also because paper production was limited in Europe and had to be used efficiently. The screw-type press should be regarded as an invention by Gutenberg. However, it required at least two people to operate, producing 300 sheets per day, while in China only one person was needed to print more than a thousand sheets per day, making the Chinese way several times as efficient as the European press.

---

[98] Oswald J C. *A History of Printing: Its Development Through 500 Years* [M]. New York, 1928.

In summary, it is evident that Gutenberg's movable type printing technology still followed the principles and basic operating procedures of the metal movable type technology invented in China, but he adapted it to European conditions in his own way, thus innovating on the traditional Chinese technology to make it suitable for the Latin cultural area and Christian world where phonetic alphabets were used. There is no shortage of inventions in his technology, such as the printing press. Before him, other Europeans had conducted early experiments, but his techniques were most systematic and widely used, establishing him as the founder of European metal movable type technology. His technology was later improved and became the starting point for the development of modern printing in the world. He achieved success by absorbing and summarizing the existing results and technical experiences of his contemporaries in Europe and Chinese ancients in this field. While he was casting type, his compatriots were using Chinese technology to manufacture gunpowder and cast cannons, and Germans were actively participating in the Renaissance movement, providing material preparation for the rise of new ideas, culture, science and technology in Europe.

For a long time, there has been a prevailing viewpoint in the West that metal movable type was invented independently by Gutenberg without any external influence, attributing the invention of movable type printing to him regardless of the historical existence of ancient East Asian type technology. This Europocentric view is still reflected in some works by nonprofessionals who generally lack knowledge of the printing history of both East and West and have little understanding of Chinese printing history and Sino-European relations, focusing only on Europe. They are unwilling to acknowledge that metal movable type evolved from non-metal movable type, which in turn evolved from woodblock and copperplate printing. All of these techniques had already come into existence in China in the East Asia long before Gutenberg's birth. Compared with China, Europe was hundreds of years late in developing these printing forms. In 1954, Joseph Needham, in his French publication *Dialogue Entre l'Europe et l'Asie* ("Dialogue between Asia and Europe"), called on some arrogant Europeans to learn more about Asia and engage in dialogues with Asia in order to truly understand Europe and the world. Now it seems that this suggestion was absolutely correct. Naturally, people from Asia, Africa, and Latin America should also manage to understand Europe and America. Residents from all continents in the global village should seek mutual understanding and talk to each other.

As time goes on and research by scholars from both East and West deepens, the notion that Europe independently invented metal movable type has been questioned by Western experts, who have turned their attention to East Asia. At the "International Symposium on Printing History in East and West" held in South Korea from September to October 1997, European printing historians sought to understand and appreciate East Asian printing culture. Dr. Eva Hanebutt-Benz, Director of the Gutenberg Museum in Mainz, discussed at the symposium the historical context of European metal movable type, pointing out:

> Considering these historical events, one inevitably faces two questions. First, did Gutenberg know about the achievements of East Asian movable type printing that had existed since the 12th century? This question is difficult to answer... But considering the extent of contact between East Asia and Europe during the 12th and 13th centuries, I believe that those travelers on the Silk Road should have known about movable type, and even if they did not introduce this knowledge in writing, they would have spread it orally. Therefore, I cannot imagine that Gutenberg had never heard of this method for printing books. I think it was this awareness that motivated him to seek an appropriate way to solve his problems and adapt to the situation he faced in his country.[99]

Professor Henri-Jean Martin, a printing historian at the University of Paris, also spoke at the symposium:

> "The Eastern printing technology and Asia's leading inventions in this field have always been of interest to French and other European printing historians. Therefore, forty years ago, I invited Madame Roberte Guignard, an expert on Chinese studies, to write a chapter on the history of printing in the East for the book *L'apparition du Livre* ("The Coming of the Book") which I co-authored with Lucien Febvre.... In an attempt to refute the long-held Europocentric view that had dominated the thinking of non-specialists, I somewhat recklessly suggested that, compared with the great gods of wisdom in East Asia, Gutenberg and Waldorfger as inventors, who spread the printing technology, were no longer creators of this field."[100]

Martin also told us that before he left Paris for this conference in East Asia, he received a letter from Professor Wolfgang von Stettenmüller, a historian of technology at the University of Nuremberg in Germany, and a paper commissioned to be distributed to the attendees. Regarding the earliest metal movable type printing in Europe, it is stated in the letter: "I (Stettenmüller) believe that the West must have mastered the technology from the East. In this regard, I emphasize the role of Emperor Charles IV at the court in Prague. At that time, Prague was a major destination for Eastern silk arriving in Europe, and information may have subsequently been transmitted from Prague to German commercial and industrial cities such as Nuremberg, Strasbourg, and Mainz."[101]

In Martin's work *Histoire et Pouvoirs de L'écrit* ("The History and Power of Writing", 1988) and Stromer's latest research *Von Turfan zum Karlstein* ("From Turfan to Karlstein", 1997), we have noted their progress in tracing the historical origins of Gutenberg's technology. This shows that after World War II western experts have gradually abandoned the Eurocentric perspective in their research and have made efforts to explore the Chinese background of European metal movable type printing and yielded new results. As a representative from China attending this international symposium, I was pleased to hear their presentations, read their works, and have friendly conversations with them. After the symposium, I also expanded on Stromer's

---

[99] Hanebutt-Benz E. *Features of Gutenberg Printing Process* [C]. Ch'ongju: International Forum on the Printing Culture, 1997.

[100] Martin H J. *The Development, Spread and Impact of Printing from Movable Type in the 15th and 16th Centuries* [C]. Seoul: International Symposium on Printing History in East and West, 1997.

[101] Martin H J. *The Development, Spread and Impact of Printing from Movable Type in the 15th and 16th Centuries* [C]. Seoul: International Symposium on Printing History in East and West, 1997.

work by drafting an article titled *Von Khan baligue zum Mainz* (From Khanbaliq to Mainz),[102] which constitutes the main content of this section.

As previously mentioned, between 1434 and 1444, while working on gem-cutting and mirror-making in Strasbourg, Gutenberg suddenly changed his profession and secretly began printing experiments, undoubtedly spurred by some external factor. More than a decade before Gutenberg's efforts, Procopius Waldfoghel had experimented with metal movable type in Avignon, which Gutenberg possibly learnt about, because one of his partners, Hans Riffe, had a relative named Walter Riffe, a goldsmith in Strasbourg, who was Gutenberg's neighbor and worked with him. During that period, Walter Riffe frequently visited Avignon.[103] By that time, Waldfoghel had taught his technique to some people and his products were already sold, making the technique semi-public. When Walter Riffe learnt of this, he mentioned it inadvertently, which got Gutenberg's attention. Seeing potential in this industry, Gutenberg brought in a Frankfurt goldsmith Dene to conduct experiments, but initial results were not as expected. The transition from failure in 1444 to success in 1450 can be attributed to his extensive travels from 1444 to 1448, during which time he learnt about the work of other European forerunners and their problems, leading him to figure out the secret to his eventual success.

When acknowledging Gutenberg's historical contributions, we cannot ignore the early work of other Europeans. But where did the ideas and techniques for movable type printing they relied on come from? To answer this question, we need to, as the aforementioned German and French experts have noted, trace back to the information brought back from China by travelers who took the Silk Road. This information, directly or indirectly, found its way to Gutenberg and other European pioneers; otherwise, it is hard to imagine how they could have quickly started out of nothing their own movable type experiments in the first half of the fifteenth century. We know that Europe, like China, went through several stages from manuscripts to metal movable type, including papermaking, handwritten manuscripts, woodcut, wooden movable type, and metal movable type. However, the transition from woodblock to wooden movable type and then to metal movable type in Europe took less than a hundred years, while it took China about 600 years. This is because Europe drew on China's experience. Europe's woodblock and wooden movable type techniques are exactly the same as those of China, and its metal movable type technique was based on the same principles as used in China, with only variations in materials and some tools used. Since this technique was complicated and could not be copied exactly, it had to be adapted to local conditions.

However, the early metal movable types in China and Europe were essentially the same in shape. In 1468, Ulrich Zell (1440–1505), a German disciple of Gutenberg's, published the Latin version of ***Cantica de Puella Dulci*** ("The Song of the

---

[102] Pan Jixing. Von Khan baligue zum Mainz: The Chinese Background of Gutenberg's Technical Activities [J]. *History of Chinese Science and Technology* (now The Chinese Journal for the History of Science and Technology), 1998, 19 (3): 21–30.

[103] Martin H J. *History and Power of Writing* [M]. Cochrane L G, tr. Chicago: University of Chicago Press, 1994: 202.

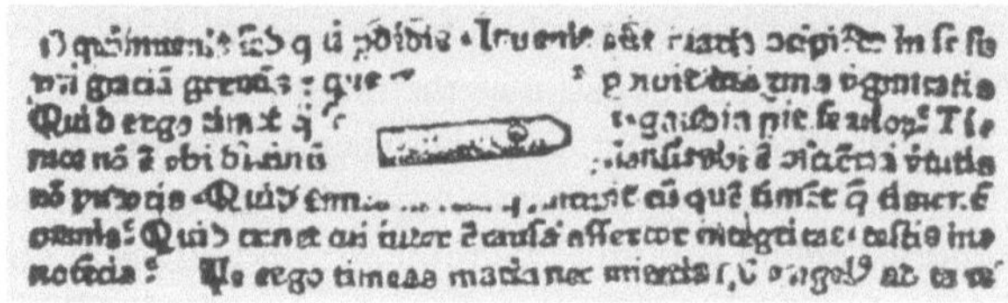

Fig. 8.26 Image of the movable type in *Cantica de Puella Dulci* published in Cologne in 1468

Sweet Maid") in Cologne in1468. On one page, a lead type was mistakenly placed horizontally during typesetting and was printed as such (Fig. 8.26), clearly showing that during Gutenberg's time, the movable types were long-cube solid castings with a small hole in the body,[104] similar in appearance to Chinese metal movable types in the twelfth-fourteenth centuries, and the typesetting method was also similar. This indicates that the Europeans of that time were using technology developed by the Chinese 400 years earlier to design their metal moveable type. Just as the early European metal guns resembled those of the Yuan dynasty in appearance, both designed and cast using the same technology. Given the frequent exchanges between China and Europe during that period, this similarity cannot be considered a mere coincidence.

Based on a wealth of known historical facts, western historians of science have summarized a law of technological diffusion: from the first millennium AD to the rise of modern science, i.e. more than one thousand years from the first century to the sixteenth century, the more complex a technology was, the less likely it was to be reinvented in different regions within a certain period of time. When a technology is developed in one country, the emergence of similar technology in another country can only be explained by technological diffusion, which can be triggered by even a hint of an idea.[105] Dr. Joseph Needham exemplifies this with printing, "Regarding the spread of printing, I am pleased that Gutenberg knew about Chinese movable type, or at least heard about it." This happens to align with the views of the aforementioned German and French scholars. Getting the general suggestion that Chinese people were able to create movable type with woodcuts or cast metal type for printing books, Gutenberg and his contemporaries could have experimented with typecasting and typesetting in their own way and achieved a series of developments, not to mention the possibility that they might have heard more than that.

At present, it is still difficult to identify the travelers who transmitted information about printing technology. However, their presence was undoubtedly evident among those who travelled between China and Europe during that time. Several examples can illustrate the existence of such stream of travelers. From 1298 to 1307, the Italians Giovanni da Monte Corvino and Pietro da Lucalonga, together with the German Arnold de Cologne, published religious images in Beijing with Chinese assistance. In 1307, the Frenchman Gerard, the Italian Andrew da Perugia, and the Spaniard Peregrine de Castello met in Beijing and then went to Quanzhou for missionary work. Andrew stayed in China for nearly 30 years and returned to Europe in 1336.

[104] Oswald J C. *A History of Printing: Its Development Through 500 Years* [M]. New York, 1928.

[105] Needham H J. *Science and China's Influence on the World* [M]. // DAWSON D. The legacy of China. Oxford, 1964: 234.

The Italian Odoric de Pordenone (1266–1331) and the Irishman James set out east in 1315 and arrived in Quanzhou in 1322, where they met Andrew and others. From 1322 to 1328, they travelled to Jinling (now Nanjing), Hangzhou, and Yangzhou, then lived in Beijing for three to four years before returning to Europe in 1330.[106] In his *Itinertarium de Orientalium* (1330), Odoric mentioned that in Venice he met many merchants who had been to China. In 1238, Giovanni de Marignolli (ca. 1290–1357) from Florence was sent by the Pope to China from Avignon, leading a group of 50. They entered the interior through Xinjiang in 1341, and arrived in Beijing in 1342 with 32 people remaining. He was received by Emperor Shundi of the Yuan dynasty and presented with fine horses. He lived in Beijing for four years, and then travelled from Hangzhou to Quanzhou before returning to Avignon in 1353.[107]

In 1355, when Holy Roman Emperor Charles IV was crowned by the Avignon Papacy, he heard that Marignolli had traveled to China and invited him to reside in Prague, the capital of Bohemia. Marignolli was tasked with compiling *Chronicles of Bohemia*, which included his experiences in China. Charles IV was keen to develop trade with the East, and Prague became a terminal for Chinese silk and other goods bound for Europe and also a transit point for information before being further distributed to other regions. Many people who returned from China lived in Avignon and Venice, and without exception, they entered and exited through Xinjiang on the overland Silk Road or through Quanzhou and Guangzhou on the maritime Silk Road, traveling to cities such as Hangzhou, Suzhou, Nanjing, and Beijing. Along the way, there were woodblock, wooden movable type, and metal movable type printing centers and book markets. Information about printing knowledge was thus transmitted orally through the two north and south transportation routes between the East and the West to Bohemia and Italy, where it further spread to other industrial and commercial cities, thus enabling Gutenberg and his contemporaries to learn about the new technologies. In 1546, Italian historian Paolo Giovio (1483–1552) wrote in his Latin book *Historia sui Temperis* ("Histories of His Own Times"), published in Venice in 1546:

> Metal type printers in Guangzhou (canton typographos artifices) printed history and ritual books on long folio paper in the same way as we did.... Therefore, we can easily believe that this technique, which was so beneficial to culture, had already reached Europe through Scythas or Sythia and the Grand Duchy of Moscow even before the Portuguese arrived in India (in the 14th century).[108]

The Scythas or Sythia mentioned in the text is the name of an ancient country located at the intersection of Asia and Europe, between the Caspian Sea and the Black Sea. Here it refers to Armenia under the Mongol Ilkhanate, while the Grand Duchy of Moscow refers to Russia (Russ or Ancient Russia) controlled by the Mongol Golden

---

[106] Cordier H. *Les Voyages du Frère Odoric de Pordenone* [M]. Paris, 1891. Yule H, Cordier H. *Cathay and the Way Thither* [M]. London: Hakluyt Society Publication, 1914.

[107] Moule A C. *Christians in China before the Year 1550* [M]. London, 1930.

[108] Jovius P. *Historia sui Temperio: 1546* [M]. Venezia. 1558: 161. Carter T F, Goodrich L C. *The Invention of Printing in China and its Spread Westward* [M]. 2nd ed. New York: Ronald, 1955: 159, 164–165.

Horde. These regions, close to Europe, were located on the south and north transportation routes through which Chinese movable type printing technology spread to the West. The Armenian people believed in Christianity, and their king Hayton I (1224–1269), after surrendering to the Mongols in 1246, sent his brother on a diplomatic mission to the Yuan court through Xinjiang,[109] where he was received by Emperor Dingzong, and later he wrote a travelogue. Armenians were skilled in metal-working and other crafts, often travelling to Europe for living. From the meaning of Paolo Giovio's Latin words, it is clear that he explicitly stated that European metal movable type printing technology was introduced from China. Some translations are inaccurate, so we have re-translated and specifically cited the original text. Giovio's account can be cross-referenced with the records of his contemporary Spanish scholar Juan Gonzeles de Mendoza (1540–1620), who wrote in Chapter 16 of his Spanish work *Historia del Gran Regno de China* ("History of the Great Kingdom of China") published in Rome in 1585:

> In the opinions of most people, the invention of European metal movable type printing began in 1458 and was completed by the German Gutenberg... However, the Chinese firmly believe that this printing technique was first initiated in their own country, and they revere the inventor as a sage. Obviously, this printing technique was transmitted to Germany via Russia and the Grand Duchy of Moscow many years after it was used in China. This is certain and it may have been transmitted by land. Some merchants who arrived in China from Arabia Felix via the Red Sea might have brought back books. This laid the initial foundation for Gutenberg, who was historically regarded as the inventor. It seems clear that this invention was passed on to us by the Chinese, and they deserve the credit for it.[110]

In this first European monograph on China, Mendoza not only referenced various early Western accounts but also cited the original account by the Spanish Martin de Rada (1533–1578), *Narrativo de Mision a Fukien*. Martin was proficient in Chinese and, in the third year of the Ming Wanli era (1575), traveled from Luzon (now the Philippines) to Fujian Province with the Mexican Geronimo Marin, purchasing a large number of Chinese books in Quanzhou, Zhangzhou, Fuzhou and other places. In 1576, Marin returned to Europe with these books and the manuscript of Martin's travelogue, which he presented to King Philip II of Spain. In Martin's travelogue, he wrote about a conversation with a local official in Fujian:

> When this Chinese official heard that we also had script for printing and printed books as they did, he was very surprised because they had been using this technology for hundreds of years before us.[111]

The word "script" used by Martin refers to handwritten metal movable type, so his conversation with the Chinese official was still about metal movable type printing. During the Ming dynasty, Fujian was well known for its development of copper movable type printing, and publications were generally labeled "*tongban*

---

[109] Boyle J A. The Journey of Het'un I, King of Little Armenia [J]. *Central Asiatic Journal*, 1964 (9).

[110] De Mendoza J G. *The History of the Great Empire and Mighty Kingdom of China* [M]. PARKER, tr. London: Hakluyt Society, 1853: 131–134.

[111] Boxer C R. *South China in the 16th Century* [M]. London: Hakluyt Society, 1953.

huozi", meaning copper movable type editions, which are still preserved today. Martin acquired such editions in Fujian, which is why he discussed metal movable type with the local official. Contemporary research by German, French, and Chinese scholars has confirmed that the records by the Italian Paolo Giovio and Spanish Juan Gonzeles de Mendoza over 400 years ago are correct: knowledge of metal movable type printing was brought to Europc by travelers who visited China in the thirteenth and fourteenth centuries, thereby laying the initial foundations for the development of metal movable type by Janszoon, Waldfoghel, and Gutenberg. The three to four hundred years' gap between China's casting of copper movable type in the Northern Song dynasty (eleventh to twelfth centuries) and the European casting of lead movable type during Gutenberg's time (the fifteenth century) also supports the Fujian local official's statement to Martin that Chinese metal movable type predates European metal movable type by hundreds of years. The Chinese background to Gutenberg's technical innovation is becoming progressively clear. The knowledge likely spread along two routes in the south and north, with a greater possibility of spreading along the northern overland Silk Road route. This was the safe route connecting Asia and Europe, stepped out by the Mongol army during their third westward campaign (1253–1258), and was also the primary route for the transmission of Chinese inventions such as gunpowder, woodblock printing, and other technologies to the West.

8.3.6 Early Development of Metal Movable Type Printing in Europe and America

In 1455, the year when Gutenberg published the deluxe edition of *42-Line Bible*, his contract with Fust expired. As he was unable to repay his debts, his printing shop was ruled to be owned by Fust. Starting from November 6, 1455, Fust controlled the shop and employed existing technicians and workers, including Peter Schöffer (about 1425–1502), a young German who had graduated from the University of Paris and was skilled in calligraphy. He was responsible for making all the handwritten Gothic typeface for casting, and later became Fust's son-in-law and heir. Together, they published numerous books and improved the metal typeface, layout design, and casting technique. In August 1457, they published *Psalter* ("Book of Psalms"), a folio for use during mass in church (Fig. 8.27). This is the first book in Europe printed in two colors, vermilion and black, and for the first time with such information as the printer's name, the year and place of publication as well as the manufacturer's emblem designed by Schöffer. This deluxe movable type edition with large typeface (20 or 24 lines per page) was reprinted many times, indicating its impressive sales. In 1459, they published *Rationale Divinorum Officiorum* by Bishop Duranti (1237–1295), followed by *Constitution* in 1460. In 1462, a folio of *42-Line Bible* was reprinted in small-sized type or "round Gothic script", making it easy for reading.

Fig. 8.27 The vermilion and black color edition of *Psalter* jointly printed and published by Fust and Schöffer in 1457[112]

After parting ways with Fust, Gutenberg received support from other investors and established another printing press on the outskirts of Mainz in 1456. Albert Pfister (1400–1465), who had worked under Fust, also came along and became an important assistant. In the same year, they published a one-sheet *Almanac of 1457*. Around 1459, they printed *36-Line Bible*, and in 1460, they published *Catholicon*, a popular book of Latin grammar and theological dictionary in one large volume of 748 pages. In 1462, Mainz experienced civil unrest and Fust's factory was destroyed. Printers took refuge in various cities such as Strasbourg, Cologne, Bamberg, and Nuremberg, spreading the movable type technology to various parts of Germany and even to other European countries. Before the unrest, Gutenberg's assistant Pfister had already established a printing press in Bamberg in 1458. In the same year, another Gutenberg's collaborator, Johann Mentelin (1410–1478), established a printing press in Strasbourg, and in 1466, he published the German edition of *Bible*. They left Mainz due to Gutenberg's shortage of funds and business depression. Gutenberg spent his

[112] Oswald J C. *A History of Printing: Its Development Through 500 Years* [M]. New York, 1928.

later years in poverty and debt. After the unrest, Fust rebuilt a new factory, which was managed by Schöffer after Fust's death.

In 1466, Gutenberg's disciple Zell started his publishing business in Cologne and produced 200 titles of books in 40 years. Alongside Mentelin in Strabourg, Heinrich Eggestein (?-1483) was also active in printing books. In 1470, Jodocus Pflanzmann published an illustrated German *Bible* in Augsburg, and in the same year, Arnold der Hoernen in Cologne published a book with a title page. In 1473, Anton Koberger (?–1513) established a printing press in Nuremberg, employing one hundred workers and publishing 236 titles of books. His most famous work was *Nurnberg Chronicle* published in 1513, a representative work of German movable type and woodblock, which had 596 pages and 1809 illustrations engraved on 645 blocks, including portraits, landscapes, maps, and more. The books printed in Europe during the fifteenth century are often referred to as "incunabula" by printing historians, derived from Latin words "in" (in) and "cunabula" (cradle), indicating that the technology of that time was still relatively immature. However, this term does not apply to China, as in the Ming dynasty, copper movable type printing had already reached a very mature stage of development so that Europeans were astonished when they saw it, and this is one of the reasons that Mendoza and others believed that European metal movable type printing technology originated from China.

After the first group of printers were trained in various places in Germany during the fifteenth century, they relied on their own skills to travel around, even venturing abroad and settling in foreign lands, becoming "tourist-printers". In 1465, Conrad Sweyenheim and Arnold Pannartz, who had been employed by the Fust printing press in Mainz, went to Italy and established the country's first printing factory in the village of Subiaco near Rome. Within two years, they printed *Latin Grammar*, *Speeches* by the ancient Roman scholar Marcus Tallius Cicero (106–43 BC), and religious literature. The required lead type, printing press, and ink were brought by carriage from Germany. In 1467, the factory was moved to the city of Rome, where they published a large amount of literary works within five years. To meet the needs of Italian readers, they used the Roman font, as used for typesetting *Historia Roma* published in 1469 (Fig. 8.28).[113] It is close to modern printing style but quite different from the handwritten Gothic script used by Germans like Gutenberg.

Gutenberg and Fust's prints imitated the font of ancient manuscripts, and when Fust sold books in France he refrained from claiming them as printed books for fear of provoking resistance from the guild of scribes. After the development of movable type printing, those who had previously worked as scribes for a living changed their profession, and printing factories began using printed fonts to print books, marking a significant advancement in typefaces.

In 1469, Johannes and his brother Vendlin from Spira, Germany, established a printing press in Venice and they also printed books using Roman typeface, which gradually gained popularity and eventually evolved into the modern Western printing font. In 1474, Dominic de Pistoia, an Italian, completed the Ripoli Press in Florence. The account books (libri contabili) used by the press from 1474 to 1483 are preserved

---

113 Oswald J C. *A History of Printing: Its Development Through 500 Years* [M]. New York, 1928.

Fig. 8.28 *Historia Roma* published by Conrad Sweyenheim in Rome in 1469

in the National Library of Florence, providing original records for our understanding of the types of raw materials used, prices, and production conditions at that time. In 1468, Berthold Ruppel, who had worked with Guttenberg, established a printing press in Basel, Switzerland (Fig. 8.29), marking the beginning of printing in Switzerland. In 1491, another German, Johann Froben (1460–1527), also established a printing press in Basel. Froben was a scholarly printer proficient in Latin, Greek, and Hebrew, and besides publishing *Bible* in Latin and Greek, he also published non-religious academic works, including works of the famous humanist Desiderius Erasmus (1466–1536). Altogether, he published 256 titles of books. The books published by Froben were known for their academic rigor and exquisite binding (Fig. 8.30). After his death, his son continued his business. Therefore, the earliest European countries other than Germany that developed metal type printing were Italy and Switzerland.

When King Charles VII of France heard about Gutenberg's creation of movable type for printing books, he sent Director of the Mint, Nicholas Jenson (1420–1480), to Mainz in 1458 to learn the technique. Jenson stayed at Fust's factory for many years and mastered the technique, but when he returned to France, Charles VII had

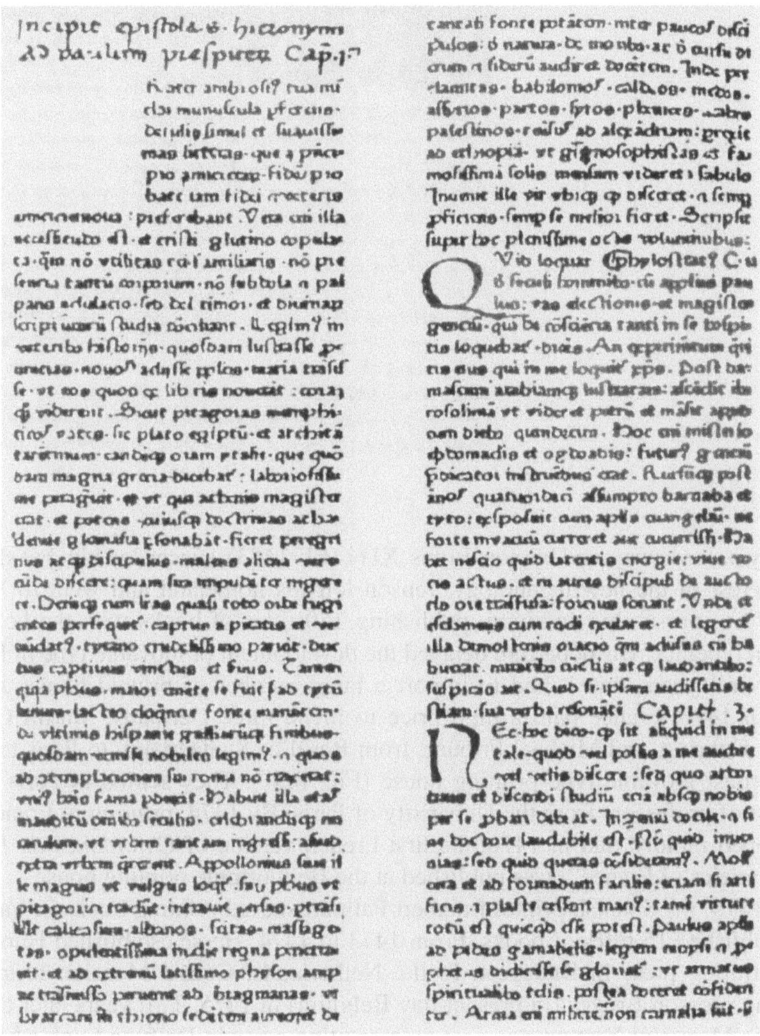

Fig. 8.29 The Latin version of *Bible* published in Basel, Switzerland by Ruppel in 1468

ΑΡΡΙΑΝΟΥ ΕΠΙΣΤΟΛΗ

Fig. 8.30 The Greek version of *Bible* printed from metal movable type, published by Froben in Basel, Switzerland in 1533

already passed away, and his son, Louis XI (1461–1483), succeeded him but showed no interest in the new technology. Jenson left his homeland and went to Venice in 1470 to pursue his ambition, publishing 150 titles of books with the beautiful Roman typeface. His departure delayed the development of movable type in France for several years, and it had to import a large number of printed books at high cost. In 1470, France paid a high price to invite three Germans, Martin Crantz, Ulrich Gehring, and Michael Friburg, from Basel of Switzerland, to Paris to build the earliest movable type printing house (Fig. 8.31). They settled in Paris at the request of professors from the University of Paris. By 1474, printing in France had developed rapidly, and in 1476, the first French version of *Chroniques de France* ("Chronicles of France") was published at the Bonhormme printing house.

In 1475, the Spanish printer Lambert Palmart started printing books in Valencia and published 15 titles of books. From 1473 to 1474, Hesse established Hungary's first printing press in Budapest. In the Netherlands, Colard Mansion founded a printing press in Bruges (in present-day Belgium) in 1475. Two years later (1477), Van der Meer and Yemantszoen set up a printing press in Delft and published *Old Testament*. In 1481, Van der Goes established another printing press in Antwerp.

Fig. 8.31 The title page of *Adagiorum Collectanea* ("Collection of Proverbs") published in Paris in 1513 (The picture shows the interior of the printing house)

William Caxton (1422–1491), the earliest printer in England, was initially a textile merchant who got to know Colard Mansion when he was doing business in Bruges and he learned the printing technique in Mansion's printing press. He translated the medieval romance *The Recuyell of the History of Troye* ("History of Troy") from Latin into English and published it in cooperation with Mansion in 1475. It was the first book published in English. In 1476, Caxton returned to London and established England's first printing press at Westminster. In 1477, he published the Pope's Indulgence, and also the English version of *The Dicta and Sayings of the Philosophers* in bold Gothic typeface. Altogether, he printed more than 30 titles of works (Fig. 8.32).

In Northern Europe, Denmark had its earliest printing press set up in 1482 in the city of Odense by Snell. Portugal saw its first printing press in Lisbon in 1489, established by Toledana. In Russia, the movable type printing press was built in Moscow in 1563 by Ivan Fedorov (?–1583) and P. T. Mstislavetz, commissioned by Ivan IV (1530–1584). After publishing some books, the printing press was destroyed, and then in 1589 a new factory was built to continue printing (Fig. 8.33). Prior to this, in 1551, Powell established a printing press in Dublin where he published *Book of Common Prayer* (The Second Prayer Book of Edward VI), which is the earliest printed book from movable type in Ireland. In 1508, King James IV issued a printing license, with which Edinburgh, the capital city of Scotland, began to have a printing press which was operated by Andrew Myllar and Walter Chepman. They published works such as Geoffrey Chaucer's *Maying and Disporte*.

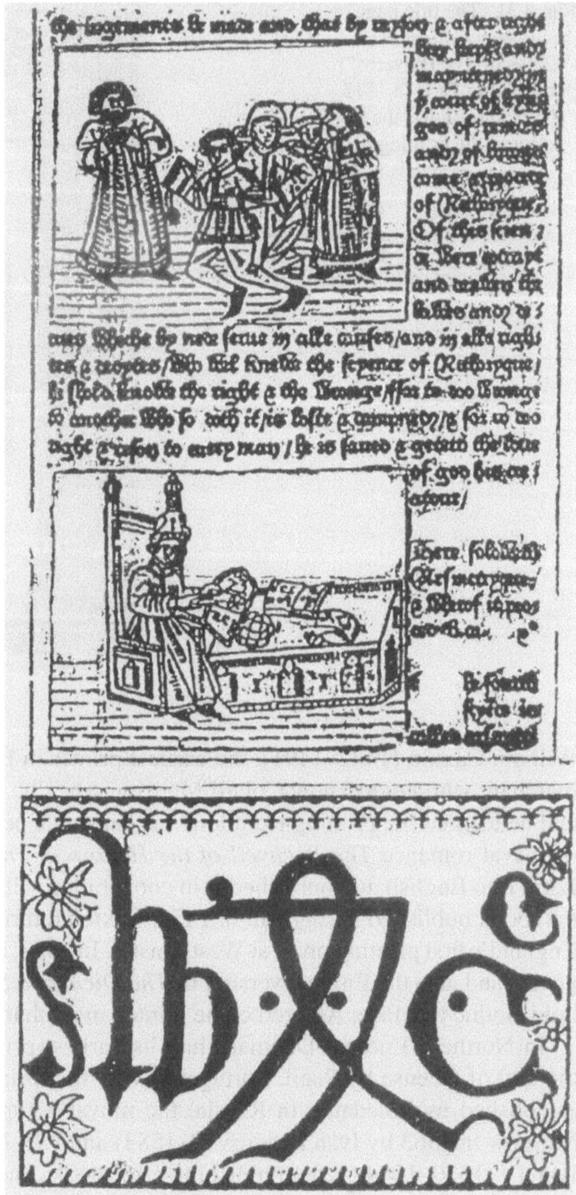

Fig. 8.32 The Pope's Indulgence published by William Caxton in London in 1477[114] (Housed in the British Library)

In about half a century from the successful casting and printing of books by Johannes Gutenberg in 1450 to 1500, a total of 250 printing presses spread throughout Europe, producing 22,000 different titles. Assuming 300 books per title, 6.6 million

[114] The illustrations are woodcuts, the text is in Gothic English metal type, and the image below shows the factory logo.

Fig. 8.33 A Russian theological book printed from metal movable type published in Moscow in 1618

books were published in Europe during this period,[115] ushering in a new era of metal movable type printing. By the sixteenth century, countries such as Italy, France, Germany, the Netherlands, and Switzerland had become major printing powers in Europe with printing presses in many major cities, and printing output doubled from the fifteenth century. Woodblocks and wooden movable type gradually faded from the historical stage, but printmaking as an art continued to develop. We can still see the influence of Chinese technology in Gutenberg's time or the "incunabula period", but from the sixteenth century to the arrival of modern machine printing industry, a period of transition, printing technology was localized in various European countries and improvements or innovations were made on Gutenberg's technology, but the original manual craft mode remained unchanged, still at the early stage of printing (Fig. 8.34).

After the discovery of the New World, European countries began to immigrate to the Americas, and with them came the technology of printing. The first printing press in the New World was established by the Spaniard Juan Pablos in Mexico City, the capital of Mexico, in 1539. In 1540, he published a book on Catholic doctrine, which is now held by the Hispanic Society in the United States. In 1584, a Spaniard

[115] Shoji, Sensui. *Chronology of the History of World Printing Culture* (インサツ ブンカシ) [M]. Tokyo: ブックドム社, 1936: 63.

Fig. 8.34 European printing press in the seventeenth century[116]

named Ricardo set up the New World's second printing press in Lima, the capital of Peru. The earliest printing press in the United States was established in Cambridge, Massachusetts, in 1638, before the country's independence. The required equipment and lead type were shipped from England by the British clergyman Jesse Glover, accompanied by Stephen Day (1594–1668) and his son, as well as the entire Glover family. Although the equipment and type intended for use by the Harvard Society (predecessor of present-day Harvard University) arrived, Glover died during the journey. The printing press was owned by Glover's wife and operated by Stephen Day. The earliest publications were *Freeman's Oath* in 1639 and *Bay Psalm Book* in 1640, both printed with the old-style screw press. One surviving publication is *An Almanack for the Year of Our Lord 1647*. However, the printer's name on the title page is Matthew Day, indicating that he was the printer, and his father was the operator (Fig. 8.35).

When printing presses were established in America, paper production was not yet locally available. It was not until 1690 when the first paper mill was established in Philadelphia, Pennsylvania that the printing industry had foundations for development. Benjamin Franklin (1706–1790), a scientist and the founding father, made significant contributions to the development of papermaking and printing in America. His elder brother James brought back movable type and printing equipment from England and established a printing press in Boston in 1717. Benjamin became an apprentice in the factory and five years later he moved to Philadelphia, establishing his own printing press in 1728. He published books, magazines, and

[116] Sandermann W. *Die Kulturgeschichte des Papiers* [M]. Berlin: Springer-Verlag, 1988.

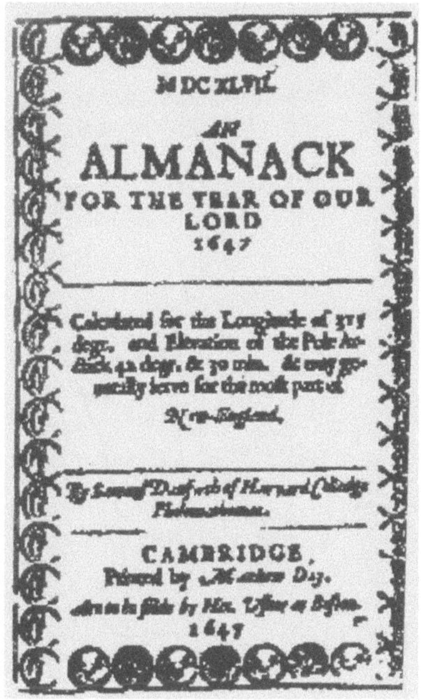

MDCXLVII
AN
ALMANACK
FOR THE YEAR OF OUR
LORD
1647

Calculated for the Longitude of 315 degr. and Elevation of the Pole Arctick 42 degr. & 30 min. & may generally serve for the most part of New-England.

CAMBRIDGE,
Printed by Mathew Day.
1647

Fig. 8.35 The title page of *An Almanack for the Year of Our Lord 1647* printed in Massachusetts before American independence

newspapers, having great influence among the masses. In North America, Canada's printing press was established by Bostonian John Fisher in Halifax in 1752, and he launched the newspaper *Gazette* in the same year. In 1795, printing technology was introduced to Sydney, Australia. By the eighteenth century, movable type printing had spread throughout Europe and the Americas. The diagram below (Fig. 8.36) illustrates the transmission routes of Chinese printing technology.

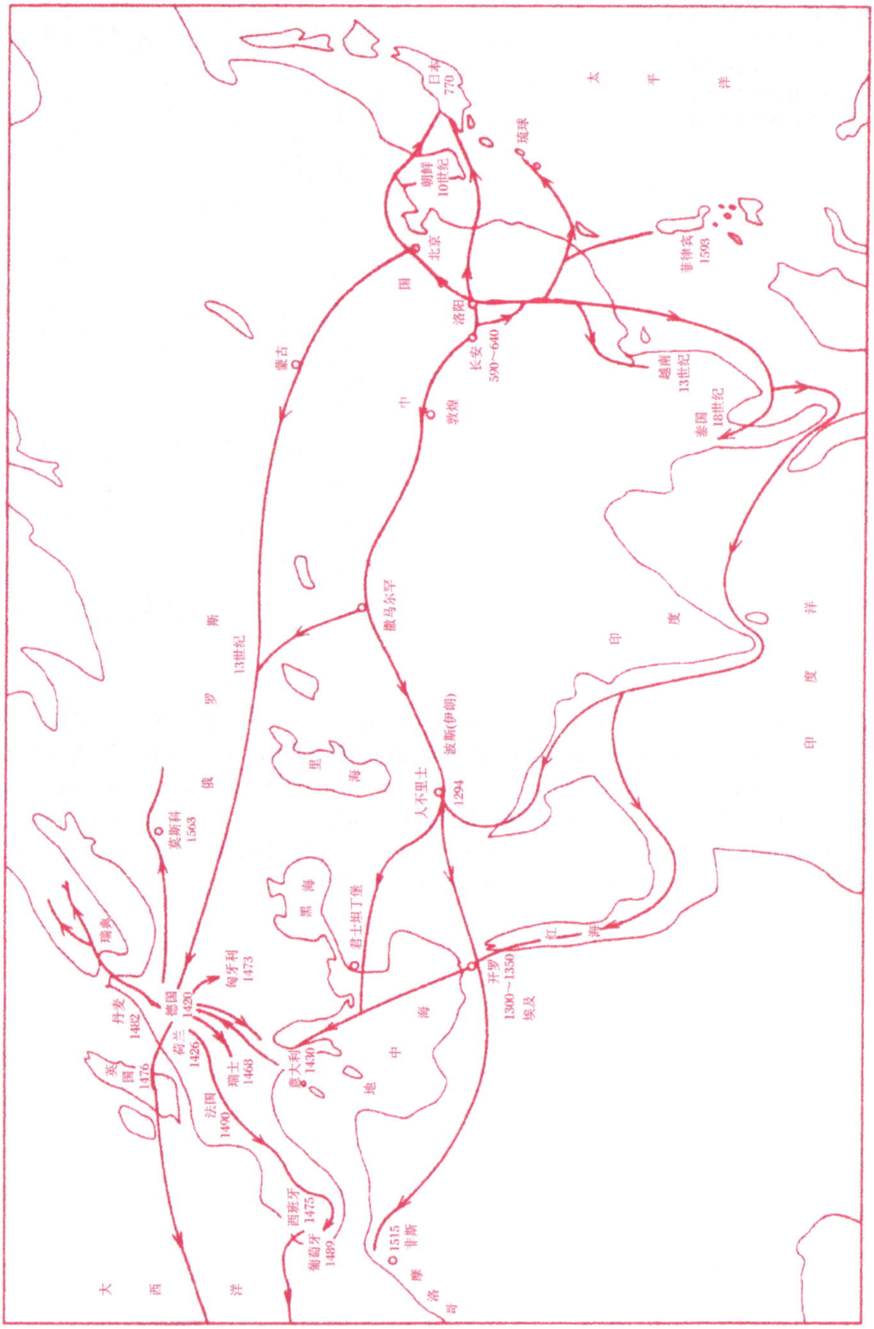

Fig. 8.36 Diagram of the transmission route of Chinese printing technology (Drawn by Pan Jixing)

Chapter 9
Influence of Paper and Printing Techniques on the Development of World Civilization

9.1 Role of Paper in Promoting the Development of Chinese and Foreign Cultures

9.1.1 Revolutionary Invention of Paper in the History of Human Writing Materials

The creation and use of writing is an important milestone in the history of human civilization. Before the invention of writing, people communicated ideas and conveyed information through face-to-face conversations within a close distance, which were then spread to others through word of mouth. This kind of transmission, solely relying on memory, was slow and could not be recorded. To improve this situation, various ethnic groups used visual aids such as drawings or knot tying to convey conceptual information over longer distances. These methods also allowed for the recording of events, effectively compensating for the limitations of oral transmission. However, it posed a challenge to expressing more complex ideas, and there was also uncertainty about whether others could accurately grasp the intended meaning of the drawer or knot-tier. Then, human beings created characters as symbols for expression of thoughts and language. The earliest form of writing was abstract and simplified drawings (pictorial writing), which later evolved into hieroglyphs. Through further simplification, deformation, and phonetic transcription, the hieroglyphs have developed into ideograms and phonetic writings, which are the characters used today in various countries in the East and West with a history of four to five thousand years so far.

The emergence and continuous improvement of characters enables human beings to fully express what they think and wish to say. Transferring characters onto writing materials allows for the conveyance of information over long distances, making it understandable to others and capable of being passed down to future generations.

P. Jixing, *History of Printing Techniques in China*,
https://doi.org/10.1007/978-981-96-6119-0_9

This expands the space and time for exchanging ideas and transmitting information, thereby documenting human activities in reliable historical records. With the invention of characters, human beings are able to create various classical texts and develop spiritual civilization. Only after a nation has developed its own characters can it be considered a civilized one with its own culture. The development of human civilization and the accumulation of spiritual culture rely on the transmission of classics recorded on various written materials across generations; otherwise, culture continuity would be interrupted. Historically, classics take different forms due to the diversity of the mediums, and they also change over different periods. In general, historical classics of various ancient civilizations in both the East and West are mainly recorded on materials such as stone, wood, clay, bark, oracle bones, metal, papyrus, parchment, pattra, silk, and others, which can be divided into the following three categories:

The first category is heavy and hard materials. Examples include the bronze bells (*zhong*) and tripods (*ding*) with cast inscriptions from the Chinese Shang and Zhou dynasties, and the codex and covenants engraved on bronze in the Western Roman Empire. Besides bronze, other metals like lead and iron were also used. In China, Egypt, and other countries, inscriptions were also carved onto stone tablets for permanent preservation. The Assyrians and Chaldeans engraved their cuneiform characters onto clay tablets, which were then baked into hard bricks. Some ethnic groups used wooden slips as writing materials, such as the bamboo and wooden slips used by the Chinese. These materials are heavy, cumbersome, unable to be rolled up, and take up a lot of space, thus difficult to store. However, they have the advantages of being sturdy and durable. Metal and stone, clay bricks, and oracle bones are fundamentally not suitable for writing; only bamboo and wooden slips are suitable for writing and recording, being easily obtainable and long lasting. These materials were used in China for a long time from the Spring and Autumn Period to the Jin dynasty (eighth century BC to fourth century).

The second category is light and brittle materials, mostly from the plant kingdom. For example, ancient Egyptians, Arabs, and Europeans used papyrus, while India and other South and Southeast Asian countries used palm leaves for writing and recording. Some ancient peoples in both the East and West also used tree bark as a writing material. While not as sturdy as the first category of materials, this category is light, capable of holding many characters, easy to carry when bound into volumes with strings, easy and cheap to produce. These characteristics kept them in use for a long time. For example, papyrus was used from the fifteenth century BC to the tenth century, and in India pattra was used for over a thousand years. However, they have drawbacks including brittleness, incapacity of being rolled up, poor foldability, rough surfaces, low whiteness, and thickness, which naturally results in a large volume when many sheets are bound together.

The third category of materials are light and flexible. For instance, the silk fabric used in China for writing since the Spring and Autumn period and the Warring States period is a textile made from silk worms; Europeans used parchment made from sheepskin as a writing material since the beginning of the Common Era. Both of these materials come from the animal kingdom and have the advantages of being white in color, smooth in texture for ink absorption, and having a large surface area, thus suitable not only for writing but also for painting, and they are flexible. In terms of sturdiness and durability, they surpass the second category of materials, and they take up less space when bound as a book, thus easy to carry. These high-quality materials were used for over a thousand years. Their main disadvantage is high cost, which limits their use to a small segment of society, preventing their popularization among the general public.

Among the three categories of materials mentioned above, only silk, bamboo and wooden slips, parchment, papyrus, and pattra are suitable for writing to create cultural classics. However, with the progress of culture and increased demand for books, their limitations became increasingly apparent. For example, writing a Latin Bible required over 100 sheepskins, which was an enormous cost. During the Han dynasty, a bolt of silk was priced equivalent to 360 kg of rice, thus the saying "the poor could not afford plain silk", indicating that ordinary people could not afford to use silk for writing. Although bamboo and wooden slips were inexpensive, writing a book required tens of thousands of slips, a weight for several people to carry. The limitations of these writing materials eventually became a barrier to sustainable cultural development and called for new materials to replace them. One new material is paper. China invented hemp paper made from broken hemp rags in the early Western Han dynasty (second century BC), and then in the Eastern Han dynasty (early second century AD) bark paper was made from woody bast. By the fourth century, paper had become the primary writing material in China, and from the end of the Tang dynasty in the eighth century, bamboo paper was made from bamboo stalks.

Compared with the aforementioned three categories of ancient writing materials, paper has the following obvious advantages:

1. Its surface is smooth and white, with good ink receptivity and a large format to accommodate many characters, and it is usable on both sides.
2. It is light and flexible, resistant to folding, capable of being rolled up and bonded, thus easy to carry and store.
3. It is cost-effective, with raw materials available worldwide, capable of being made anywhere, and it has a long life span.
4. It has a wide range of uses, suitable for writing, painting, printing, and packaging. Paper can be further processed into industrial, agricultural, military and daily products.

It can be said that paper is a universal material that surpasses all previous writing and recording materials in human history. The emergence and use of paper represent an epoch-making revolution in the history of carriers of written language and dissemination of images and texts. For over 2000 years, as a universal material used around the world, paper has played a significant role in advancing human civilization. We believe that it will continue to do so for a long time to come, even after the arrival of the second millennium.

From the perspective of world history, previous writing materials were only used in some specific regions without being promoted worldwide. The reason why paper is superior to other writing mediums and used worldwide is that all the other materials only undergo simple mechanical processing of raw materials without change in their composition, form, or natural properties. Paper, on the other hand, is made by chemically extracting effective components (plant fibers) from the raw materials, making them pure by removing other impurities, and then going through a series of mechanical processes. In the process of papermaking, raw materials undergo both morphological changes and chemical changes in composition, with the treatment of raw materials reaching the microscopic level of matter, namely, the molecular level. Paper is the product of deep processing of natural raw materials at the molecular level. This is why it still retains its vitality even today after 2000 years.

After inventing paper, China did not monopolize its use but shared it with all of humanity. During the Wei and Jin dynasties, papermaking technology was first introduced to neighboring countries including Goguryeo, Baekje, Silla, Japan, and Vietnam. During the Tang dynasty, it was spread to the Indian subcontinent, Central Asia, West Asia, and North Africa's Arab regions. In the twelfth century, it reached Europe through the Arabs, and in the sixteenth and seventeenth centuries, it was brought to the Americas through Europe, completing its millennium-long journey around the world. Wherever paper appeared in the world, it immediately became a strong competitor to other ancient writing materials used locally, gradually pushing them out of the historical stage. Whether the regions were Buddhist, Islamic, or Christian, people preferred to use paper for writing in different languages and they also re-transcribed on paper classic texts previously written on other materials for permanent preservation.

Early paper documents in both Eastern and Western countries were handwritten copies, including both new works and ancient works transcribed from bamboo or wooden slips, silk, papyrus, parchment, and pattra. While ancient books written on other materials gradually disappeared over time, paper manuscripts have been passed down through generations. Paper manuscripts have made an indelible historical contribution to preserving human cultural heritage and ensuring its perpetuity. They have enabled ancient cultures to continue and flourish without interruption. Since the Han dynasty, most pre-Qin scientific and cultural classics have been preserved in the form of paper books and volumes. The works of ancient Greek and Roman scholars, Indian Sanskrit classics, and works by Arab ethnic groups have also been preserved thanks to paper manuscripts, allowing people to gain spiritual enlightenment from them, inherit their essence, and promote the revival of academic studies in the future. It can be said that paper manuscripts are the sacred flame that disseminates human civilization, and their widespread use in various countries has greatly promoted the development of culture and education.

9.1.2 Role of Paper in Promoting the Development of Chinese Culture

The development of papermaking technology during the Han dynasty greatly advanced the development of education in the first place. After Emperor Wu of Han ascended the throne, he decreed to "dismiss the hundred schools of thought and revere only Confucianism". The Confucianism founded by Confucius was transformed into the official ideology and continued for more than 2000 years until the end of the Qing dynasty. The Western Han court established the highest academy *Taixue*, known to Westerners as "Imperial Academy", in the capital city of Chang'an. The academy had scholars of the Five Classics as teachers and representatives of official academic authority, teaching students Confucian classics such as *Yijing* ("Book of Changes"), *Shujing* ("Book of Documents"), *Shijing* ("Book of Songs"), *Liji* ("Book of Rites") and *Chunqiu* ("Spring and Autumn Annals"). Initially, there were 50 students, but no limit was set to the number of students recommended by local officials. Those who passed the examinations were awarded official positions. Later, enrollment gradually increased, reaching 200 during the reign of Emperor Xuan, 1000 during the reign of Emperor Yuan, and 3000 during Emperor Cheng's reign. In the Eastern Han dynasty, the Imperial Academy in the capital city of Luoyang had more than 30,000 students, with 240 schoolhouses and a total of 1850 rooms, making it the largest higher education institution in the world at that time.[1]

In addition to the central *Taixue*, the Han government also established public schools in various counties and districts, recruiting students widely. At the same time, the trend of private teaching by scholars was prevalent. *The Biographies of Confucian Scholars* in the *Book of Later Han* listed some cases of private lecturers recruiting students; sixteen scholars taught nearly 74,000 students. One example is Cai Xuan, a native of Runan (now Shangcai, Henan), who was "well-versed in *Five Classics* with a regular following of a thousand disciples, his teachings being recorded by sixteen thousand people".[2] The total number of students in schools across the country was at least tens of thousands to millions, and the textbooks used were mostly paper scrolls of sutras. The flourishing education cultivated a large group of intellectuals for society, who not only engaged in the study of Confucian classics but also conducted research on science and technology and created literary and artistic works. In this process, various pre-Qin ancient texts were carefully annotated and circulated continuously, and a large number of new works came out. The imperial court also repeatedly sent people to search for books among the people. Experts and scholars were organized to systematically revise books in the imperial library to provide standardized texts. Since the Han dynasty, the number of works significantly increased compared to that in the pre-Qin period, and the content discussed laid the foundation for academic research in subsequent dynasties.

---

[1] Meng Xiancheng, Chen Xuexun, et al. *Sources on the History of Education in Ancient China* [M]. Beijing: People's Education Press, 1961: 144–154.

[2] Fan Ye. *Book of Later Han*: *Vol. 66*, Biography of Jia Kui / / *Twenty-five Histories*: *Vol. 2*. Compact edition. Shanghai: Shanghai Classics Publishing House, 1986: 912.

The *Book of Han: Treatise on Literature* lists the catalog of works at that time, totaling 678 authors with 14,994 scrolls, including works on Confucian classics, elementary studies (language and writing), Confucianism, Daoism, *Yin-Yang*, *Mingjia* (Logicians/School of Names), Mohism, *Zongheng* (Diplomacy), *Zajia* (Miscellaneous School), *Nongjia* (School of Agriculture), *Bingjia* (School of Military), *Yijia* (School of Medicine), Astronomy and Calendrical Science, Legal Codes, Mechanics, Literature, Arts, and various other works representing the hundred schools of thought. With the introduction of Buddhism in the late Han dynasty, Buddhist texts were added. During this period, the traditional system of ancient science and technology also took shape, generating excellent new works such as *Zhou Bi Suan Jing* (the oldest mathematical classic in China), *Jiuzhang Suanshu* ("Nine Chapters on the Mathematical Art"), *Ling Xian* ("Mystical Laws"), *Bencao Jing* ("Classic of Materia Medica"), *Shanghan Lun* ("Treatise on Cold Pathogenic Disease"), and *Fan Shengzhi Shu* ("Fan Shengzhi's Book"). The philosopher Wang Chong (27–97), who integrated Confucianism and Taoism and was well-versed in various schools of thought, pointed out the inaccuracy and fallacy in Confucianism in his work *Lun Heng* ("Discourses Weighted in the Balance") (83): "Confucians' interpretation of *Five Classics* often distort the truth. Earlier Confucians did not know the ins and outs, only producing false theories, and the later Confucians believed in the words of their predecessors, following the old ways.... Hence, false theories continued to spread, while the truth was obscured and the essence of *Five Classics* has been lost." (*Zheng Shuo* "Correcting Falsities") This critique served as a challenge to the official ideology. He also criticized theological teleology, Yin-Yang calamity theory, various superstitions, popular divination and astrology studies, and religious tendencies of Taoism. The publication of this book further disseminated the materialist thought and rational critical spirit, playing a great role in promoting intellectual liberation.

The literary forms of Han *Fu* and poetry have had a long-term impact on later literature. They use rhymed language to depict things and human emotions and to reflect social reality as well. For instance, the poem *Qi Ai Shi* ("Seven Laments") by Wang Can (177–217), one of the Seven Scholars of Jian'an, portrays the pain that years of continuous warfare brought to people: "Stepping outside, I see nothing but white bones covering the plain. Along the road, there is a starving woman who abandons her child in the grass. Hearing the wailing, she moves on alone with tears unable to return." Such lines undoubtedly resonated in society. During the Jin dynasty, the literatus Zuosi (250–305) gained widespread acclaim for his *Yong Shishi* ("Epic Poetry"), particularly the sixth verse which reads: "Though the noble deem themselves high, to me they are mere dust; though the humble consider themselves lowly, to me they weigh a thousand catties." His *San Du Fu* ("Rhapsodies on the Three Capitals"), which he conceived over a decade, was eagerly copied and circulated among affluent families, so that "paper in Luoyang became expensive", which illustrates how quickly handwritten works could spread in society. Liu Xie (ca. 466–538), a poor man with no social status in the Liang dynasty, wrote a monumental work *Wenxin Diaolong* ("The Literary Mind and the Carving of Dragons", 500), which was not taken seriously after completion. He then carried his manuscript and, pretending as a peddler, handed it to Shen Yue (441–513) when he saw him leaving in a carriage. Shen Yue read it and regarded it as "profound in literary theory", thus attracting attention from the public. This book summarized the literary phenomena of previous eras, proposing that literature should benefit society, and emphasizing the importance of both form and content, with content being more important. It criticized the pursuit of ornate literary style, established the earliest literary theory system, and provided systematic comments on writers and works of all times, exerting a profound impact on later generations.

After the Han dynasty and through to the Six Dynasties period, the number of paper books increased sharply. Xie Lingyun (385–433) of the imperial secretariat in the Liu-Song dynasty, compiled *Four-Part Catalogue* in 421, which lists 64,582 volumes of books. Compared with *The Book of Han: Treatise on Literature*, the number of books had increased 4.3 times over the past 338 years. Shen Yue in the Liang dynasty had a passion for collecting books in his later years, and he alone had a collection of 20,000 volumes of books. So, it is recorded in history that during the reign of Emperor Wu of Liang dynasty (502–549), "Within the four borders, every household had books on history and literature." The surge in the number of classic books often accompanied the arrival of a climax in cultural and scientific development, which was directly related to the development of papermaking. This was true of the Han dynasty and so of the Wei, Jin, and Northern and Southern dynasties. While both paper and bamboo slips were used in the Han dynasty, the Wei, Jin, Northern and Southern dynasties were the period when paper manuscripts dominated. Since the twentieth century, a large number of handwritten books from this period have been discovered in places like Turpan and Shanshan in Xinjiang and Dunhuang in Gansu, including both Buddhist scriptures and non-religious works. Therefore, the phrase "Within the four borders, every household had books on history and literature" must refer to paper books.

The Wei, Jin, Southern and Northern dynasties were a period of enrichment and improvement of China's traditional scientific and technological system. The types and quantity of classics published during this period exceeded those of the Han dynasty. Representative works from this period include *Qimin Yaoshu* ("Important Arts for the People's Welfare") by agricultural expert Jia Sixie, *Jiuzhang Suanshu Zhu* ("Annotations on the Nine Chapters on Mathematical Art") by Liu Hui, *Daming Calendar* and *Zhuishu* ("A Mathematical Treatise") by mathematician Zu Chongzhi, *Zhitu Liuti* ("Six Principles of Cartography") by cartographer Pei Xiu, *Shuijing Zhu* ("Commentaries on the Water Classic") by geographer Li Daoyuan, *Bencaojing Jizhu* ("Collective Commentaries on the Classic of Materia Medica") by Tao Hongjing, *Mai Jing* ("Pulse Classic") by Wang Xi, *Zhenjiu Jiayi Jing* ("Classic of Acupuncture and Moxibustion") by Huangfu Mi, and *Zhou Hou Bei Ji Fang* ("A Handbook of Prescriptions for Emergency") by Ge Hong. The authors of these outstanding works made significant achievements in their respective fields, opening up new directions and reaching the world's most advanced level in many aspects. For example, Zu Chongzhi is the first to determine the value of πbetween 3.1415926 and 3.1415927 in the world. Jia Sixie's *Qimin Yaoshu*, which summarizes agricultural techniques and experience of the middle and lower Yellow River regions, is a classic work on

Chinese agriculture, discussing cultivation techniques and tools for crops, oil plants, vegetables, and fruit trees as well as animal husbandry and veterinary medicine, agricultural product processing and sidelines, and paper and ink manufacturing. The book emphasizes the idea of adapting agriculture to local conditions and the use of artificial selection, and introduces methods like fruit tree grafting, castration of poultry and livestock, and microbial fermentation. Wang Xi (courtesy name Shuhe) summarized previous literature on pulse diagnosis and wrote a systematic monograph on pulse studies; he proposed 24 types of pulse conditions, providing the theoretical basis for pulse diagnosis. Cartographer Pei Xiu proposed six principles for mapmaking: *fenlu* (scale), *zhunwang* (orientation), *daoli* (distance), *gaoxia* (topography), *fangxie* (inclination), and *yuzhi* (straightness), i.e. methods for indicating scale, orientation, distance and topography, which essentially align with the scientific principles of modern cartography and occupy an important position in the history of world cartography.

After Buddhism was introduced to China during the Han dynasty, it developed greatly in the Wei, Jin, Southern and Northern dynasties. This was because during this period, vast regions of the Yellow River basin were plagued by warfare and disasters, leaving people hopeless and without a way out. Buddhism advocates karma and reincarnation as well as the concepts of heaven and hell with fair rewards and punishments after death, providing people with spiritual solace. Buddhism also has profound philosophical teachings that attract the literati's interest. On the other hand, monks from the Western Regions and India continuously came to China, where they preached, translated Buddhist scriptures, and recruited believers. Chinese monks also travelled to India in search of the Dharma and brought back Buddhist scriptures to introduce to their countrymen. These activities received support from the rulers, and temples were built all over the country, leading to a significant increase in the number of monks. Let's take some examples. The monk Dharmaraksa (240–315) of Gandhara origin traveled to Dunhuang, Chang'an, Luoyang and other places during the reigns of Emperor Wu and Emperor Min of the Jin dynasty (265–316). Along the way, he translated 154 Buddhist scriptures totaling 309 scrolls, mostly Mahayana sutras. The monk Budhochinga (232–348) from Kucha (now Kuche, Xinjiang), who was proficient in Chinese and could recite Buddhist scriptures of millions of words, came to Luoyang at the end of the Western Jin dynasty (310). Later in 319, he was welcomed by the Zhao Emperor Shi Le into Chang'an, where he took on 100 disciples. One of his disciples, Dao'an (314–385), preached in Xiangyang for 15 years, teaching *Fang Guang Prajnaparamita Sutra*. In 379, he was invited to Chang'an by the Former Qin Emperor Fu Jian who frequently consulted him on state affairs. He taught thousands of monks and supervised the translation of *Agama Sutra* and other works. Dao'an made great contributions to the integration of Buddhism, a foreign culture, with China's indigenous culture.

In 401, the monk Kumarajiva (344–413) from Kucha, who was well-versed in both Mahayana and Hinayana Buddhist Sutras, was invited to Chang'an by Emperor Yao Xing of Later Qin. He and his disciples translated over 300 volumes of Buddhist Sutras, including *Great Prajna Sutra*, *Diamond Sutra*, *Vimalakirti Sutras*, *Amitabha Sutra*, and the *Great Treatise on the Perfection of Wisdom*

(Mahāprajnāpāramitāśāstra), producing clear and smooth texts. He had 5000 disciples. The South Indian monk Bodhidharma (unknown—528) sailed to Guangzhou during the reign of Emperor Wu of Liang and was welcomed to Jiankang (modern-day Nanjing). Later, he went to Shaolin Temple on Mount Song in the Northern Wei dynasty, starting the transmission of Chan Buddhism. Chinese monks also traveled to the Western Regions and India to seek Buddhist scriptures and Dharma. For example, Faxian (337–422) of the Eastern Jin left Chang'an in the third year of Long'an (399) and traveled to various countries in the Western Regions and India to seek Buddhism. He returned to Guangzhou in the eighth year of the Yixi era (412) via Sri Lanka and Java after 14 years of travel to 30 countries, bringing back many Sanskrit Buddhist scriptures. The following year, he translated the scriptures in Jiankang and wrote *A Record of Buddhist Kingdoms* (412), documenting his experiences. During the Northern Wei dynasty, in the first year of the Shengui era (518), Hui Sheng and Song Yun were sent to the Western Regions and India to seek scriptures. In the third year of the Zhengguang era (522), they returned with 170 Mahayana Buddhist texts and published *Hui Sheng's Travelogue.*

According to *Bianzheng Lun* ("Essays of Disputation and Correction") and other works by Falin (572–640), in the Eastern Jin dynasty, there were 1768 Buddhist temples and 24,000 monks and nuns. During the time of Liang in the Southern and Northern dynasties, there were 2845 temples and 82,700 monks and nuns, increasing the number of temples by more than 1000 and more than doubling the number of monks and nuns compared to the Eastern Jin dynasty. The *Book of Wei: Treatise on Buddhism and Taoism* mentions that in the first year of Taihe (477) in the Northern Wei dynasty, there were 6478 Buddhist temples and 77,258 monks and nuns. During the Yan Chang period (512–515), the number of temples increased by more than 100%, to 13,727, and at the end of the Eastern Wei, "there were over two million monks and nuns and more than 30,000 temples". A large number of Buddhist scriptures circulated in society, as evidenced by the manuscripts from this period found in the Dunhuang caves. After Buddhism was sinicized, it became one of the three major religions in China along with Confucianism and Taoism, and as an integral part of traditional Chinese culture its content was enriched, which was manifested in many aspects such as language, literature, art, architecture, philosophy, and craftsmanship. Moreover, it became a social factor advancing the development of papermaking and printing technologies. During the Wei, Jin, Southern and Northern dynasties, Buddhism spread from China to Korea and Japan.

As writing materials changed, Chinese writing styles also evolved. The small seal script, popular during the Qin dynasty, ended the situation of varied scripts in the Warring States period. The small seal script was simpler and easier to write than the pre-Qin era's large seal script. During the Han dynasty, the clerical script was generally used, which further simplified the small seal script. Throughout the Wei, Jin, Southern and Northern dynasties, the *kaili* script (a combination of the regular script and clerical script) was commonly used, which was more convenient to write than the clerical script and became the early form of the regular script that has been used since the Sui and Tang dynasties. The round strokes of the small seal script evolved into horizontal and straight strokes. For faster writing, the running and cursive scripts were then developed. The evolution of Chinese writing styles is directly related to the widespread use of paper. Writing on paper is not only more convenient than on other materials, but it also allows the brush to move more freely, exhibiting the aesthetic beauty of the characters. Due to the prevalence of copying books, calligraphy became an independent art form. Before the invention of paper, painters used to work on silk and cloth, but after paper became available, they gradually became accustomed to creating on paper. The Jin dynasty saw a new stage of development in calligraphy and painting. Notable figures such as the calligrapher Wang Xizhi (321–379) and the painter Gu Kaizhi (348–409) led the trend of using paper for creations and exerted a profound influence on later generations. Whether for meticulous coloring, small script, or freehand brushwork depicting flowers, birds, landscapes and figures, paper can produce better artistic effects, especially the rendering effect for ink-splashed landscapes, which is unmatched by silk. Paper-cutting and paper flowers are folk arts developed from the Jin dynasty, which are used for indoor decoration.

The increased production and use of paper also had an impact on politics, economy, military affairs, daily life, and customs, bringing about new changes. Paper was more convenient than any other writing materials for making official documents, laws, proclamations, certificates, books, imperial edicts, diplomatic documents, meeting records, and archives, which greatly improved work efficiency from the imperial court to all levels of government, and promoted construction of political power. Papermaking is a low-cost and high-income industry because raw materials are cheap and easy to obtain, leading to the establishment of paper mills in various regions and the development of local industry, commerce and transportation. Contracts, account books, and bills made of paper ensure the normal operation of social and economic order. Paper is also an ideal packaging material for many commodities, and its use helps promote sales of commodities. Kites for transmitting information, lanterns of various colors for sending signals, waterproof maps, paper armor for self-defense, umbrellas for rain protection, and paper gunpowder barrels for military use make marching and fighting more convenient. The use of napkins, toilet paper, and menstrual paper has significant implications for health care and hygiene, marking a historical change in human living habits. Before the Qin and Han dynasties, the deceased's personal belongings, including metal currency (copper coins), were often buried in tombs. After the Northern and Southern dynasties, some physical objects were replaced with paper ones, such as paper-made shoes, hats, chariots and horses; paper money was used in place of copper coins so that the burial style became frugal.

9.1.3 Contribution of Paper to the Development of Arabic Culture

The changes in China brought about by the appearance of paper manuscripts and paper products also occurred in other Sino-sphere countries such as Korea, Japan, and Vietnam. After the spread of papermaking technology to the West, it also made significant contributions to the cultural development in the Arab Islamic countries and the European Christian world. The Arabs were mostly nomadic tribes originally living on the Arabian Peninsula, who experienced social changes brought about by economic development in the sixth to seventh centuries. In 622 AD (the first year of the Islamic calendar), the Islamic prophet Muhammad (570–632) established a theocratic state in Medina (now northwest Saudi Arabia); in 631 AD he unified the various tribes on the Arabian Peninsula and continued to expand outward. After the prophet's death, his battle companion and father-in-law Abu Bakr (576–634) was elected as the leader known in Arabic as "Khalifah", the highest ruler who held administrative, military, and religious powers. After the establishment of the Umayyad dynasty (661–750) by the fifth Khalifah Muawiya, Khalifah became hereditary. The Arab Empire was a multi-ethnic state, with its territory including culturally advanced regions such as Egypt, Syria, Mesopotamia, and Persia. The Islamic holy book *Quran* and the Arabic language were powerful tools for unifying the empire. Arabic culture, created by people of various ethnic groups within the empire and also incorporating certain elements from Greek, Chinese, and Indian cultures, developed over a long period and shone brightly in the medieval world.

Early Khalifahs, including those of the Ummayad dynasty, were mainly devoted to conquering surrounding areas, showing more military prowess than cultural governance. Since the Abbasid dynasty (750–1258) established by Abu'l-Abbas (about 721–754), more emphasis was placed on cultural construction, and the introduction of papermaking technology from China greatly boosted the development of Arabic culture. Prior to this, the writing material there was still papyrus used by ancient Egyptians and it was mainly available to a few people because the vast majority were illiterate. During this period, the empire capital moved from Damascus in Syria to Baghdad in Iraq. After the second Khalifah al-Mansur (about 715–775) consolidated the new dynasty's rule, he rewarded learning and developed production, transforming Baghdad into a political, industrial, commercial, and cultural center. His successors, Harun al-Rashid (about 764–809) and al-Ma'mun (786–833), continued to pursue this policy with fruitful achievements.

During the Ummayad dynasty, schools were established within mosques, albeit in limited numbers, and they mainly trained religious talents. Since the Abbasid dynasty, the education system stepped on track, with aristocratic children attending palace schools and commoners attending mosque-affiliated schools. In the ninth century, there were 30,000 mosques of different sizes in Baghdad alone, and such affiliated schools existed throughout the country. In addition, there were primary schools known as "kuttāb" or "maktab" in Arabic, and some private schools were run in people's homes. Primary education began at the age of 7, with a 5-year curriculum focusing on the study of *Quran*, Arabic grammar, calligraphy, arithmetic, horseback archery, and swimming. For further studies, students attended advanced courses such as Quranic studies, astronomy, history, law, etc. at schools known as "madrasah", equivalent to advanced vocational schools, which became state-run schools after the eleventh century. Graduates became scribes, judges, teachers, and officials. The total number of such schools was 238.[3] Madrasahs were concentrated in cities like Cairo, Baghdad, Damascus, and Jerusalem, with tens of thousands of enrolled students. Damascus, Cairo, and Baghdad were all centers of papermaking in the Arab empire, supplying not only local need for paper use but also exporting to other places or countries. Thus, all the teaching materials used in schools everywhere should be paper manuscripts, and writing on paper makes the emergence of calligraphy and the evolution of fonts possible. The role of paper manuscripts in promoting the development of Arabic culture and education is evident.

During the reigns of the Abbasid rulers al-Mansur, al-Rashid and al-Ma'mun (eighth-ninth century), scholars and scientists from various regions were invited to Baghdad to engage in teaching and research, and they were also instructed to translate various ancient Greek and Indian works into Arabic. Keeping in mind Prophet Muhammad's hadith: "Seek knowledge even if you have to go as far as China", the rulers made efforts to introduce Chinese scientific and technological knowledge. In particular, the House of Wisdom (Bait al-Hikmah) in Baghdad established by al-Ma'mun in the early ninth century became a comprehensive national institution integrating scientific research, higher education, and academic translation, with similar functions to a scientific academy and a university in nature. The institution had a well-stocked library and a well-equipped astronomical observatory. First-class scholars from various fields worked there and they also engaged in teaching activities. The scientific courses offered included medicine, astronomy, mathematics, philosophy (including ethics), and law. The required paper, ink, pens, and daily necessities were funded from the government budget. Grand and sumptuous, the building inside the House of Wisdom resembles a palace.

Hunayn ibn Ishaq (808–873), who once served as Chief Librarian of the House of Wisdom, was a scientist and translator born in Iraq. He was proficient in Arabic, Syriac, and Greek, and had excellent medical skills. His son, Abu Yaqub Ishaq ibn Hanayn, inherited his knowledge, and they both were summoned to serve as court physicians in Baghdad before being sent to work at the House of Wisdom. They led the translation of medical works by Greek physicians Galen (130–200), Hippocrates (460 BC–370 BC), and Dioscorides (40–90), as well as *Elements* and optical works by Euclid (330 BC–275 BC), mathematical works by Archimedes (287 BC–212 BC), *Almagest* by Ptolemy (85–165), and philosophical works by Aristotle (384 BC–322 BC) and Plato. Hunayn ibn Ishaq himself also wrote books such as *Medical Questions and Answers* and *Introduction to Medicine*. The Arabic translations of these ancient Greek scholars' works were widely circulated within the Arab empire and became reference books for scholars' research and teaching in various schools.

The mathematician and astronomer Al-Khwarizmi (Abu Jafar Muhammed ibn-Musa al-Khwarizm, 780-approximately 850), who served as Chief Librarian at the House of Wisdom, was a famous Persian scholar, knowledgeable about Greek and Indian sciences and innovative personally. Originally from Khwarizm in Central Asia, Al-Khwarizmi was highly valued by Khalifah al-Ma'mun, to whom some

[3] Totah K A. *The Contribution of the Arabs to Education* [M]. 2nd ed. New York: Columbia University Press, 1926.

of his works were dedicated. His book *Kitab fi 'l-Jabr wa 'l-Muqabalah* ("The Compendious Book on Calculation by Completion and Balancing"), completed in 820, is the first monograph on algebra written in Arabic, drawn from ancient Indian texts and mainly discussing methods for solving linear and quadratic equations, though quadratic equations are not found in ancient Indian literature. His another work *Kitab al-Hisab al-Hindi* ("The Book of Indian Arithmetic") introduces the decimal algorithm based on the Indian mathematical system. His *Zij al-Sindhind* ("The Astronomical Tables of Sindhind") is the first Arabic astronomical table influenced by Ptolemy and Indian astronomy, forming the basis for the later European *Toledan Tables*, and it includes calendar, planetary motion, and eclipse calculations. His *Kitab Surat al-Ard* ("The Book of the Shape of the Earth"), written around 817, is a geography book based on Ptolemy's *Geography*, but with new developments including accurate data on meridians (latitude and longitude) measurements for the Arab region and other areas. Al-Farghani, an astronomer who worked at the House of Wisdom's observatory, participated in celestial observations and wrote *Jawami* ("Outline of Celestial Studies"), providing a comprehensive explanation of Ptolemaic astronomy and having a broad impact on Europe in later years.[4]

Arabic medicine developed its unique features by combining theory with practice on the basis of absorbing the achievements of Greek, Indian, Persian, and Chinese medicine. In order to practice medicine, physicians had to pass examinations, and by the tenth century, Baghdad alone had over 1000 practicing physicians. Some hospitals also trained apprentices, functioning as medical schools. The Persian medical scientist and philosopher Abu Bakr al-Razi (866–925), director of Baghdad's largest hospital with both theoretical knowledge and clinical experience, authored over 100 works. Among them, *Kitab al-Hawi fi Tibbi* ("The Comprehensive Book of Medicine"), consisting of 30 volumes, is a classic in Arab clinical medicine; his *Al-Judar wa al-Hashah* ("On Smallpox and Measles") is a valuable document on smallpox. Experts believe that al-Razi was influenced by the ideas in *Zhouhou Beiji Fang* ("A Handbook of Prescriptions for Emergency", 341) by the Chinese physician Ge Hong (284–363), as al-Razi had interactions with a Chinese doctor who could speak Arabic. Al-Razi's *Kitab al-Mansuri* ("Book of Secrets in Medicine") is a compilation of Greek medical books, including those of Hippocrates and Galen, but he made rational criticisms of these authorities. Besides, he wrote *Kitab sirr al-Asrar* ("The Secret of Secrets", 912) (Fig. 9.1), a book on alchemy with some formulas, the knowledge of which came from China. Philosophically, Al-Razi was a rational skeptic, refusing to recognize any authority and being critical of the teachings of Galen and Aristotle. He believed that anyone could achieve correct understanding through practical experimentation and theoretical thinking, and that religious leaders were unnecessary and religions were actually harmful because they incited hatred and warfare. He thought that prophets were deluded by demons and that there was no creation ex nihilo, maintaining that evil had already triumphed over good in the world. He also upheld thorough egalitarianism. Al-Razi was a unique thinker in the

[4] Sarton G. *Introduction to the History of Science*: *Vol. 1, Vol. 2, Vol. 3* [M]. Baltimore: Willians & Wilkins Co., 1927, 1931, 1947.

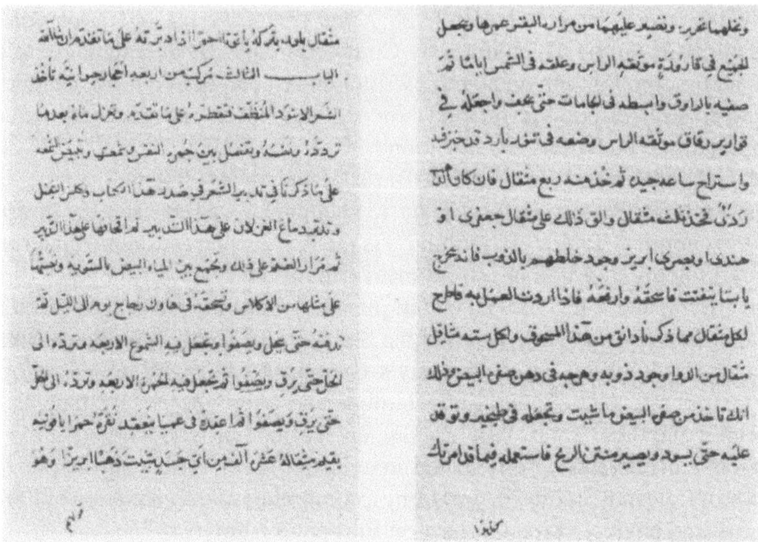

Fig. 9.1 Arabic text page from al-Razi's *The Secret of Secrets* (912) (Taken from Karikov, 1957)

history of Islamic thought, and his humanistic ideas and religious beliefs made him the target of sharp attacks by his contemporaries.

Another physician and philosopher with equal renown to al-Razi was Ibn Sina (980–1037) from Bukhara in Central Asia, also known by his Latin name Avicenna. He had extensive travel experience and wrote over 100 works, including essays and treatises covering multiple scientific fields. One of his widely circulated medical books was *Al-Qanun fi al-Tibb* ("Canon of Medicine"), a five-volume work with over a million words. The book summarizes medical knowledge from various countries in the East and West, incorporating his own thoughts and practical experience. Covering anatomy, etiology, diagnostics, medication, obstetrics and so on, this book is considered an encyclopedia of medicine. Pulse diagnosis, an important part of Chinese medicine, is introduced in the book, and some Chinese herbal medicines are also discussed in the medication section. Ibn Sina's major philosophical work is *The Cure for Ignorance*, also known as *Al-Shifa* ("The Book of Healing") in Arabic. In fact, this is an encyclopedia divided into four parts: logic (equivalent to Aristotle's *Organon*), physics, mathematics (geometry, arithmetic, and astronomy), and metaphysics. His philosophical views mainly derived from the Aristotelian school of thought, but also had elements of Neoplatonism. He criticized astrology and, based on personal experiments, refuted the alchemists' claims of turning base metals into gold.

The Fatimid dynasty (909–1171) established by the Arab nobility in Egypt roughly stood side by side with the Abbasid dynasty. During the reign of Khalifah Abu-Ali Mansūr al-Hakim (996–1021), an institution similar to the House of Wisdom and a library were also established in Cairo. The Later Umayyad dynasty (756–1036) established in Spain had a comparable institution in its capital Cordova in 970. These institutions were modeled after the House of Wisdom in Baghdad and thus functioned as universities. Ibn Yunus, who worked in Cairo, was an astronomer, mathematician and a poet as well. His main works include an overview of astronomy and astronomical tables, entitled *Al-Zij al-Hakimi al-Kabir* ("The Great Hakim Astronomical Tables") which provides observational data from previous scholars and the author's own observations spanning 175 years (829–1004), including solar and lunar eclipses, equinoxes, planetary motion theories and calculation methods. Al-Zarqali (1029–1100), who worked at the Cordova observatory, made numerous astronomical instruments with explanations of their structures, including the distinctive Arabic astrolabe. These contents are all included in his *Toledo Tables* (1080), in which the Ptolemaic astronomical system was revised.

The humanities in Arab culture also shone brightly. In addition to the aforementioned philosophical works, there is *Kitab Akhbar al-Rasul w- 'al-Muluk* ("The History of Prophets and Kings") written by the historian Al-Tabari (838–923) of the Abbasid dynasty, a 13-volume chronicle of the Arab world from ancient times to 915 AD, commonly abbreviated as *Annals* in Western translation versions. Another historian, Al-Masudi (896–956) born in Baghdad, wrote a comprehensive 30-volume Arabic chronicle, condensed into *Muruj al-Dhahab wa Maadin al-Gawhar* ("The Meadows of Gold and Mines of Gems"), an encyclopedia of history. Ibn-Khaldūn (1332–1404), born in Tunisia, completed a comprehensive historical work in 1382, which includes three parts: (1) "Muqaddimah", a treatise on history and philosophy; (2) history of the Arabs and neighboring peoples; and (3) history of North African Muslims and Mecca. Geographic works in Arabic are also abundant, represented by *Kitab al-Masalik wa 'l-Mamalik* ("The Book of Roads and Kingdoms", 885) by Ibn-Khordadzbeh (ca. 820–912), which records the journeys and business travels from Baghdad to various regions of the Arab Empire as well as to India, China, and other countries. One of the most famous literary works is *Alf Layla wa Layla* ("One Thousand and One Nights"), also known as *The Arabian Nights*, completed in the tenth century. The book is based on Persian stories from the sixth century, and incorporates tales and fables from places like Greece, India, Egypt, and Hebrew, reflecting social life and customs of various ethnic groups in the Arab world and showcasing their imagination. This work has long been renowned in world literature.

Before the seventh century, the Arabs under the rule of Byzantine and Persia did not attract much attention worldwide. The various tribes were constantly engaged in internal strife, culturally far inferior to surrounding regions such as Egypt, Persia, and Babylon. Since Muhammad founded Islam and unified the Arabian Peninsula in the seventh century, the Arab Islamic state gradually grew stronger, and taking advantage of the long-term war and mutual weakening of Byzantium and Persia, it quickly expanded its territory to the surrounding areas. In less than a hundred years, it had established the Islamic Caliphate, whose territory exceeded that of the Roman Empire at its peak. Although the caliphs forced people within the empire to convert to Islam, use Arabic and pay taxes, they did not reject the advanced cultures of the conquered regions; they embraced their strengths to offset their own weaknesses. After the Abbasid dynasty, there was a focus on cultural development. In

Fig. 9.2 The Hulwan library in Baghdad in the eleventh century, housing 200,000 volumes of books (Collection of the National Library of Paris)[5]

addition to the local cultures of various ethnic groups, Greek, Indian, and Chinese cultures were also widely absorbed. Within a century, the light of Arab culture radiated to the surrounding areas, attracting endless Christians and Jews to Baghdad, Cairo, and Cordova to study various disciplines. These places had rich collections in their libraries (as shown in Fig. 9.2), and schools and bookstores were everywhere, producing many talented individuals. The role of paper in promoting development of Arab culture and education is obvious. When paper produced by paper mills in various parts within the territory was exported to Europe, large amounts of European gold coins flew into Arabia, increasing its fiscal revenue. Arab culture had a significant impact on Europe in the later Middle Ages (especially the eleventh to fourteenth centuries) and also played a communicative and mediating role between Eastern and Western cultures.

9.1.4 Role of Paper in the Development of European Culture

The history of Europe is older than Arab history, but while Arab culture was flourishing, Europe entered the dark era of the Middle Ages. With the division of Rome in 395 and the fall of the Western Roman Empire in 476, ancient Greek civilization gradually faded away and European culture declined. After Christian countries recovered in the tenth century, feudalism developed, but frequent wars left no time for

[5] Taken from *Al-Hariri's Assemblies*, painted by Yahya al- Wasit in 1237.

the development of education and culture, knowledge progressing slowly. Coarseness and ignorance were common among the upper classes of society. Knights were usually illiterate, even unable to write their own names, and many kings and emperors lacked reading abilities. Henry IV (1050–1106) was praised for being able to read letters.[6] The vast majority of peasants and artisans were illiterate, and no one cared about their education. People were poor and ignorant. Knowledge was only held by a very small number of church clergy, who used Latin as a written language that was incomprehensible to the masses. Schools aiming at training missionaries made students rote-memorize the Latin Bible. *Vulgate*, the Latin version of Bible, revised by Saint Jerome (342–420) in 405 and *De Civitate Dei* ("City of God") by Aurelius Augustinus (354–430) became the official philosophy for ruling people's thoughts and theological preaching, and they were also used as main textbooks at church schools. Even this kind of education was a privilege enjoyed by only a few, and the masses were excluded.

In the Middle Ages, European schools were strictly controlled by the church they served. They were often located in monasteries, and future priests were trained in the "seven liberal arts", namely Latin grammar, rhetoric, logic, arithmetic, geometry, astronomy, and music. Rhetoric was intended for training preaching skills, logic for arguing theological propositions, music for training hymn singing, astronomy for calculating religious holidays, geometry was precisely for describing geography, plants and animals when annotating the Bible, and arithmetic was simply for calculating. These subjects were all tools and accessories for promoting religion. Dogmatic indoctrination was used in teaching and independent thinking was not allowed. The Bible and annotations of the Church Fathers were absolute authorities only to be believed in. The purpose of acquiring knowledge was to deepen religious faith. Christianity promoted otherworldliness, asceticism, and obscurantism, which reinforced its ideological tyranny, thereby stifling creative thinking and blocking the path of knowledge advancement, a major cultural regression compared with ancient Greece. Rulers sometimes established palace schools to cultivate imperial management talents, but only a few aristocratic children were admitted, and the subjects they learned were limited to a few topics such as law. Medieval literature was predominantly religious, such as hymns, prayers, Christian stories and Acts of the Apostles; dramas were superstitious and tasteless. Folk literature, on the other hand, belonged to oral literature consisting of tales in dialects, so the spiritual life of society was quite poor.

However, during the eleventh to fourteenth centuries, the situation changed. The Crusades (1096–1291) launched by the Roman Pope and Western feudal lords exposed Europeans to the more advanced Arabic culture in the Islamic world than in the Christian world and introduced it to Europe. Through the medium of the Arabs, they also introduced some scientific and technological inventions from China, such as papermaking, gunpowder and firearms, compass and knowledge of magnetism, and alchemy. On the other hand, due to the development of commerce and overseas trade

[6] Kosminskii E A. *The History of the Medieval World*: Part 2 [M]. Translated by Wang Yijin. Shanghai: Kaiming Bookstore, 1947.

in European cities, there appeared citizens engaged in handicrafts and commerce, or a class of bourgeoisie, a new class in feudal urban economic development and a new social force for promoting scientific and cultural development. In line with this, secular schools also emerged and gradually evolved into universities teaching students multidisciplinary knowledge; a group of deviant scholars who dedicated themselves to academic research separated from church intellectuals. They initially broke through the ideological shackles set by the church authorities, awakened the masses with rational knowledge, and lit up the guiding light in the darkness for the arrival of a new era. All these new developments occurred after the establishment of paper industry in Europe, just as we see in East Asia and the Arab regions, because it was paper that made these changes possible and accelerated their development process. The reason why Chinese scientific inventions that had an impact on European social development could be introduced through the Arab regions was that the westward expedition of the Mongolian army in the thirteenth century opened up the once-blocked passage between the East and the West, providing convenient conditions for the movement of people and goods as well as cultural communication.

In 1085, when the European Crusaders captured Toledo, a Muslim-ruled city in Spain (south of Madrid), they discovered a large number of Arabic manuscripts, including translations of Greek works. Archbishop Raimundode Penafort (1176–1275) established a translation agency and recruited people who knew Arabic to translate them. The translation work reached a peak between 1125 and 1280, when the Italian Gerard da Cremona (1116–1187) alone translated 80 Arabic works, including those by Aristotle, Ptolemy, and Arab scholars like Ibn Sina. Spaniards and Sicilians cooperated with Arabs and Jews to translate books of Euclidean geometry, Arabic algebra, astronomy, alchemy, medicine, and more into Latin.[7] Later, they searched for Greek manuscripts left behind by the Byzantines, including *Complete Works of Aristotle*, and translated them directly from Greek into Latin. The appearance of these new Latin translations astonished Europeans, revealing to them a world of ancient Greek intellectual culture that they had long forgotten and felt unfamiliar with, as well as the fresh brilliant treasure of Arabic culture and the nascent scientific and cultural civilizations of China and India. These works were spread and copied, explosively increasing people's knowledge and enabling them to find new fields of study and absorb new ideas and inspirations to ultimately revive and utilize ancient knowledge for contemporary use.

After the eleventh century, secular urban schools and universities appeared one after another to meet the needs of the emerging urban middle class. What was taught in Latin was no longer limited to theology, but also included law, medicine, and literature; contents of astronomy and mathematics became richer. There were secular scholars among the teachers. These were public universities managed by rectors elected by students, such as universities of Padua and Bologna in Italy. Another type of university was founded by the church, such as University of Paris, Oxford, and Cambridge. By the fourteenth century, there had been more than 40 universities in

[7] Sarton G. *Introduction to the History of Science*: *Vol. 2* [M]. Baltiomore: Williams & Wilkins Co., 1931: 832–833.

Europe, cultivating many scholars. Each university had a library, where teachers and students obtained much philosophical and natural science knowledge from the translations of Greek and Arabic works, recognizing the absurdity of Catholic theology and consequently doubting the existence of God. This terrified the church authorities, and thus Albert the Great (Albertus Magnus, ca 1200–1280), a German-born theologian and professor at the University of Paris, attempted to use the newly introduced scientific knowledge and Aristotelian doctrines to serve theology, claiming that "science is but a preparation for faith". Albert the Great's student, Thomas Aquinas (ca.1225–1274), systematically developed Albert's approach in his *De Summa Theologica* ("A Systematic Compendium of Theology").

Aquinas used philosophical methodology to demonstrate theological propositions, arguing that truth resided firstly in reason and secondly in things, with God being the reason and the highest truth, and faith in God surpassed rationality. He employed the necessity of a first mover for the motion of celestial bodies to prove the existence of God. This theological system, pioneered by Aquinas and others, was further developed and disseminated within the Church's scholastic tradition, hence referred to as scholasticism. Realism in scholasticism represented by Aquinas is seen as a reactionary and regressive response to Greek philosophy and Arabic science introduced to Europe through Arabic literature in the late Middle Ages, reflecting the feudal forces' dying struggle in ideology.[8] On the other hand, we also observe that the empirical spirit in Arabic and Chinese sciences inspired European scholars to engage in scientific experiments, leading to the establishment of critical rationalism against scholastic philosophy and the Church. Some even philosophically critiqued Aquinas's fallacies. For instance, Pierre Abélard (1079–1142), a professor at the University of Paris, advocated that faith must be based on knowledge, promoting free discussion and opposing the Church's supreme authority. He asserted, "Doubt is the path to inquiry," "Inquiry leads to truth," and "To believe, one must first understand,"[9] putting forward tit-for-tat battle cries against the mainstream views of scholasticism, thus being branded as a "heretic" by the Church.

William de Ockham (1295–1349), an Englishman from Ockham, launched a movement, denying Aquinas' claim about the existence of God as the first mover. Taking action at a distance as an example, he argued that a moving object does not require continuous physical contact with the mover, as a magnet can move an iron rod without contact between the two. Roger Bacon (1214–1294) from Oxford, who knew Arabic and meticulously researched various scientific fields including knowledge of the newly introduced Chinese gunpowder, provided detailed explanations in his three major works. He believed that true scholars should understand natural sciences through experimentation, so he personally conducted optical and chemical experiments. He advocated that the only way to verify the correctness of previous assertions should be observation and experimentation, and therefore he urged, "Do not be ruled by dogmas and authorities any longer, but look at this world." His

[8] Hong Qian, et al. *A Brief History of Philosophy* [M]. Beijing: People's Publishing House, 1957.

[9] Danpier W C. *A History of Science and Its Relations with Philosophy and Religion* [M]. 4th ed. Cambridge: Cambridge University Press, 1958: 80, 90.

contemporary, Pierre de Maricourt or Peter Peregrinus (1205–1275), persisting in conducting a series of experiments in magnetism, said in his *Epistola de Magnete* ("Letter on the Magnet") (1269) that researchers in magnetism must be diligent in hands-on work to correct their misunderstanding. Roger Bacon believed that he gained knowledge of natural sciences through experimentation, which brought him wisdom and peace.

The new change that emerged in medieval European literature is marked by the publication of the Italian Dante Alighieri's *Divina Co-mmedia* ("The Divine Comedy", ca. 1307). This political and philosophical poem depicts the story of the author being led by a Roman poet in a dream to wander through the three realms of Hell, Purgatory, and Heaven. The work metaphorically reflects the arduous journey of experiencing the hardships of the real world and striving for an ideal state. The author places the ideal ruler in Heaven, and the Pope in Hell. The pursuit of ideological liberation and knowledge and the call for absorption of classical culture and leniency towards heretics as expressed in the work are the sprout of humanistic thought. It was written in the Tuscan dialect of central-western Italy, paving the way for later European literature during the Renaissance to be created in vernacular languages. Meanwhile, the criticism of the Church and scholastic philosophy by such figures as Abelard and William sowed the seeds for the future religious reform; the experimental spirit and serious pursuit in natural sciences by Roger Bacon and Peter et al. became a precursor to the upcoming scientific revolution. Through the efforts of several generations, Europe had completely shaken off its previous stagnation in science, technology, and culture by the fifteenth century, presenting itself with a new face. Although there were still social factors hindering progress, the forces driving societal advancement were rapidly gathering and growing. Dante said in *Divina Commedia*, "*Segui il tuo corso, e lascia dir le genti* (Follow your own road and let people talk)". This quote inspired many to challenge the old forces, even if they were criticized and persecuted. With this fearless spirit, European progressives faced up to reality and the future.

9.2 Contributions of Printing Techniques to the Development of World Civilization

9.2.1 Contributions of Printing Techniques to the Development of Education and Science in China

Although manuscripts written on paper were superior to those on any other material, they shared a common drawback: each book had to be copied by hand word by word, and each copying effort could only produce one manuscript. As the number of books increased, people had to spend a lot of time and effort copying books. For the Chinese who use ideographic characters, the beauty of Chinese characters comes with the large number of characters and strokes, making writing even more laborious. Moreover, the mistaken use of words due to similarity in word form often occurred in the process of copying, thus confusing readers. The emergence of printing technology relieved millions of people from the tedious task of copying books. Tens of millions of copies with the same content and font can be printed at one time from the same printing plate. With standardized proofreading, there are fewer errors and omissions; texts are clear and easy to read and the price is cheap. Printed books circulate more quickly than handwritten books and spread to various corners of society on a larger scale, becoming a powerful medium for disseminating ideas, knowledge, and information. The role they played in past societies is similar to that of television and the internet in today's world. The emergence of printing was another epoch-making revolution in the history of human communication of images and texts. Starting in the Sui dynasty and flourishing in the Tang dynasty, printing technology gave wings to the development of Chinese education, culture, and various academic disciplines, and brought about Renaissance to the East of the Old Continent ahead of time. For limited space, we discuss only the Tang and Song dynasties here.

Early printed materials in China were mostly Buddhist scriptures and images intended for the common people to use, as Chinese Buddhists, driven by their faith, were enthusiastic about replicating a large number of Buddhist texts. Printed scriptures were considered to have the same spiritual power and function as handwritten sutras and mantras, and people preferred printed texts so they did not have to copy sutras and mantras dozens or even hundreds of times. The Buddha required believers to repeatedly write the same mantra many times to accumulate merit, while Confucius, the founder of Confucianism, advised his disciples to "learn and practice from time to time", suggesting that one copy of each book was enough and that making more copies was unnecessary. Therefore, there are reasons that Chinese Buddhists made a special contribution to the development of printing, as it enabled Buddhism to achieve greater development after the Sui and Tang dynasties. On the other hand, publishers found that by using the method of publishing Buddhist scriptures to produce other secular readings needed by the masses, such as dictionaries, phonetics and other linguistic tools, fortune-telling books and almanacs, they could also find a market; rulers also quickly realized that publishing Confucian classics such as *Nine Classics* and *Three Histories* for use in schools could provide standard teaching materials, cultivate future officials at all levels with a unified ideological system, and strengthen the ideological rule of official philosophy across the country. Therefore, from the Tang dynasty onwards, all dynasties set up official printing houses at both the central and local levels to publish various books. Printing technology greatly promoted the development of China's education as well as the establishment and improvement of the Imperial Examination System after the Tang dynasty.

Taking the Tang and Song dynasties as an example, the number of schools and enrolled students as well as learning contents all exceeded those of previous dynasties.[10] The Tang dynasty established the *Guozijian* (Imperial Academy) in the capital to manage educational affairs, with all schools in the capital under its jurisdiction.

---

[10] Meng Xiancheng, et al. *Sources on the History of Education in Ancient China*: Part 2 [M]. Beijing: People's Education Press, 1961.

There were six institutions, among which *Guozixue* and *Taixue* were the highest, equivalent to ancient universities, mainly teaching Confucian classics, literature and history. *Simenxue* was a higher education institution exclusively for children of high officials and foreign students, teaching similar subjects as *Guozixue* and *Taixue. Lvxue* (Legal Studies), *Shuxue* (Calligraphy Studies), and *Suanxue* (Mathematical Studies) were specialized schools teaching law, calligraphy, astronomy and calendar calculations, respectively. Each school had *boshi* (a prestigious academic title given to accomplished scholars who had teaching positions in ancient China's educational system, equivalent to present-day professors), who were referred to as *boshi* of a certain subject, such as *Suanxue boshi*. Under the *boshi* were teaching assistants and instructors. Students typically entered school at the ages of 14–19, except for *Lvxue*, where the age range was 18–25. The number of students in each school was generally fixed: 500 students in *Taixue*, 300 in *Guozixue*, 1300 in *Simenxue*, and 30 in *Shuxue* and *Suanxue* respectively, and 50 in *Lvxue*, totaling 2210 students. The *Taiyi* Bureau (Imperial Medical Office) recruited medical students separately and it had a pharmacy garden and departments for physicians, acupuncturists, masseurs, and pharmacists, each having *boshi* as teachers. The *Taiyue* Bureau (Imperial Music Office) had two departments for *Yue* (instrumental music) and *Shengyin* (vocal music), and also enrolled students. The Court Painting and Calligraphy *boshi* also recruited artistic talents for painting training. The *Hongwen Guan* and *Chongwen Guan*, institutes for text correction, also admitted students. Therefore, the total number of students in the capital city of the Tang dynasty definitely exceeded 3000. According to *New Book of Tang: Biographies of Confucian Scholars*, in the sixth year of Zhenguan reign in the Tang dynasty (632 AD), there were 1200 districts of school houses in the capital, with a total of 3200 students. In addition, there were 8000 students enrolled in *Simenxue*, including overseas students from Silla, Baekje, Goguryeo, and ethnic minority students from Gaochang (Xinjiang) and Tubo (Tibet). In the second year of Tang Longshuo reign (662), another *Guozijian* was established in the Eastern Capital (Luoyang), with the same setup as in the Western Capital (Chang'an) but with fewer students. Therefore, the number of students enrolled in the "universities" in the two capitals exceeded 10,000 per session.

In addition to central-level schools, there were also local schools during the Tang dynasty, known as *Fuxue* (provincial-level schools), *Zhouxue* (prefectural-level schools), and *Xianxue* (county-level schools). The enrollment sizes varied, with *Xianxue* admitting 20–40 students, *Zhouxue* 40–60 students, and *Fuxue* 80–100 students. Some *Fuxue* and *Zhouxue* even had *boshi* of medicine and teaching assistants to train students. In the 29th year of the Kaiyuan era (741), there were 328 prefectures and 1573 counties across the country, with nearly 86,000 students enrolled in local schools each year and a total of nearly 100,000 students, including students from the two capitals (Tang dynasty had a "dual capital system" with the Western Capital at Chang'an and the Easter Capital at Luoyang). These students were all from state schools where the government provided unified teaching materials. Besides, there were private schools and village schools everywhere. According to Volume 35 of *Tang Huiyao* ("Economic and Political Structure of Tang Dynasty"), in the 26th year of the Kaiyuan era (738), an edict was issued to "set up one school in each village in all prefectures and counties, with selected teachers to teach students". The "*New Book of Tang: Biographies of Confucians*" records that Ma Jiayun from Wei Prefecture established a private school at Bailushan, attracting thousands of students. In the text, Wang Tong (canonized as Wenzhongzi) taught in the area between the Yellow and Fen rivers (now Shanxi Province), and his students were too numerous to count, nearly a thousand, including famous figures like Fang Xuanling. Wang Gong from Hua Prefecture taught in his community, with hundreds of disciples. Private schools were an important supplement to official schools, and many learned scholars or resigned, secluded officials served as teachers. The massive education network from the central to the local level during the Tang dynasty laid foundations for the cultural prosperity. In the first half of the Tang dynasty, handwritten manuscripts and printed books coexisted, but in the second half, printed books gradually increased. During the Five Dynasties period (tenth century), the central *Guozijian* started to engrave and print *Nine Classics* and issued them to various places, and then the number of handwritten textbooks and reference books in schools at all levels gradually decreased. On this basis, the Northern Song dynasty further developed the trend, ultimately making woodblock printing popular in society. The movable type printing also appeared at the same time so that teachers and students mainly used printed books, especially the standard editions issued by *Guozijian* to the whole country, which was a revolution in the history of teaching materials with profound impact on education. Teachers and students did not have to spend much time and energy copying various books, but instead they could focus on teaching and learning. The same was true for the vast majority of intellectuals in society, as printed books relieved their physical burden of copying books and allowed them to concentrate on research.

The education system of the Song dynasty generally inherited from the Tang dynasty but had new developments. The central government established *Guozijian* to oversee national education affairs, and under its jurisdiction were higher education institutions including *Guozixue* in the capital, admitting 200 children of officials ranked seventh grade or higher, *Taixue*, enrolling 2400 children of lower-rank officials and civilians, *Simenxue*, which had the same target group as *Taixue*, and *Guangwen Guan* with 2400 students. The number of students in higher education institutions exceeded 10,000, but as the population of the Song dynasty was smaller than that of Tang, there were relatively more university students in the Song dynasty, and more civilian children enrolled, which is noteworthy. The local official schools included prefectural-level schools (each with 53 students) and county-level schools (with 30–50 students), totaling about 64,000 students nationwide. Primary schools prevalent throughout the country were mostly privately run family schools called *jiashu*, where children started school at age 8. The *Guozijian* also had publishing functions, providing textbooks and reference books for schools at all levels throughout the country. According to incomplete statistics, it published 256 kinds of dictionaries,

Confucian classics, history books, medical books, and works of various schools of thought.[11]

Private lectures, which were more advanced in teaching content than private schools, prevailed in various regions. These lectures were often conducted by scholars, and the teaching quality sometimes surpassed that of official institutions. For example, the philosopher Lu Jiuyuan (1139–1193), who was a *jinshi* (nominee who passed the highest level of imperial examinations), was so popular that "every time he lectured, the area outside was filled with shoes; the elderly leaned on canes to listen." Yang Jian, Yuan Xie, Shu Lin, and others were all his students. Another *jinshi*, Zhang Zai (1020–1077), gave lectures in Guanzhong (Shaanxi), attracting many followers, and his teaching was known as "Guan Learning". Zhou Dunyi (1017–1073), a native of Daozhou (Dao County, Hunan), retired from his official post and built a hall in his hometown where he engaged in writing and lecturing. Cheng Hao, Cheng Yi, and others all studied under him; known as "Master Lianxi", his teaching was referred to as "Lian Learning". However, one major characteristic of the Song dynasty was the rise of larger-scale academies (*shuyuan*). Initially privately operated, these academies later gained support from officials and were granted with court plaques, school-owned land, and books, with academic officials appointed to oversee them, thus turning to be semi-official and semi-private. The famous Four Great Academies are Bailu Academy in Jiangzhou (Jiujiang, Jiangxi), Songyang Academy in Xijing (Dengfeng, Henan), Yuelu Academy in Tanzhou (Changsha, Hunan), and Yingtian Academy in Jiangning (now Nanjing). These academies had educational rules and regulations, and provided accommodation for students, with dozens to over a hundred dormitories. Besides, they had rich collections of books, and received imperial grants of engraved editions of *Nine Classics* and other books. Later, there appeared more academies such as Mingdao Academy, Danyang Academy, Ziyang Academy, Hedong Academy, Heshan Academy, Tianmen Academy, Kaoting Academy, Xuancheng Academy, etc., spreading across the country, and in level they were equivalent to the government-run prefectural schools. These academies cultivated a large number of talents for the country and had a profound impact on later generations. The academies of the Yuan, Ming, and Qing dynasties were all established based on the Song dynasty model. The various government-run prefectural schools and academies in the Song dynasty were also active in publishing, producing over 300 kinds of books, more than those published by *Guozijian*. They integrated publication and education, enhancing each other, which is another distinctive feature of the period.

With the highly developed education system in the Song dynasty, "it was shameful for a five-foot child not to be able to talk about literature and writing", and a significant proportion of the population was literate. For instance, in the twelfth century, with an average national population of 30 million people, 200,000 *juren* (nominees who passed provincial-level imperial examinations) were produced. In the thirteenth century, the population halved, but the number of *juren* doubled to 400,000. Assuming

[11] Needham J. *Science and Civilisation in China*: *Vol. 5*. Cambridge: Cambridge University Press, 1985: 379.

that there were 20 candidates for one *juren*, the number of people participating in the imperial examinations in the twelfth century would be 4 million, accounting for 13% of the total population and this proportion continued to increase in the thirteenth century. At that time, China was the only country in the world with so many intellectuals, therefore, there was inevitable prosperity in all academic fields in the Song dynasty. Science and technology, with the Four Great Inventions as the backbone, reached a new peak, which can be seen from the works and publications by Song scholars. The 40-volume W*ujing Zongyao* ("Complete Essentials for the Military Classics", published in 1044 and reprinted in 1231), written on imperial orders by Zeng Gongliang (999–1078) and other scholars, is a large military science encyclopedia. It provides detailed descriptions and studies of various cold weapons, fortress building techniques, warships and chariots, gunpowder and firearms, compass and magnetism, accompanied by illustrations. Shen Kuo's *Mengxi Bitan* ("Dream Pool Essays", 1088) is an encyclopedic academic monograph covering mathematics, astronomy and calendrical science, magnetism and compass, movable type printing, optics, geology, and medicine, to which he made original contributions. Joseph Needham regarded this book published in 1163 as a milestone in the history of Chinese science.

The *Nongshu* ("Book On Agriculture") by Chen Fu (1076–1154) first summarizes the cultivation techniques used in the rice regions in the south of the Yangtze River, and it also discusses land use planning, proposing the idea that proper management and sufficient manure could keep soil perpetually new and fertile, and unprecedently includes sericulture techniques. The *Ju Lu* ("On Citrus Fruit", published in 1178) by Han Yanzhi (1131–1206) records the varieties, cultivation, pest prevention and storage of citrus in Yongjia (Wenzhou, Zhejiang Province). It is the world's first monograph on citrus, highly regarded by Western scholars. The *Xi Yuan Lu* ("Collected Cases of Injustice Rectified", published in 1247), written by the former judicial officer Song Ci (1186–1249) based on forensic practice and theoretical research, is the world's first systematic forensic medicine work. It covers all contents of forensic science, having a profound impact on later generations, and has been translated into various foreign languages. Medical officer Wang Weiyi (987–1067), who was ordered by the court to summarize ancient acupuncture and moxibustion techniques, accurately determined acupuncture points and led the casting of a bronze acupuncture figure to help explain them. He authored *Tongren Shuxue Zhenjiu Tujing* ("Illustrated Manual of Acupuncture Points on a Bronze Figure", 1027). The life-sized bronze figure was an important teaching tool, considered as a national treasure at the time. During the Song dynasty, herbal medicine experienced significant development. In the early Song period (974), Emperor Taizu ordered Ma Zhi (ca. 935–1004) and others to annotate the manuscript *Tang Bencao* ("Tang Herbal") and publish it under the title *Kaibao Bencao* ("Materia Medica of the Kaibao Era"), making contributions to the preservation of ancient herbal texts. The scientist Su Song (1019–1101) and others were commissioned again to revise herbal texts and compiled *Bencao Tujing* ("Illustrated Classic of Materia Medica", 1061) based on a national survey of medicine, which was published the following year. There were 103 new herbal medicines added in the book and 923 illustrations of medicinal plants attached. With strong scientific significance, it is the earliest illustrated edition of a pharmacopoeia on herbs, representing the highest level of pharmacology in the world at that time. Tang Shenwei (1056–1163) further supplemented herbal books and completed *Zhenglei Bencao* ("Classified Materia Medica from Historical Classics for Emergency", 1108), also an illustrated edition, which encompasses the essence of Chinese materia medica prior to the Northern Song dynasty. The *Bencao Yanyi* ("Extension of the Materia Medica", 1116) published in 1119 by Kou Zongshi (ca. 1071–1146) has its own unique characteristics with the form of notes to supplement the deficiencies of earlier materia medica. During the Song dynasty, commerce and overseas trade flourished, leading to the inclusion of numerous foreign medicinal materials in Chinese materia medica. The multitude of herbal books published during this dynasty is dazzling and too numerous to list. Classical works of traditional Chinese medicine such as *Huangdi Neijing: Suwen* ("Inner Canon of the Yellow Emperor: Basic Questions"), *Zhu Bing Yuan Hou Lun* ("Treatise on the Pathogenesis and Manifestations of All Diseases"), *Maijing* ("Classic of the Pulse"), and *Zhenjiu Jing* ("Canon of Acupuncture and Moxibustion") previously circulated only in handwritten copies, but after the Song dynasty, they were published in printed editions and reached thousands of households.

The Song dynasty was the golden age for the development of mathematics, and especially in the field of algebra, with its achievements far ahead of the rest of the world. Jia Xian (1005–1065) put forward *Kaifang Zhuofa Benyuantu* (Original Principle Diagram for Square Root Extraction) in his book *Huangdi Jiuzhang Suanfa Xicao* ("Detailed Exposition of the Nine Chapters on the Mathematical Art of the Yellow Emperor", ca. 1050), listing the binomial coefficients of whole powers in a tabular form, known as the "Jia Xian Triangle". His "method of multiplicative increment and root extraction" creates a way to find the positive roots of any high-order power. Qin Jiushao (1202–1261), in his book *Shushu Jiuzhang* ("Mathematical Writings in Nine Sections", 1247), presented the "method of positive and negative square root extraction" to solve the problem of numerical solutions to high-order equations, and his *Da Yan Shu* (Dayan Solution) provides a method for solving the simultaneous linear congruence equations.[12] One of the facilitators for the advancement of mathematics in the Song dynasty is the compilation and publication, for the first time in history, of *Suanjing Shishu* ("Ten Mathematical Classics"), which reflects the mathematical achievements over a thousand years from the Han dynasty to the Tang dynasty. This work was reprinted several times in the Southern Song dynasty, thus popularizing mathematical knowledge. In astronomy, from 1010 to 1106, five large-scale star position observations were conducted with much higher precision than the previous ones. Based on these observations, a star map with 1464 stars was created, which was attributed to the development of advanced astronomical instruments. Su Song's *Xin Yi Xiang Fa Yao* ("New Treatise on the Astronomical Clock", 1092) documents the structure of the Kaifeng Astronomical Observatory along with 47 design plans, which he and the engineer Han Gonglian built from 1086 to 1092 under the imperial decree. The observatory was 35 feet high (about 12 m), consisting

[12] Qian Baocong. *History of Mathematics in China* [M]. Beijing: Science Press, 1964: 144–167.

Fig. 9.3 Water-powered Astronomical Clock Tower[14]

of three levels, with the top level for an armillary sphere observing celestial bodies, the middle level for a celestial globe demonstrating the daily motion of celestial bodies, and the lower level for a time-reporting mechanism. The three parts were connected by transmission devices and gears which were driven by water flowing from a clepsydra, rotating the gear and activating the armillary sphere, celestial globe, and the time-reporting mechanism simultaneously. It is worth noting that the most important component of the modern mechanical clocks, the linkwork escapement, was already installed in the time-reporting mechanism, making the water-powered time-reporter the direct predecessor of the astronomical clocks in the world.[13] Fortunately, this Chinese invention has become known to the world by virtue of the circulation of publications (Fig. 9.3).

The Song dynasty also made various achievements in the field of technical sciences. For example, the architect Li Jie (ca. 1060–1110) compiled *Yingzao Fashi* ("Treatise on Architectural Methods", 1100) under imperial decree, which was published in 1103. The book consists of 34 volumes with 357 sections and 3555

[13] Needham J, Wang L, De Solla Price D J. *Heavenly Clockwork: The Great Astronomical Clocks of Medieval China* [M]. Cambridge: Cambridge University Press, 1960.

[14] Wang Zhenduo. *Restoration of the Water-powered Astronomical Clock Tower in the Song Dynasty* [M] // Wang Zhenduo. *Archaeology of Science and Technology Series*. Beijing: Cultural Relics Press, 1989: 238–273.

entries, systematically summarizing the architectural achievements and engineering management experience of previous generations. It provides detailed descriptions of the technical and operational standards for 13 different types of work in the industry, along with labor and material quotas. The book also includes 193 architectural drawings, especially reflecting the new achievements and technical experience of builders since the Northern Song dynasty. This book is rare in its breadth of content, systematization, and scientificity of narration. At the same time, in 1103, the tower-building master Yu Hao from Zhejiang authored a three-volume work entitled *Mu Jing* ("Timberwork Manual"), a specialized treatise on architecture. Additionally, there were two books detailing the technology of making ink: *Mo Pu* ("Ink Manual", 1095) by Li Xiaomei (1055–1115) and *Mo Jing* ("Book of Ink", 1100) by Chao Guanzhi (1050–1120), the former being an illustrated edition. Wang Zhuo (1115–1175) wrote *Tang Shuang Pu* ("Sugar Frost Recipes", 1154), explaining the techniques of making sugar from sugarcane. During the Song dynasty, the study of "recipe compilation" was quite developed, and many of these works were related to science and technology. In addition to the works mentioned above, there were also *Lizhi Pu* ("Lychee Recipes"), *Tong Pu* ("Tung Tree Recipes"), *Ju Pu* ("Chrysanthemum Recipes") and more.

9.2.2 Development of Neo-confucianism, Literature and Historical Studies in the Song Dynasty Promoted by Printing Technology

During Northern and Southern Song periods, printed editions of exegetical works on *Nine Classics* by Confucian scholars from the Han, Wei, and Six dynasties circulated widely in society. Continuing in this direction, Confucianism would hardly have any new development. Therefore, the Song scholars shifted their focus from the traditional emphasis on the literal interpretation of texts to a philosophical interpretation of meanings and principles in Confucian classics. With traditional Confucian ethics as the core and incorporating the theoretical thinking of Buddhism and Daoism, they expanded their exploration from human principles to heavenly principles, i.e., the origin of the universe and the development of nature. They introduced the concept of "*li* (principle)", holding that "*li*" was not only the highest principle of human society but also the fundamental root of all things in nature. That is why the Confucianism developed during the Song dynasty is called *Lixue* ("Study of Principle"), or "Neo-Confucianism" in the West. Zhou Dunyi, the founder of Neo-Confucianism, proposed in his book *Tai Chi Illustrations* ("Explanation of the Diagram of the Great Ultimate", 1060) a theory about the creation of the universe and transformation of all things, using concise language and with dialectical thinking elements. Zhang Zai developed Zhou Dunyi's theory by putting forward the proposition "*tai xu* is *qi*"

(The ultimate void is nothing but *qi*)[15] in *Zhengmeng* ("Correcting Youthful Ignorance", 1076), explaining the transformation of all things through the aggregation and dispersion of *qi*. By the time of Cheng Hao and Cheng Yi, *li* had become the highest philosophical category, with the proposition that "all things only come from one heavenly principle (*li*)", indicating that everything in the world came from *li*. Going against the feudal hierarchy system and ethical norms was just going against the heavenly principle and therefore was unacceptable. With this, the transition from Confucianism to philosophical and abstract Neo-Confucianism was completed.

Zhu Xi, a great scholar of the Southern Song dynasty, summarized and integrated various schools of Neo-Confucianism since the Northern Song dynasty and established a complete and rigorous system of Neo-Confucian thought, encapsulating the highest achievements of Neo-Confucianism. Zhu Xi was knowledgeable of a vast array of subjects including classics and history, philosophy, Buddhism and Daoism, astronomy, and geography, and he had particular expertise in natural science, in which he made great achievements. The core of his philosophical system is the theory of *tianli* (heavenly principle), centering on the concepts of *li* (principle) and *qi* (vital force). *Tianli* or *li*, also known as *taiji* (the great ultimate), is the root of all things in the universe, and *li* and *qi* are interdependent. "Where there is *li*, there is *qi*, but *li* is the source." All things, different as they are, stem from "one principle with various manifestations", meaning that each thing has its own principle which is a manifestation of the heavenly principle (Volume 1, *Zhuzi Yulei* ("Classified Dialogues of Master Zhu")). While all things unified through *li* differ from one another, they are interrelated with each other. Zhu believed that things move through either "*hua*" (quantitative change) or "*bian*" (qualitative change). He advocated "*Gewu Yi Zhizhi*" (investigating things to extend knowledge), stressing knowledge-action unity (Volume 117, *Zhuzi Yulei*). "In terms of sequence, knowledge comes first; in terms of importance, action weighs more", emphasizing the importance of seeking knowledge from practice. His philosophy is generally objective idealism but contains considerable elements of simple dialectics. Moreover, what was described as "natural philosophy of organism" by Joseph Needham has had a positive impact on the development of natural sciences.[16] When Zhu Xi applied his philosophy of *tianli* to human society, *tianli* became manifested as feudal order and ethical morality. The publication of works by Neo-Confucian masters contributed significantly to the development of Neo-Confucianism.

Zhu Xi was diligent in teaching and writing. Based on his philosophical system, he annotated *Sishu* ("Four Books"), *Yijing* ("Classic of Changes"), and *Shijing* ("Book of Songs"), and he also compiled and published *Zizhi Tongjian Gangmu* ("Compendium of Comprehensive Mirror to Aid in Government") and other works. In the late Southern Song dynasty, the imperial court recognized Zhu Xi's teachings as

---

[15] A view on the origin of the world proposed by the Northern Song philosopher Zhang Zai. He believed that "*qi*" is the material origin of all things in the universe, and "*Tai Xu*" is the natural state of *qi*. When *qi* aggregates, it takes on tangible forms as all things, and when it disperses, it becomes intangible, which is "*tai xu*".

[16] Needham J. *Science and Civilisation in China*: *Vol. 1* [M]. Cambridge: Cambridge University Press, 1954.

the orthodox Confucianism, and thereafter, the Song dynasty's Neo-Confucianism became the officially endorsed philosophy. Neo-Confucianism dominated the intellectual world of China for 700 years, and its basic principles penetrated all social strata, serving as standards for judging good and evil and playing a role in stabilizing social order. Neo-Confucianism was a major subject in school education and imperial examination essays during the Yuan, Ming, and Qing dynasties. Due to this highly developed social ideology, Chinese feudalism was stronger, lasted longer, and had more far-reaching ideological influence than Western feudalism. Neo-Confucianism also exerted a long-term ideological influence on the Sinosphere such as the Korean Peninsula, Japan, and Vietnam. It should be noted that traditional Confucianism, including the Neo-Confucian philosophical system, is vast and profound, containing both essence and dross. Today, we should promote its essence and discard its dross instead of wholly affirming or rejecting it. Moreover, it is important to point out that Chinese Confucian classics, after being introduced to Europe by Western Jesuit missionaries in the seventeenth century, had a beneficial impact on the European intellectual world, particularly on the 18th-century Enlightenment scholars. The French thinker and literary giant François-Marie Voltaire (1694–1778) even hung a portrait of Confucius in his room. The German philosopher Gottfried Wilhelm Leibniz (1646–1716) was undeniably influenced by Zhu Xi while developing his organic view of nature. This also proves that there are valuable intellectual insights worth learning in Song dynasty's Neo-Confucianism.

The flourishing publication industry in the Song dynasty also promoted the prosperity of historiography and literature. Historical works in the form of chronicles and separate accounts of important events emerged continuously, such as the *Old Book of Tang*, *New Book of Tang*, *Old History of the Five Dynasties*, *New History of the Five Dynasties*, *Zizhi Tongjian* ("Comprehensive Mirror to Aid in Government"), *Tongjian Jishi Benmo* ("Historical Events in Their Entirety: A Rearrangement of Zizhi Tongjian"), *Zuozhuan Shilei Benmo* ("Zuo's Commentary"), *San Chao Beimeng Huibian* ("Collected Records of the Northern Alliances during Three Reigns"), *Wenxian Tongkao* ("Comprehensive Textual Research of Historical Documents"), *Tongzhi* ("General Treatises"), and others. Works on epigraphy and archaeology include *Jigulu* ("The Collection of Antiques"), *Jinshi Lu* ("Reflections on Things at Hand"), and *Bogu Tu* ("Illustrated Antiquities"). Extensive publication of local chronicles began in the Song dynasty. Many historical works from previous dynasties were also published during the Song dynasty, and the number of historical works published during this period may have exceeded the total number of books listed in *Hanshu: Yiwen Zhi* ("Literature Records in Book of Han"). As a result, the trend of lecturing on history in schools and academies became prevalent, and scholars usually discussed both classics and history. The Song dynasty also saw popularity in the historical storytelling scripts (*huaben*) which were used as a base by urban storytellers for performance. Generally written in popular language, historical storytelling scripts, also known as *pinghua*, are the initial form of ancient vernacular fiction. They typically narrated the rise and fall of various dynasties and wars recorded in historical books, such as *Wudai Shi Pinghua* ("Popular Stories of the Five Dynasties").

During the Song dynasty, the creation of *ci* poetry (a Chinese poetic form) in the literary world reached its heyday, surpassing the late Tang and Five Dynasties in terms of quantity, quality, and content, making it on par with Tang poetry in literary history. Early representative figures include Yan Shu (991–1055) and Ouyang Xiu (1007–1072), known for their graceful language. Liu Yong (ca. 984–1053) depicted the lives of Bianjing citizens in his *ci* poetry, excelling in using plain description and a large amount of colloquial language. Su Shi (1037–1101) explored a wide range of themes for his *ci* with a bold and unconstrained style, opening up new avenues for the development of Song *ci* poetry. Female poet Li Qingzhao (1084–ca.1155), who experienced drastic changes during the Northern and Southern Song periods, wrote about her leisurely sentiments in her early works while later on, she focused on reminiscing about her homeland and lamenting her fate. Xin Qiji (1140–1207) of Southern Song expressed his patriotic concerns and aspirations for reclaiming lost territories in his works, creating the "Jiaxuan style", which represents the highest achievement of Southern Song *ci* poetry. Chen Liang (1143–1194), influenced by Xin Qiji, also expressed passionate patriotism. The modern scholar Tang Guizhang's *Quan Song Ci* ("Complete Collection of Song *Ci*", Commercial Press, 1940) consists of 300 volumes, including works from 1330 Song *ci* poets with a total of 1990 poems. Literary giant Ouyang Xiu promoted the prose style of Han Yu and Liu Zongyuan from the Tang dynasty, advocating for a quaint and plain literary style while opposing the *pianwen* style (parallel prose style), thus fostering a new literary trend. His writings shun classical allusions and fancy language, instead they convey the sorrows of the common people and expose the absurdities of the ruling class. He also devoted himself to cultivating young writers including Wang Anshi (1021–1086) and Su Shi, who both received his guidance and assistance.[17]

With the efforts of such figures as Ouyang Xiu, Mei Yaochen (1002–1060), Wang Anshi, and Su Shi, the prose genre dominated the literary world. Wang Anshi was also an outstanding statesman, and his achievements in poetry and prose were in line with his advocacy of political reform. He cared deeply for the people and expressed his innovative political ideas in his prose, believing that literature should "serve society". Su Shi's prose was free and unrestrained, not overly focused on crafting phrases. He advanced that writing should flow naturally, "constantly moving where it should and stopping where it must". His achievements reflected further development of literature in the Northern Song period, distinguishing literature from the teachings of Confucianism. Although Song poetry evolved from Tang poetry, it also broke new ground in content and form. Ouyang Xiu's poetry, absorbing characteristics of Han Yu's prose, created the "poetic discourse" style, while his student Wang Anshi swept away the "*xikun* style", which was still used in the early Song dynasty and characterized by the use of allusions and ornate language, and presented a vigorous and concise style to reflect reality. Su Shi's poetry was characterized by the use of prose and discussion, with surging momentum. The poetry of Southern Song poet Lu You (1125–1210) overflowed with patriotic fervor as found in his personality,

[17] Lu Kanru, Feng Yuanjun. *A Compendium of Chinese Literature* [M]. Beijing: The Writers Publishing House, 1957.

and before his death, he left behind the lines: "When the Royal army recover the heartland in the North, do not forget to let your father know in the family sacrifice", a verse that has been passed down through generations. The heroic anti-Mongol figure of the late Song dynasty, Wen Tianxiang (1236–1283), left behind a poem before his noble sacrifice: "What man is ever immune from death? Leave me with a loyal heart shining in the pages of history", marking Song poetry and *ci* with a deep imprint of the era.

9.2.3 Influence of Printing Technology on European Education and Scientific Revolution During the Renaissance

As mentioned earlier, by the fourteenth and fifteenth centuries, Europe had already moved beyond the stagnation it experienced in culture and science, and was exhibiting a new face. This new momentum was strengthened by the introduction of printing technology, especially movable type printing from China. Metal movable type is well-suited to the phonetic writing systems used by Europeans, allowing for the mass production of books without investing too much. School textbooks were no longer expensive manuscripts which were replaced with printed books. Early printed materials, apart from religious texts, were primarily school textbooks and language reference books, followed by humanities and science works. By the sixteenth century, the number of printed materials with content of science and technology surged. These emergent inexpensive printed books made it possible for more people to read and become literate, creating a greater demand for books at the same time, which in turn further stimulated the development of education and science. The emerging urban bourgeoisie were eager to send their children to school; scholars could also quickly publish their research results; craftsmen found it necessary to become literate to understand technical methods and illustrations in printed guides. Upper-class nobles also felt that being ignorant was incompatible with their status. The development of urban industry and commerce, along with technological improvement and the introduction of new technologies such as gunpowder and compass, constantly raised problems that needed to be solved theoretically, thus turning some cities into centers for academic study.

Between the eleventh and fifteenth centuries, some universities already emerged in Europe, such as those in Paris, Oxford, Cambridge, Bologna, Naples, Padua, Prague, Krakow, Vienna, Salamanca, Ferrara, and Saint Andrews. By the year 1500, there were already 65 universities throughout Europe, and thereafter, the number increased rapidly. In the sixteenth century, universities were widespread in countries such as Italy, Germany, France, Belgium, the Netherlands, Switzerland, Austria, Poland, England, Spain, the Czech Republic, and Denmark, with Italy, Germany, France, and England having more universities. Previously, there were relatively few universities in Germany, and Germanic people went to Italy and France for higher education, but later, Germany caught up and nearly matched Italy in the number of universities. It is worth noting that the cities where universities were located were often centers of printing and publishing.[18] Germany, being the earliest European country to develop metal movable type printing, established universities in major printing centers such as Augsburg, Nuremberg, Cologne, Strasbourg, Frankfurt, Munich and Berlin. Other countries exhibited similar patterns. For instance, printing centers like Rome, Venice, and Florence in Italy, Paris and Lyon in France, Basel and Zurich in Switzerland, Louvain in Belgium, and Vienna in Austria all had universities, with a large number of scholars gathering there.

The so-called universities in the early days were established according to the model of the madrasah in the Arab Empire. They were institutions affiliated to religion, mainly for teaching the Bible. Mathematics and astronomy served religious purposes, and scholars had far less freedom in research than the Arab teachers, making these institutions quite different from modern universities. After the fifteenth century, the situation changed. Teaching of mathematics, astronomy, and medicine in European universities was strengthened and scholars conducted more in-depth research, thus the quality of published textbooks and reference books naturally improved. With the increasing urban population, medicine became a sought-after profession and many universities offered medical majors, making it a trend to study medicine, which stimulated the development of human anatomy and related subjects. Those specializing in theology also developed an interest in natural sciences and engaged in amateur research. Talents cultivated by these universities later became active participants in the Scientific Revolution and founders of modern science.

The Scientific Revolution began with astronomy, and the study of astronomy was stimulated by ocean voyages and geographical discoveries. As the manuscripts of the travelogue of the Venetian explorer Marco Polo (1299) spread around, Europeans' horizons were broadened and they learned about the rich, highly developed material civilization of China thousands of miles away and the abundant natural resources of India, which aroused their desire to travel far to seek fortune and wealth. The introduction of the compass navigation and ocean-going shipbuilding technology from China made it possible for Europeans to engage in long-distance sea voyages in the fifteenth century. The Italian explorer Christopher Columbus (1451–1506), who had read *The Travels of Marco Polo* and been captivated by the wealth of the East described in the book, was informed by the Italian astronomer Paolo dal Pozzo Toscanelli (1397–1482) that sailing straight westward across the Atlantic would take him to the East. In 1492, with the support of the Spanish king, Columbus crossed the Atlantic and opened a new route from Europe to the New World of the Americas, ushering in a series of geographical discoveries. In the process of travelling by sea, sailors needed to find directions, determine the ship's position in the ocean, know seasonal changes and phases of the moon and the sun, and compile accurate planetary ephemerides for navigation. Observations from ocean voyages also provided new data. The major development of long-distance sea voyages raised new problems

[18] Martin H J. *The French Book: Religion, Absolutism and Readership*: 1585–1715 [M]. Saenger P, Saenger N, tr. Baltimore: Johns Hopkins University Press, 1996: 4.

for astronomy, which could not be resolved with the outdated translations of the Ptolemaic astronomical system. Knowledge innovation was needed in astronomy.

As early as the first half of the fifteenth century, Austrian astronomer Georg von Peuerbach (1423–1461) from the University of Vienna compiled the *Tablae Eclipsium* ("Table of Eclipses") based on precise observations and strict theoretical calculations. This work was published in 1459 and used for 200 years. In 1454, he completed *Theoricae Novae Planetarum* ("New Theories of the Planets"), commonly known as *Handbook of Astronomy*. Repeatedly published and used as a textbook, it elaborated in detail the Ptolemaic theory about the movement of planets, which was later refuted. His German student Johannes Müller (1436–1476), also known as Johann Regiomontanus, studied astronomy at the University of Vienna. They worked together to annotate new translations of Ptolemy's works. From 1471 to 1475, while staying in Nuremberg, Müller established an observatory with the help of his wealthy merchant friend Bernhard Walther (1430–1504), and he conducted systematic astronomical observations. Müller also ran a printing press specializing in publishing astronomical and mathematical works. In 1475, he published a nautical almanac for the years 1475–1505, which was used by navigators including Christopher Columbus. Müller also completed his teacher's unfinished work, producing *Epitoma Almagesti Ptolemaei* ("Summary of Ptolemy's Almagest"), but he expressed doubts about the geocentric theory and criticized Ptolemy's lunar theory. He was an astronomer and mathematician who also ran a publishing business. Meanwhile, the German philosopher Nicolaus Cusanus (1401–1464), despite being a cardinal, advanced in his book *De Docta Ignorantia* ("On Learned Ignorance", 1440) that the Earth revolves around the sun and also rotates on its own axis. He also proposed that the universe was infinite, with everything in motion and having complexity beyond human comprehension. The movement of celestial bodies observed from other planets is the same as seen from the Earth. People thought the Earth was stationary, but actually it is in motion like all celestial bodies. These discussions lacked astronomical evidence.

The accumulation of observation data and sporadic expression of ideas by predecessors laid the foundation for the heliocentric theory advocated by the great Polish astronomer Nicolaus Copernicus (1473–1543). Copernicus studied astronomy at the University of Krakow and traveled to Italy in 1497 to study for ten years at the universities of Bologna, Padua, and Ferrara, during which he participated in astronomical observations and recorded his observations. In 1506, Copernicus returned to Poland to teach in Frombork, and he continued his observations from a platform on the church roof. As early as 1502, while studying in Italy, Copernicus had doubts about the Ptolemaic theory which stated that the Earth was the stationary center of the universe and that the Sun and planets revolved around it. Copernicus claimed that the Earth and other planets moved along orbits around the Sun, which is known as the heliocentric theory. After returning to Poland, he continued to ponder this problem and verified it with observation data kept by himself and his predecessors, and also through mathematical calculations. From 1510 to 1515, he wrote the initial manuscript of his astronomical system entitled *De Hypothesibus Moturum Coelestium a se Constituti Commentariolus* ("Commentariolus on the Hypotheses of Celestial Motion", hereafter referred to as "Commentariolus"), which only circulated among a small circle of trusted friends. In 1539, the Austrian astronomer Georg Joachim von Lauchen (1514–1576), who later changed his name to Rheticus, came to Poland to study under Copernicus and became his loyal follower. In 1541, Rheticus sent a summary of Copernicus's "Commentariolus" to a publisher in Gdansk, Poland, and had it published as a booklet entitled *De Narratio Prima de Libris Revolutionum Copernici* ("The First Account of the Book on the Revolutions by Nicolaus Copernics").[19]

Rheticus had previously urged Copernicus to turn his "Commentariolus" manuscript into a book for publication, and his suggestion was adopted. Around 1530, Copernicus completed the manuscript of *De Revolutionibus Orbium Coelestium* ("On the Revolutions of the Orbs of the World"), but he continuously revised it for the sake of rigor and caution. More importantly, he was concerned that publishing it would result in persecution by the church, so he hesitated. In this context, only a summary of his work was published in 1541. Copernicus's book was not published until his later years, in 1543, in Nuremberg, Germany, where on his deathbed, he finally saw the printed copy (Fig. 9.4). The Beijing Library holds the second edition of this book, published in Basel in 1566 and printed on hemp paper. Copernicus's *De Revolutionibus Orbium Coelestium* presents a new perspective on the cosmos; with observational data, mathematical calculations, and strict logical reasoning, it overturned the Ptolemaic geocentric theory that had ruled Western thought for more than 1000 years and had been integrated into the religious system of scholastic philosophy. By repositioning the Earth as one of many planets rather than the center of the universe, the book fundamentally changed people's view of the cosmos. It challenged the ecclesiastical authority from the perspective of natural phenomena, hence liberating natural science from theology. The dissemination of scientific works acted as a silent cannon, tolling the death knell of feudal ideology, but natural science also underwent baptism by fire in its struggle against feudalism.

In 1584, Italian astronomer Giordano Bruno (ca. 1548–1600) published *Cena de le Ceneri* ("The Ash Wednesday Supper") in Italian, promoting Copernicus's heliocentric theory. In the same year, he also published *De l'Infinito Universo e Mondi* ("On the Infinite Universe and Worlds"), asserting that the universe is infinite and stars are scattered throughout endless space, thereby rejecting the theory that stars are scattered on a crystalline celestial sphere centered around the Sun. In 1593, he was imprisoned and tried by the Roman Catholic Church for his beliefs, but he remained steadfast, insisting that the Sun was the center of human galaxy. He was burned at the stake in 1600. However, scientific revolution had its successors. In 1632, the Italian scientist Galileo Galilei (1564–1642) published *Dialogo dei due Massimi Sisemi del Mondo* ("Dialogue Concerning the Two Chief World Systems") in Italian, which presents a debate between followers of Aristotle and Copernicus, with the former clearly being no match for the latter. In 1633, Galileo was forced to abjure his belief in heliocentrism, tortured by the Inquisition in Rome and condemned to

[19] Symposium on Copernicus [C]. Proceedings of the American Philosophical Society, 1973, 117 (6).

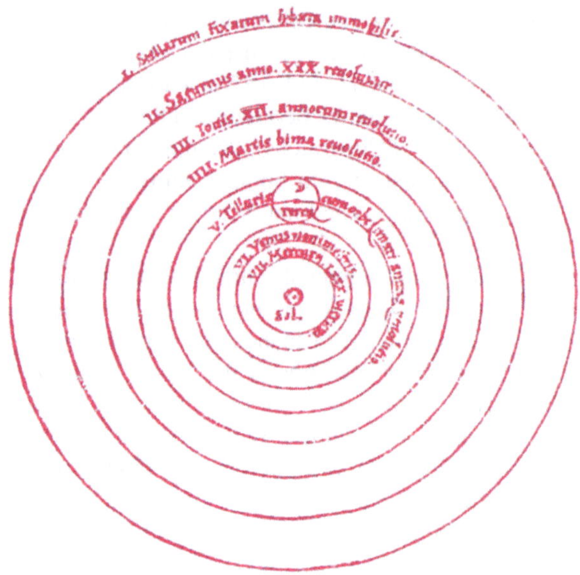

Fig. 9.4 A diagram of Copernicus's heliocentric system in *De Revolutionibus Orbium Coelestium*

life imprisonment for the rest of his life. Despite being forced to sign a verdict, he supposedly muttered "Eppur si muove" ("And yet it moves"). His works were spirited out of the country and circulated abroad. In 1638, Galileo published *Discorsi a due Nuove Scienza* ("Dialogues Concerning Two New Sciences"), in which he introduced his experiments on falling bodies from the Leaning Tower of Pisa and proposed the law of falling bodies in mathematical terms. He observed celestial objects with a telescope and discovered, among other things, sunspots, the moons of Jupiter, and mountains on the moon, establishing his place as a leading figure in the advancement of Scientific Revolution.

Breakthroughs were also achieved in the field of medicine. Andreas Vesalius (1514–1564) from Flanders (now Belgium) developed a keen interest in anatomy while studying medicine at the University of Paris in 1533, and he collected bones from graveyards for study. He earned a doctorate in medicine in 1537 from the University of Padua and served as a professor of anatomy and surgery; he persisted in performing surgical dissections personally and teaching on the spot, which made him popular. Not bound by the theories of ancient medical authorities, he based his research on his own dissections, particularly excelling in his studies of bones, veins, abdomen, and brain. In 1543, he published *De Humani Corporis Fabrica* ("On the Fabric of the Human Body") in Basel, which includes beautiful illustrations. It is a representative work of medicine and anatomy during the Renaissance. Vesalius pointed out that the ancient physician Galen had mistakenly applied observations from animal dissections to humans, which needed correction.[20] Then in 1628, William Harvey (1576–1657), an Englishman who studied medicine at the University

[20] Hall A R. *The Scientific Revolution* [M]. 2nd ed. London, 1962.

of Padua, published *Exercitatio Anatomica de Motu Cordis et Sanguinis in Animalibus* ("Anatomical Exercise on the Motion of the Heart and Blood in Animals"). In this illustrated work, Harvey presented his theory of blood circulation, demonstrating that blood flows continuously from the constantly contracting and expanding heart to the arteries through the veins, and then back to the heart, and this cycle repeats. This theory refuted Galen's belief that blood flowed from the liver to the rest of the body without circulation. Harvey proved it incorrect and pointed out that anatomists "should rely on experiments, not solely on books, and they should learn from nature, not from philosophers", advocating a new spirit of the era. His research laid foundations for physiology and provided medicine with a scientific theoretical basis.[21]

The greatest achievements in technology during the Renaissance were manifested in the fields of mining, metallurgy, casting, chemical engineering, and mechanical engineering, as reflected in the extensive designs and illustrations created by the Italian painter and engineer Leonardo da Vinci (1452–1519). The included cannons, crossbows, cranes, water pumps, spinning machines, bicycles, flying machines, rollers, and rotary machines all feature new concepts that can be found in da Vinci's notebooks, which, unfortunately, were not published in time. His sketches might not be implemented due to insufficient funding. However, Agostino Ramelli (1530–1590), an Italian engineer who had learned from Leonardo da Vinci, published *Le Diverse et Artificiose Machine* ("The Various and Ingenious Machines") in Paris in 1588. This book, in both French and Italian, includes 195 illustrations with explanations for machines like devices that convert rotary motion to linear motion, piston water pumps, spiral cranes, and windmill-driven vertical grist mills, all of which were later used in production. Another Italian metallurgist, Vannoccio Birringuccio (1480–1539), published *Pyrotechnica* ("Pyrotechnics") written in Italian, a ten-volume treatise on metallurgy and casting, which also discusses gunpowder and firearms. The *De re Metallica* ("On the Nature of Metals") published by the German Georg Agricola (1490–1555) in Basel is a representative work on technology during the Renaissance, covering topics such as ore vein exploration, mining techniques and equipment, metal smelting and separation, and testing, as well as fabrication of aqua valens (inorganic acids), glasses, and various inorganic salts. The book also has sections on mine management and contains 295 illustrations, reflecting the highest level of European mining technology at that time. Thereafter, a series of theoretical breakthroughs were achieved successively in fields of astronomy, physics, chemistry, mathematics, and others. Up to the seventeenth century, the publication of *Philosophiae Naturalis Principia Mathematica* ("The Mathematical Principles of Natural Philosophy") by Isaac Newton (1643–1727) in 1687 marked the completion of Scientific Revolution and the rise of modern science in Europe.

---

[21] Harvey W. *Anatomical Exercise on the Motion of the Heart and Blood in Animals* [M]. Translated by Huang Weirong. Beijing: The Commercial Press, 1962.

9.2.4 Rise of European Humanism and National Literature During the Renaissance Promoted by Printing Technology

Printing technology in Europe spurred a great development in education and secularization of knowledge, broke the Church's monopoly on knowledge and catalyzed the Scientific Revolution, thereby triggering a chain reaction of massive changes in the fields of philosophy and literature. Medieval Christianity advocated for God as the center, God being supreme as the Creator of the universe and humanity, while humans were born sinful and insignificant. Freedom and happiness in the afterlife "Heaven" were only achievable through complete devotion to God and His earthly representative, the Church. The Church used divine authority to oppress human rights, seeking to maintain feudal rule and subjugate people to the authority of God and the Church. However, in the fourteenth and fifteenth centuries, capitalism began to develop in Western Europe, and the emerging bourgeoisie demanded that they be free from the shackles of feudal rule and they sought justifications for their own rational existence and development. So, under the banner of reviving classical culture, a new cultural movement was initiated, which they called "Renaissance". The real purpose of this social movement, however, was not simply to revive ancient Greek and Roman cultures, but to create an entirely new anti-feudal culture that suited the needs of the bourgeoisie. Therefore, the term "Renaissance" does not fully express the essence of this movement. It can be said that this movement, which swept through Europe at that time, was not limited to the cultural sphere; it was followed by economic, political, and military actions, ultimately aimed at overthrowing the feudal system and establishing a capitalist system.[22]

The ideological system of the Renaissance movement is humanism, a term derived from the Latin word *humanus* (human). Humanism, opposed to Church theology, advocates for centrality of humanity, celebrating the nobility of human nature, elevating human authority, and extoling human value and dignity. It advocates for human rights in opposition to divine rights and seeks to deprive the Church of its authority. Italy is the birthplace of the Renaissance. As early as the fourteenth century, Giovanni Boccaccio (1313–1375) from Florence exposed hypocrisy, greed, and cruelty of church officials and feudal aristocrats in his *Decameron* (1353), directly targeting the Roman Catholic Curia. This book promotes the idea that all humans are born equal, opposes differentiation based on social status, and advocates for liberation of individuality. Francesco Petrarch (1304–1374), a poet from Florence, was the first to propose the opposition between human sciences and theology and he exposed the evils of papal rule. In his *Conzoniere*, a sonnet collection written in Italian, he depicted love with a human touch and anti-feudalism overtones, reminisced about the glory of ancient Rome and yearned for the unity of Italy.

[22] Zhu Huan. *Medieval History of the World* [M]. Changchun: Jilin People's Publishing House, 1981.

During the fifteenth and sixteenth centuries, the current of humanism spread with even greater momentum and on a broader scale to become a prairie fire, with printing technology acting as an accelerant. Italian philosopher Pietro Pomponazzi (1462–1524) refuted Aquina's belief in the immortality of the soul in his work *De Immortalitate Animi* ("On the Immortality of the Soul", 1516), arguing that sensations are caused by external objects and form the basis of rational understanding. As a result, he was persecuted, and the Pope issued an order to burn his writings. Italian linguist Lorenzo Valla (1406–1457) published his work *De Falso Credita et Ementito Constantini Donatione* ("On the Donation of Constantine") in 1440, exposing an 8th-century forgery of historical documents by the Roman Curia through rigorous textual research. The Curia falsely claimed that the Roman Emperor Constantinus I (280–337) had transferred the jurisdiction of four dioceses outside of Rome and temporal power over the western regions of the empire to the Roman Catholic Church. In the Middle Ages, the Pope used this "Donation of Constantine" as a weapon in the fight against secular regimes to demand power from the other four dioceses and kings in Western Europe. Valla's textual research denied the legitimacy of papal territories and the historical basis of the Pope's temporal power. He also used his linguistic knowledge to reveal that the doctrine, regarded as divine "revelation" by the Church, was actually the result of errors made in the translation of the Bible. His book tore off the "holy" veil of the Roman Curia and exposed its deceitful nature to the world.

According to Valla's research, the Roman Catholic Church had tampered with the translation of the Greek *Bible* ("Ta Biblia") into Latin. To give people access to the original Greek text, Desiderius Erasmus (1467–1536), a humanist from the Netherlands who was well-versed in Greek, Latin, and other languages, published the first Greek *New Testament* in Basel in 1516, along with his own Latin translation, breaking the Church's monopoly on the translation of scriptures. In 1509, he also published *Encomium Moriae* ("Praise of Folly") in Basel, in which he used the metaphor of a "fool" to expose the sins of feudal rule and the deceit of the Church towards people, portraying the popes, bishops, monks, and scholastic philosophers as a group of greedy, lecherous worshipers of stupidity, making a significant contribution to the European religious reform. In his work *Diatribe de Libero Arbitrio* ("Discourse on Free Will", 1526), he advocated for individual freedom and human liberation, opposing the Church's asceticism. He called for secular rule, rejecting theocratic dictatorship and supporting the establishment of a rational church. Since after the mid-fifteenth century, many works by humanists were published by printing houses in cities such as Florence, Venice, Basel, Lyon, Paris, and Barcelona. Although the Church issued bans on these books and made known lists of banned books, the propagation of new ideas throughout Europe could not be stopped. Publishers placed books in sealed barrels and spirited them to various locations via carriages or ships (Fig. 9.5).

During the medieval period, the Christian church was hostile to the classical Greek and Roman secular culture, whose emphasis on worldly life and pursuit of happiness conflicted with the teachings of the Pope. As a result, literature during this time was generally boring with religious tones. During the Renaissance, some humanists not only published their own works but also edited and published many classical

Fig. 9.5 Shipping books

works, much like the restoration of ancient Roman buildings and sculptures. This was also a form of revolt against the church ideology, and people made efforts to search for ancient manuscripts to replace the distorted texts of the medieval period. Especially valuable were the manuscripts brought by Greeks to Italy after the fall of the Byzantine Empire's capital Constantinople in 1453, and these manuscripts were steadily published after being edited. For example, the pastoral poem *Bucolics* by the ancient Roman poet Vergil (70–19 BC) and his didactic poem *Georgics* praising farmers' life and loyalty were printed in 1470. The epic poems *Iliad* and *Odyssey* by the ancient Greek poet Homer (ninth-eighth century BC) were respectively published in 1488 and 1504. The Latin version of Aristotle's complete works came out in 1469, and the Greek version was published between 1495 and 1498. Plato's complete works were published in Latin in 1483 and the Greek version in 1513. Ancient language courses were offered at European universities, and in 1530, the "College of Three Languages" (Latin, Greek, and Hebrew) was established in Paris and became another center of humanism. The Italian humanist historian Leonardo Bruni (1369–1444), who translated works of Plato and Aristotle, discovered in his translation that the main theories of scholastic philosophers were built on distorted texts of Aristotle.

Besides, the breeze of humanism also swept into the field of European literature and brought about new changes.[23] The Frenchman Pierre de Ronsard (1524–1585) studied medicine and authored several medical textbooks. He was also a humanistic lyric poet and organized a seven-member literary group "La Pléiade", advocating for the restoration and use of French in literary creation and opposing writing in Latin and other foreign languages. He wrote in French works such as *Odes* (1550), *Amours de Cassandre* (1552), and *Hymnes* (1556), reflecting the awakening of national

[23] Anon. *Vozrozhdenie, Bol'shaya Sovetskaya Entsiklopediya* [M]. Translated by Wang Yizhu. Beijing: People's Publishing House, 1955.

consciousness and the strengthening of patriotic ideas. The French satirical writer François Rabelais (1494–1553), who had a background in natural sciences, used folk tales as prototypes to create the images of ideal monarchs, the Giant Gargantua and his son Pantagruel, in his French novel *Gargantua and Pantagruel* published in two parts (1533 and 1535). His works satirize feudalism, and expose the darkness of the Church as well as the decadence of scholastic philosophy and medieval education. He propagated humanist views on politics, education, and ethics and proposed the creed "Fais ce que voudras" (Do what you will), reflecting the demand for liberation of individuality.

During the English Renaissance, Thomas More (1476–1535), despite his noble background and important positions in the cabinet and parliament, criticized the Catholic Church and feudal rule and thereby criticized the bourgeoisie itself in his work *Utopia* published in Latin in 1516. The ideal society depicted in this book, like the one portrayed in the *Civitas Solis* ("City of the Sun", 1623) by Italian philosopher Tommaso Campanella (1568–1639), features public ownership of the means of production, equality for all, and collective labor. This utopian socialism represents the most radical idea in humanism, but the English and French authors who proposed this idea were both persecuted. The English satirist Thomas Nash (1567–1601) wrote the novel *The Unfortunate Traveller* (1594), a work of realism that portrays the lives of people from various professions and classes in the historical context of the time. Meanwhile, William Shakespeare (1564–1616), who had limited knowledge of Latin and even less of Greek but abundant social experience from his lower-class upbringing, became the greatest playwright and poet of the English Renaissance. He allowed the talent of the English people to be fully expressed in drama, creating a literary form with a new flavor of the era. His plays, different from Italian drama, has English characteristics. Unlike other playwrights, Shakespeare not only wrote scripts but also acted, directed, and managed theaters. He was a versatile and prolific writer, producing 37 plays and 154 sonnets.

Shakespeare's plays create many distinctive and archetypal characters, primarily depicting the conflicts between various social forces during the collapse of the English feudal system and the rise of capitalism. With an evidently humanistic flavor, his plays advocate for liberation of individuality and oppose feudal constraints and divine rule. Drawing materials from secular life, Shakespeare wrote plays in easily understood English for the masses, with genres including historical plays, comedies, and tragedies. Representative historical plays are *Henry* IV(1597–1598), *Richard* III(1592–1593), etc.; comedies include *A Midsummer Night's Dream* (1595) and *Twelfth Night* (1601) etc.; tragedies are represented by *Romeo and Juliet* (1595), *Hamlet* (1600–1601), *Othello* (1604), and *King Lear* (1605–1606) etc.. Shakespeare's historical plays praise national unity, oppose feudal division, and support royal power, demonstrating humanistic political ideals. His comedies are filled with an optimistic atmosphere, praising friendship and love, advocating freedom and equality, and highlighting humanistic societal morality. He used the tragic endings of just and virtuous characters to denounce feudal powers and criticize the greed and cruelty of merchants during the period of primitive capital accumulation and materialism in society.

The Renaissance movement in Spain began in the sixteenth century after the establishment of universities when new Italian culture was introduced and humanistic ideas spread, resulting in the emergence of a group of famous European literary figures. One of the early representatives is Miguel de Saavedra Cervantes (1547–1616). As a young man, Cervantes travelled to Italy and experienced hardships before returning to Madrid in 1580 to work on plays and novels. His most significant work is the novel *El Ingenioso Hidalgo Don Quijote de la Mancha* ("Don Quixote") published in 1605. It was written in the Castilian dialect of central Spain (Castellano), a language that was later used throughout the country and became what is now known as Spanish. The novel depicts the adventures of an outdated knight errant Don Quixote and his squire Sancho travelling around and performing chivalrous deeds but encountering numerous setbacks. Believing that there are demons everywhere, Quixote attacks with his lance, resulting in many absurd incidents, and it is only before his death that he finally awakens to the truth and rejects his previous beliefs. The novel exposes the tyrannical, cruel, and hypocritical nature of the feudal nobility, reveals people's tragic situation and criticizes the medieval knight literature that glorifies feudalism. Don Quixote's behavior, though comical and ridiculous, embodies feudal moral values and the longing for freedom, happiness, and social equality. He typifies a conflicting personality, reflecting the complex psychology at the transition from old to new eras. Sancho, a peasant character, overcomes his selfishness while traveling with his master, and later becomes a governor of an island, implementing honest and fair governance and reflecting the author's humanistic political ideals. Sancho realizes Don Quixote's unfulfilled dreams, demonstrating people's wisdom and the power to overcome darkness and change societal status quo.

During the Renaissance, some scientific and literary works were written and published not in Latin but in various national languages such as Italian, French, Spanish, and English, which had historical significance. In medieval Europe, manuscripts were primarily written in two ancient languages, Greek and Latin, which were respectively the official languages of the Byzantine and the Roman Empire. After the Great Schism of Christianity in the eleventh century (1054), Latin became the official language promoted by the Roman Catholic Church in Western European countries. At that time, few people in Western Europe understood Greek, and the use of Latin was limited only to a minority of clergy and nobles; the vast majority of people did not understand this language, leading to a significant disconnection between the official language and the mother tongue and grammar used by most people in various countries. After the fall of the Western Roman Empire in 476, Europe was in a state of continuous division. The Charlemagne Empire (751–843), after its collapse, split into three embryonic nation-states of present-day Germany, France, and Italy. In the eleventh century, Western Europe had more than 20 separate kingdoms, each using a different language. The formation of various ethnic states necessitate the development of national cultures because a nation without its own culture has no future, and this requires the use of suitable writing for its own language. At that time, Latin literature was mostly religious works, exclusively for the Church. People in various countries could not understand or read these works, and they were excluded from the cultural sphere, making it impossible to develop national cultures.

After the arrival of printing technology, a large quantity of popular reading materials were published in national languages, and scholars had their new thoughts and discoveries published in their mother tongue, which certainly made them easy to be accepted by the public. Literary works written by authors in their own national languages were popular with the common people. So, the true history of European literature began with the appearance of printed books in the fourteenth and fifteenth centuries. Literary works published in different countries provided models for writing in standard Italian, French, English, Spanish and German, and also reduced the language barriers caused by the use of different dialects within the same country. These written languages gradually became the unified literary language in their respective countries. Writers and scholars in different countries refined and standardized their national languages. Through printed materials, they fixed and enriched vocabulary, improved grammar, sentence structures, spelling, and pronunciation principles. These languages were continuously refined in use, and eventually had the function of philosophical discussion akin to Greek and Latin, even surpassing them in many regards. This development helped raise national and ethnic awareness in various European countries and led to the development of national cultures. Only with their own literary languages could nations truly have their own culture and cultural classics. In the process of forming various ethnic literary languages in Europe, printing played the role of a "midwife", and by the sixteenth century, literary works in Latin had ceased to exist.

9.3 Political and Economic Effects of Printing Techniques on the East and West

9.3.1 Influence of Printing Technology on Political Systems in the East and West

Science, philosophy, literature and art, education, law and institutions, etc. are ideological reflections of social politics and economy, serving them and in turn influencing and driving their development. Printing, as a product of technological revolution in document reproduction, also has a significant impact on politics. In China, it was initially used by the ruling class to consolidate feudal rule. In the second century BC, Emperor Wu of the Han dynasty "dismissed the hundred schools of thought and revered only Confucianism", hereafter promoting Confucian ideology throughout the country. During the Five Dynasties period in China, the imperial court promulgated engraved copies of the Confucian *Nine Classics* to strengthen the ideological control of the official philosophy over the people. After the Song dynasty, the *Guozijian* inherited this tradition from the Five Dynasties period and published numerous Confucian classics and works by Neo-Confucian scholars as textbooks and reference books for schools at all levels. To provide political officials to the central and local government departments, the Sui dynasty initiated the open examination system for the imperial court to select officials. This system did not consider candidates' social status or rely on recommendations from local authorities. Instead, it awarded official ranks and positions based on examination results. This is the imperial examination system (known as *keju*), which is a significant improvement on the ancient system of selecting officials.

The imperial examination system underwent development during the Tang dynasty, and examinations were divided into two types: *changke* (regular exams) and *zhike* (special exams hosted by the emperor). The *changke* exams were held annually, including more than 50 categories, such as those for the degrees of *xiucai* (lower-level scholar), *mingjing* (master of classics), *jinshi* (advanced scholar), *junshi* (distinguished scholar), *mingfa* (master of law), *mingzi* (master of calligraphy), and *mingsuan* (master of calculation), with *mingjing* and *jinshi* being the most common for candidates. Since the exam for *jinshi* was the most difficult to pass with a very low success rate of one in a hundred, it was highly valued by scholars. Candidates for the *changke* exams came from schools in the capital or from prefecture and county schools, and they were sent to *Shangshu Sheng* (Department of State Affairs) for the exams. Those who were not from the official schools had to pass a preliminary exam at the prefectural or county level before they could be sent to *Shangshu Sheng*. The examination in the *Shangshu Sheng* was called provincial examination and *libu* (Ministry of Rites) examination. The *zhike* exams were set up by the emperor on an ad hoc basis, comprised of a hundred different subjects such as virtue and propriety, straightforwardness and expostulation, multiple talents, etc.. The candidates were current officials, successful candidates from the regular exams, or ordinary people. These exams, which included writing essays and composing poetry, were personally overseen by the emperor. The Song dynasty reformed the imperial examination system by establishing the palace examination system, the highest level examination conducted inside the palace by the emperor after the *libu* examination. The top scorer was titled *bangshou*, while the second and third places were called *bangyan*. During the Southern Song dynasty, the top scorer was renamed *zhuangyuan*, and the second and third places were respectively referred to as *bangyan* and *tanhua*. After the palace examination, official positions were directly awarded. The *changke* exams were divided into three levels: provincial exam, *libu* exam, and palace exam. Those who successfully passed the palace exam were granted the title of *jinshi*. For the generous treatment from the imperial court, the exam for *jinshi* had the most candidates and later it became known as the *jiangxiang* exam (examination for generals and ministers).[24]

The well-developed printing technology in the Song dynasty brought cheap printed reference materials and textbooks for readers participating in the imperial examinations, which strongly promoted the implementation of the examination system. The imperial examination system provided an opportunity for ordinary

---

[24] *New Book of Tang*: *Vol. 44, Records of Election* [M] // *Twenty-five Histories*: *Vol. 6*. Compact edition. Shanghai: Shanghai Classics Publishing House, 1986: 4254. *History of Song Dynasty*: *Vol. 155, Records of Election* [M] // *Twenty-five Histories*: *Vol. 7*. Compact edition. Shanghai: Shanghai Classics Publishing House, 1986: 5639.

literate people to compete for official titles through examinations and be selected for government positions at various levels. The Chinese civil and scholar-official class, known as "mandarinate" to Westerners, were highly educated intellectuals selected through the examination system, and they were a product of the literary and cultural traditions. The imperial examination system was a great innovation in the civil service examination system, which was later emulated by other countries in both the East and West. Dr. Joseph Needham described the scholar-official class as "non-hereditary and almost rare social elite", and although the Chinese feudal system ruled by the literati appeared weak, it was actually stronger than the European feudal system ruled by the military nobility, and it prevented the feudal system from being jeopardized by industrial and commercial capitalism more effectively. This is because the imperial examination system is more reasonable than the feudal selection system of European aristocracy. The latter did not have a reasonable reference for selection in that the hereditary noble offspring were not necessarily competent but could still occupy high positions without exams and selection.[25] Literary giants like Ouyang Xiu, Wang Anshi, and Su Shi were appointed to local and central government positions after they successfully passed the exam for *jinshi*. They did not rely on heredity for their appointments but on their abilities demonstrated in the examination, and their descendants were required to do the same, a situation rare in Europe. This illustrates that the civil official feudal system was more advanced and stronger than the military aristocratic feudal system, capable of absorbing significant inventions such as printing, gunpowder, and the compass, thereby enriching and sustaining itself. In contrast, the Western feudal system could not withstand the social impact brought by these inventions.

Statistical data show that during the Song dynasty, there were over 40,000 *jinshi* and a total of 200,000 and 400,000 *juren* respectively in the twelfth and thirteenth centuries. There were also a large number of people who had the same qualifications but failed to pass the exams. In addition to being proficient in *Thirteen Classics*, literature and history, these scholars also had different interests in natural sciences and technology, and engaged in long-term observation and research, which undoubtedly provided a talent resource and driving force for the high development of academic research. Regions with the most *jinshi* were also places with the highest production of printed materials and the most developed education. For example, the five provinces of Liangzhe (present-day Zhejiang), Fujian, Sichuan, Jiangnan (Jiangxi), and Jiangnandong (present-day Jiangsu) in the Song dynasty produced a total of 24,172 *jinshi*, accounting for 84% of the total, and during the same period, these five provinces printed 1168 types of books, nearly 90% of the total number of printing publications (1303 types) in the country. In contrast, Guizhou Province in southwestern China had only 103 *jinshi* and printed the fewest books (2 types).[26] Major printing provinces were also major papermaking provinces, and the proportional relationship between

---

[25] Needham J. *China and the West* [M] // Jungk R, Mcmullin E, Needham J, et al. *China and the West: Mankind Evolving*. New York: Humanistics Press, 1970: 19–34.

[26] Tsien T H. *Paper and Printing* [M] // Needham J. *Science and Civilisation in China: Vol. 5*. Cambridge: Cambridge University Press, 1985: 379–380.

book publishing and the imperial examination system clearly shows the contribution of printing to the imperial examination system. Conversely, implementation of the imperial examination system stimulated the development of printing, publishing, and education, forming a mutually reinforcing relationship.

From a political perspective, compared to the systems of selecting officials in previous dynasties, the imperial examination system has the following characteristics: First, it consoliated the power of selecting officials from the local to the central authority, strengthening centralized governance, and it gave ordinary landowners and commoners opportunities to participate in the political power, thereby expanding the social base of the ruling group. Second, it linked studying, taking exams, and holding official positions together, providing a way for a wide range of intellectuals to improve their social status and living conditions, thus attracting more people to join this group. This encouraged them to focus their energy on studying for the imperial examination instead of causing uprisings or rebellions, which helped to maintain social stability. Third, it addressed the previous problem of focusing only on family background and moral conduct while neglecting knowledge and talent; the use of a uniform national standard to select talented individuals for government positions enhanced the quality of officials at all levels and strengthened government functions. As a result, the imperial examination system lasted for more than 1000 years in China until the nineteenth century when it was abolished in the late Qing dynasty and replaced by modern examination systems.

The imperial examination system in the Tang and Song dynasties had a significant impact on both the East and the West. According to *Chronicles of the Three Kingdoms* (1145) by Jin Fushih (1075–1151), Volume 10, after the unification of the Korean Peninsula by the Silla dynasty, in the fourth year of King Yeonseong's reign (788), a three-tiered educational system modeled on the Tang imperial examination system was established to select officials for the state after examination. Those who had studied *Chunqiu Zuoshi Zhuan* ("Spring and Autumn Annals"), *Liji* ("Book of Rites"), *Wen Xuan* ("Literary Selections"), *Lunyu* ("The Analects of Confucius"), and *Xiaojing* ("Classic of Filial Piety") were considered the first tier; those who had studied *Quli* (part of *Liji*), *Lunyu*, and *Xiaojing* were the second tier, and those who had studied only *Quli* and *Xiaojing* were the third tier. Candidates were divided into first, second, and third tiers based on their areas of study and examination results, and then were awarded corresponding official positions. "Those who are well-versed in *Five Classics*, *Three Histories*, and books of various schools of thought will be promoted exceptionally." There were Silla students who studied in China, took part in Chinese examinations and were awarded the title of *jinshi*. According to *Election Annals, Volume 77,* in *History of Goryeo* (1451) by Zheng Linzhi (1395–1468), in the ninth year of King Gwangjong's reign (958), the Goryeo dynasty started the imperial examination system, testing poetry, prose, odes, contemporary policy essays for selecting scholars and experts in fields like Confucian studies, medicine, and divination. This system continued through the Joseon dynasty and was practiced for over 1000 years. According to *Đại Việt sử kí toàn thư* ("Complete History of Dai Viet", 1479) by Ngo Si Lien (1439–1499), in the fourth year of Empeoror Renzong's Taining reign in the Vietnamese Ly dynasty (1075), a decree was issued to select scholars well-versed in classics and three examinations of Confucian studies were held. That was the beginning of implementing the imperial examination system in Vietnam. Subsequent dynasties followed suit and continued to select scholars through examinations until the early twentieth century when Vietnam abolished the imperial examination system. The Nara dynasty in Japan established the Daigaku-ry (a set of educational institutions) with divisions for the study of classics, music, law, calligraphy, and mathematics. Students were required to study Confucian classics as well as books such as *Jiuzhang Suanshu* ("The Nine Chapters on the Mathematical Art") and *Sunzi Suanjing* ("Master Sun's Mathematical Manual"), and were awarded eight-rank official titles after passing examinations. This bears similarities to the imperial examination system.

The modern civil service examination systems in Europe and America are directly influenced by China's imperial examination system. Firstly, in 1696, the French Jesuit missionary in China Louis Daniel le Comte (1655–1728) introduced the Chinese examination system and pointed out its four benefits: (1) selecting capable young people through examinations regardless of their family background and based only on their scholarly achievements, thus inspiring them to strive forward; (2) cultivating a spirit of advancing learning, promoting societal respect for knowledge and enhancing cultural refinement of Chinese youth who were motivated to excel in state examinations; (3) preventing greed, moral decline, intellectual emptiness and indulgent behaviors; (4) gathering talented individuals from all corners of the country to remove corrupt officials and select more suitable successors.[27] In 1735, another French Jesuit missionary Jean Baptiste du Halde (1674–1743) published in Paris a compilation of Jesuit communications from China, which included reports on the imperial examination system. He noted that the Chinese emperor established the examination system to select talented people nationwide through state examinations and award them official positions. The examinations were held every three years, and those who passed the highest-level exam were given the title of *jinshi* and entered the Hanlin Academy. The emperor selected the Hanlin Academy scholars to serve as ministers, the prime minister, and tutors to the crown princes. Hanlin Academy scholars also engaged in writing and were highly respected in society.[28]

In the eighteenth century, the French Enlightenment thinker Voltaire (Francois-Marie Arouet, 1694–1778) praised China's imperial examination system in his work *Essai sur les Moeurs* ("Essay on Manners", 1756) after reading the above-mentioned reports on the topic. He mentioned that the Chinese government was composed of six ministries—Personnel, Revenue, Rites, Justice, War, and Public Works, and officials were divided into nine ranks, all appointed only after passing several strict examinations. Therefore, he believed that China's administrative organization was the best in the world. Francois Quesnay (1694–1774), a French physiocratic economist, published a long serialized article from March to June, 1767 entitled *Despotisme de la Chine* ("Despotism of China") under the pen name A. M. in the publication of his school of thought "*Ephérméide du Citoyen*". He pointed out that, different from

[27] Le Comte L D. *Nouveaux Mémoires Sur l'état Présent de la Chine* [M]. 3éd. Paris, 1698: 61–62.

[28] Du Halde J B. *Description de l'Empire de la Chine*: *Vol. 2* [M]. Paris: Le Mercier, 1735: 28.

the West, China did not have hereditary nobility and official positions were obtained based on ability and merit. Quesnay highly praised China's imperial examination system, saying that candidates of any background could compete with their abilities, and even sons of artisans could become governors after passing the examinations. Quesnay believed that China pursued enlightened monarchy and was worth emulating by Western countries of the time. In 1793, George Thomas Staunton (1781–1859), who visited China as deputy to the British envoy George Macartney (1737–1806), also introduced and praised Chinese imperial examination system in Chapter 12 of his work *An Authentic Account of an Embassy from the King of Great Britain to the Emperor of China* (1797).

Under the influence of China's imperial examination system, France established its first civil service examination system in 1791. Although it was suspended ten years later in 1801, it was reinstated in 1840. In the eighteenth century, some British writers also praised the Chinese examination system and advocated for the implementation of a similar system in England. In 1806, the British East India Company first introduced a civil service exam system, and thereafter many Britons referred to the Chinese model as an argument for the establishment of a universal civil service system in the UK. For instance, the British diplomat in China Thomas Taylor Meadows (1815–1868) introduced the imperial examination system in his book *Desultory Notes on the Government and People of China* in 1847 and he pointed out that the reason for the enduring existence of the Chinese Empire was that the government was composed of capable and merit-based officials. They advocated for the establishment of a comprehensive and open civil service examination system in the UK to improve administrative work. These calls led to the British government establishing a committee to study the issue. In 1853, a report was presented to the Parliament, and the UK officially implemented the civil service examination system in 1855.

The British civil service examination system undoubtedly had a demonstrative effect on the establishment of a similar system in the United States, but China's influence remains apparent. For example, in 1868, Senator Thomas A. Jenckes from Rhode Island first proposed to Congress the establishment of a civil service examination system and his proposal included a chapter on China's imperial examination system. In the same year, Boston's renowned cultural figure Ralph Waldo Emerson (1803–1882) delivered a speech at a gathering in Boston hosting the Qing Empire envoys Zhi Gang and Sun Jiagu, where he praised China's imperial examination system and urged the US Congress to pass Senator Jenckes' proposal. He remarked in his speech: "There is one point about China's political system which interests us, and I suppose everyone present remember the bill presented to Congress twice by Mr. Jenckes of Rhode Island that civil officials should first pass the exam to obtain academic qualifications before they can hold positions. In terms of correcting abuses, the Chinese are indeed ahead of us, and also ahead of Britain and France. They attach great importance to education, and in this respect, they are also ahead of us. This is a source of honor for China."[29]

---

[29] Teng S Y. Chinese Influence on the Western Examination System [J]. *Harvard Journal of Asiatic Studies*, 1943 (7): 267–312.

Emerson's speech in Boston aroused responses, with some people agreeing and others opposing. Just as in the UK, many people in the US who obtained vested interest from the old system strongly opposed this new idea. Some argued that using examinations to decide on suitable candidates for administrative positions was a Chinese-style or foreign-style system, and therefore "un-American". Due to the obstruction of conservative senators, Jenckes' legislative proposal was not passed by the US Congress until 1883. Professor Derk Bodde, a contemporary sinologist at the University of Pennsylvania, believes that the Chinese imperial examination system has two major advantages: First, it is open to all members of society without exception, making it the most democratic way to select government officials in the pre-modern world. Second, it ensures that people in government must have a higher education background. Bodde noted: "Today, the civil service examination system is actually accepted by all democratic countries. More and more people enter government agencies based on their personal abilities rather than political favoritism. The result is that political corruption so pervasive a hundred years ago ceases to be visible. The civil service examination system is undoubtedly one of the most precious intellectual gifts that China has given to the West."[30]

In its later stage, especially during the Ming and Qing dynasties, the Chinese imperial examination system and its related educational system were more closely intertwined and more complete. After receiving elementary education, students who passed the exams were called *tongsheng*, eligible to take the three-tiered county, provincial, and national exams. The *tongsheng* who studied in county or prefectural schools and passed the exams were conferred the title of *xiucai* and became eligible for the provincial exams. County-level exams were administered by provincial or local officials and were held twice every three years. The provincial-level exams, held every three years in provincial capitals, were run by officials appointed by the imperial court as principal and deputy examiners, and those who passed were awarded the title of *juren*, with about 100 *juren* admitted per province. Those who earned the title of *juren* traveled to the capital the following year to take the national-level exam *huishi* (the metropolitan examination), which was held every three years and presided over by the Ministry of Rites. The emperor appointed members of the Hanlin Academy as the chief examiner, and those who passed were awarded the title of *gongshi*, with 100–400 people receiving the honor and as backup candidates for the palace examination. The palace examination was the highest level held inside the imperial palace and was personally overseen by the emperor. Those who passed were known as *jinshi*, typically 100–200 in number. Those who obtained the titles of *juren* and *jinshi* were eligible for direct official appointments, with some *jinshi* entering the Hanlin Academy for several years before being appointed to key positions after further examination. The titles of *xiucai*, *juren*, and *jinshi* are equivalent to the Western academic degrees of Bachelor, Master, and Ph.D., respectively. After the Ming dynasty, however, the shortcomings of the imperial examination system became evident, primarily manifested in the rigid and dogmatic examination content,

---

30 Bodde D. *Chinese Ideas in the West* [C]. Washington, D.C.: The Committee on Asiatic Studies in American Education, 1948.

confined to *baguwen* (eight-legged essay) based on Confucian interpretations and lacking practical application and knowledge relevant to the real world; the education system remained focused on Confucian classics and history, neglecting science and technology, hence hindering the development of science and culture and failing to meet the needs of the times. In contrast, the Western civil service examination system selectively adopted rational elements from the Chinese imperial examination system and adapted them to its own characteristics instead of copying everything. The establishment of this system in the West contributed to the improvement and strengthening of capitalist regime construction, whereas in China, the degeneration of th system made the feudal system further waning.

The influence of printing technology on religion and the resulting political consequences were exhibited differently in China and the West. In China, since the Tang dynasty, printing was initially used to publish Buddhist scriptures, serving the development of Buddhism. Like Confucianism, Buddhism was harmless to feudal rule. Throughout history, except for a few emperors who opposed Buddhism in favor of Taoism, most rulers supported Buddhism. Soon after the founding of the Song dynasty, Emperor Taizu ordered the publication of *Tripitaka*, and thereafter, Buddhism and Taoism continued to develop along their own paths without major changes or causing social unrest because they were embraced by society and had a relatively broad mass base. This was true even of dynasties established by ethnic minorities, such as the Liao, Jin, Yuan, and Qing. The Korean Peninsula and Japan in East Asia were slightly different: Buddhism reached its heyday during the Goryeo dynasty, followed by the promotion of Confucianism over Buddhism and the decline of Buddhism during the Yi dynasty, while in Japan Buddhism began to stagnate and secularize during the Edo period. Chinese Buddhism developed steadily all the time, and after the Song dynasty, it infiltrated with Confucianism, resulting in the emergence of Neo-Confucianism and new Confucian schools as mentioned previously.

9.3.2 Printing Technology and European Reformation Movement

In Europe, the impact of printing technology on religion is vastly different from that in China, due to different societal backgrounds. Germany was the earliest country in Europe to develop printing, especially metal movable type printing, in the fourteenth and fifteenth centuries, but it was also the region with the strictest control by the Roman Catholic Church. After the fourteenth century, the national states of England and France grew stronger politically and economically, and the monarchy could no longer tolerate the Curia's tyranny and wealth accumulation so that the ties between the local churches and the Curia gradually weakened. Germany, however, was in a state of division, lacking a unified and powerful government, so it became a "docile cow" of the Roman Church. Not only did it have to pay onerous taxes to Rome but a large amount of land, farms, forests, and pastures also became the property of the Church. Popes, bishops, and monks all lived on the exploitation of citizens and farmers. The earliest publications of Guttenberg's printing press were the Vatican-sanctioned Latin Bible by Hieronymus or Jerome (342–420) and the Indulgence issued by Pope Nicholas V (1477–1455) in 1454. Believers paid for the indulgences to obtain the Pope's pardon for "sin and punishment", which was actually a means of collecting money from the masses.

Therefore, it is obvious that in Europe, printing was initially used to serve the Curia's purpose of ideologically controlling the masses and economically exploiting them. In this context, large-scale publication was allowed. However, when a powerful consortium in Southern Germany published the first Bible (probably in German) for citizens in 1479, the Cologne Church immediately reported to the Pope, and the Roman authorities first imposed heavy taxes on the publisher as an economic sanction, and then ordered that publishers in this large German city should obtain permission from the Church before publishing books. On May 4, 1515, the Pope extended the book censorship system to the entire Christian world. Books not in line with the Curia's taste were considered "heretical" and all the publishers, buyers, and readers were to be punished.[31] With the continual publication of anti-church humanistic works, the Pope's list of "banned books" grew longer and longer, and the struggle of control and counter-control continued. Eventually, this struggle erupted in the relatively obedient Germany, resulting in the Reformation movement. It was a large-scale anti-feudal social and political movement launched under the banner of religious reform by the emerging bourgeoisie in 16th-century Europe, directly targeting the Roman Catholic Church, the main pillar of the feudal system, headed by the Roman Pope.

The Reformation first occurred in Germany because Germany was the country where the most indulgences were printed and distributed, resulting in the most suffering and the strongest resistance. The money the Pope extracted through this method in Germany was more than that of the Emperor. Citizens believed that a large amount of money flowed into Rome, damaging the development of the German economy and their vital interests. The lower-class peasants hated the Catholic Church even more. The publication of the Dutch humanist Erasmus's satirical work *Moriae encomium* ("Praise of Folly") in Basel in 1509 had the greatest influence on Germany. It was widely circulated and reprinted more than 20 times in just a few years. Ulrich von Hutten (1488–1523), a German humanist and thinker of the knight class, also one of the authors of the *Epistolae Obscurorum Virorum* ("Letters of Obscure Men", 1517), fiercely attacked the Roman Pope and the Catholic Church, advocating for the establishment of a monarchy built on the pillar of the knight class. He translated works by the Italian Lorenzo Valla that exposed the Curia's forgery of documents, and he also expressed similar thoughts in his poems written in German, hoping that one day Germany could unify and become a strong nation.

---

[31] Martin H J. *Histoire et Pouvoirs de l'écrit* [M]. Paris: Librarie Académique Perrin, 1988: 252–253. Martin H J. *History and Power of Writing* [M]. Cochrane L G, tr. Chicago: University of Chicago Press, 1994: 267.

The trigger for the Reformation occurred in October 1517 when Pope Leo X (1513–1521) sent the ignorant and vulgar German clergyman Johannes Tetzel (1465–1519) to Germany as a special envoy to sell a large number of printed indulgences under the pretext of raising funds for the construction of St. Peter's Cathedral. He proclaimed in the streets to the public, "As soon as the coins sound in the box for indulgences, the soul will immediately rise from purgatory to heaven." This deceitful tactic caused great anger among the people. Martin Luther (1483–1546), then a theology professor at the University of Wittenberg, was outraged and wrote *Ninety-Five Theses,* exposing the fraudulent fundraising tactics of indulgences. The next day (October 31, 1517), he nailed the theses to the door of the Augustinian Church in Wittenberg, calling for a debate to clarify right from wrong. In the *Ninety-Five Theses,* Luther argued on the basis of religious principles that the Pope could not forgive any punishment other than those imposed by himself or prescribed by the canon law. The Pope's indulgence was merely an announcement of God's mercy, not applicable to the souls in purgatory; every believer could receive Christ's forgiveness of sins, and even without the Pope's aid, Christ and the saints would make amends for them.

With his rational religious arguments, Martin Luther overturned the claim that only the Pope and the Roman Curia had the power of atonement, proving that the act of selling indulgences as a way to absolve believers of sins had no legal basis. This was a heavy blow to the Roman Curia and a liberation of thought for the masses of believers. The torch Luther ignited quickly grew into a prairie fire against the Pope. His *Ninety-Five Theses* became a common programme for everyone, with the entire German nation joining the movement. The theses were printed and spread throughout Germany within two weeks and across Europe in four weeks, as fast as an angel delivering Christ's gospel, as described by people at that time.[32] Luther himself believed that "printing is God's supreme and greatest gift, allowing the Gospel to spread far and wide."[33] In June 1519, during a theological debate in Leipzig with the Pope's representative Johannes Maier Eck (1486–1543), Luther moved beyond discussing the indulgences to challenge the Pope directly. He believed that the Pope was not God's representative, and decisions made by the Curia were not necessarily correct, such as the 1414 Council of Constance's wrongful declaration of Jan Hus (1372–1415) as a heretic, thereby denying the Pope's supreme authority. To systematically promote his ideas, Luther published three booklets in August 1520: (1) *An den Christlichen Adel der Deutschen Nation* ("To the Christian Nobility of the German Nation"); (2) *De Captivitate Babylonica Ecclesiae Praeludium* ("On the Babelonian Captivity of the Church"); and (3) *Von der Freiheit eines Christenmenschen* ("Concerning Christian Liberty").

In the first booklet mentioned above, Luther called for the German nobility to unite against the Pope and liberate Germany. He demanded that the Pope return the rightful

---

32 Snyder L I. *Documents of Germany History* [M]. New Brunswick, N. J.: Rutgers University Press, 1958: 62.

33 Black M H. *The Printed Bible* [M] //*Cambridge History of the Bible*: *Vol. 3.* Cambridge, 1963: 408.

Fig. 9.6 The title page of the Bible translated by Martin Luther from Greek to German (Taken from the Wittenberg edition of 1533)

freedom and property that belonged to Germany to make the imperial authority truly substantive. Luther proposed an end to taxes paid to the Pope, revocation of the Pope's power to appoint bishops in Germany, and rejection of any authority that placed the Pope above the Emperor. In fact, this was Luther's political programme and Germany's declaration of independence from Roman control. The second and third booklets were Luther's religious programme, focusing on fighting against the Pope's control over various national churches and the church's ownership of property. He advocated for simplifying the complex rituals in religious ceremonies and replacing Latin with the national language, and allowing believers to read the German version of the Bible (Fig. 9.6). Luther believed that the Bible was the highest criterion of faith, denying the church's absolute authority to interpret doctrines. He emphasized that individual believers could communicate directly with God without the need for the intermediary of priests, calling for the Church and doctrines suitable for the monarchy. These ideas became the theoretical basis for the Protestantism which separated from the Catholic Church. To quickly spread the booklets and create a strong public opinion, they were published in quarto format (18 cm $\times$ 22 cm to 25 cm $\times$ 33 cm) and included some cartoons. So, for the first time, German printers ensured the successful spread of Protestantism.[34]

---

[34] Martin H J. *The French Book: Religion, Absolutism and Readership*: 1585–1715 [M]. Saenger P, Saenger N, tr. Baltimore: John Hopkins University Press, 1996: 12.

On June 15, 1520, Pope Leo X issued a papal bull declaring Martin Luther's teachings as heretical, and ordered his works be burned and his ecclesiastical membership be revoked. With the support of the masses, Luther also declared that the Pope was an impenitent pagan and his orders were anti-Christian. On December 20 of the same year, Luther publicly burnt the Pope's bull in Wittenberg and supported the confiscation of church property by the princes. In early 1521, under the Pope's directive, the ruler of the Holy Roman Empire ordered the arrest of Luther. Protected by the Elector of Saxony, Luther took refuge in Wartburg and survived. In 1522, Luther returned to Wittenberg, translated the Greek version of the Bible into German and established a new Protestant church based on his own teachings. Later, Luther's Protestantism (Lutheranism) developed in Germany and Nordic countries. Under his influence, the Frenchman John Calvin (1509–1564) established a radical sect (Calvinism) in Geneva, which gained ground in Switzerland, France, the Netherlands, and Scotland, while in England, there appeared Anglicanism. This led to a tripartite situation of Protestantism, Catholicism, and Eastern Orthodoxy, greatly reducing the territory controlled by the Roman Catholic Church. Influenced by the Reformation, the Peasants' War (1524–1525) led by the lower-level pastor Thomas Münzer (ca. 1490–1525) broke out in Germany with the "Twelve Articles" proposed to feudal lords as the struggle program, which was printed and circulated throughout Germany. This time, printing was used to serve the Peasants' War. The political consequences of the Reformation and the Peasants' War dealt a heavy blow to feudalism and the rule of the Roman Catholic Church in Germany, awakening the German nation and laying the preliminary foundation for later bourgeois revolution and the formation of a unified nation-state, with impacts extended to other European countries.

9.3.3 *Economic Effects of Printing Technology in China and Abroad*

The economic impact of printing technology is no less significant than its impact on thought, culture, education, science, religion, and politics, both in China and abroad. Therefore, it is worth a brief discussion here. Since the Song dynasty, the printing industry has become a continuously thriving sector in the national economy with a stable market. Unlike industries such as mining, metallurgy, shipbuilding, and machinery, products of the printing industry have a vast and nationwide customer base. The sales volume remains high, unaffected by other social factors. Therefore, printing houses and booksellers specializing in printed products always have business to do. Printing houses often cluster in specific regions of certain provinces, forming printing centers, such as Jianyang in Fujian, Hangzhou in Zhejiang, Chengdu in Sichuan, Yangzhou and Nanjing in Jiangsu, etc., where products are distributed to other cities. The printing industry, often located near paper-producing areas, also stimulates the development of paper industry, thereby fostering local economic prosperity and the development of commerce, water and land transportation. Bookstores sometimes concentrate in certain districts of a city, forming bookstore streets to facilitate customer purchases. As early as the Tang dynasty, bookstores already appeared in Chang'an, Luoyang, and Chengdu. During the Song dynasty, cities such as Kaifeng, Hangzhou, Luoyang, Suzhou, Quanzhou, and Guangzhou had bookstores as well. The famous painting scroll *Qingming Shanghe Tu* ("Along the River during the Qingming Festival"), which depicts the scene of Kaifeng in the Northern Song dynasty, even includes the bookstore.

After the Tang and Song dynasties, printed books remained one of the important commodities in China's export trade. Export destinations included the Korean peninsula, Japan, Vietnam, and other regions. These countries also sent people to China to purchase books, and the trading volume was enormous. For example, according to Volume 34 of *History of Goryeo*, in the first year of King Chungseok's reign (1314), emissaries were sent to Jiangnan in China to purchase books, and in only one place, Nanjing, over 100,000 volumes of scriptures were purchased for 150 ingots (equivalent to 62.5 million *guan*) of *baochao* (paper currency). Additionally, in Volume 5 of the same book, it is noted that in the 18th year of King Hyeonjong's reign (1027) Song merchant Li Wentong shipped 597 volumes of books to Goryeo. Therefore, over 1000 years ago, an international market for printed products was already formed within the East Asian Sinosphere and it continued up to the Qing dynasty, primarily for the export of Chinese books. After merchants shipped books to foreign ports, they then transported local goods back from the ports, promoting the development of foreign trade and simultaneously expanding the sales market for books. Some coastal booksellers had their own fleets organized by clan names, and they took advantage of monsoons to navigate between domestic and foreign port cities. Countries connected to China by land, such as Korea and Vietnam, also imported books through overland trade routes.

Similar situations also occurred in Europe. In the fifteenth century, Germany was a major printing power in Europe and once kept the technology of movable type printing secret from the outside world. It exported large quantities of Latin Bible, religious images, Latin grammar, and playing cards to neighboring countries, earning substantial foreign gold coins. Printing centers quickly emerged throughout the country, boosting the great development of papermaking industry as well. The Guttenberg printing house once had to use some parchment due to shortage of paper supply, and after paper mills were established in various places, printing was done entirely on paper, reducing production costs and book prices. After the sixteenth century, the printing industry in Italy, the Netherlands, France, Switzerland, and England also developed rapidly and became one of the pillar industries. From 1450 to 1500, 250 printing houses in Europe produced 25,000 types of books, totaling 5 million copies. Production doubled in the sixteenth century, and bookstores became common in some cities, mainly commercially and culturally developed cities. From the very beginning, European publishing industry has been closely linked with banks. Publishers obtain loans from banks to establish printing factories, and as the printing and publishing industry rapidly develops, bank business expands greatly. The distribution of newspapers in various places provide important information for the development of industry and commerce, allowing business owners to know market trends at home and abroad without leaving home, and to promote their products through newspapers.

Manufacturers often hire people to design trademarks and disseminate them in society through various forms of advertising or flyers to promote their products, maintain reputation, and establish brands. This helps to promote the development of industry and commerce. The practice of printing advertisements and trademarks originated in China. The earliest surviving physical evidence is a copper printing plate used for advertising Liu's Needle Shop in Jinan Prefecture during the Northern Song dynasty, now housed in the National Museum of China. The trademark for their steel fine sewing needles was a white rabbit, and the signboard was hung in front of the needle shop. Similar practices could be found in other shops in cities like Kaifeng and Hangzhou. In the printed books from the Song and Yuan dynasties, it's common to come across trademark designs and advertising text used by various shops. In Europe, trademarks and advertisements first appeared in printed books in the fifteenth century, and later newspapers and magazines with larger circulation provided even better functions for publishing advertisements. Manufacturers also printed product names, trademarks, product performance, and advertising text on packaging materials to distinguish their products from similar products produced elsewhere and to highlight their own product promotions, thereby attracting customer attention. This business strategy continued from Song through to the Qing dynasty, bringing new business to printing houses. Such use of printed materials for industry and commerce started in Europe as well after the fifteenth century. Today, such printed materials are almost ubiquitous in urban and rural areas around the world, and the economic benefits they bring are self-evident.

In economic history, the Tang dynasty had two initiatives that had profound impact on later generations: First, in the fourth year of Emperor Dezong's Jianzhong era (783), the government issued a uniform printed ticket, also known as *yinzhi* (printed paper), to taxpayers when collecting property and income taxes, with the taxpayer's name, tax item and amount filled in as a proof of tax payment. This was the first use of printed materials for financial management, which improved work efficiency and effectiveness, and standardized financial operations. Second, during the early Yuanhe era of Emperor Xianzong (806–820), the government issued standardized exchange vouchers known as *feiqian* (flying money), allowing foreign merchants who travelled to the capital for trade to first exchange the coins they received from selling goods, and then retrieve their coins with the vouchers in their home province without having to carry large amounts of coins on the return journey. The issuance of *feiqian* made trade more convenient and promoted the development of the commodity economy. The Northern Song dynasty inherited and developed these systems. In the eighth year of the Qingli era (1048) of Emperor Renzong, the government issued *yanchao* (salt notes), which allowed merchants who had paid in cash to the government for a certificate (*yanchao*) to legally trade salt. In the seventh year of Emperor Shenzong's Xining era (1074), merchants were required to buy tea from the government and pay taxes, and then they received a permit called *chashou* (tea receipt), allowing them to legally trade tea in other regions.[35] During the Northern Song, the government also uniformly printed land and housing deeds for civil use as transaction and tax certificates when buying and selling real estate. To prevent local officials from over-printing and selling these deeds for personal gain, each deed printed from movable type was numbered with a character from *Qianziwen* and recorded in the official register. These economic measures adopted by the Northern Song government led to a significant increase in the national treasury's revenue and also benefited merchants.

As commodity economy developed significantly during the Song dynasty, it was inconvenient for merchants to carry a large amount of minted coins when traveling out for trade. They needed a currency smaller in size, lighter in weight, and higher in value to replace the coins. The *feiqian* printed in the Tang dynasty provided ideas for the issuance of paper money. Thus, in the fourth year of Zhenzong's Dazhong Xiangfu era of the Northern Song dynasty (1011), in Sichuan where heavy iron coins were used, wealthy merchants issued exchange vouchers printed on paper, known as *jiaozi*, which became the precursor of paper money. In the first year of the Northern Song Tiansheng era (1023), the government set up the Yizhou *Jiaozi* Office in Sichuan and issued official *jiaozi*, with iron coins as reserve funds, for use in Sichuan. The *jiaozi* had a validity period of three years, and were exchanged for new ones when the period expired. The denominations ranged from one *guan* (1000 *wen*) to ten *guan*, each numbered with a character from *Qianziwen*. Starting from 1039, all contents on the *jiaozi* were printed. The amount of each issuance was set at 1.256 million *guan*; besides Sichuan, *jiaozi* was also used in Shaanxi. The official issuance of *jiaozi* in 1023 marked the beginning of paper money in China, the earliest country in the world to issue paper currency. The use of paper money is a revolutionary innovation in the history of currency development, promoting the development of commodity economy and laying foundations for the establishment of the later capitalist financial system, and it has far-reaching significance.

When issuing *jiaozi*, the Song government developed a relatively complete set of regulations or systems for paper currency, including currency design and printing, issuance and circulation, exchange methods, and reserve fund provision. These provided references and lessons for subsequent dynasties. Initially, the total amount of *jiaozi* issued was strictly controlled without excessive printing and issuance, thus ensuring the stability of currency value. Later on, as military expenditures for the resistance against the Jin increased and the government faced financial deficits, there was overprinting and issuance, resulting in the devaluation of *jiaozi*. In the first year of Emperor Huizong's Daguan era (1107) the *Jiaozi Wu* (*Jiaozi* Office) was transformed into *Qianyin Wu* (*Qianyin* Office) and *jiaozi* was renamed as *qianyin* with changes to the face design. Issuance and use of *qianyin* continued until 1234. The Southern Song issued *huizi* from 1160 to 1279 with copper coins as the base currency circulating in southeastern regions such as Hangzhou, Huai, Zhejiang, Hubei, and Jingxi; corresponding laws were also enacted. The coexisting Jin dynasty imitated

---

[35] *History of Song*: *Vol. 181, The Chronicles of Food, Silks Clothing or Currency:* Part 2 [M] / / *Twenty-five Histories*: *Vol. 7*. Compact edition. Shanghai: Shanghai Classics Publishing House, 1986: 5741.

the Song system and issued *jiaochao* in 1154, but with indefinite circulation period, which was a reform in monetary history. In the first year of the Zhongtong era (1260), Emperor Shizu of Yuan issued *baochao* based on silver, with "two *guans*" equivalent to one tael of silver, having indefinite circulation period. After unifying the country, the Yuan government prohibited the use of copper coins, and *baochao* was prescribed by law as the only legal currency in the country. Before the Yuan dynasty, paper money still had some nature of exchange vouchers and was used only in certain regions. It was not until the Yuan dynasty that *baochao* truly functioned as non-convertible paper money, which was a major event for China and even for the whole world. Moreover, Yuan *baochao* was also used in regions such as the Korean Peninsula during the Goryeo dynasty and Central Asia under Yuan Empire, making it an international currency similar to toady's US dollars.

The Yuan dynasty developed a more complete monetary system with sound institutional settings, strict administrative and legal systems, and ample reserves. Additionally, the establishment of stabilization funds for trading gold and silver helped maintain currency value. All these measures were later adopted by numerous other countries. The Yuan Empire was also an unprecedentedly powerful empire spanning Asia and Europe. It opened up the once blocked passage between Asia and Europe with military forces, ushering in a new era of east–west communication and cultural exchange. Various Chinese inventions including paper money spread to the West. The first region west of China that printed paper money was the Mongolian Il-Khanate (1260–1353). In 1294, Il-Khanate's fifth ruler Gaykhatu Khan (1240–1295) took a major economic measure during his reign, ordering the issuance of paper money based on the Yuan dynasty's *baochao* system. The paper money was printed in Tabriz, capital of Il-Khanate, in Persia (present-day Iran), with Mongolian, Chinese, and Arabic characters on the note. Due to lack of experience in the first issuance, insufficient gold backing and local resistance to the new form of currency, the issuance ultimately failed. However, the printing of paper money was successful and set a precedent.

The Chinese method of printing paper money became increasingly sophisticated during the Yuan dynasty, and the issuance of *baochao* during this period drew the attention of Europeans. In 1253, French King Louis IX dispatched his Franciscan missionary Guillaume de Rubrouck on a mission to China, and upon his return in 1255, Rubrouck recorded the Yuan dynasty's *baochao* in his report *Itinerarium ad Orientales*. Later, Italian traveler Marco Polo also described in his travelogue (1294) that in Khanbalique, i.e. Dadu during the Yuan dynasty (present-day Beijing), banknotes were printed on mulberry paper in rectangular shape, with various denominations and official seals, and anyone found counterfeiting was executed. This paper money circulated throughout the country, and people could use it to purchase any goods, including gold, silver, and jewelry, and no shop was allowed to refuse it. Marco Polo also noted that when foreign merchants arrived to trade, the Great Khan would summon 12 experienced and knowledgeable people to determine a fair price for the goods plus a reasonable profit, and that merchants were paid in *baochao*, which was the only accepted payment method. If foreign merchants could not use this paper money in their home countries, they could use *baochao* to buy other Chinese products suitable for their markets and ship them back home. When the received paper money was worn out from use, it could be exchanged for new banknotes with a 3% fee. "All the military troops of His Majesty the Emperor are paid with this paper money, which is considered to be of the same value as gold and silver. For these reasons, it can be said that the Great Khan has more extensive power over treasures than any monarch in the world."[36] Marco Polo's firsthand account provided a relatively accurate introduction to the Yuan dynasty's paper money system and offered a ready-made success model for European countries to implement this system.

During the Yuan dynasty, Italian merchants from Venice, Genoa, and Florence were keen on trade with China, which spurred economic prosperity in these regions. In 1340, the Florentine Francesco Balducci Pegolotti (1280–1355) wrote a trade guide titled *Practica della Mercantura* ("Practice of Commerce", also known as "Merchant's Handbook"), in which he dedicated Chaps. 1–3 to explaining how European merchants should conduct trade in China. Pegolotti pointed out that upon entering China, they needed to exchange their silver ingots for paper money named *baochao* (*balishi*), available in three denominations, and this money was "accepted throughout the country and could be used to buy silk and other goods. The paper money was on par with silver coins, without extra payment despite being made of paper."[37] The book also discusses how much silk and satin could be bought with a certain number of *baochao* exchanged from European silver coins, allowing for calculations of the exchange rate between *baochao* and European coins. Pegolotti's records and his observations of merchants traveling between China and Europe indicate that Marco Polo's account of China's paper money issuance is not a fabrication.

In the fourteenth century, some European merchants who used paper money in China discovered that a banknote with a face value equivalent to 10 Byzantine gold coins (Bezants) weighed less than a single Byzantine gold coin, and yet this light and secure new form of currency could be used to buy any goods in any market in China. Once the information about this paper money reached Europe and when the time was ripe, it began to be imitated. After the fifteenth century, with the rapid development of European commodity economy and trade, people increasingly found the inconvenience of using traditional minted coins. After the sixteenth century, paper money issued in accordance with the Chinese model began to be officially used in some European countries, which had a significant impact on the development of European banking industry.[38] To meet the needs of the development of commercial capitalism, primitive capital accumulation, and market expansion, banks were established in various parts of Europe, such as Bank of Venice (1580), Amsterdam Bank (1609), Hamburg Bank (1619), Bank of England in London (1694), and Banque

[36] Polo M. *The Travels of Marco Polo* [M]. Translated by Li Ji. Shanghai: The Oriental Book Company, 1936: 159–161. Komroff M. *The Travels of Marco Polo* [M]. New York: Grosset & Dunlap, 1936: 137–140.

[37] Yule H. *Cathay and the way thither*: *Vol. 3* [M]. London: Hakluyt Society, 1914: 134–171. Zhang Xinglang. *Compilation of Historical Materials on the Transportation between China and the West*: *Vol. 2* [M]. Beiping: Beiping Publishing House, 1930: 327–331.

[38] Temple R. *China: Land of Discovery* [M]. Wellingborough: Patrick Stephens, 1986: 117–119.

Générale of Paris (1716), and they became financial centers in their respective countries. "Bank" in Italian is called "banco", and its original meaning is "counter" because the bank's predecessor was the counter set up by money operators in the market to receive customers. The English word "bank" and the French "banque" both derive from the Italian word "banco". After the establishment of banks, paper money was subsequently issued in some countries such as Sweden (1661), the United States (1690), France (1720), and Germany (1806), and by the eighteenth and nineteenth centuries, banks had spread to many countries. The establishment of banking and credit systems based on paper money was a necessary part for the development of the capitalist economic order. Western paper money, from design to security measures, issuance management, market circulation, and other procedures, directly drew on the valuable experience of 600 plus years provided by China. Ultimately, modern financial systems in various countries have been fundamentally established on this basis.

Concluding Remark

As discussed above, the direct impact of printing technology on China, surrounding Asian countries and European countries was a significant boost to the development of education and the popularization of knowledge. This led to the emergence of generations of intellectuals who engaged in research and creation in different fields of natural and humanistic sciences, publishing works with new ideas and achievements or reinterpreting ancient works, thereby triggering a series of social effects. However, due to differences in social backgrounds and cultural traditions between China and foreign countries, the resulting effects, manifestations, and scope of influence are varied. Therefore, it is advisable to make a brief summary here.

In China, printing began during the period of great unification in the Sui and Tang dynasties when the country was powerful. By the Song dynasty, printing technology had undergone significant development. At that time, the feudal system was mature with comprehensive cultural relics and classics, prosperous economy, and well-developed cultural education, producing a wealth of scholars, writers, and scientists. Tang poetry and Song *ci* dominated the literary world, while prose became increasingly refined, and music, fine arts and drama entered a new realm, achieving a form of renaissance well ahead of the rest of the world. The Song dynasty saw the printing of numerous scientific classics from previous dynasties as well as many new achievements in its own era. Chinese science and technology, centered around the Four Great Inventions, were at the forefront of the world at that time. Buddhism also experienced great development, but it was merely a religious belief voluntarily chosen by the masses, rather than a state ruling ideology. Buddhist temples were not tyrannical and extortionate like the Roman Catholic Church causing resentment among the people. China's official philosophy was Confucianism, which became more philosophical after the Song dynasty, absorbing achievements of natural sciences and developing into *Lixue* ("Study of Principle") or "Neo-Confucianism", and thus it sustained feudalism more effectively. The organic view of nature in Neo-Confucianism continuously promoted scientific development. Science, philosophy, literature, and religion developed steadily and orderly in China without causing any societal shocks, but

P. Jixing, *History of Printing Techniques in China*,
https://doi.org/10.1007/978-981-96-6119-0

only bringing about academic prosperity. The well-experienced feudal governance in China had mechanisms for self-adjustment, capable of assimilating any major invention and using it for the empire's benefit. Although the Song dynasty had a developed commodity economy, the sprout of bourgeoisie was not strong enough to challenge feudalism as a social force.

Conversely, when printing technology reached Europe, it coincided with the period when the feudalism was declining and capitalism was developing rapidly. The early bourgeoisie, having accumulated a large amount of wealth, were dissatisfied with their existing social status and threw themselves into the anti-feudal struggle and eventually established their own political power. The important pillars of feudal rule in Europe were the Roman Catholic Church and the papacy. They held authority over secular regimes, owning territories in some countries, collecting exorbitant taxes and levies. As the representative of God, the Pope had absolute ideological authority and the privilege of interpreting the Bible and doctrines, ruling people's thoughts with medieval scholastic philosophy and forbidding dissenting voices, which were regarded as "heretical" and were persecuted. The Church suppressed human rights with divine power, promoted obscurantism and asceticism, and urged people to endure pain so that their souls could ascend to heaven after death. The Church and feudal forces infringed on the interests of the bourgeoisie and restricted its development. Therefore, any anti-feudal struggle inevitably targeted the Church first.

The Renaissance movement that began in the fourteenth century adopted humanism as its ideological banner in opposition to the theology of the Church. Humanists advocated human-centeredness, praising human value and dignity, and demanded liberation of individuality, freedom, and equality. They exposed the hypocrisy and greed of church authorities and advocated reason as a replacement for divine revelation. This was an ideological weapon used by the bourgeoisie to oppose feudal restraints and to fight for their own political and economic status. As humanistic works were widely disseminated in society in printed forms, they created powerful public opinions that could challenge the authority of the Pope and the Church and undermine their ruling foundations. With the anti-feudal struggle entering a new active phase, the cries of a few turned into a mass movement participated in by millions of people. Humanistic ideas were reflected in various fields including philosophy, political thought, literature, art, and science. Writers in various countries insisted on writing in their own national languages, a move of political significance. Their published works helped to enhance national and state consciousness among the people of various countries and to advance national cultures, enabling national languages to function in philosophical discussion like Greek and Latin, thus promoting the formation of national states.

In the struggle against feudalism and theology, the field of natural science became a new battleground. In medieval Europe, science was subordinate to the Church. With the development of printing technology in the fourteenth and fifteenth centuries, the number of universities increased dramatically, and the teaching of medicine and science was included. The introduction of ancient Greek, Arabic scientific works and Chinese inventions opened up Europeans' scientific vision and stimulated their research efforts. This necessitated breaking through the ideological barriers of theology and opposing it. The publication of Copernicus' *De Revolutionibus Orbium Coelestium* ("On the Revolutions of the Orbs of the World") in Nuremberg in 1543 marked the beginning of the Scientific Revolution. With scientific evidence, the book overthroew the geocentric theory that had dominated European thought for a thousand years and had been integrated into its religious system by scholastic philosophy. Repositioning the Earth from the center of the universe to one of the ordinary planets revolving around the sun caused a fundamental change in people's worldview, shaking the theoretical foundations of theological rule and liberating natural science from theology. The struggle between science and theology was intense; books about Copernicanism were banned by the Pope, and Bruno, who advocated the heliocentric theory, was burned at the stake. Nevertheless, there were always successors to a revolutionary cause. After several generations of effort by scientists such as Galileo, a series of theoretical breakthroughs were achieved in disciplines including astronomy, mathematics, physics, chemistry, anatomy, and physiology in the seventeenth century. Modern natural science, characterized by systematic experimentation and mathematization of natural hypotheses, rose in Europe and laid the theoretical foundation for the Industrial Revolution of the eighteenth century. With the help of science and technology, capitalist production developed rapidly, and the bourgeoisie gained political power as well.

Finally, in the sixteenth century, a religious reform movement broke out in Western European countries, which was a large-scale rebellion from within the Catholic Church, directly targeting the Pope. The tyranny and greed of the Roman Curia had long incited anger among secular authorities and people in various countries. Germany, most strictly controlled by the Roman Curia, suffered the most and thus had the strongest resistance. In 1517, theology professor Martin Luther nailed his "Ninety-Five Theses" to the door of the Wittenberg church to expose the Pope's fraudulent act of peddling mass-printed indulgences for enrichment without legal or theological basis, sounding the horn of the Reformation. The "Ninety-Five Theses" quickly spread throughout Germany and all of Europe in the form of printed flyers, garnering widespread response. Luther also published three booklets promoting his religious reform program, calling for Germany's independence from papal control, opposing the Pope's control of churches in various countries and their possession of lands, advocating for simplification of religious rituals, and replacing Latin with national languages. He denied the Pope as God's representative, advocating for direct communication between believers and God without the priest as an intermediary, and he rejected the Church's authority to interpret doctrine. Regardless of papal persecution, Luther established the Protestant church based on his own teachings. Later, Protestantism developed in Germany, France, and other countries, reducing the Roman Curia's control over dioceses. The Reformation also triggered the Peasants' War in Germany, dealing a heavy blow to the feudal powers and the rule of the Roman Curia.

The Renaissance, the Scientific Revolution, and the Reformation were intellectual and cultural movements against feudalism and religious authority, paving the way for the emerging bourgeoisie to ascend to the historical stage; the printing technology provided a powerful means for mobilizing the masses and shaping public opinions. Once mobilized, the masses could resort to concrete actions. With the mastery of gunpowder technology, the bourgeoisie acquired powerful armed forces to establish a political regime for their own will, and the roar of cannons tolled the death knell for European feudalism. During the sixteenth and seventeenth centuries, the bourgeois revolution in some European countries achieved success on both cultural and military fronts, marking an unprecedented and progressive revolution in human history. The great geographic discoveries accomplished through compass navigation and the consequent flourishing ocean voyages helped capitalism to explore overseas colonies, world trade, and world markets, thus opening a new chapter in modern world history and sparking off countless changes that altered the face of the world. In 1620, the English thinker Francis Bacon (1569–1626) wrote in his book *Novum Organum* ("New Instrument") about the role of inventions beneficial to humanity:

> First, it should be said that among all human activities, making significant discoveries is the noblest, as judged by all previous generations who bestowed upon inventors sacred honors. Those who have contributed to the state, such as founders of cities and empires, legislators, saviors of the state from long-standing calamities, and those who have overthrown tyrants, are only granted the title of heroes. If these two groups of people are compared properly, it is undeniable that the judgment of the ancients is fair. This is because discoveries benefit all humanity, while the achievements in human affairs affect only specific regions. The latter may last only a few generations, whereas the former endure throughout eternity.[1]

Bacon's discussion of inventions and discoveries here refers to China's Four Great Inventions. Although he was unaware of their origins in China at the time of writing, he clearly recognized their significance to the world and he stated:

> Second, the power, efficacy, and consequences of inventions are fully visible, as demonstrated by the three relatively recent inventions that were unknown to the ancients and of obscure origin: printing, gunpowder, and the magnetic compass. These three inventions have transformed the face of the world and the state of things: the first in academia, the second in warfare, and the third in navigation. From these, countless changes have arisen, so much so that no empire, no sect, no celebrity seems to have exerted as much power or influence on human affairs as these mechanical inventions.[2]

The discussion here of the three inventions of printing, gunpowder, and the compass should actually have included papermaking, because printing cannot exist without paper, and papermaking technology had already been introduced to Europe in the twelfth century well before the Renaissance. In the nineteenth century, Karl Marx pointed out in his work *Die Maschinen. Die Anwendung der Naturlicher Kraft und die Wissenschaften* ("The Machines. The Application of Natural Force and the Sciences", 1863):

> Gunpowder, the compass, and printing technology—these three inventions heralded the arrival of the bourgeois society. Gunpowder shattered the feudal knightly class, the compass

---

[1] Bacon F. *Novum Organum* [M] // Ellis R L, Spedding J. *Bacon's Philosophical Works*. London: Routledge, 1905.

[2] Bacon F. *Novum Organum* [M] // Ellis R L, Spedding J. *Bacon's Philosophical Works*. London: Routledge, 1905.

> opened up the world market and established colonies, while printing became a tool for Protestantism, and more generally, a means of scientific revival and the greatest lever for creating necessary prerequisites for spiritual development.[3]

Finally, the twentieth-century British scholar John Desmond Bernal (1901–1971) believed: "For many centuries, China has been one of the great centers of human civilization and science... It is already evident that during the Western Renaissance, in the transition from ancient Greek abstract mathematical science to modern mechanical mathematical science, China's technological contributions—the compass, gunpowder, paper, and printing—played a role, and possibly a decisive role."[4]

[3] Marx. Die Maschinen. *Die Anwendung der Naturlicher Kraft und die Wissenschaften* [M]. Beijing: People's Publishing House, 1978: 67.

[4] For the preface written by Bernal J D for the Chinese translation of *Science in History*, see *Science in History* by Bernal J D [M]. Translated by Wu Kuangfu et al.. Beijing: Science Press, 1959.

Postscript

Ten years ago, my father completed the manuscript of this book, but for various reasons, it was not published in time. This book is not only his third work on the Four Great Inventions of ancient China (on papermaking, gunpowder, and printing, in that order), but also the last research achievement of his life. Sadly, he did not live to see it published. He dedicated his whole life to the study of ancient Chinese civilization, deeply amazed by the wisdom of the ancients. I admire his perseverance!

As the manuscript was completed a long time before it was edited for publication, and with my father's passing away, this book inevitably has inadequacies in terms of content and references. This acknowledgement invites readers to show understanding and consideration for any limitations they might encounter in the book. It is hoped that this book serves as a "stepping stone" for young people to cultivate love for science and explore knowledge, blending the wisdom of the ancients with modern technology to make contribution to the greatness and glory of China.

Pan Feng
October 8, 2021

P. Jixing, *History of Printing Techniques in China*,
https://doi.org/10.1007/978-981-96-6119-0

Epilogue

The publication of this book is a joint effort made collaboratively by six Chinese scholars, chief translator Prof. Chen Jiliang (陈纪梁), and co-translators of Prof. Mu Yuanyuan (慕媛媛), Lecturer Chen Fumei (陈馥梅), associate Prof. Tao Wei (陶伟), associate Prof. Song Binbin (宋彬彬) and associate Prof. Su Tao (苏涛). I would like to extend my sincere thanks to them on behalf of my father.

My thanks also go to the project leader of the University of Science and Technology of China Press Yao Shuo (姚硕), and editors of Yu Xiumei (于秀梅) , Yang Xinqi (杨昕琦), Li Panfeng (李攀峰), Huang Chengqun (黄成群) and Han Jiwei (韩继伟) for their conscientious work and ready help. They have coordinated the necessary tasks to make the publication of this book possible among three parties, the author, the translators, and Springer Nature.

I am grateful to Divyalakshmi Ganesan, project manager at Springer Nature. During the translation of this book, she did a careful and meticulous editing to ensure the quality of the book. The book is jointly published by Springer Nature and the University of Science and Technology of China Press. As the daughter of the author of the Chinese version of this book, I would like to express my sincere thanks to them all.

Pan Feng (daughter of Pan Jixing)
2025-06-16

P. Jixing, *History of Printing Techniques in China*,
https://doi.org/10.1007/978-981-96-6119-0

MIX
Papier aus verantwortungsvollen Quellen
Paper from responsible sources
FSC® C105338

If you have any concerns about our products,
you can contact us on
ProductSafety@springernature.com

In case Publisher is established outside the EU,
the EU authorized representative is:
Springer Nature Customer Service Center GmbH
Europaplatz 3, 69115 Heidelberg, Germany

Printed by Libri Plureos GmbH
in Hamburg, Germany